Advances in Spatial Science

Springer

Berlin
Heidelberg
New York
Hong Kong
London
Milan
Paris
Tokyo

Titles in the Series

C. S. Bertuglia, M. M. Fischer and *G. Preto* (Eds.)
Technological Change, Economic Development and Space
XVI, 354 pages. 1995. ISBN 3-540-59288-1
(out of print)

H. Coccossis and *P. Nijkamp* (Eds.)
Overcoming Isolation
VII, 272 pages. 1995. ISBN 3-540-59423-X

L. Anselin and *R. J. G. M. Florax* (Eds.)
New Directions in Spatial Econometrics
XVX, 420 pages. 1995. ISBN 3-540-60020-5
(out of print)

H. Eskelinen and *F. Snickars* (Eds.)
Competitive European Peripheries
VIII, 271 pages. 1995. ISBN 3-540-60211-9

J. C. J. M. van den Bergh, P. Nijkamp and *P. Rietveld* (Eds.)
Recent Advances in Spatial Equilibrium Modelling
VIII, 392 pages. 1996. ISBN 3-540-60708-0

P. Nijkamp, G. Pepping and *D. Banister* (Eds.)
Telematics and Transport Behaviour
XII, 227 pages. 1996. ISBN 3-540-60919-9

D. F. Batten and *C. Karlsson* (Eds.)
Infrastructure and the Complexity of Economic Development
VIII, 298 pages. 1996. ISBN 3-540-61333-1

T. Puu
Mathematical Location and Land Use Theory
IX, 294 pages. 1997. ISBN 3-540-61819-8

Y. Leung
Intelligent Spatial Decision Support Systems
XV, 470 pages. 1997. ISBN 3-540-62518-6

C. S. Bertuglia, S. Lombardo and *P. Nijkamp* (Eds.)
Innovative Behaviour in Space and Time
X, 437 pages. 1997. ISBN 3-540-62542-9

A. Nagurney and *S. Siokos*
Financial Networks
XVI, 492 pages. 1997. ISBN 3-540-63116-X

M. M. Fischer and *A. Getis* (Eds.)
Recent Developments in Spatial Analysis
X, 434 pages. 1997. ISBN 3-540-63180-1

R. H. M. Emmerink
Information and Pricing in Road Transportation
XVI, 294 pages. 1998. ISBN 3-540-64088-6

P. Rietveld and *F. Bruinsma*
Is Transport Infrastructure Effective?
XVI, 384 pages. 1998. ISBN 3-540-64542-X

P. McCann
The Economics of Industrial Location
XII, 228 pages. 1998. ISBN 3-540-64586-1

L. Lundqvist, L.-G. Mattsson and *T. J. Kim* (Eds.)
Network Infrastructure and the Urban Environment
IX, 414 pages. 1998. ISBN 3-540-64585-3

R. Capello, P. Nijkamp and *G. Pepping* (Eds.)
Sustainable Cities and Energy Policies
XI, 282 pages. 1999. ISBN 3-540-64805-4

M. M. Fischer and *P. Nijkamp* (Eds.)
Spatial Dynamics of European integration
XII, 367 pages. 1999. ISBN 3-540-65817-3

M. M. Fischer, L. Suarez-Villa and *M. Steiner* (Eds.)
Innovation, Networks and Localities
XI, 336 pages. 1999. ISBN 3-540-65853-X

J. Stillwell, S. Geertman and *S. Openshaw* (Eds.)
Geographical Information and Planning
X, 454 pages. 1999. ISBN 3-540-65902-1

G. J. D. Hewings, M. Sonis, M. Madden and *Y. Kimura* (Eds.)
Understanding and Interpreting Economic Structure
X, 365 pages. 1999. ISBN 3-540-66045-3

A. Reggiani (Ed.)
Spatial Economic Science
XII, 457 pages. 2000. ISBN 3-540-67493-4

D. G. Janelle and *D. C. Hodge* (Eds.)
Information, Place, and Cyberspace
XII, 381 pages. 2000. ISBN 3-540-67492-6

P. W. J. Batey and *P. Friedrich* (Eds.)
Regional Competition
VIII, 290 pages. 2000. ISBN 3-540-67548-5

B. Johansson, Ch. Karlsson and *R. R. Stough* (Eds.)
Theories of Endogenous Regional Growth
IX, 428 pages. 2001. ISBN 3-540-67988-X

G. Clarke and *M. Madden* (Eds.)
Regional Science in Business
VIII, 363 pages. 2001. ISBN 3-540-41780-X

M. M. Fischer and *Y. Leung* (Eds.)
GeoComputational Modelling
XII, 279 pages. 2001. ISBN 3-540-41968-3

For information about further volumes please see end of the book

Stan Geertman
John Stillwell
Editors

Planning Support Systems in Practice

With 181 Figures
and 38 Tables

Springer

Dr. Stan Geertman
Utrecht University
Faculty of Geographical Sciences
Heidelberglaan 2
3508 TC Utrecht
The Netherlands

Professor John Stillwell
University of Leeds
School of Geography
Leeds LS2 9JT
United Kingdom

ISBN 3-540-43719-3 Springer-Verlag Berlin Heidelberg New York

Library of Congress Cataloging-in-Publication Data applied for
Die Deutsche Bibliothek – CIP-Einheitsaufnahme
Planning Support Systems in Practice / Stan Geertman; John Stillwell Ed.. –
Berlin; Heidelberg; New York; Hong Kong; London; Milan; Paris; Tokyo: Springer, 2003
(Advances in Spatial Science)
ISBN 3-540-43719-3

Springer-Verlag Berlin Heidelberg New York
a member of BertelsmannSpringer Science + Business Media GmbH

http://www.springer.de

Printed in Germany

Cover design: Erich Kirchner, Heidelberg

SPIN 10881482 42/2202-5 4 3 2 1 0 – Printed on acid-free paper

Preface

Planning Support Systems: Technologies that are Driving Planning

Michael Batty

Centre for Advanced Spatial Analysis (CASA), University College London,
1-19 Torrington Place, London WC1E 6BT, United Kingdom

I had always thought the term 'Planning Support Systems', abbreviated to PSS, had been coined by the father of land use modelling, Britton Harris, in his article 'Beyond Geographic Information Systems: computers and the planning professional' published in the *Journal of the American Planning Association* in 1989 (Harris 1989). Until I asked him, that is. In a response to a paper he gave to the Urban and Regional Information Systems Association (URISA) in the summer of 1987, he told me that someone in the audience who he cannot quite remember, actually coined the term, referring to 'planning support systems' as that constellation of digital techniques (such as GIS) which were emerging to support the planning process. In fact, the predecessor term 'decision support systems' (DSS) from which this unknown originator obviously defined PSS by analogy, was coined as far back as the late 1970s in the management literature for a loose assemblage of techniques, usually computer-based, which aided management decisions. The term slowly entered the geographical lexicon as 'spatial decision support systems' (SDSS) and this is probably first attributable to Lew Hopkins and Mark Armstrong who used it in a paper published in *AutoCarto 7* in 1985 (Hopkins and Armstrong 1985). What makes these terms significant in contrast to earlier computer-based planning processes is the assumption that support systems are not usually highly structured as are computer models or optimisation techniques but simply loosely coupled assemblages of computer-based techniques which form a toolbox into which decision-makers can dip in the quest to manage and tame their decision problems. In this sense, they reflect the times in which we live and the dominant way we currently conceive of planning and management, government and control.

The history of computers in planning mirrors the evolution of land use planning itself during the last 100 years since it first became institutionalised as a function of government in the west in the late nineteenth century. The optimism and idealism that characterised the early responses of planning pioneers to the evils of the industrial city with their emphasis on new towns, garden cities and green belts gradually gave way to more corporate, rational thinking by the mid twentieth century. Scientific rationality, it was argued, was the way in which the great visions of people like Howard and Geddes, Benton MacKay and many others might be realised in the context of increasingly complex urban agglomerations. This consolidation of thinking through rational planning was intimately bound up with the use of computers and by the 1960s, early applications of computation were quite consistent with the top-down methods used to produce and implement physical plans. Since then, the picture has changed radically. The role of central planning in the modern state has been split asunder as society at large has reacted against centralised control. In the last 30 years, physical planning has become a much more bottom-up, decentralised activity as strategic planning has all but been abolished in western nations. This again is entirely consistent with what has been happening in the development of new technologies, with computers being used in many diverse ways to support these emerging decentralised forms of planning. In short, PSS reflect a toolbox approach with the emphasis on providing the planner and decision-maker, the public at large, indeed anyone interested in fashioning spatial solutions to contemporary problems, with tools to explore, represent, analyse, visualise, predict, prescribe, design, implement, monitor and discuss issues associated with the need to plan (Batty 1995).

There has been another subtle change in the use of new technologies which is also reflected in the way PSS are maturing. As the editors point out in their introduction, PSS began as a collection of tools to support 'planning for people' but increasingly they are becoming toolboxes to be used in 'planning with people'. There is no grand model or even vision of an ideal planning support system and thus the emphasis is increasingly on tools which can be used to communicate rather than generate solutions to planning problems. Indeed, as soon as the focus changes towards communication, the support offered by techniques that comprise PSS broadens to include everything from raw data to methods of analysis, simulations and models as well as procedures for generating optimal designs through everything from computer aided design to group problem-solving. These trends are clearly evident in this volume which represents one of the most complete collections of articles in this area to date. Although it comes hard on the heels of the edited reader published by Brail and Klosterman (2001), it is entirely complementary, covering many new applications in theory and practice as well as focusing on the way the field might develop over the next 20 years. This shift to communicating plans from generating them is quite evident in this book with much of the emphasis on the development of online communications spinning off from desktop applications. The use of the Web is very evident in many of the contributions here while an increasing focus on finer scale applications – on development and design at the more micro scale – is also clear.

It is not easy to provide a framework for PSS due to the breadth of the field and its diversity of applications. Yet despite the general problem of classifying where PSS begins and ends, the editors do a very skilful job in collecting together a series of papers which bound the field rather well. Moreover, the most active research and practitioners of computer technologies in planning are writing here. One of the great things about this book is that it provides a defining moment for PSS: not only does it focus on the way theory and applications are beginning to reflect practice, it describes real progress in showing how researchers and practitioners alike are beginning to use the toolbox that is being added to at an ever increasing rate, in applications that reflect contemporary planning concerns. In this way, planning is beginning to reflect its science: the use of techniques is starting to affect the very act of planning in terms not of what we plan and where but more to the point, how we plan and with whom. New information and communications technologies are changing the very way we think about planning while at the same time enabling us to plan better. PSS are in the vanguard of such change.

There is a real wealth of detail contained in the pages that follow. The editors have done a magnificent job in presenting some of the best applications of PSS in theory and practice not only in terms of the state of the art but also in a more prospective way. Let me point very briefly not only to some of the achievements reported here but also to the signposts to the future that are charted in this volume. What is most encouraging is the fact that PSS span the different scales of planning, from national systems such as the desktop migration modelling and visualisation system produced for the UK's Department of Transport, Local Government and the Regions to the community based systems implemented through the Web for urban design and for eliciting public responses such as those developed for Pilsen in Chicago and Verona, near Madison, Wisconsin. Furthermore, this range of scales is complemented by applications which span many different styles and types of planning from strategic to the local, from land use planning to urban design, and from urban and regional monitoring to more environmentally based systems where resource conservation is critical. Tools and techniques of course drive such systems and the useful articles collected together in Part 2 help give insights into how PSS are built from the bottom up. The other feature of these systems is how GIS is becoming less significant. Although many of the systems reported here do use desktop or internet GIS to deliver maps, there is pragmatism about the technologies and their use which is much wider than GIS *per se*. Increasingly, multimedia systems, CAD, the development of spreadsheet-like applications and specific purpose-built software characterise this field. In fact, for those of us who think that planning must always be specific, community focussed and in this sense independent of any formulaic or procedural templates, what is so encouraging about the PSS landscape sketched here is that so many of the systems that are being built are sensitive to the local context. Although there are of course general principles, these are implicit rather than explicit. The field is marked by one-off applications, tailored to the specific contexts in question and in this sense represent responses to local conditions. Far from this being an exercise in how big brother is watching you, the kinds of systems described here give real hope for a democratic planning built from the bottom up.

The articles collected here present as good a picture of the field as you are likely to find. They contain the keys to the past and the keys to the future and we can all look forward to a wealth of new applications of PSS built around the experiences reported in this volume.

References

Batty, M. (1995) Planning support systems and the new logic of computation, *Regional Development Dialogue*, 16, (1): 1-17.

Brail, R. K. and Klosterman, R. E. (eds) (2001) *Planning Support Systems: Integrating Geographic Information Systems, Models, and Visualization Tools*, ESRI Press, Redlands, CA.

Harris, B. (1989) Beyond Geographic Information Systems: computers and the planning professional, *Journal of the American Planning Association,* 55: 85-90.

Hopkins, L.D. and Armstrong, M.P. (1985) Analytic and cartographic data storage: a two-tiered approach to Spatial Decision Support Systems, *Proceedings of the Seventh International Symposium on Computer-Assisted Cartography: Auto-Carto 7*, American Society for Photogrammetry and American Congress on Surveying and Mapping, Falls Church, VA, pp. 283-292.

Contents

INTRODUCTION

1 Planning Support Systems: An Introduction

Stan Geertman[1] and John Stillwell[2]

[1] URU/Nexpri, Faculty of Geographical Sciences, Utrecht University, Heidelberglaan 2, 3508 TC Utrecht, The Netherlands

[2] School of Geography, University of Leeds, Leeds, LS2 9JT, United Kingdom

Abstract: This introductory chapter begins with an initial discussion of some of the trends that have characterised planning in recent years and that provide the context for the development and application of Planning Support Systems (PSS). Given the focus of the book on demonstrating the application of PSS in practice, it was necessary to create an inventory of systems either in prototype form or in use in different planning environments. Many of the entries in the inventory have been extended into more detailed contributions for this book. The latter part of the chapter is devoted to outlining the structure of the book and its contents.

1.1 Introduction

Planning is an activity orientated towards the future that links *"scientific and technical knowledge to actions in the public domain"* (Friedmann 1987). It normally takes place through formal discourse between the groups, organisations and individuals concerned with particular public or private sector initiatives or proposals. Urban and regional planning encompasses a wide range of activities – many of which are responses to particular societal problems – that impact at different spatial scales. Traditionally, distinctions have been drawn between the physical planning of urban structure and the strategic planning of regional entities. However, the diversity of planning applications now extends from spatial planning at the European level, through the European Spatial Development Perspective (ESDP), for example, to the development of plans at the very local level for neighbourhood renewal and community regeneration. In each case, attention is increasingly given to economic, social and environmental considerations in view of the imperative to ensure sustainability. In each context, planning tends to be characterised by the involvement of a plurality of frequently conflicting interests, an explicit future orientation, an emphasis on strategic choices as well as operational decisions (Friend and Jessop 1969) and a necessity to recognise the long term impact of policies and plans.

It is inevitable that the processes of planning, and the agencies involved, will change over time as governments seek to make improvements to their national systems on grounds of efficiency and to confront the pressures of new circumstances and priorities. Within Europe, for example, there is plenty of evidence to indicate that planning has become more integrated ('joined-up'), holistic, region-oriented, partnership-oriented, strategic, scenario-based, as well as being more interactive and participatory in nature (Stillwell *et al.* 1999, Geertman and Stillwell 2000). The corollary of this has been that the complexity of the planning tasks has grown extensively due to: a greater involvement in planning processes of more parties with diverging interests; the increasing pace of development that shortens the time available for plan preparation and design; the need for better insights into real world processes that a more holistic approach to planning demands; the necessity for better information on which to monitor progress and base scenarios; and the difficulties associated with enhancing participation, collaboration and partnership. As a result, requests for support to fulfil a range of planning tasks in an appropriate way have grown rapidly, stimulating the demand for consultancy services from private companies and research institutions and encouraging the development of new instruments that take advantage of the rapid advancements that have taken place in computer and information technologies.

There is no denying that planners have always wanted computer-based tools to enhance their analytical, problem-solving and decision-making capabilities, but it should also be recognised that planning practice has never effectively embraced many of the methods, techniques and models developed over time in the research laboratory. These tools and their applications have changed through the years too (e.g. see Batty and Densham 1996). In the 1960s and 1970s, computer-based planning tools, including the early generations of Geographical Information Systems (GIS) and Decision Support Systems (DSS), were used primarily for data management, modelling and strategic planning support. Although these systems and models offered huge opportunities to support planning, they were not readily adopted by the planning profession. Lee (1973) has identified the reasons for failure of various large-scale models developed in the 1960s and 1970s, not least of which was their complicated nature and their 'data-hungriness'. The 1980s and 1990s have witnessed a change in emphasis towards the application of geotechnology to the more routine management and graphical display of spatial datasets. Newer generations of GIS with their extensive database management and display capabilities have attuned to these developments. However, more sophisticated analysis, simulation and modelling have remained 'on the back-burner' as far as planning practice is concerned despite some major developments in new multi-media and virtual reality technologies as well as a new generation of computationally intensive modelling (simulation, neural networks, fuzzy modelling) methods, many of which have enormous application potential.

The picture continues to alter. One important dimension of this in current planning practice is the change from 'planning for' to 'planning with' the people whose lives are affected by a new development or policy. This particular emphasis requires new tools that can support the communicative aspect of current planning practice. On the other hand, geotechnology in its broadest sense also continues to change: increased availability of affordable geodata and geoinformation; more user-friendly software and hardware; enhanced capabilities of software tools; exploitation of the Internet and the World Wide Web as a channel for data capture, analysis and dissemination; *et cetera*. All these trends in geotechnology extend its fields of application, its user groups and its audience. At the same time, however, the changing nature of planning draws attention to the current dearth of well-designed and well-tested planning support tools. At the moment, it can be argued that most of current proprietary tools that do exist are far too generic, too complex, too inflexible, incompatible with most planning tasks, oriented towards technology rather than problems and too focused on strict rationality (see, for example, Couclelis 1989, Ottens 1990, Klosterman 1994, Worrall 1994, Nedovic-Budic 1998, Geertman, 1999).

As a consequence of these trends and the shortcomings of the existing geotechnology, a range of new or renewed tools to facilitate 'new' planning practice has materialised under a new generic term, Planning Support Systems (PSS). These systems include geotechnology-related instruments that are primarily developed to support different aspects of the planning process, including: problem diagnosis, data collection, enhancing participation, spatial and trend analysis, data modelling, visualisation and display, report preparation, collaborative decision-making, *et cetera*. In the field of scenario-building and alternative plan formulation, for example, a variety of conceptual or operational prototypes of new planning support tools can be found. These range from the electronic conference board rooms (Group Decision Support Systems) discussed by Laurini (1998) and the GIS-supported collaborative decision-making tools outlined by Nyerges and Jankowski (1997) to WWW-based mediation systems for cooperative spatial planning (Gordon *et al.* 1997, Shiffer 1992) and support tools for different planning tasks during the planning process (Hopkins 1998, 1999, Kammeier 1999, Klosterman 1999b, Singh 1999, Geertman 2002).

Despite these developments and the ongoing progress made over the last decade since Scholten and Stillwell (1990) reported on the state of the art in GIS at the end of the 1980s, the general picture seems to be that PSS still remain a set of purpose-specific tools, providing for only a limited range of operations (Bishop 1998). In fact, it is the uncertainty surrounding our current knowledge about the development, adoption and implementation of PSS in different planning contexts throughout the world that has led to the construction of an inventory of PSS and the preparation of this book. The inventory is discussed in Section 1.3 and the chapter structure in Section 1.4. Beforehand, some clarification of the nature of PSS *vis à vis* other information system categories is attempted.

1.2 PSS, GIS and SDSS

Planning Support Systems (PSS) involve a wide diversity of geo-technology tools (geographical information and spatial modelling systems) that have been developed to support public or private planning processes (or parts thereof) at any defined spatial scale and within any specific planning context. In particular, they are developed to support the derivation and evaluation of alternative futures. Inevitably PSS are closely related to two other more well-known systems, Geographical Information Systems (GIS) and Spatial Decision Support Systems (SDSS), and therefore it is important to clarify what makes each of these distinctive.

Although the term 'planning support system' itself is fairly recent, the underlying concept goes back to the 1950s. Harris (1960, 1989) has long argued for an approach to planning that combines 'sketch planning' – the rapid and partial description of alternatives – with state of the art modelling of the implications of these alternatives. According to Harris and Batty (1993), the concept of PSS refers to an appropriate model for combining a range of computer-based methods and models into an integrated system that can support the planning function. More precisely, a PSS forms a framework in which three sets of components are combined: the specific planning tasks and problems at hand; the system models that inform the planning process through analysis, prediction and prescription; and the transformation of basic data into information which in turn provides the driving force for modelling and design (cyclic process). In this description, the planning profession and its tasks and needs are stated explicitly. Alternatively, Klosterman (1997, 1999a) and Brail and Klosterman (2001) more recently, describe PSS as information technologies that are used specifically by planners to perform their unique professional responsibilities. In their opinion, PSS have matured into a framework of integrated systems of information and software, which brings together the three components of traditional decision support systems – information, models, and visualization – into the public realm. Hopkins (1999) believes that the search for an appropriate role for information technology in planning must begin not with the technology but with the planning problems. In his view, a PSS can and should be based on a common set of objects inherent in urban development processes (e.g. like actors, activities, flows, facilities) and on views (e.g. like sketches, models, scenarios) and associated tools that share these common objects so as to support a full range of planning tasks.

It is our view that PSS, GIS and SDSS are categories that are not mutually exclusive but have certain distinguishing characteristics. PSS are a subset of geotechnology-related instruments that incorporate a suite of components (theories, data, information, knowledge, methods, tools) that collectively support all of or some part of a unique planning task. They aid the planning process by providing integrated environments usually based on multiple technologies and a common interface. Therein, it is not so much the technology, with its capabilities and restrictions, that dictates the support function of the PSS, but it is primarily the specific needs of the planning context in which the PSS is designed to operate. One typical example of a recent PSS with all these characteristics is *EUROSCANNER* (Stillwell and Scholten 2001), a system for simulating future land use in different parts of Europe. *EUROSCANNER* consists of components to capture different types of data, to store and manage databases, to compute land suitability measures, to model land use change, to develop scenarios and to visualise and evaluate alternative simulations of future environments. These components are grounded in theoretical frameworks both for identifying the driving forces behind land use development and for adopting a particular modelling approach.

GIS, on the other hand, can be broadly regarded as general-purpose systems, comprising tools for the handling of geo-referenced data, and applicable for a wide diversity of tasks and problems in various settings. Typically GIS provide generic solutions whereas PSS are specific and focused on the tasks of the application. This implies the handling of aspatial and nonspatial data as well as spatial data; historical data sets as well as projections; qualitative information as well as quantitative; implicit and semi- or even ill-structured knowledge as well as explicit and well-structured knowledge; and the modelling, design, and analysis of dynamic spatial data and information as well as the management and presentation of static spatial data or information. Each different planning situation will have its own planning procedures, underpinning theory, data, information, knowledge, tools, methods, presentation requirements, *et cetera* and will therefore require a customised PSS. On many occasions, a proprietary GIS will form part of a PSS, given the useful set of functional capabilities that it provides.

Spatial Decision Support Systems (SDSS) are *"explicitly designed to support a decision research process for complex spatial problems. SDSS provide a framework for integrating database management systems with analytical models, graphical display and tabular reporting capabilities, and the expert knowledge of decision makers"* (Densham 1991, p. 404). SDSS have much in common with PSS since their computer-based components such as the database, user-interface and modelling software can often be found in PSS too. Moreover, SDSS are designed explicitly to handle semi- or ill-structured problems, which is also the task of most PSS. Both SDSS and PSS can be considered task-specific systems rather than being applicable in a diversity of situations. In contrast to these similarities, PSS will have to pay particular attention to long-range problems and strategic issues and may be designed explicitly to facilitate group interaction and discussion. SDSS, on the other hand, are generally designed to support shorter-term policy-making by isolated individuals and by business organizations (Clarke 1990). Furthermore, they are intended primarily to support executive decision-making (Klosterman 1997).

In summary, PSS are not equivalent to GIS or SDSS, although elements of both the latter are frequently found within PSS. PSS are not a radically new form of technology that will replace existing planning tools. They take the form of *information frameworks* (after Klosterman 1997) that integrate the full range of information technologies useful for supporting the specific planning context for which they are designed.

1.3 A PSS Inventory: Context, Aims and Content

An assessment of planning practice at the beginning of the twenty-first century suggests, rather depressingly, that the adoption and use of geotechnology tools (geographical information and spatial modelling systems) is far from widespread and far from being effectively integrated into the planning process (Stillwell *et al.* 1999). Planners and designers remain, at best distrustful, or at worst downright antagonistic toward highly systematic and computer-based models (Harris 1998, 1999).

Klosterman (1998) suggests that the tools for planning support are no more developed than they were ten years ago. Although prospects may seem exciting, it is more likely that the adoption of new tools and the take-up of computer applications in planning for the next 25 years will remain disappointing. The reasons for this situation are primarily associated with the diversity of analytical tasks which planners perform, the relatively small market for public sector software and the expense of developing and supporting commercial software. As a consequence, analytical tools for planning purposes will continue to lag far behind those developed for other professions such as transportation engineering and other areas of government. Nedovic-Budic (1998) has shown that within planning practice, whilst the quality of information generated with GIS technology has improved, GIS is consistently underemployed for more sophisticated analytical and modelling exercises, and its impact on planning decisions remains low.

In general, we conclude that many planners now have access to the geodata and meta-geoinformation facilities of their organisations, and many are proficient in using their GIS tools to perform spatial queries and to generate thematic maps. Sadly, however, progress towards the use of GIS beyond these basic activities to help solve key planning problems through more sophisticated analyses, has been very limited (Geertman 1999). The percentage of planners who consider their geotechnology as an intrinsic and indispensable tool for performing their job properly (as financial experts use their spreadsheet software and as medical specialists use their ECG technology) remains remarkably low (Geertman 2001).

The desire and need to enhance all types of planning practice provides our overall rationale for wanting to undertake a review of the field and to identify innovative examples of geotechnology application and 'best practice'. The mechanism through which this was carried out involved the assembly of an inventory of PSS. Over the last few years, certain individuals or groups, based at scientific, research or planning institutions around the world, have been involved in the development, testing and implementation of a range of PSS. However, no comprehensive picture of the extent of these developments has been constructed and no assessment of the success of their adoption has been undertaken. As a result, the different groups of researchers and developers themselves may well be undertaking a substantial amount of work that duplicates or overlaps with work undertaken elsewhere. At the same time, the planning community as a whole has little idea of where to look in the academic or planning literature for good examples of *Planning Support Systems in Practice*, details of new PSS instruments, or advice and support for PSS. This is problematic for the potential consumers (planners) at a time when there is an increasing need for geotechnology support. Moreover, the developers of PSS have to prove the worth of their products in real-world planning situations. Thus, underlying the initiative to produce an inventory and to publish a book has been the assumption that a greater degree of access to relevant information will lead to improved planning decisions (Shiffer 1995).

Based on this rationale, an Internet-based inventory of PSS was established in order to collect together details of PSS in development as prototypes or as commercial products, as well as those PSS now implemented and operational in planning practice. Entries to the inventory *(www.nexpri.nl)* were made between June 2000 and June 2001 and about 50 contributions were received from 20 countries including the Netherlands, UK, USA, Italy, Columbia, Australia, New Zealand, Russia, South

Africa, Malaysia, Sweden, Finland, Czech Republic and Hong Kong. The entries represent a wide diversity of instruments that are being used in a variety of planning tasks. This diversity relates to differences in aims (e.g. area-based or generic tools), goals (e.g. monitoring, scenario building), functional capabilities (e.g. sketching, analysis, presentation), content (e.g. tools, methods, data, information, knowledge, models, theories), structure (e.g. toolbox versus integrated system), technology (e.g. stand-alone, Intranet, Internet systems), looks and feels (e.g. amateurish versus professional interfaces), *et cetera*. In summary, two main groups of PSS in practice can be identified: those dedicated to analytical and/or design tasks and those attempting to improve presentation and/or communication.

Several entries in the inventory are PSS that support the phase of modelling, analysis and/or design in the planning process and possess a diversity of tools to perform simulations, sketching, evaluation, *et cetera*. Thus, the concept of a 'planner's toolbox' is appropriate as a defining characteristic of these PSS. Most of the examples involved a set of loosely coupled tools, with some theoretical and methodological meta-information ensuring their correct application, focused on performing some specific planning tasks (e.g. monitoring, sketching, simulation) by professional planners with both substantive, technological and methodological knowledge. It is the functional capabilities and their correct application within the planning process that are of prime importance, rather than their user-friendliness. Moreover, whilst their application is usually area-independent, it is specific to a particular planning phase or task (e.g. the SPARTACUS project).

In contrast, another group of PSS focus primarily on facilitating the visualization, communication and dissemination of a diversity of geographical information to a restricted audience of professionals or to a much wider public audience. Characteristic of the majority of these systems is their attractive presentation functionality and the slick Internet-based technologies upon which they are based. Moreover, these PSS are mostly area-specific but planning phase-independent, in that they are oriented to and applicable within diverse planning contexts (e.g. Great Lakes Information Network (GLIN) *http://www.great-lakes.net/*).

In general, it was the purpose of the inventory to gain an overview of and insight into the diversity and innovative state-of-the-art of PSS in practice and the experiences that have been gained within different planning contexts. Most of the entries in the inventory are contemporary applications; many remain in a phase of perpetual development; others are in prototype form or have recently been released as commercial products; some have recently been implemented in planning practice. As a consequence, factual experience with their application and use in real-world planning situations has not always been illuminating.

Following careful scrutiny of the inventory entries, we selected contributions from 15 different countries covering a wide range of application areas for inclusion in more detailed form as chapters in this book. The contributions were selected on the basis of four criteria. Firstly, they should explicitly describe the application of the PSS in a real world planning practice context. Where has it been applied? What were the experiences of this application? How has or will the experience be translated into the PSS itself and/or into later applications? Thus, the accent is put on the first S for Support in conjunction with the P of Planning in PSS, instead of on the second S for System, as is the case of other concurrent books on planning or decision support. This focus emphasises that the search for an appropriate role for geotechnology in planning must be planning-led rather than technology-led, and must strive to adapt available technologies to the particular needs of the planning profession. Secondly, contributions should explicitly refer to instruments (including tools, data, knowledge, information, methods, theories, meta-information), which may be applicable and translatable in principle to a wide diversity of cases and/or planning practices. Thirdly, they should not entail once-and-for-all stand-alone research projects without instrumental outcomes that can be applied for support in subsequent planning practices. Fourthly, they should fit into the general PSS description given previously. Thus, applications of MapInfo desktop GIS for instance are not really PSS since MapInfo is a generic toolbox, not identified for supporting specific planning tasks, although an ArcIMS application for facilitating the communication of planning proposals with the wider public can be classified as being appropriate.

1.4 Structure of the Book

The book has been divided into five parts, the contents of which are summarised in the remainder of the chapter. There is always some danger in imposing structure on a wide range of contributions but we have chosen a five part categorisation based on the purpose or use of each PSS. Part 1 is a collection of contributions that illustrate PSS that are used specifically for disseminating information to the public for enhancing participation. Part 2 contains a series of contributions that outline and exemplify a variety of PSS tools for undertaking particular integration, analysis or visualisation activities. Thereafter, Parts 3, 4 and 5 are dedicated to the detailed presentation of examples of individual PSS in action to support strategic planning, land use and infrastructure planning and environmental planning respectively.

Part 1
Systems and Technologies for Enhancing Participation in the Planning Process

2 Interactive support systems for participatory planning
Stan Geertman and John Stillwell

3 Public participation via on-line democracy
Richard Kingston, Andrew Evans and Steve Carver

4 Web-based tools and interfaces for participatory planning and design
Kheir Al-Kodmany

5 Community engagement in land use planning through web-based technologies
Steve Ventura, Tracy Miller and Glen Barry

6 Tools for community design and decision-making
Ken Snyder

Chapter 2 is the first of five that constitute Part 1 and that consider how participation in the planning process can be enhanced using PSS. The chapter initially elaborates on the increasing importance of participation and how this trend, in combination with other trends and with developments in geotechnology, has resulted in increased use being made of interactive PSS to provide opportunities for the public to take a much more direct and interactive role in contemporary planning of various types at different spatial scales. Two web-based systems developed by pioneer local authorities in the UK are used to illustrate how the Internet is being used as a channel for disseminating statistical information about localities and for stimulating public interaction at a time when neighbourhood regeneration is high on the political agenda. In addition, two PSS used in the Netherlands are considered: *I-pl@n* is an internet-based support tool for communicating information about development plans and *SketchGIS* is a PSS designed for use in the process of spatial scenario-building.

Two case studies of the use of the Internet and the WWW as a means for carrying out public consultation using GIS-based PSS are reported in Chapter 3. Firstly there is an exercise to identify the views of local residents in the village of Slaithwaite in West Yorkshire on how they would like to see their village develop. Secondly, there is a case study of woodland planting in the Yorkshire Dales National Park. A two-stage approach is followed involving GIS-based modelling to identify areas best suited for regeneration of natural forest cover, and user-focused participatory mapping methods to identify which of the suitable areas are most acceptable for regeneration. The chapter makes some very useful recommendations on accessibility to the technology, GIS and IT understandability, data and copyright issues, trust in the system and response legitimacy.

Developments in web-based systems to improve participation are probably more advanced in the USA than in Europe. Chapter 4 explains some of the new tools that have been developed at the University of Illinois at Chicago to encourage public participation in the local planning process. The context of the work is the Mexican-American community of Pilsen, south of the University. Specifically, the chapter introduces six prototype web interfaces that invite users to read maps of their community and then make selections and type comments in order to provide feedback into the planning process. Each of the prototypes highlights a different tool or feature for using online maps: to navigate a large map area online and use different zooming methods; to select areas spatially using a grid-based system or by freehand drawing; and to allow users to create their own maps using freehand drawing technology. In particular, this project lays out some of the potential contributions and limitations related to enabling two-way communication and participation over the web.

A web-based planning toolkit called *Planning Analyst* is outlined in Chapter 5 that has been constructed for the town of Verona in Dane County, Wisconsin, to allow citizens to participate in community-based, land use decision-making. The toolkit supports land use allocation, impact assessment and visualisation. Areas within the community can be examined in three-dimensional detail and future growth scenarios can be generated. Two particular features of the system – accessible at the Shaping Dane web site – are the Internet map serving software called *WebGIS* for viewing, zooming and panning and querying spatial data; and the construction of a specialised search engine or 'planning portal' for facilitating full text searching of Internet content for relevant planning data and information. Foreseeable future developments are suggested that entail a tool to electronically survey citizens and convey the results, and the development of an orthobrowser for providing images on limited band-width Internet connections.

The American perspective is continued in the final chapter in this part of the book with an elaboration in Chapter 6 of the results of four national meetings on tools for community design and decision-making that were organized by the US Department of Energy. The prime intention of these meetings was to bring together professional experts, tool providers and community practitioners to elaborate on their experiences in making use of tools that help communities to improve their decisions about land use, resource use, community development and community design. This chapter highlights key outcomes of these meetings, some key principles of effective PSS and an overview of many of the tools presented at these meetings. It provides a valuable overview of tools already in use for community design and decision-making in the US and some experiences in working with them. This chapter provides a useful bridge to the second part of the book that deals with tools that support other phases of the planning process.

Part 2
Tools for Supporting the Planning Process

7 The Planner's TOOLBOX: a web-based support system for sustainable development
Jukka Mikkonen, Mika Ristimäki, Kari Oinonen and Henning Sten Hansen
8 STEPP: a strategic tool for integrating environmental aspects into planning procedures
Gerrit Jan Carsjens, Ron van Lammeren and Arend Ligtenberg
9 FLOWMAP: a support tool for strategic network analysis
Stan Geertman, Tom de Jong and Coen Wessels
10 Geovisualisation for planning support systems
Bin Jiang, Bo Huang and Vit Vasek
11 Key to Virtual Insight: a 3D GIS and virtual reality system
Gert van Maren
12 Cellular automata and multi-agent systems as planning support tools
Paul Torrens
13 The application of case-based reasoning in development control
Anthony Yeh and Xun Shi
14 Fuzzy algorithms to support spatial planning
Dmitry Kurtener and Vladimir Badenko

Chapter 7 is the first of a sequence of chapters in Part 2 that provides more detailed insights into particular tools developed for supporting the planning process in different ways. It presents a PSS for Sustainable Development (PSSD) for the Baltic Sea region that consists of a suite of GIS-based computer-aided tools to promote sustainable spatial planning and development in this region of northern Europe. It offers examples, theories, methods, indicators, tools and scientific documents of relevance for sustainable spatial planning and development. With the help of these it will be possible to take into account more effectively the natural environmental and socioeconomic factors plus their combined effects during the various stages of regional development and spatial planning.

Environmental sustainability is the central focus also of Chapter 8. Local authorities have responsibility for creating a sustainable environment that, in practice, means finding a sustainable balance between the influences of present and proposed human activities on the urban and rural environment. This chapter describes a GIS-based Strategic Tool for integrating Environmental aspects in Planning Procedures (*STEPP*). It is intended to support the work of local authorities, in particular to improve the quality of their physical plans and the communication between disciplines and stakeholders that are dealing with environmental and spatial planning. From educational and professional applications, it is apparent that *STEPP* represents a fast, efficient and representative tool for the analysis of the environmental aspects in a planning process. The chapter concludes that present planning practice is not well equipped to deal with real-time participation in the planning process.

Chapter 9 moves the focus from environmental planning to network planning. It describes a GIS-based PSS known as *Flowmap* and illustrates ways how this instrument can support the process of finding solutions for so-called higher-order planning activities. *Flowmap* is (part of) a PSS dedicated to perform network analysis. After a discussion of the support instrument itself, its objectives, development, capabilities and users, special attention is paid to the current and potential functionality of the instrument. Thereafter, its actual application is illustrated with the help of three different real-world planning case studies. Each of these illustrates different aspects of the capabilities of *Flowmap* and provides some details of user experience.

One important area of rapid technical innovation over the last decade has been the field of visualisation. Two chapters on this subject are included in Part 2. Chapter 10 demonstrates that there is now a range of tools that can be used for geovisualisation within PSS. Several prototype geovisualisation tools for use in the context of urban and environmental planning are described. Attention is focused primarily on two potential roles of geovisualisation tools for PSS: data exploration and collaboration. The prototypes that are described integrate recent developments in Internet GIS, geospatial virtual environments and multi-agent simulations. In Chapter 11, a prototype VR environment called *K2-vI* is outlined together with its application in different planning contexts. *K2-vI* is a very powerful tool that allows users to visualise, manipulate and analyse two-dimensional and three-dimensional spatial data within a virtual reality (VR) environment. The software uses the multi-view mode approach where each view mode has a preferred VR display system. The application runs on a

PC but can also make use of high tech VR systems such as Virtual Workbench and the CAVE in order to maximize the data visualisation. *K2-vI* is a breakthrough in VR software since it allows direct interaction with three-dimensional GIS and CAD data from within the VR environment.

Spatial interaction modelling has long been associated with PSS and Chapter 12 reviews traditional land use and transport models based on gravity principles before arguing that a new wave of models based on artificial intelligence methods have much potential for use in planning contexts. Two particular simulation approaches, cellular automata models and multi-agent systems, are outlined that have the capacity to reproduce real world landscapes in a very realistic way, although the challenges to implement and operationalise these models are considerable.

The authors of Chapter 13 indicate the difference between Case-Based Reasoning (*CBR*) and rule-based reasoning which is the method adopted by most Knowledge-Based Systems (KBS) and demonstrate the advantages that the *CBR* approach has for use as a PSS for development control. The chapter provides a comprehensive explanation of the CBR system that has been built for processing planning applications in the Planning Department of Hong Kong. The system integrates a CBR shell (*ESTEEM*) with the ArcView GIS package, providing the user interface. Because the system is case-based and not rule-based, it is also applicable to other cities in dealing with development applications.

The final chapter in this part of the book outlines an approach based on the synthesis of fuzzy modelling applications with GIS. The resulting GIS Fuzzy Modelling approach (*GISFM*) is a powerful tool for scenario analysis taking into account technological, economic, ecological and social factors. The software is intended for operation with conventional MapInfo and oriented for workers in the fields of agricultural engineering, land use and ecology. The chapter outlines the main steps in the approach, including the derivation of the membership function of a fuzzy model, before describing three examples: the application of the approach to study soil contamination near St Petersburg, to assess land drainage near St Petersburg, and to estimate the sustainability of land in the Volga delta.

Part 3
Support Systems for Strategic Planning

15 A migration modelling system to support government decision-making
Tony Champion, Glen Bramley, Stewart Fotheringham, James Macgill and Philip Rees

16 Spatial decision support systems for petrol forecourts
Mark Birkin, Peter Boden and Julia Williams

17 An urbanisation monitoring system for strategic planning
Jan-Jaap Harts, Kees Maat and Henk Ottens

18 An on-line planning support system to evaluate urban and regional planning scenarios
Christopher Pettit, Tung-Kai Shyy and Robert Stimson

Chapter 15 is the first chapter of four that deal with strategic planning support, the theme of Part 3. A prototype internal migration modelling system called *MIGMOD* is demonstrated that has been built for the Department of Transport, Local Government and the Regions (DTLR) in the UK, to assist them in understanding the impacts of changes in various economic indicators on the volume and patterns of internal migration within the United Kingdom. This chapter explains the structure of the PSS that contains three modelling stages: the historical modelling of out-migration from a set of 100 origin areas; the historical modelling of the flows from each of these areas to 99 destinations; and the predictive modelling of the flows in the system on the basis of assumptions about the values of indicators and parameters in future. The chapter illustrates what the user interface looks like and how it works. The way in which the PSS operates to deliver projected flows is illustrated for one set of scenario assumptions.

Whilst Chapter 15 represents an example of the develoment of a public sector PSS for planning at the national level by central government, the PSS presented in Chapter 16 exemplifies the type of PSS that is used in the private sector. It describes a PSS known as *Site2001*, which has been developed by GMAP Ltd for the retail petroleum industry. The Internet version is known as *Forecourt Planner*. The chapter explains the components of the system including the GIS capability, site assessment facilities, market area planning applications and the analysis modules. Various applications of the PSS are considered including capital allocation, sales forecasting, investment modelling, format optimisation and representation planning. Examples of more advanced modelling applications such as price modelling, ratings optimisation and advanced location modelling are also described.

The two remaining chapters in Part 3 both involve PSS that are used in urban and regional planning. Chapter 17 describes a system called the *UPDD Monitor* that is used for monitoring the pattern, density and diversity of urban land use. The project attempts to build a PSS that enables national physical planning policies to be monitored and evaluated *ex post*. A classification of urban environment types is developed that is based on the values of the three elements of pattern, density and diversity. Moreover, a methodology for detecting change over time is presented and examples are given of how the *UPDD Monitor* is used in the Netherlands to show areas where metropolitan centres are shifting, where service areas are growing and where business estates are emerging. The final chapter describes the development of a web site that incorporates a suite of planning support tools to undertake spatial analysis at different scales. Current functionality allows users to undertake shift-share and multicriteria analyses. The site has been built at the regional level for the Wide Bay-Burnett region in Queensland in Australia and within it for the rapidly growing Hervey Bay local government area. It is intended that by providing access to multi-scaled analytic tools, the user can better understand the performance of a region and the implications of regional planning policy for settlement and land use patterns under an urban growth scenario. Several examples are used to show how these analytic techniques are or can be applied in a planning context.

Part 4
Support Systems for Land Use and Infrastructure Planning

19 Models for assessing the effects of community change on land use patterns
Cary Gaunt and Laura Jackson

20 The New Jersey (USA) Growth Allocation Model: development, evaluation and extension
James Reilly

21 Using an operational planning support system to evaluate farmland preservation policies
Richard Klosterman, Loren Siebert, Mohammed Ahmadul Hoque, Jung-Wook Kim and Aziza Parveen

22 Development of the Klang Valley regional planning support system
Ruslan Rainis, Kamalruddin Shamsudin, Mohd Nazari Jaafar and Hadi Harman Shah

23 The LiNC Viewer: an information system designed to assist with land reform
Marilene Heunis, Tania Smith and Elizabeth Gavin

24 The SPARTACUS system for defining and analysing sustainable urban land use and transport policies
Kari Lautso

25 An infrastructure potential cost model for integrated land use and infrastructure planning
Sharon Biermann

Land use examples are amongst the most frequent applications of PSS across the world. In the first chapter out of seven on the theme that deals with support systems for land use and infrastructure planning (Part 4), an overview is given of the most important simulation instruments for modelling future land use changes. It appears that many potential clients for land use change models, such as city and county planners, community groups and environmental agencies, need better information on the features, strengths and limitations of various model packages. Because of this growing need, the U.S. Environmental Protection Agency (EPA) has developed a selective inventory and evaluation of 22 leading land use change models currently in use or under development. The EPA's Office of Research and Development (ORD) initiated this effort in order to improve its ability to assess and mitigate future risk to ecological systems, human health, and quality of life. The findings of the research are summarised in Chapter 19.

The following chapter describes the New Jersey Growth Allocation Model known as *GAMe* that is a PSS, designed for interactive use, that identifies the implications of the application of certain trend-based or plan-based scenarios on land development. It allows users to explore various land use policies in light of both the existing conditions in their town and region and in light of the values and goals of the model user. The tool simply takes user-selected policy choices and forecasts how these policies would affect the location of growth (or decline) and the costs associated with the resulting land use pattern. The chapter explains the structure and tools of the PSS and contains some useful reflections on the problems that have confronted the developers of the systems that have led to revisions.

Chapter 21 describes the use of a fully operational, commercially available PSS, *What if?*™, to evaluate alternative policies for preserving farmland in a rapidly urbanizing area in the midwestern United States. It begins with a description of the study area, Medina County Ohio, and the issues facing this rapidly developing area. The *What if?* model is explained and the data and procedures required to apply the model to Medina County are outlined. The chapter concludes by considering the implications which the study has for efforts to preserve farmland and accommodate growth in the County.

The geographical focus shifts from the USA to Malaysia in Chapter 22 where a PSS developed for the Klang Valley to assist in the identification of optimum land use allocation is presented. The system, known as *KV-RPSS*, allows alternative land use maps to be generated for scenarios defined by the user. Demand for land is generated from population projections and land suitability is determined using multicriteria analysis; the system then allocates the most suitable land for the use required in the optimum way. Examples are shown of the software interface and of the results of three different scenarios – development clustered around existing built up areas, development linked to transport stations and interchanges, and more dispersed development. In addition to the scenario generation module, the system also contains modules for impact analysis and monitoring purposes. Examples of each of these are provided. The chapter concludes with some comments on the strengths and limitations of the system, some discussion on the implications for using the system in practice and some ideas about future developments.

The *LiNC Viewer*, described in Chapter 23, is a web-based viewer to assist departmental workers in effective and efficient land reform in South Africa. The main benefit of the *LiNC Viewer* to land reform officials is immediate access to spatial information concerning the location of a property or properties in relation to infrastructure and services, which relates to the feasibility of the project. Resources in terms of cost and time involved in travelling to the sites of potential land reform can thus be saved. This chapter describes and illustrates the steps followed to design and implement the *LiNC Viewer*, data available on the system, its functionality and future development plans. In general, the response of users to the system has been positive but difficulties were encountered in developing the LiNC Viewer as well as in integrating its use into the land reform process.

Chapter 24 presents a synopsis of *SPARTACUS* (System for Planning and Research in Towns and Cities for Urban Sustainability), a strategic urban planning system based on a state-of-the-art land use/transport model (*MEPLAN*) combined with a set of urban sustainability indicators, the GIS based Raster method to calculate values for some of the indicators, the *MEPLUS* database and presentation module as well as the decision support tool *USE-IT*. The *USE-IT* module is for evaluating the results of policy options based on weighted indicator values representing social, environmental and economic effects of the policies. The *SPARTACUS* system has been used for assessing options for urban sustainability policies in a systematic way in three cities, Helsinki, Bilbao and Naples. The results show that the level of sustainability can be increased in the test cities in a cost-effective way by using a variety of policies, especially pricing and regulatory policies and their combinations. Recommendations based on these case studies have been given for sustainable urban policies and future research.

The last chapter in Part 4 explains the development of a potential cost model for strategic 'bulk' infrastructure that allows decision-makers to ensure that the full costs of alternative infrastructure plans are identified. The theoretical basis of this work is found in threshold analysis in which urban development is thought to encounter thresholds based on physical or man-made limits. The existence of these thresholds is indicated by a steep rise in the marginal cost curves associated with further development. The costing methodology is explained in detail in the chapter and there are case study examples of the application of the methodology in plan generation and plan evaluation in the Greater Pretoria Metropolitan Council area.

Part 5
Support Systems for Environmental Planning

26 The Environmental Information System: a data analysis and presentation tool
Flavio Bonfatti, Paola Monari, Andrea Aime and Silvia Ascari
27 DESIMA: a DEcision Support tool for Integrated coastal zone MAnagement
Marieke Eleveld, Philippe Carreau, Wolfram Schrimpf and Ardy Sieger
28 WadBOS: integrating knowledge to support policy-making for the Dutch Wadden Sea
Guy Engelen, Inge Uljee and Kees van de Ven
29 A planning support system for policy formulation in water resources rehabilitation
Mohammad Ali Sharifi and Erasmo Rodriguez

Part 5 contains four contributions about PSS developed for use in environmental planning. The first of these outlines the ISOLA project. This EU-funded (LIFE-programme) project is aimed at providing a set of methods and tools to support environmental planning in medium-sized towns. The resulting piece of software – the Environmental Information System (*EIS*) – supports three fundamental environmentally oriented activities: eco-balance; eco-plan; and eco-management. In the chapter, the requirements, architecture, process, methodology and planned improvements are presented. Moreover, the different functionalities are illustrated with examples from the Italian city of Modena.

Coastal erosion and the effects of coastal engineering – such as altering ports and constructing sea defences and the impact of oil spills – are just some of the many problems that management authorities in Europe's coastal regions have to face. The Marine Environment Unit of the European Commission's Joint Research Centre supports these authorities through the development of methods, tools and systems that integrate distributed coastal data and information. The contribution in Chapter 27 introduces a distributed information system, named *DESIMA*, that was specifically created to support integrated coastal zone management. Key requirements for the design and implementation were that large data sets remain with data providers and that their data formats need not be changed. Furthermore, users had to have real-time access to data and models through a user interface that allows interpretation of various formats on a single PC or workstation using a regular browser. To meet these requirements, the Common Object Request Broker Architecture (CORBA) was selected as a basis for the design of the *DESIMA*-demonstrator, whereas Java was used for the user interface. Two applications, a coastal defence for the town of West Bay in the South of England and an oil spill scenario in the Gulf of Lion in the Mediterranean, demonstrate how a user can benefit from the *DESIMA* concept by using data and software that are implemented and maintained elsewhere.

Chapter 28 describes the development of a PSS for the Wadden Sea, an important estuarine system in the northern part of the Netherlands. It is part of a larger system extending into northern Germany and western Denmark. In the Netherlands, the Wadden Sea has been declared a protected natural site because of the important ecological functions that it fulfils. At the same time, the sea has an important economic function. Fishing, recreation, transportation and gas exploitation are among its main economic activities. The management of the different activities and functions of the sea is distributed over a great number of institutions, ranging from the local to the European. In order to streamline the process of decision-making, it was decided in 1996 to initiate the development of a PSS. The aim of this system was to gather the knowledge available about the sea and to link it as part of an integrated model with a view to make this knowledge available in an operational form usable for policy and decision-making. Today, *WadBOS* exists in its second version. It consists of an integrated model representing the ecological and the economic functions of the sea. The constituting sub-models represent processes operating at different time scales and at different spatial scales. The *WadBOS* system relies heavily on GIS information for its inputs, but its models need economic, demographic and ecological data from other sources. The chapter documents in detail the development process of *WadBOS*, its functions and capabilities, and its application in the policy-making process.

In the final chapter of the book, a PSS is described that helps in the formulation and evaluation of the impacts of different policy instruments aimed at reducing the extraction from groundwater for agricultural use. Western Mancha is a low populated region in Spain characterized by severe dryness. Since 1970, in some years the volume of extractions (discharge) has been greater than the recharge of the aquifer system. The progressive lowering of the water table in the aquifer has reduced the nearby National Park wetland area by more than 60 per cent of its original area, and has jeopardized the supply of drinking and irrigation water. To improve the situation, a plan to reduce extractions was carried out. This was a problem that involved decision-making at two levels. At one level, policy-makers had to decide on the policy having most effect on rehabilitation in the region, in the face of uncertainty due mainly to the farmers' response to the proposal. At another level, farmers had their own decision-making problem: how best to respond to the new policy environment, given their own objectives and constraints. In order to solve this problem, the uncertainty was to be reduced by modelling the farmers' reaction to various policy decisions. Using this system, policies were generated and their performance evaluated in each case. The system included three models: a water balance model for simulating the average yearly recharge of the aquifer system; a mixed integer programming model to simulate the reaction of the farmers towards changes in the present subsidy schemes; and a multicriteria analysis model to support evaluation of the policy instruments. All three models are described, as well as their application in western Mancha.

Acknowledgements
This book on *Planning Support Systems in Practice* could not have been completed without support of many individuals. As editors, we would like to thank all the authors (more than 70 in total) from different countries all over the world for their contributions both to the inventory and to the book. Second, we would like to thank our colleagues in the Graphics Units in School of Geography at the University of Leeds (UK) and in the Kartografisch Laboratorium of the Faculty of Geographical Sciences at Utrecht University (NL) for their tremendous efforts in preparing the manuscript in camera ready form, financed by both our institutions. We are particularly grateful to Alison Manson, Ton Markus and Adam Davenport for their immense contributions.

References

Batty, M. and Densham, P. (1996) Decision support, GIS, and urban planning (at: *http://www.geog.ucl.ac.uk?~pdensham/SDSS/s_t_paper.html*).

Bishop, I.D. (1998) Planning support: hardware and software in search of a system, *Computers, Environment and Urban Systems*, 22(3): 189-202.

Brail, R. and Klosterman, R. (eds.) (2001) *Planning Support Systems: Integrating Geographic Information Systems, Models, and Visualization Tools*, ESRI-Press, Redlands.

Clarke, M. (1990) Geographical information systems and model-based analysis: towards effective decision support systems, Chapter 15 in Scholten, H. and Stillwell, J. (eds.) *Geographical Information Systems for Urban and Regional Planning*, Kluwer, Dordrecht, pp. 165-175.

Couclelis H. (1989) Geographically informed planning: requirements for planning relevant GIS, Paper presented to the 36th North American Meeting of Regional Science Association, Santa Barbara.

Densham, P.J. (1991) Spatial Decision Support Systems, in Maguire, D.J., Goodchild, M.F. and Rhind, D.W. (eds.) *Geographical Information Systems: Principles and Applications*, Longman, London, pp. 403-412.

Friedmann, J. (1987) *Planning in the Public Domain; From Knowledge to Action*, Princeton University Press, Princeton.

Friend, J. and Jessop, W.N. (1969) *Local Government and Strategic Choice: An Operational Research Approach to the Processes of Public Planning*, Tavistock Publications, London.

Geertman, S. (1999) Geographical information technology and physical planning, in Stillwell, J., Geertman, S. and Openshaw, S. (eds.) *Geographical Information and Planning*, Springer Verlag, Heidelberg, pp. 69-86.

Geertman, S. (2001) Planning support by PSS: an inventory, Paper presented at the CUPUM-2001 Conference, Honolulu.

Geertman, S. (2002) Participatory planning and GIS: a PSS to bridge the gap, *Environment and Planning B: Planning and Design,* 29(1): 21-35.

Geertman, S. and Stillwell, J. (2000) Geoinformation, geotechnology and geoplanning in the 1990s, *Working Paper 00/01*, School of Geography, University of Leeds, Leeds.

Gordon, Th., Karacapilidis, N., Voss, H. and Zauke, A. (1997) Computer-mediated cooperative spatial planning, in Timmermans H. (ed.) *Decision Support Systems in Urban Planning*, E. & F. N. Spon, pp. 299-309.

Harris, B. (1960). Plan or projection: an examination of the use of models in planning, *Journal of the American Institute of Planners*, 26(4): 265-272.

Harris, B. (1989) Beyond Geographic Information Systems: computers and the planning professional, *Journal of the American Planning Association*, 55: 85-90.

Harris, B. (1998) Optimization and design, *Environment and Planning B: Planning and Design; Anniversary Issue*, pp. 23-28.

Harris, B. (1999) Computing in planning: professional and institutional requirements, *Environment and Planning B: Planning and Design*, 26: 321-331.

Harris, B. and Batty, M. (1993) Locational models, geographic information and planning support systems, *Journal of Planning Education and Research,* 12: 184-198.

Hopkins, L.D. (1998) Progress and prospects for planning support systems, *Environment and Planning B: Planning and Design, Anniversary Issue*, pp. 29-31.

Hopkins, L.D. (1999) Structure of a planning support system for urban development, *Environment and Planning B: Planning and Design*, 26: 333-343.

Kammeier, H.D. (1999), New tools for spatial analysis and planning as components of an incremental planning-support system, *Environment and Planning B: Planning and Design*, 26: 365-380.

Klosterman R.E. (1994) International support for computers in planning, *Environment and Planning B*, 2: 387-391.

Klosterman, R.E. (1997) Planning support systems: a new perspective on computer-aided planning, *Journal of Planning Education and Research*, 17(1): 45-54

Klosterman, R.E. (1998) Computer applications in planning, *Environment and Planning B: Planning and Design, Anniversary Issue*, pp. 32-36

Klosterman, R.E. (1999a), Guest editorial: New perspectives on planning support systems, *Environment and Planning B: Planning and Design*, 26: 317-320

Klosterman, R.E. (1999b) The What if? Collaborative planning support system, *Environment and Planning B: Planning and Design*, 26: 393-408.

Laurini, R. (1998) Groupware for urban planning: an introduction, *Computers, Environment and Urban Systems*, 22(4): 317-333.

Lee, D.R. (1973) Requiem for large-scale models, *Journal of the American Institute of Planners*, Washington, May: 163-178.

Nedovic-Budic, Z. (1998) The impact of GIS technology, *Environment and Planning B: Planning and Design*, 25: 681-692.

Nyerges, T.L. and Jankowski, P. (1997) Adaptive structuration theory: a theory of GIS-supported collaborative decision making, *Geographical Systems*, 4(3): 225-259.

Ottens, H. (1990) The application of geographical information systems in urban and regional planning, in Scholten, H.J. and Stillwell, J. (eds.) *Geographical Information Systems for Urban and Regional Planning*, Kluwer, Dordrecht, pp. 15-22.

Scholten, H. and Stillwell, J. (eds.) (1990) *Geographical Information Systems for Urban and Regional Planning*, Kluwer, Dordrecht.
Shiffer, M. (1992) Towards a collaborative planning system, *Environment and Planning B: Planning and Design*, 19: 709-722.
Shiffer, M.J. (1995), Interactive multimedia planning support: moving from stand-alone systems to the World Wide Web, *Environment and Planning B: Planning and Design*, 22: 649-664.
Singh, R.R. (1999) Sketching the city: a GIS-based approach, *Environment and Planning B: Planning and Design*, 26: 455-468.
Stillwell, J., Geertman, S. and Openshaw, S. (1999) Developments in geographical information and planning, Chapter 1 in Stillwell, J., Geertman, S. and Openshaw, S. (eds.) *Geographical Information and Planning*, Springer Verlag, Heidelberg: 3-22.
Stillwell, J. and Scholten, H. (eds.) (2001) *Land Use Simulation for Europe,* Kluwer, Dordrecht.
Worrall L. (1994) The role of GIS-based spatial analysis in strategic management in local government, *Computers, Environment and Urban Systems*, 18(5): 323-332.

PART 1

SYSTEMS AND TECHNOLOGIES FOR ENHANCING PARTICIPATION IN THE PLANNING PROCESS

2 Interactive Support Systems for Participatory Planning

Stan Geertman[1] and John Stillwell[2]

[1] URU/Nexpri, Faculty of Geographical Sciences, Utrecht University, Heidelberglaan 2, 3508 TC Utrecht, The Netherlands

[2] School of Geography, University of Leeds, Leeds, LS2 9JT, United Kingdom

Abstract: Central to the focus of this chapter is the recognition that public participation in planning is increasing at a tremendous pace. One of the consequences of this trend is the necessity to develop new and user-friendly planning support systems (PSS) which can be accessed by all those interested in the development of their regions or localities. The imperative is for interactive and/or participatory PSS, dedicated to the broad diversity of actors involved. In this context we seek to illustrate in this chapter some examples of these PSS designed to disseminate information as widely as possible and to facilitate participation in contemporary planning practice. We conclude that, although it is still too early to evaluate the benefits of these new geotechnological systems, the first signs are promising.

2.1 Introduction

More and more planning support systems (PSS) of different levels of sophistication are being used at different stages in the planning process. Contributions in this book reflect the diversity of PSS that is apparent in the wider literature (see, for example, Gordon *et al.* 1997, Nyerges and Jankowski 1997, Hopkins 1998; Batty *et al.* 1999, Kammeier 1999, Klosterman 1999, Singh 1999). However, although the range of examples continues to expand, planning practice itself still remains underprovided with tools for planning support. Various authors have detected this discrepancy. Klosterman (1998), for example, states that the prospects for planning support are exciting but that the practice will be disappointing for the next decade at least. In his view, this deficiency is due primarily to the wide diversity of the analytical tasks that planners perform, the relatively small market for public sector software, and the costs of developing and supporting commercial software. Nedovic-Budic (1998) too demonstrates the underutilization of GIS tools in planning practice, especially for more sophisticated analytical and modelling requirements.

At least two further reasons can be identified for this underutilization. Firstly, there is still a deep-rooted resistance in planning practice to more quantitative modelling and analysis methods that some regard as the focus of PSS. This may be a function of the need to have transparent and easily understood methods underpinning policy and planning proposals, but it may be partly due to the lack of attention paid in planning education to quantitative methods of analysis, design and evaluation and the way these methods can be applied to good effect. Besides the resistance voiced by the less mathematically educated students, this lack of will to embrace more sophisticated methods may be seen as a consequence of the historical frustrations associated with the very grand aspirations of the so-called quantitative revolution in the social – including the geographical – sciences during the 1960s. Lee's requiem for large-scale models illustrates many of the shortcomings that were apparent in the urban models of this era (Lee 1973). In many parts of the world, the newer generation of planners with skills in GIS, spatial analysis and modelling is still in the process of overhauling previous generations lacking proper awareness of the benefits of new technologies or the willingness to try them out (Geertman 1999).

Secondly, underutilization can be ascribed to the characteristics of many PSS, whose development has been driven primarily by the potential capabilities of the technology itself (supply-directed) instead of the quest to provide solutions to real world problems or to encourage participation (demand-directed) (Geertman 2001).

Thus, in this chapter we begin with a brief discussion of trends in the practice of planning itself, the developments that have occurred therein and the emerging and still unfulfilled needs for support systems. Important among the trends has been greater emphasis on more openness, accountability and participation in the planning process. For too long, the manipulation and interrogation of information has been perceived as the responsibility of the practitioners without the means to involve the public or even to disseminate their results more widely. In Section 2.3, we take a closer look at the opportunities for the adoption of support systems in different contexts that involve participation. In Section 2.4, two examples are presented that illustrate PSS developed by selected pioneer local authorities in the UK to facilitate access to quantitative data that relate to particular neighbourhoods and that are used to underpin community policy. Two generic systems used in the Netherlands, *I-pl@n* and *SketchGIS* are then considered in Section 2.5 that are designed respectively to communicate development plans and to facilitate the process of scenario-building. Both of these latter PSS attempt to address the need to improve participation in planning practice and some initial experiences with the use of *SketchGIS* in Dutch planning are also provided. Concluding remarks are presented in the final section.

2.2 Key Themes in Contemporary Planning Practice

During the last decade of the twentieth century, planning practice changed significantly. In Europe and, to a large extent, in North America too, planning became more holistic in scope, more strategic and scenario-oriented in content and more interactive in nature (for an elaboration of these trends, see Stillwell *et al.* 1999).

The boundaries between different types of planning were redefined in the 1990s. Traditional planning has tended to be undertaken through the activities of different planning professions (environmental; infrastructure; economic; social); contemporary planning is much more holistic, exemplified in the UK by central and local government preoccupation with 'joined-up' administration and policy-making. Region-oriented, integrated land use planning, as well as economic, environmental and transportation planning, manifest characteristics of the trend towards a more holistic and comprehensive approach to strategic planning. Integrated transport planning, for example, has become a means for developing sustainable policies that balance the freedom to travel with the need for a healthy environment and support for a competitive economy while assisting urban and regional regeneration. Along these lines, new strategies for regional economic development, innovation, integrated transport and land use, sustainable environments and enhanced quality of life are being formulated and implemented. These are typified by the strategy documents prepared by the new regional development agencies and authorities for regional governance that have emerged in the UK under 'New Labour'. More generally, integration is no longer restricted to plan design, but applies to the overall planning process, including the specification of goals, programmes, financial arrangements, implementation, monitoring and evaluation.

At a more technical level, the last decade has seen a shift towards the use of scenarios as a means of giving concrete expression to spatial proposals for the future and a move towards communicating these proposals to those concerned in a more effective fashion. Spatial scenarios (e.g. the construction of a new airport, the location of a new industrial estate or the development plan for a region or locality) offer various views of the future, based on different assumptions or underlying trends and on what might be the best or, at least, the satisfactory spatial outcome(s). Scenario planning reflects the idea that planning should offer inspired visions of the future. This kind of strategic scenario planning activity can be seen as a reaction to the more procedural and instrumental orientations typical of planning in the 1970s and early 1980s. One corollary is the shift in emphasis towards broader strategic programmes and away from an approach that focuses on smaller projects (for example, see Bryson and Einsweiler 1988). This project-to-programme shift has been particularly apparent in the restructuring of European Union funds for regional development during the last two decades.

The trend that we focus on in this chapter is that towards improved openness, transparency and participation in the planning process. This applies to planning at different levels by different organizations. In particular, central, regional and local government planning authorities have begun to embrace Internet technology as a means of communication. Most agencies now have web sites that contain key information and data sets that can be accessed, interrogated and often downloaded by whoever is interested. The World Wide Web has become an extremely efficient channel for transferring data across the Internet because of its visual capabilities and its relatively advanced multi-media tools. The use of the Internet to allow the public to exploit GIS capabilities and play an enhanced participative role in the decision-making process has been well documented by Carver and Peckham (1999).

Traditionally, planners prepare preliminary plans. The public and other parties affected by the outcomes of the plan are consulted through public inquiries or hearings. This consultation has usually taken place when the plan is almost in its definitive stage and when the range of alternative possibilities has been narrowed down. In the new era, the public and various private agencies are collaborating much more in the development of planning projects and policy initiatives. In fact, public-private partnerships (PPPs) have become a familiar approach for many development schemes. More generally, all the 'stakeholders' with an interest in the plan are invited to participate in the planning process right from the start. In this way, planning is no longer an exclusive task of government agencies, local authorities or private companies of professional planners. Rather, it is an organizational/financial/responsive agreement between the public and private partners involved. At least in part, satisfactory implementation of the plan depends on successful co-operation between all parties concerned (Jankowski *et al.* 1997).

2.3 Participatory Planning

Generally speaking, the lack of transparency and the complexity of public administration are felt to be responsible for the slow and laborious procedures that tend to characterize many planning processes. Participatory planning is seen as a means to overcome some of the problems and facilitate the processes inherent in contemporary planning practice. However, for public participation this often means a transformation from confrontation to collaboration. *"The process will be defective so long as politicians believe that they have an exclusive writ to represent the needs of the public, so long as planners believe that they have the exclusive prerogative of studying and understanding the city and deciding what should be done with it, or of presenting plans to the public and interpreting their reactions, or so long as the public believes that it is being disregarded or mistreated, and in each case feels that the only way to defend particular interests is through intransigent dispute with other interests and with planning bodies"* (Harris 1999, p. 326). As a consequence, what is required is for all participants involved in the participatory planning process to adopt a particular attitude that results in a network of mutually interdependent actors striving for a common image of reality, for consensus on the problems that need to be dealt with and for a solution to the particular problems that have been prioritised.

2.3.1 Situations for Participatory Planning

Participatory planning is not necessarily the most appropriate planning method to deal with real world problems in all instances. Its appropriateness depends first and foremost on the unique combination of problem and type of planning. Table 2.1 is a simplification of reality but it does indicate that participatory planning is most appropriate for handling so-called 'unstructured problems' where there is less agreement concerning the goals and values as well as the knowledge required to solve the problems. As a consequence, each individual's perception of the problem must be taken into account explicitly. Mutual exchange of ideas, knowledge, perceptions, viewpoints *et cetera,* is of foremost importance (*planning as a learning process*) and for this situation, a participatory planning process is highly suitable. In the situation where problems can be classified as less-structured, participation is restricted because there will be no discussion concerning the goals and values that have been agreed. Instead, participants will concentrate their discussions on the pros and cons of each of the alternatives, which means that the process is much more a *planning by negotiation* instead of mutual learning. Another situation in which participatory planning is actively restricted and only partially appropriate is where there is less agreement concerning goals and values, so there is the fear of escalation of discussions as a consequence of conflicting goals and *planning is a pacifying process*. Finally, there is the situation wherein planning is restricted to the regulation of 'well-structured' planning problems. This implies *planning by regulation* and in this case, participatory planning is not seen as the most appropriate planning method.

Table 2.1: Cross-classification of policy problems and types of planning

	less agreement concerning goals and values	**full agreement concerning goals and values**
less agreement concerning knowledge needed	unstructured problems planning as a learning process very appropriate for pp	less-structured problems planning as negotiation partially appropriate for pp
full agreement concerning knowledge needed	not-to-structure problems planning as pacifying partially appropriate for pp	well-structured problems planning as regulation inappropriate for pp

note: pp = participatory planning

Source: Drogendijk and Duijn (1999).

Thus, the type of policy problem is very important in dictating the appropriateness of the planning method. Complex societal and political problems are more suitable for participatory planning but this is not the case for complex technical problems, where the relationship between action and effect is unclear (Drogendijk and Duijn 1999). In short, the use of participatory planning is most appropriate when the policy problem is clearly defined, not abstract, preferably focused on a certain area and pre-defined scale, and involves many different parties with conflicting interests. However, in essence, the classic approach and the participatory planning method are not necessarily mutually exclusive. In many cases, they may be complementary, with each predominating in certain phases of the planning process. Thus, every situation and problem will call for a unique mix of participatory and classic planning elements.

In fact, several authors (including Alexander 1998) make a strong plea for complementary methods. In the words of Voogd and Woltjer, *"both can complement each other in varying combinations as the quality of planning processes and its outcomes are contingent upon the particular context of space, time and people."* (Voogd and Woltjer 1999, p. 851).

The reasons for embarking on a participatory planning process can be very diverse (Harris 1999). On the one hand, these reasons might be largely instrumental (content-oriented) in that the process is expected to yield some advantage. For instance, citizens can contribute directly to the knowledge and innovation needed within the planning process. They can help define and reconcile values, offer new information about urban conditions, provide new insights into relationships and thus help predict the consequences of planned actions, and bring in new ideas on measures for public action and on ways to orchestrate such actions. On the other hand, the reasons might be cultural (process-oriented). The expected/seen advantages might take the form of better working relationships with the parties involved or more tolerance of each other's standpoint. It is expected that people with diverse interests who will work together to analyze and solve problems are the more likely to reduce than exacerbate conflicts over values. Both kinds of motives should be explicitly taken into account in the participatory planning process. From practice, while is seems that participatory planning increases the support for the ultimate decision, it does not appear to reduce the amount of time and expense involved, and rarely does it lead to win-win situations.

2.3.2 Requirements for Participatory PSS

At least three characteristics of the (participatory) planning process influence the suitability of its supporting instruments: the diversity of participants involved; the increased interaction between the participants and between the participants and the supporting instruments; and the characteristics of the planning method, planning process and policy context.

A participatory planning process is frequently characterized as involving a wide range of actors with a great diversity of interests, backgrounds, status, *et cetera* (e.g. public agencies, politicians, civil servants, private-sector firms, pressure groups, civilians, personally involved individuals, technical or scientific advisers). Each participant will bring different goals, tasks, skills and limitations to the planning process. Moreover, they will all have different perceptions of the need for and the availability, accessibility and quality of data, information and knowledge. New information will be continually used to update personal knowledge and differences between factual information and experience will prompt a profound questioning of the validity of the data. In addition, participants will possess a variety of levels of skill that will restrict some but allow others to fulfil their requests for data, information and knowledge. Those with information and communication technology (ICT) skills are at an advantage. Some people prefer pictures and diagrams to written text; some possess advanced computational and programming skills, while others are computer illiterate. In response to this situation, planning support instruments must endeavour to meet the differing requirements and skill levels of the participants concerned. This means that the instruments should be at least transparent, understandable and user friendly for all the participants. Moreover, they should be adaptable to the situation at hand, i.e. to the problem under investigation.

Human interaction with analytical tools is frequently abstract in nature, both for inputs (e.g. computer commands) and for outputs (e.g. numerical data). It is sometimes argued that this level of abstraction may cloud the public's understanding of potentially beneficial projects and thereby contribute to misunderstanding in the planning process (Forester 1989). On the other hand, some media such as maps tend to be a useful way of allowing data to be visualized and interpreted effectively. Nevertheless, maps can also present a biased perspective, oversimplify reality (leave out too much detail) or over-impress the audience (e.g. a fancy presentation concealing a weak argument). Thus, acceptance of the inputs (e.g. reliable, up-to-date data) and outputs (e.g. meaningful modelling results) of a PSS is of utmost importance to fulfil a support function. In response to the obstacles in human interaction with computer tools, transparency is seen as one of the most important prerequisites for support tools and for the way in which they are handled.

It is also essential to consider the type of planning problem when considering the appropriate PSS. In fact, the nature of the planning problem dictates the choice of an appropriate planning method and its supporting instruments. Moreover, the planning problem is itself influenced by the characteristics of the planning or decision-making process and the characteristics of the policy context in which the problem should be solved (Geertman 1999). Each planning or decision-making process is subject to a certain time constraint, degree of participation, publicity, *et cetera*. These factors influence the potential for and restrictions on the use of the technology. Likewise, the characteristics of the policy context (e.g. democracy or dictatorship) influence the preferred technology (transparent or black-box) and the way in which it is used.

Instruments that are developed to support the planning process need to take these factors or pre-requisites outlined above into consideration. In the next two sections of the chapter, we present some examples of PSS that have been developed to enhance planning participation through either the improved access of individuals to information of different types or through the availability of generic PSS or their application in different circumstances. In Section 2.4, we start from the premise that poor information breeds ignorance and that ignorance leads to poor policy-making, resulting in failed interventions and deeper problems. Consequently, it is necessary to provide better information to help diagnose problems more accurately, to develop 'smart' strategies to deal with the causes rather than the symptoms, and to evaluate the outcomes that result from selected policy responses to find out whether the policies are working or not. We present two selected examples of PSS developed by individual local authorities in the UK to exemplify good practice in terms of information dissemination and the encouragement of participation. Thereafter, in Section 2.5, two different PSS instruments are described, both of which support overall participation into the planning processes. The first system, *I-pl@n*, is especially relevant for communication between civil servants via an intranet or with the wider public via the Internet. The second PSS, *SketchGIS*, is designed more specifically for facilitating communication within participatory planning sessions.

2.4 Interactive Access to Online Vital Statistics in the UK

The measurement and monitoring of what changes, by how much and for whom is it important is a current preoccupation with policy-makers and practitioners throughout the UK, where the paucity of statistical indicators for small areas puts great significance on the results of the Census of Population that is conducted at the beginning of each decade. Between censuses, the amount of reliable evidence that is available on demographic, economic and social indicators for small areas is very limited, let alone on measures of environmental sustainability. Yet, it is at this spatial scale where planning and policy-making has its impact, especially in connection with the wide range of initiatives that government agencies have implemented for neighbourhood regeneration. At national level, the Office for National Statistics (ONS) has begun to develop a Neighbourhood Statistics Service (NeSS) (*http://www.statistics.gov.uk*) that provides data across the country via the Internet on electoral populations and benefit claimants at ward level. This complements the existing National Online Manpower information System (NOMIS) that provides small area labour market data (*http://www.nomisweb.co.uk*).

At a local level, a number of local authorities in the UK have begun to develop web-based PSS that enable information about neighbourhoods to be accessed and progress towards meeting targets for regeneration to be evaluated. We consider two systems by way of illustration. *Oi4B* stands for Open Information for Birmingham and is a website that was developed on behalf of City Pride, the strategic body that monitors the overall progress of the city. The City Pride Manager co-ordinates a steering group for *Oi4B* that includes City Pride, Birmingham City Council and Birmingham Health Authority. This site, whose home page is shown in Figure 2.1, is open to anyone wanting access to the information and the acronym '*Oi*' cleverly draws attention to the inequalities that exist across the city. City Pride has been confronted with the problems of agreeing a number of outcomes for the city, defining indicators to use for measuring progress towards those outcomes, disseminating this information to the public and raising the debate about the underlying issues. They are conscious of the limitations of the information available. In particular, the reliance on old, unreliable and inconsistently collected information on ethnicity is a serious limitation which the partners will address when the results of the 2001 Census become available in 2003, although this will not provide the year-on-year information which is required for monitoring change on a regular basis.

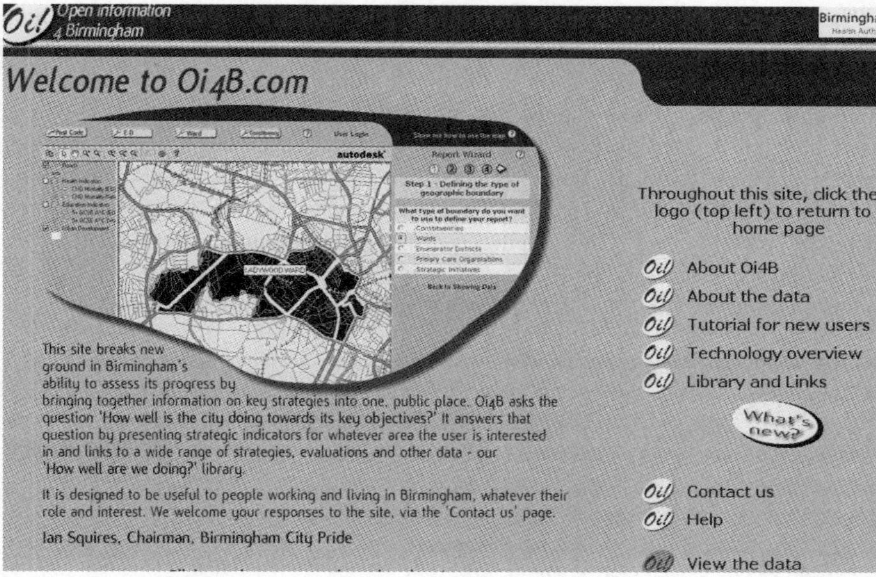

Fig 2.1: The *Oi4B* home page (*http://www.oi4b.com/*)

Reducing inequalities in localities has become an important focus for Government policy initiatives. The Department of Transport, Local Government and the Regions (DTLR) has a Neighbourhood Renewal Unit focused on tackling deprivation through the bending of main Departmental programmes such as the police and health services, to deal more specifically with the most deprived areas (*http://www.neighbourhood.dtlr.gov.uk/*). All the relevant Departments now have minimum floor targets to meet, which means that services will be judged on the areas where they are doing worst, and not on the national average.

Oi4B currently displays the Government's floor targets for neighbourhood renewal, covering health, education, employment, community safety and housing. These targets include:

- elimination of substandard social housing by 2010, and its reduction by a third by 2004;
- ensuring that no area has burglary rates three times higher than the national average by 2005;
- ensuring that at least 25 per cent of pupils in every school and 38 per cent in every local authority area achieve five or more GCSEs at grades A* to C;
- reducing by at least 10 per cent the gap in mortality rate between manual groups and the general population by 2010; and reducing by at least 10 per cent by 2010 the gap between the fifth of areas with the lowest life expectancy and the national average; and
- raising employment in the 30 local authority districts within the worst labour market problems and narrowing the gap between these and the overall rate.

The *Oi4B* website provides access to the indicators that monitor progress towards these targets through a library of material. City Pride is attempting to make evaluations more transparent by placing them in the public domain and in so doing, creating a culture of openness. Users can ask for information about the data (where it comes from, how it is defined and how reliable it is); request an overview of the application; request how to use the map to find a particular area (neighbourhood); display information on a map of Birmingham; and produce a report combining several pieces of information using the 'report wizard'. The area that is displayed on the map can be updated with 'map controls' that allow the user to zoom in and out and to move left or right as indicated in Figure 2.2. There are 'goto' buttons to jump to particular areas.

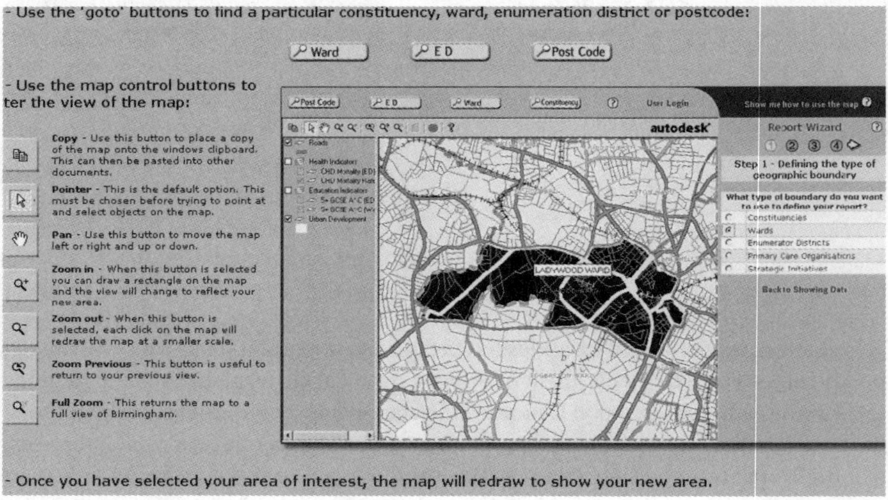

Fig 2.2: The mapping functionality available at the *Oi4B* website
Source: Oi4B website

The *Oi4B* web site is built up from HTML web pages, AutoDesk's Mapguide and map server applications, cold fusion scripts and an underlying database that contains both the information provided to support the indicators and also the GIS metadata, sourced from the Ordnance Survey and Birmingham City Council.

A second example is the web-based PSS developed by the City of Bradford Metropolitan District Council that also uses Autodesk's Mapguide software and whose *raison d'etre* comes as a response to the growing demand within the local authority for statistical information to support bids for funding, particularly European, and also from community groups who also require quantitative information to support their bids for resources. The system is part of Bradford Community Statistics Project (*http://www.bcsp-web.org*) whose function is to make statistical data more open to community, voluntary and campaigning groups. Thus the system has been developed to improve dissemination and broaden the use of geo-referenced data; to facilitate access to the data via an intuitive user interface; to combine 1991 Census data with other geographically referenced data; and to provide 'boundary-free' data (Thomasson 2000).

The latter aim is a response to one of the classic problems in geographical analysis – the so-called Modifiable Areal Unit Problem (MAUP) considered by Openshaw and Rao (1995) which indicates that the spatial configuration of zone boundaries will effect the geographical patterns of a particular census variable. The Bradford project recognises that a user's area of interest is unlikely to align with existing zone boundaries such as those of enumeration districts (EDs) used to report the results of the 1991 Census. Thomasson emphasises the painstaking task of trying to split existing EDs to derive an area that is consistent with the user's area, and the problems that this involves because of the likelihood that population counts are not spread evenly across each ED. The simple but nevertheless effective solution devised by the project has been to 'spread out' the census count to postcode centroids of residences within the ED. Thus, in an ED with a population of 500 residents and 250 postcode centroids, each point would be assigned a population of two people. However, this allocation is further adjusted according to the number of residential properties within each postcode. One of the important tools that is available for the user of the Bradford system is an on-screen boundary drawing tool that enables a user to draw a specific area of Bradford MDC on a detailed large scale map as indicated in Figure 2.3.

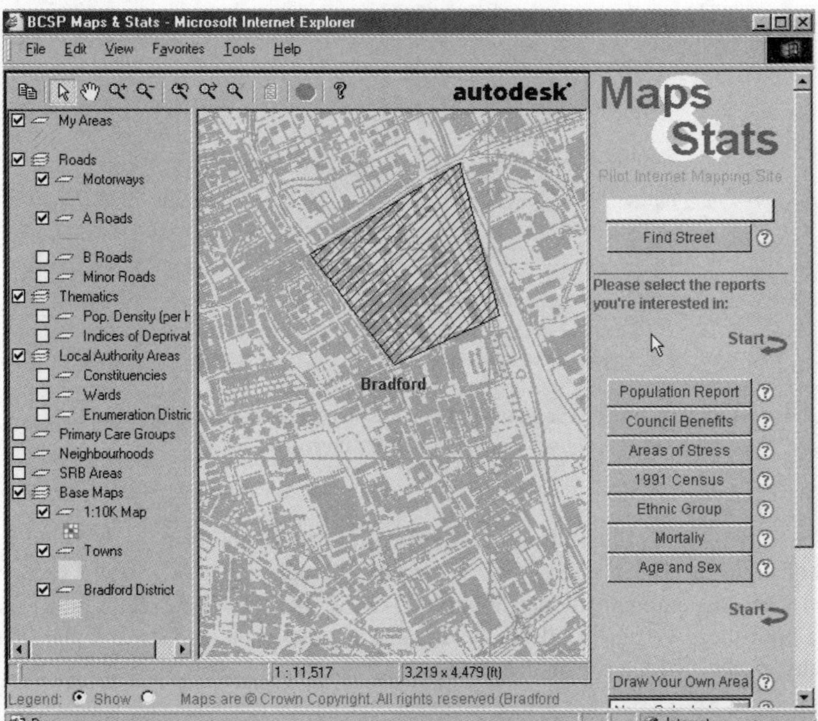

Fig 2.3: User-defined polygon, Bradford internet mapping site

Once the boundary has been 'digitised' by tracing over the background map, the user has the option of pressing buttons embedded in the browser that link to programs that take the raw data associated with each of the points in the user's area, analyse that data and present the data in the form of a report. Summary calculations can therefore be undertaken on the fly and reports also presented on the screen. In addition to this facility of generating reports for user-defined areas, the system allows the user to create selections of wards, neighbourhoods or constituencies. Reports can then be generated that illustrate the area profiles relating to demographic variables such as population by age and sex, mortality and ethnicity, as well as other stress and deprivation indices. Figure 2.4 illustrates a report on ethnicity (from the 1991 Census) for an area comprising three wards of Bradford. The ethnic structure of the selected area is compared with the ethnic structure of the whole of the Bradford District.

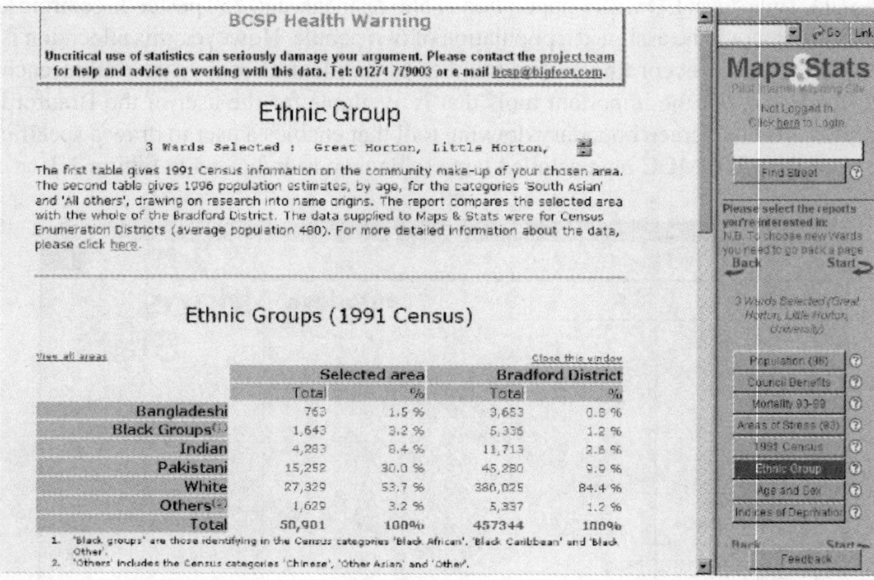

Fig 2.4: Ethnicity profile for a user-defined area based on wards in Bradford
Source: http://www.bcsp-web.org/mapguide_site/maingeo.cfm

These two examples of English metropolitan local authority web-based PSS demonstrate the type of efforts that are currently being made to provide the user with open access to community or neighbourhood-based data sets. There seems little reason why both these systems will not be adapted to give users access to some of the results of the 2001 Census in due course.

2.5 Interactive Support Tools in the Netherlands

In the Netherlands, there have been a number of attempts to involve the general public in physical planning. One early example of the use of computer technology to achieve this objective was the use of a system known as RIA (Regional Information by Automation), created by the National Physical Planning Agency in the late 1980s to disseminate information relating to the Dutch Fourth Memorandum on Physical Planning (Scholten and Meijer 1988). RIA was a system accessible from 100 or so PCs in public places throughout the Netherlands that gave individuals the opportunity to see information about the national physical plan in the form of maps, tables and text using a very simple and user-friendly interface.

More recently, several other instruments have been developed to support participation of the public into the plan-making process of physical planning. Here we will present two Dutch examples of these instruments: *I-pl@n* and *SketchGIS*.

2.5.1 *I-pl@n*: A Support Tool for Communicating Development Plans

At the local level, since the introduction of the Law on Housing in 1901, municipalities have been monitoring ongoing changes in the physical environment and planning new developments as part of the process of formulating Development Plans (*Bestemmingsplan*). Public participation always had a relatively restricted role within the planning process, although participation has improved over time so that at the moment, draft Development Plans are subject to critical enquiry by anyone who feels it necessary to do so. The public can look at the draft plans either at the Town Hall of the municipality or at the Building Department for a period of about six weeks. The same kind of procedure operates when Development Plan have been drawn up and civilians are anxious to know if the local council will give them a building permit to make changes to their homes (e.g. to erect a sun-lounge or a garage).

In most cases, members of the public have to visit the Building Department during restricted opening hours to make their enquiries. Whilst this is restrictive for those who are working, the current practice may result in a good deal of work for the municipality officials, primarily because different (parts of) Development Plans are likely to be stored in different places. Thus, the task of gathering the necessary pieces of information together to answer a seemingly simple question may be very time-consuming.

In order to enhance the accessibility of physical planning information and to restructure its possible dispersed storage, a PSS known as *I-pl@n* was conceived as an intranet and Internet-based application with which a Development Plan and its associated information, such as the building register, could be made available via a computer screen at any computer with a network connection. The aim of the PSS was to provide access to draft Development Plans, 24 hours a day for seven days a week allowing any member of the public the opportunity to evaluate proposals, scrutinize specific developments or assess the probability of being given a building permit for a building application. It provides access in a very user-friendly way, to all relevant spatial information relating to a specific address, to the judicial rules and their exceptions, and to intended developments by governmental or private organizations. For example, a member of the public can complete a request for a building permit online or request further information from the Building Department by email. In general, *I-pl@n* increases substantially the accessibility, availability and accuracy of governmental information provision to the interested public and consequently the interaction within and between government and civilians. Moreover, for municipality officials, *I-pl@n* offers a very user-friendly way to update physical planning information as stored in a building register, for example. There, information can be kept up to date much more easily.

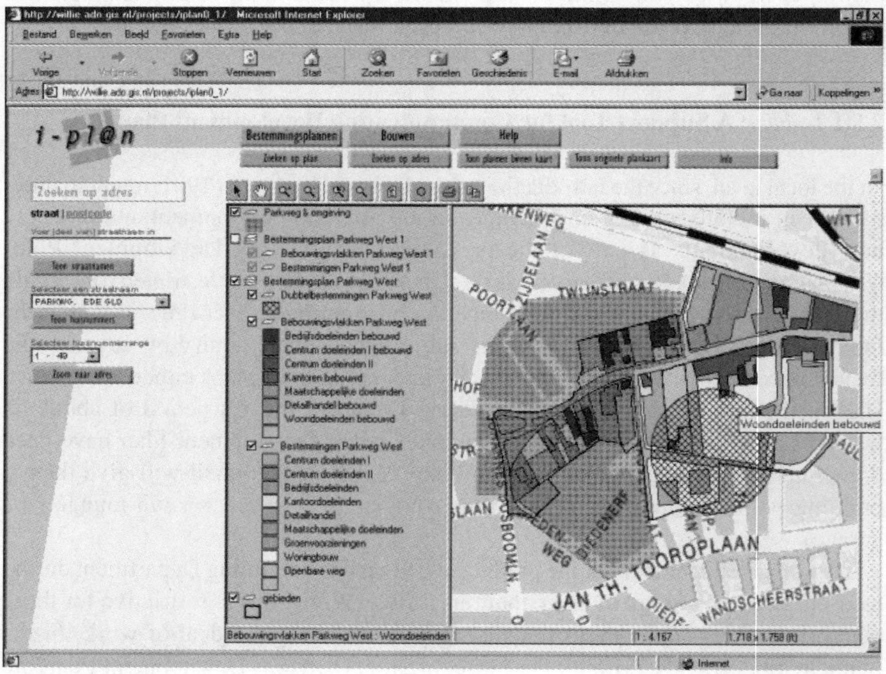

Fig 2.5: *I-pl@n* for communicating Development Plans

To work with *I-pl@n* there are two different ways to find the requested physical planning information: by entering an address or by telling *I-pl@n* the Development Plan of interest. In response, *I-pl@n* will show a map of the given street and its immediate surroundings or of the area in which the Development Plan has jurisdiction (Figure 2.5). In this map users can find the kind of development permitted for the specified address. Access is also permitted to all relevant spatial information relating to the specific address, to the judicial rules and their exceptions, and to intended developments by governmental or private organizations. Figure 2.5 shows, within *I-pl@n,* the Development Plan for a neighbourhood in which a circle indicates the area excluded for living expansion due to the location of a petrol station. The user can send back an open question for more information, reject or support the proposed plans or alterations, ask for a building permit, *et cetera.* As such, *I-pl@n* can be considered a tool to facilitate interaction and participation.

I-pl@n has been designed and built by a consortium of public sector (Utrecht University) and private sector (Nexpri; SAB; Adogis) organisations with the help of Mapguide and is currently in prototype form. Detailed information is provided at the website (*www.i-plan.nl*).

2.5.2 *SketchGIS*: An Instrument for Interactive and Participatory Spatial Scenario Building

Within the framework of a research programme called the LWI-programme (Land Water Environment Information Technology), an instrument known as *SketchGIS* has been developed and tested to support the first phase of a participatory plan-making process, the creation and evaluation of spatial scenarios about the future. The activity of scenario building is a difficult but essential step required to provide a basis from which policy initiatives are developed and implemented. The creation of visionary insights of potential or desired spatial developments is of great benefit. Spatial scenarios should encompass current and expected potentials, restrictions and bottlenecks, foreseeable and expected trends and desires for the future. Moreover, the pros and cons of each spatial scenario should be explained in the context of relevant policy goals and objectives. In this way, a comparison of the positive and negative consequences of each scenario should be made possible, allowing policy initiatives to be formulated that maximise the positive consequences and avoid the negative impacts wherever possible. The prime purpose of the *SketchGIS* instrument is to facilitate the interactive process of group scenario building (Langerak *et al.* 1999, Geertman 2001).

The *SketchGI*S planning support tool has a number of characteristics. Firstly, it has the ability to project the content of a computer screen onto a so-called 'smartboard', an electronic whiteboard of A0 format. This both enlarges the image on the screen and enables users to input commands by touching the smartboard at an appropriate location. In addition, objects can be drawn on the smartboard with the help of an electronic pencil, which are then immediately stored in the computer.

Secondly, the *SketchGIS* instrument incorporates a 'concept mapping tool' that helps participants to express, structure and communicate spontaneous thoughts during a brainstorming process. These thoughts are presented as a 'network' of concepts (e.g. goals, problems, measurements, criteria, functions, localities) that can be visualized with different symbols (boxes, lines, arrows or whatever).

Thirdly, the instrument offers regular ArcView functionality in that it projects on the smartboard all kinds of geo-information such as text documents, maps, charts, html-pages, photographic material, meta-information and 3-D images. It therefore becomes possible to show participants in the interactive planning sessions the area under study in different ways. Clearly this facilitates information provision to the various participants who might be attending one of the sessions as well as enhancing the level of communication between them.

Fourthly, with the help of sketch tools and projected geo-information, the participants can build and rebuild spatial scenarios within the area under study. They can designate areas for new housing or industry, indicate the route of a new road or railway, restrict certain attractive areas for future nature or recreational purposes, and so forth. Different scenarios can be built by departing from alternative starting-points so as to recognise different objectives such as environmental protection or sustainability, economic prosperity or attaining good living conditions. Figure 2.6 shows a sustainability scenario in which some new developments for the area under study have been designed. Finally, the spatial scenarios that have been constructed can be evaluated for their impact and/or suitability. Based on Saaty's (1977) 'principal eigenvector', a multi-criteria analysis can be performed to measure the 'goodness of fit' of the proposed spatial scenarios to the goals, criteria and restrictions of the participatory planning sessions.

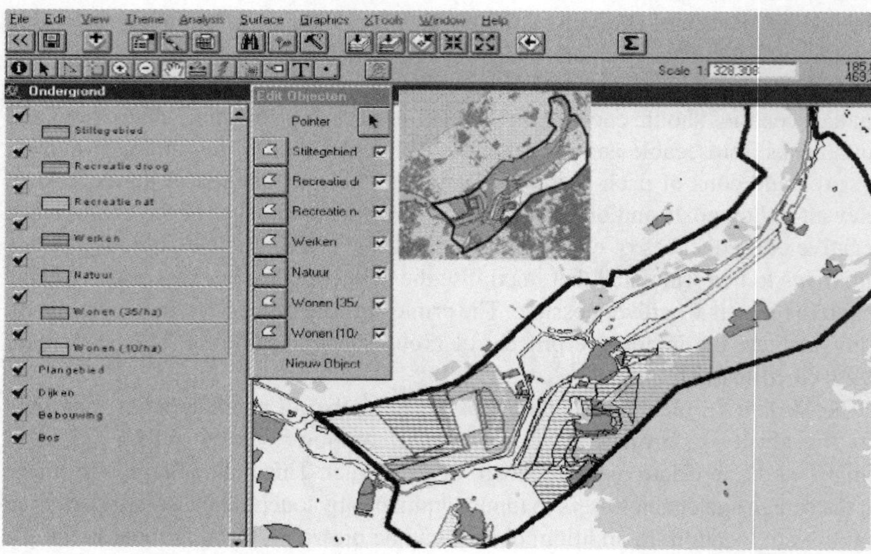

Fig 2.6: *SketchGIS* instrument for interactive and participatory spatial scenario building

The *SketchGIS* instrument has been developed by a consortium of public sector (Utrecht University; Ministry of Traffic) and private sector (NexpRI; TNO-Inro; TNO-FEL; RIKS; Cartago) organisations within the Netherlands. It incorporates

ArcView GIS, Avenue, Microsoft Visual C++, a toolkit for 3D-applications named KarmaVI, and a TOPIC prototype as a base for the concept mapping tool. *SketchGIS* is currently in prototype form. Detailed information is provided at the website (*www.nexpri.nl*).

The *SketchGIS* PSS has been applied in a variety of educational workshop settings and in some real-world planning contexts. Based on these experiences, it can be concluded that the instrument is quite flexible and easy to handle, although it still necessitates a facilitator for the interaction between participants and the computer, between man and machine. Of prime importance, however, is that it offers functionalities which ease the communication between the participants in substantial ways: proposals in the form of spatial scenarios become much clearer when they are drawn on a familiar background map on the smartboard; implications of proposals can be evaluated immediately; rebuilding of proposals in the form of revised spatial scenarios can be designed during the interactive sessions; knowledge can be captured, exchanged, combined and stored for later sessions; and by providing a meta-information service, the instrument facilitates the accessibility of available data and information.

In fact, the *SketchGIS* makes it possible to change the design process from a one-way and one-time procedure into a flexible iterative and interactive design process. In that respect, the designs of spatial scenarios should not be seen as the final product of a closed procedure, performed by some professionals. Instead, the resulting designs are intended to continuously provoke discussion within the group of participants. The designs are subject to continuous alteration until one alternative design is adopted. In that way, the design process can be characterized as an interactive process that moves from rough and tentative sketches at the outset, via a process of trial and error and discussions of consequences, to arrive at a final set of more detailed designs that all the participants feel comfortable with. For a more generalized reflection on instruments for collaborative planning like *SketchGIS*, see for example Langendorf (2001) and Shiffer (2001).

2.6 Conclusions

This chapter has reflected on a number of trends in planning practice, highlighting the importance of enhanced user participation both in terms of access to plans and policy proposals, but also to the statistical data sets that underpin them. New PSS are emerging rapidly that will transform the provision of information to all those who want to be informed. The two examples of web-based PSS developed by pioneer UK local authorities illustrate the current state of the art. We can expect many new participatory systems to emerge over the next few years as local authorities decide to use 2001 Census data to inform their local populations of the demographic and social conditions of their neighbourhoods.

In addition to these specific examples of PSS providing statistical information via the Internet, we have also reviewed two PSS developed in the Netherlands that are used to enhance knowledge of plans and facilitate scenario-building respectively. *I-pl@n* is an instrument dedicated to situations in which an unknown diversity of participants want to interact with the planning officials about Development Plan proposals. It possesses a limited but dedicated functionality that can be used without training, the only difficulty being the interpretation of the content of the system (the descriptions, the maps and the judicial rules). If a local authority chooses to simplify this information, the legal status of the text and maps will be lost and this would diminish its value.

I-pl@n is an intranet or an Internet application, so it is primarily the capacity of the server and the network that dictates its accessibility. The interaction of the user and the system has been tested with the help of some experimental subjects and resulted in some improvements. Moreover, *I-pl@n* is dedicated to supporting a democratic policy process in which participants have access to all officially approved information and are invited to influence its content as long as this information is not laid down in an officially approved report or law.

The *SketchGIS* instrument opens up data and information and facilitates the exchange of knowledge among the participants involved in a collaborative planning session where the agenda involves brainstorming ideas about scenarios of the future. The facility to visualize alternative concepts on a 'smartboard' enhances the progress that can be made in these inherently difficult stages of the planning process. Furthermore, it makes ideas transparent and offers a very effective means of communication between the participants.

Both *I-pl@n* and *SketchGIS* have been designed to fulfil specific tasks within planning practice. As with the development of PSS for planning authorities to provide statistical information for their local inhabitants, so it will be only a matter of time before the application of these tools for disseminating planning proposals and for devising scenarios will be widespread and widely appreciated.

Acknowledgements

We are grateful to Bradford Community Statistics Project and Birmingham City Pride for allowing us permission to replicate pages from their websites.

References

Batty, M., Dodge, M., Jiang, B. and Smith, A. (1999) Geographical information systems and urban design, in Stillwell, J., Geertman, S. and Openshaw, S. (eds.) *Geographical Information and Planning,* Springer-Verlag, Heidelberg, pp. 43-66.

Bryson, J. M. and Einsweiler, R.C. (1988) *Strategic Planning; Threats and Opportunities for Planners*, Planners Press, Chicago.

Carver, S. and Peckham, R. (1999) Using GIS on the Internet for planning, Chapter 20 in Stillwell, J., Geertman, S. and Openshaw, S. (eds) *Geographical Information and Planning,* Springer, Heidelberg, pp. 371-390.

Drogendijk, A. and Duijn, M. (eds.) (1999) *Handreiking voor Interactieve Planvorming, Land Water Milieu Informatietechnologie,* LWI/CUR, Gouda.

Forester, J. (1989) *Planning in the Face of Power*, University of California Press, Berkeley.

Geertman, S. (1999) Geographical information technology and physical planning, in Stillwell, J., Geertman, S. and Openshaw, S. (eds) *Geographical Information and Planning*, Springer-Verlag, Heidelberg, pp 69-86.

Geertman, S. (2001) Participatory planning and GIS: a PSS to bridge the gap, *Environment and Planning B.*

Gordon, Th., Karacapilidis, N., Voss, H. and Zauke, A. (1997) Computer-mediated co-operative spatial planning, in Timmermans, H. (ed.) *Decision Support Systems in Urban Planning,* E & F N Spon, New York, pp 299-309.

Harris, B. (1999) Computing in planning: professional and institutional requirements, *Environment and Planning B,* 26 321-331.

Hopkins, L. D. (1998) Progress and prospects for planning support systems, *Environment and Planning B,* Anniversary Issue, 29-31.

Jankowski, P., Nyerges, T., Smith, A., Moore, T. and Horvath, E. (1997) Spatial group choice: a SDSS tool for collaborative spatial decision-making, *International Journal of Geographical Information Science,* 11(6): 577-602.

Kammeier, H. D. (1999) New tools for spatial analysis and planning as components of an incremental planning-support system, *Environment and Planning B,* 26**:** 365-380.

Klosterman, R.E. (1998) Computer applications in planning, *Environment and Planning B,* Anniversary Issue**,** 32-36.

Klosterman, R.E. (1999) The What if? collaborative planning support system, *Environment and Planning,* 26 393-408.

Langendorf, R. (2001) Computer aided visualization: possibilities for urban design, planning, and management, in Brail, R. and Klosterman, R. (eds.) *Planning Support Systems; Integrating Geographic Information Systems, Models, and Visualization Tools*, ESRI Press, Redlands, pp. 309-359.

Langerak, L. *et al.* (eds.) (1999) LWI-IP3: Scenarios and visualisatie, *Land Water Milieu Informatietechnologie*, LWI/CUR, Gouda.

Lee, D.B. (1973) Requiem for large scale models, *Journal of the American Institute of Planning,* 39: 163-187.

Nedovic-Budic, Z. (1998) The impact of GIS technology, *Environment and Planning B,* 25: 681-692.

Nyerges, T. L. and Jankowski, P. (1997) Adaptive structuration theory: a theory of GIS-supported collaborative decision making, *Geographical Systems,* 4(3): 225-259.

Openshaw, S. and Rao, L. (1995) Algorithms for re-engineering 1991 census zoning systems, *Environment and Planning A*, 27: 425-46

Saaty, T.L. (1977) A scaling method for priorities in hierarchical structures, *J Math. Psychology,* 15: 234-281.

Shiffer, M. (2001) Spatial multimedia for planning support, in Brail, R. and Klosterman, R. (eds.) *Planning Support Systems; Integrating Geographic Information Systems, Models, and Visualization Tools*, ESRI Press, Redlands, pp. 361-385.

Scholten, H. and Meijer, E. (1988) From GIS to RIA, a user-friendly microcomputer orientated regional information system for bridging the gap between researcher and policy maker, in Polydorides, N. (ed.) URSA-NET Proceedings 1988, Athens.

Singh, R.R. (1999) Sketching the city: a GIS-based approach, *Environment and Planning B,* 26: 455-468.

Stillwell, J., Geertman, S. and Openshaw, S. (eds.) *Geographical Information and Planning,* Springer-Verlag, Heidelberg.

Thomasson, E. (2000) Small area statistics on-line, Paper presented at a conference on 'The Census of Population: 2000 and Beyond', at the University of Manchester, 22-23 June.
Voogd, H. and Woltjer, J. (1999) The communicative ideology in spatial planning: some critical reflections based on the Dutch experience, *Environment and Planning B,* 26 835-854.

Web Sites

Bradford Community Statistics Project: *http://www.bcsp-web.org/*
I-pl@n: http://www.i-plan.nl/
National Online Manpower Information System: *http://www.nomisweb.co.uk/*
Neighbourhood Renewal Unit: *http://www.neighbourhood.dtlr.gov.uk/*
Neighbourhood Statistics Service (NeSS): *http://www.statistics.gov.uk/*
Nexpri: *http://www.nexpri.nl/*
Open Information for Birmingham: *http://www.oi4b.com/*

3 Public Participation via On-line Democracy

Richard Kingston, Andrew Evans and Steve Carver

Centre for Computational Geography, School of Geography,
University of Leeds, Leeds, LS2 9JT, United Kingdom

Abstract: This chapter will examine the use of the Internet and the World Wide Web as a medium for carrying out public consultation and participation in environmental decision-making in the UK. Focusing on two case study examples, the authors illustrate how web-based public participation GIS (PPGIS) can facilitate and increase participation in local and regional planning decisions. The chapter illustrates how Internet technologies have been developed and applied, drawing upon research (ESRC 2000) that examined the potential and actual benefits of using on-line PPGIS. Potential and future applications are examined, making recommendations and suggesting best practice techniques in implementing a web-based public participation strategy for a local or regional planning exercise. The systems which are under development are aimed primarily for use by public bodies, particularly planning authorities, who have to undertake public consultations as part of their statutory function.

3.1 Introduction

Recently the Internet and the Web have become a popular medium for carrying out all kinds of commercial, social and governmental activities. Arguably the Internet has encroached upon and become a part of society quicker than any other previous new technologies (such as television or automobiles). Across the world, the Internet in being championed as a new democratising tool, supposedly bringing people closer together and allowing them to participate in civilised society. In the UK, vast sums of money are being spent by governments to promote and ensure that all members of society are capable of accessing the 'information society' more actively. The UK Government is committed to having all of its services on-line by 2005 (Cabinet Office March, 1999). The extent to which this may occur has recently been criticized as an extremely ambitious project (Millar 2001).

Using two case study examples, we examine the processes of providing electronic access to public participation in real environmental decision-making problems at the local and regional scale in the UK. Various map-based devices and interfaces have been tested as part of the research. Drawing on the evidence from these two case studies, we examine how the public interacted with the technology and evaluate its effectiveness against traditional methods of participation before making recommendations. Key to the research has been the successful incorporation of GIS data and functionality into interactive web-based interfaces and also its use as a public-friendly interface to experimentation in decision problems with a high degree of spatial data content. This is fundamental to facilitating better social understanding of environmental decision problems across a range of spatial scales from local to regional to national. A key aspect of these systems is that they provide a two-way flow of information between the client (the general public), and the server (the local authority).

3.2 Web-based Public Participation

Most environmental decision-making problems are inherently spatial in nature and are regularly best represented using a GIS. Some of the recent, and legitimate, attacks on GIS have been that they are complicated systems which require a high degree of knowledge to be able to use and understand them (Pickles 1995). It has also been recognised that information systems can promote and support citizen participation, particularly in public planning (Lenk 1999, p. 88) and that GIS *"can be used effectively by the community involved"* (Ceccato and Snickers 2000, p. 925). Others though, make different suggestions, highlighting the polarisation of views which currently exist towards PPGIS. On the one hand, proponents of the technology suggest concepts of freedom, empowerment, opportunity, communication and democracy (Craig 1998), while those opposing it suggest it aims to deliver control, exploitation and is elitist (Clark 1998, Pickles 1995).

It could be argued that the more traditional methods of participation are also restricted because they are geographically and temporally constrained. Also, more often than not, they tend to be dominated by forceful, influential actors causing the 'local' participants to become the marginalised, less vocal majority. Healey (1998) notes that in traditional local authority open meetings with large numbers of people and organisations attending *"the residents felt awed by the proceedings and kept silent"*. In other examples the *"residents were daunted by the 'men in suits' who seemed to know everything and could talk well"* (Davoudi and Healey 1995). Traditional public participation in environmental decision-making in the UK has a relatively lengthy history. Ever since the first Town and Country Planning Act in 1947, varying degrees of public participation have existed, although it was not until 1969 (Skeffington 1969) that widespread public participation became embedded in the process.

It is suggested in this chapter that the use of GIS information technologies, and in particular, web-based systems, could help overcome some of the criticisms by creating a more level playing field on which to conduct public debate. The advent of these on-line technologies is beginning to have serious implications for the methods by which local government and democratic processes interact with citizens and how they participate in decision-making processes across the whole spectrum of public service delivery. What this chapter intends to look at is the methods required to put in place an on-line PPGIS and to examine its benefits. Traditional methods of public participation are well documented (see for example, Thomas 1995).

There are several examples of how new forms of participation are beginning to evolve in the new 'information society'. Ghose (2001, p. 142) identifies that *"a number of programs have been developed to provide access to data and GIS analysis to citizens in local communities"*. Some of the tangible benefits of these new systems include:

- their accessibility from any place, any time, anywhere, and theoretically for any period;
- the concept of '24-7' access to systems and services;
- their cost effectiveness;
- opinions can be expressed freely and without fear;
- hierarchical information about the decision problem can be presented;
- this can increase in complexity depending on the users' interest; and
- information can be processed and responses provided with a quicker turn-around.

It is important to keep in mind that these new systems and processes should be complimentary to current methods and not entirely replace them. To illustrate these key issues, we recount the events of two case studies which we carried out to develop, live test and evaluate web-based PPGIS in real environmental decision-making problems. The experiences and findings gained from these case studies have helped to develop new methodologies in relation to the development and implementation of new forms of electronic public participation. It is suggested that the provision of access to specific decision-making problems in on-line environments will play an increasing role in the way local, regional and national planning authorities will carry out future consultation and participation exercises.

3.2.1 Local Planning and Participation

In June 1998, the West Yorkshire Village of Slaithwaite carried out a *Planning For Real* (PfR) exercise to identify the views and opinions of local residents regarding the environment in which they lived and how they would like their village to develop in the future. The whole process was organised by Colne Valley Trust (CVT), an independent community-based rural agency, which promotes and facilitates the economic, social and environmental regeneration of the Colne Valley. PfR is an idea developed and patented by the Neighbourhood Initiatives Foundation (NIF), a charitable trust based in Telford, as a means of involving local people more closely in local environmental planning problems and decision-making. This is achieved through active participation and interaction with large-scale physical models of the area that, more often than not, involves using the local community to construct the models. In this particular case, local school children helped construct a 1:10,000 scale model of the village after the initial sculpting of the underlying contours by NIF.

The physical model was used as a focus for local discussion about planning issues within Slaithwaite. Local people were invited to register their views about particular issues at a one-day event by placing flags with written comments onto appropriate locations on the model (Figure 3.1). A problem with this system is that the public are limited in what they can say due to the size of the flag which they place on the map. Typically the flag is about 4cm by 4cm, colour coded to represent a particular issue, such as red for transport or blue for housing. This is very inhibiting if you wish to make a complex comment or suggestion. The results of this exercise were then collated by NIF with the potential to feed them back into the planning process via the local authority. The PfR initiative provided the research project with an ideal opportunity to test out new methods of public participation by running a parallel exercise over the Web. A web site was set up and, in addition, Internet terminals were supplied at the event and users observed. Figure 3.2 shows the web-based interface to the PfR process developed by the research team. Using the same 2km^2 area of land around Slaithwaite, the project developed a virtual model of the village that allowed the local community to interact with a two dimensional dynamic map. The map is interactive and provides information about features on it via simple mouse-clicks, thereby giving users relatively instant answers to queries which they pose and the ensuing results.

Fig 3.1: The physical model of Slaithwaite

The Virtual Slaithwaite system is an on-line GIS facility and was arguably among the first such system available to the public that allowed a two-way flow of information (others include Carver *et al.* 1997). The web browser window consists of four frames each containing particular pieces of information (Figure 3.2). Members of the public can view a map of Slaithwaite, perform zoom and pan operations to assist in visualisation and navigation, ask such questions as 'what is this building?' or 'what is this road?' and then make suggestions about specific features identified on the map. All user input can be stored in server-side logs which is then used for future analysis and feedback into the planning process. In this manner, a community database is created, representing the range of views and feeling about planning issues in the village.

When a feature is selected, the small window in the top left hand corner of the screen displays its attributes and the 'Instructions and Help' window changes to a form which can be filled in with the comments and suggestions regarding the feature. If a building feature is selected, a form is provided allowing the input of text relating to that building. Other features, such as open spaces, the river, or the canal, provide a free-form text box allowing the user to type in their comments. This is then 'sent' to the system for future analysis. When users have finished, they exit the system and are provided with a series of questions asking them how they felt about using the system. Users also have the facility to read other users' comments. Originally there was no option to comment directly on these comments although the system in future will be adapted to allow this type of functionality.

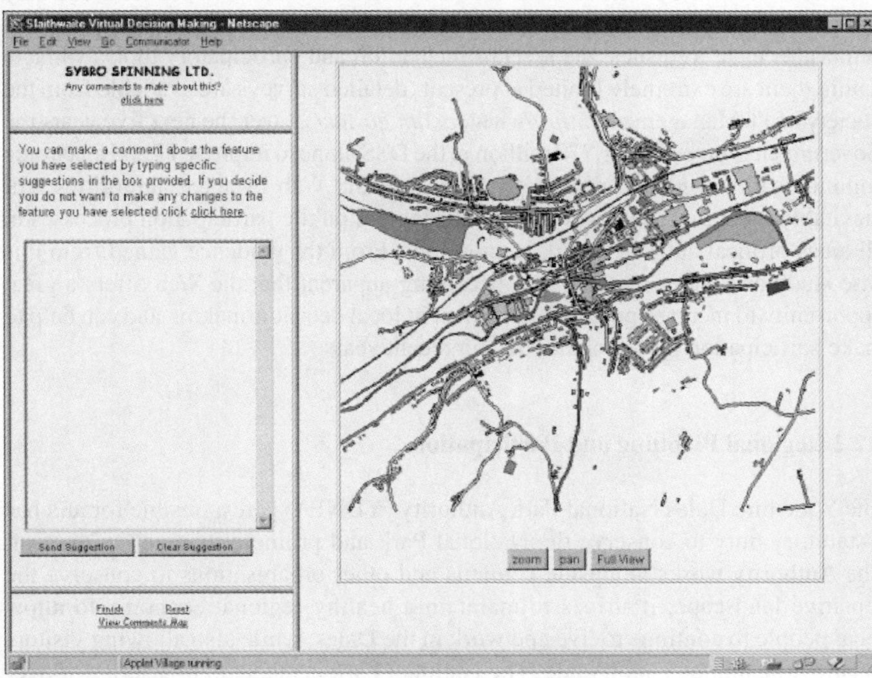

Fig 3.2: Virtual Slaithwaite
Source: http://www.ccg.leeds.ac.uk/slaithwaite/

One encouraging aspect arising from this case study was the relative ease with which the people of Slaithwaite seemed to grasp the concept of using the GIS. This appears to go against some of the work done by others in the past on the publics' interpretation of maps (see for example Keates 1996, or MacEachren 1995). One noticeable feature was the way parents and their children used the system. There were many cases where the children would operate the mouse and use the keyboard while parents and grandparents would direct them around the map getting them to click on particular features and tell them what to type in. This helps to account for the apparent skewness in the user profile that is dominated by school children. Other patterns in the response data show a propensity toward the professional and managerial occupational groups. This, together with the large 'at school' group, closely matched the demographics of the web-using public at the time (GUV 1998a, GUV 1998b, BBC 1999). Recent surveys have shown that users are fast becoming more representative of the population as a whole, particularly with the arrival of free/flat-rate Internet fees (Which? 2001). There are several government initiatives aimed at moving towards a more equitable access situation (Cabinet Office April, 2000). On the whole the public response to the system was positive, particularly with the ability to type in comments at great length, as opposed to being restricted to a short sentence, as is the case in the traditional PfR process. The ability to provide feedback and read other peoples' comments could potentially be used for conflict resolution or on-line debate and discussion about particular issues.

There is very little evidence at the moment on how local authorities view, and may react to, the possibility of on-line public participation systems. While nearly all local authorities have web sites, the level of interaction and participatory tools available within them are extremely limited at present (detailed surveys are available from the Society of IT Management, *http://www.socitm.gov.uk/*). Over the next five years the Government is investing £1,870 million in the DSS alone to improve IT and electronic communication with the public, so the use of the Web by local authorities will inevitably increase. The potential for carrying out public participation processes in all areas of local authority work is enormous. From the evidence gained from this case study and wider research it is becoming apparent that the Web offers a great opportunity to increase public involvement in local decision-making and can help to make participation more popular than in recent years.

3.2.2 Regional Planning and Participation

The Yorkshire Dales National Park Authority (YDNPA) is responsible for and has a statutory duty to conserve the National Park and promote its quiet enjoyment. The Authority works alongside councils and other organisations to conserve the sensitive landscape; it strives to maintain a healthy regional economy to allow local people to continue to live and work in the Dales; while also allowing visitors to enjoy the unique countryside. The National Park is made up of a series of over 20 dales, in total covering an area of 1,769 square kilometers. The Park lies astride the Pennines in the north of England, in the counties of North Yorkshire and Cumbria.

It was decided, in collaboration with the YDNPA, that the issue of woodland planting would be a viable and worthwhile 'decision problem' to use as an experimental case study. This provided the research with a single-issue decision problem at the regional scale. The case study generated interest from a wide range of the public; from Park residents and workers through to tourists living in other parts of the country and visiting the area because of its historic value as a National Park. The study involved collating data from a variety of sources and undertaking some manipulation to generate new data sets. Once the necessary GIS data was in place, the task of designing the web-based system began. The system needed to accommodate a wide range of user-stakeholder groups, including local residents, farmers, land-owners, park visitors and YDNPA representatives, and required them to identify areas that are deemed both suitable and acceptable for natural forest regeneration and new planting.

The case study focused on a two-stage approach involving GIS-based modelling to identify areas best suited for regeneration of natural forest cover, and user-focused participatory mapping techniques to identify which of the suitable areas are most acceptable to Park residents and users. The user is taken through a series of maps with associated (attribute) information about that particular data set and how it may be important to and effect planting. The user then has to decide whether this is important in their personal decision-making process. When the final data input is completed, the on-line system calculates the most suitable locations for tree planting based on the factors and constraints set by the user. Figure 3.3 reproduces a typical scenario. A map is produced on screen showing the results of the GIS modelling. The user is also provided with a window showing the choices that they had previously made. They are then given the opportunity to re-assess their decisions and can refine these interactively if they are unsatisfied with their decisions made previously. As the user alters the initial factors and constraints, the map changes in 'real time' accordingly.

The final version of the system was live tested over a public holiday weekend in 1999, with Internet terminals supplied at four of the National Park's main visitor centres: Hawes, Aysgarth, Malham and Grassington. Over the three days of testing, the 'Woodland Web' was used by over 200 people. From these, 125 valid responses were received. System log files have been used to reproduce the choices and weightings the individual members of the public placed on the factors and constraints. There was an observed increase in the proportion of middle-aged people who used the system from the previous case study. This is largely due to the age-structure of the people visiting the Dales and National Park visitor centres over the bank holiday weekend period. Another factor that may have caused this change was the fact that this case study was carried out twelve months after the first case study and may represent a change in peoples' ability and familiarity with information technology. Once again, however, observations of system use showed that many school students appeared very comfortable using the system while their parents made suggestions about which factors and constraints to set and the comments to make.

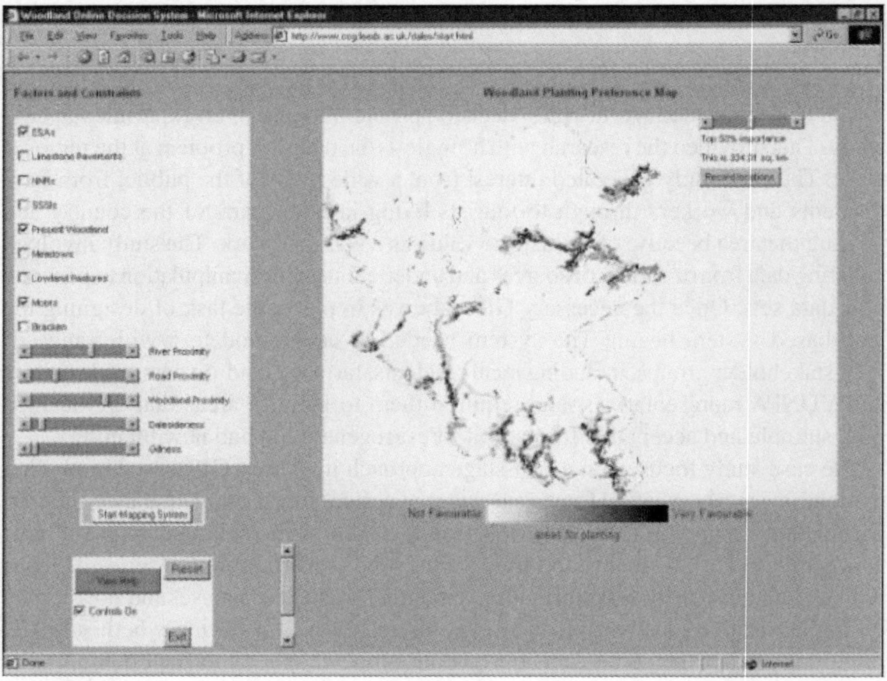

Fig 3.3: A typical users' map
Source: http://www.ccg.leeds.ac.uk/dales/

The results outlined in this section all rely on the public being truthful when they fill in the user profile at the beginning of the system. It is quite possible that some people do not fill this in correctly, and unfortunately there is no way of ensuring that correct and accurate information is collated. Out of the eight questions people were asked in the user profile, only five of these were essential. The required data fields are: whether you live in the National Park or not; your postcode; gender; age group; and occupation group. The 125 valid responses received included only seven people who actually lived within the National Park. The users living outside the National Park came from all parts of the UK as Figure 3.4 illustrates. Nearly twice as many males used the system as females, with 64.8 per cent being male and the remaining 35.2 per cent being female. Again, this reflected the demographics of Internet users at the time, and it has recently been found that there are now more female users than male (BBC 2001). Figure 3.4 has been produced using the postcode data entered into the 'Profile' page of the web site. The centroid of each postcode in the UK has been converted to a six-figure national grid co-ordinate using the Postcode Address File (Royal Mail 1997), thus allowing an approximated location to be plotted on the map using a GIS. While the vast majority of users of the system came from the Yorkshire and Humberside region a fairly dispersed spread of people from other parts of England and Wales also took part. There were no users from Scotland although a software developer from the United States found the system 'very interesting'.

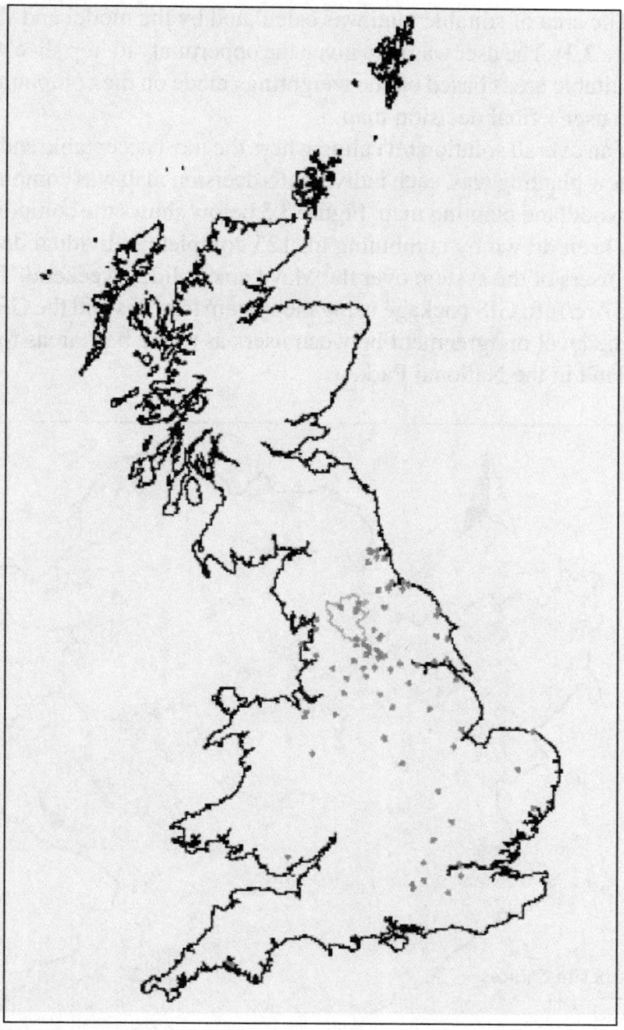

Fig 3.4: Map identifying where system users live

Fourteen individual maps were used in the system. These represented the hard-fixed constraints, user-selectable soft constraints and user-selectable weighted linear factors that were used to derive individual decision maps. Each of the fourteen maps had a short textual explanation to give the user some background to allow them to form an opinion as to whether the particular data set was important in their decision-making. User choices of constraint maps and user-specified weightings of factor maps were used to run a simple multi-criteria evaluation (MCE) model to show the suitability of selected areas for planting. This model was 'hidden' from the user, though a short explanation was provided for interested people explaining MCE techniques. The area of suitable land was calculated by the model and shown on the display (Figure 3.3). The user was then given the opportunity to 'top slice' a percentage of the most suitable areas based on the weightings made on the compound map. This then gave the user's final decision map.

To arrive at an overall solution revealing where the most acceptable and appropriate location for new planting was, each individual's decision map was combined to create a composite woodland planting map. Figure 3.5 below shows the composite decision map that has been drawn by combining all 125 complete individual decision maps generated by users of the system over the May bank holiday weekend. The map was created in the Arc/Info GIS package using the system log files and the GRID module. This shows the level of agreement between users as to the best areas for replanting native woodland in the National Park.

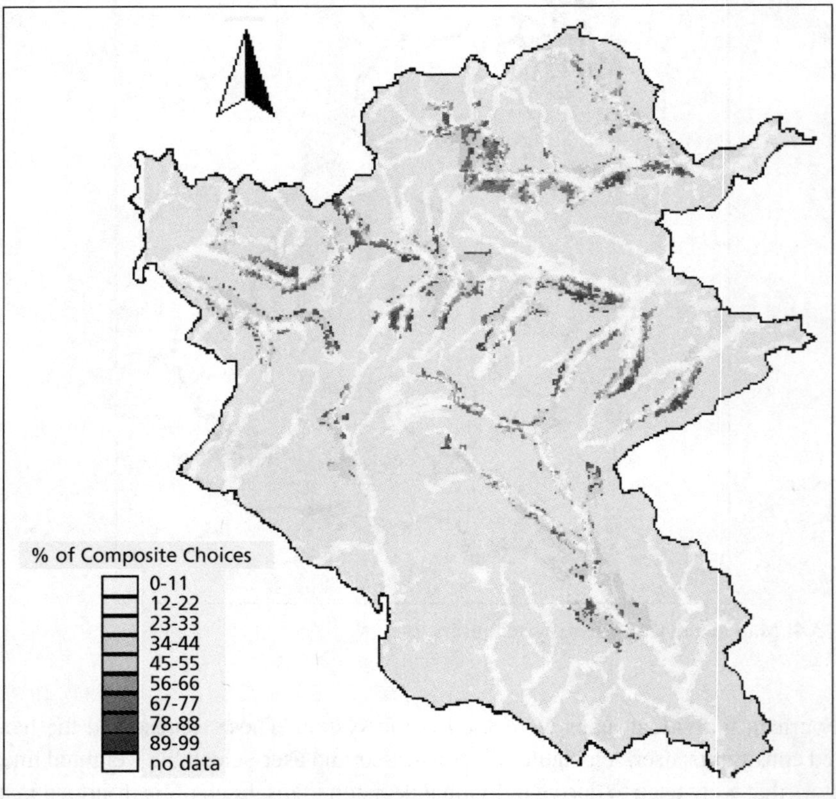

Fig 3.5: Final composite decision map

It can be seen in Figure 3.5 that there is widespread agreement among participants that woodland should be concentrated on the dale sides. The darker areas of the map identify those areas where the greater numbers of participants agreed that new planting should take place. The darkest areas of the map represented the choices of over 95 per cent of users of the woodland system. These areas are broadly in line with the favoured locations for planting suggested by the National Park Authority in their Woodland Strategy document. The expansion of current woodland along the dale sides can be clearly seen on the map. The white areas of the map represent rivers, lakes, reservoirs and roads which are generally in the lower lying areas of the National Park and were excluded from the model. The overall solution is a function of the data itself to a certain degree. This is due to the use of binary maps for certain datasets that limits the options for planting to a simple 'yes/no' scenario dependent upon the particular datasets. For example, the Sites of Special Scientific Interest (SSSI) map allowed the user to make a decision over whether to allow tree planting to occur in a SSSI or not. This makes the decision over these binary data sets relatively simplistic compared with some of the other data sets used, which involved setting levels of weighting in relation to distance to/from a road, for example.

3.3 System Structure and Components

The web sites were based around two Java™ 1.1 applets that give general PPGIS facilities. The Slaithwaite case study used a vector-based applet, while the Yorkshire Dales case study centered on a raster-based system. Both systems were initially developed for the studies, but were given sufficient flexibility that non-programming developers with a basic understanding of web pages and server-side logging can choose the data displayed, functionality used and outputs recorded. This flexibility has been utilized in a number of on-going studies.

3.3.1 Vector System Details

The vector system applet allows those with GIS knowledge to load ESRI shapefiles into a web page, map the features, and show dbf file attribute information to the user, and allows the user to pan and zoom. When a feature or coordinate is selected by the user, a web page request containing information about the location of the click, and the feature it represents, is sent as a traditional hypertext transfer protocol (HTTP) request to with the server where the page originated. This can be captured by a standard common gateway interface (CGI) script and an appropriate response initiated. In the case of the Slaithwaite study, this involved sending the user a PERL generated form to fill in detailing their comments, which appeared in a frame next to the map.

The applet was based around a set of reusable open-source Java classes (the GeoTools package) developed at Leeds for dealing with geographical data (Macgill 2001). The system is controlled from outside the Java code by the simple hypertext markup language (HTML) which embeds the applet in the web page, allowing developers to easily stipulate the files they want shown, line colours, shading, *et cetera*.

3.3.2 Raster System Details

This applet allows developers to show a composite weighted raster image in a web page. It displays a series of scrollbars or checkboxes, each of which is associated with a single graded or binary raster image respectively. Moving a scrollbar changes the degree, in real time, to which the composite display includes the associated raster image. Checking a checkbox cuts out or allows in areas of the composite display depending on the associated binary (cut-leave) raster image. The system also allows for masked areas, which show up as the colour of the applet background, allowing irregular shapes (such as the UK mainland) to be displayed.

Once a user is happy with their composite image, he/she can submit the weighting values chosen. The composite weighted image can be shown to the user as he changes the weights, or after an initial submission to test for the effect of geographical understanding. This was added to allow for assessment of not in my back yard (NIMBY) alteration of the weights in sensitive planning situations. If the user is shown the map after an initial submission, he/she can then be given the option of further experimentation with the map. In addition, the user can be presented with top-slicing or single location selection tools. Once the user is entirely happy, the data is sent to the server using an HTTP request, which is stored by a CGI script in a log file.

In the case of the Yorkshire Dales case study, the raster maps were generated using Arc/Info GRID. However, the final files pulled into the system are grayscale or black-white images in the standard web page GIF or JPEG format. This has the great advantage of opening the system up to developers without GIS knowledge. Again, the files uploaded to the system, the associated weighting controls, and the colour schemes, are all controlled outside the Java code with simple HTML.

3.4 Problems, Experiences and Recommendations

There are several issues unique to web-based participation and which differ from the more traditional methods which are usually encountered. In designing and implementing an on-line PPGIS, careful consideration of the following issues need to be addressed:

- access to the technology;
- GIS and IT understandability;

- data and copyright issues; and
- trust and response legitimacy.

Public consultation and participation in decision-making processes over the Internet is an intricate problem requiring multi-level systems depending on the characteristic of the individual user. Although it is noted that everyone should be capable of making judgments about a particular problem, it is recognised that differences in age, background, education, profession, *et cetera,* require different levels of information and interface complexity if effective interaction is to be achieved. How well multi-level systems can be engineered depends very much on the complexity of the problem to which they are addressed.

3.4.1 Access to the Technology

Access is one of the most important issues, and problem populations fall into one of three categories: those who are financially unable or who are unwilling to connect; those with technological problems; and those who are physically disadvantaged with respect to the medium.

The studies outlined above found a significant skew in the user base of the systems. However, this base accurately reflected the general Internet-using population. This has become increasingly representative of the population in general since the studies (GUV 1998a, GUV 1998b, BBC 1999). In particular, the advent of 'free' connections (especially the cheap-rate call charge only scheme of Dixon's Freeserve' service) and the subsequent introduction by British Telecom of flat-rate schemes has leveled the demographics and allowed almost any business to consider a connection for their staff. In addition to market forces, the matter is also being addressed by the UK Government with substantial amounts of investment going into IT infrastructure and provision, including their commitment to having all services on-line by 2005 (Cabinet Office March, 1999). Public access points such as cafés and community buildings have remained a slower though steady growth area, but may be the most appropriate forum to target for consultancy exercises (see Liff *et al.* 1999, for examples of the typical spaces). In the case studies above, terminals were provided at key community spaces or at community events, and this was found to be extremely worthwhile. While it may be said that there is little difference between this and the traditional PPGIS mechanisms of taking GIS to the stakeholders, the interfaces are, by nature, not expert led, and the results can be instantly fed back to the community as a whole. At present, TV web browsers tend to be incapable of running Java, though Flash is common (see below). In addition, a lack of understanding of the nature of public interest in the Web has driven most TV companies to produce 'closed garden' systems of limited content or use. This will undoubtedly change as broader access to the whole Web becomes available through PCs.

Technological limitations are likely to reduce dramatically. When the case studies above were carried out, there was still some concern that user's machines would not be fast enough to run the applets, and their connections rapid enough to download the data in a reasonable time (Lake 2000). The standard specification for home machines has been driven by the gaming industry above that necessary for most business or academic users, and is now well above the minimum estimated needs (~32MB memory, 330Hz chipset, and a 52K modem). Java 1.1 is now found on almost all standard web browsers. In addition, the development of ActionScript in Macromedia's Flash 4™ system and the opening of the language to developers in Flash 5 means Flash can be used to develop sophisticated client-server applications using HTTP requests or XML sockets. Flash is based around a relatively lightweight but robust plug-in, and is distributed with all the major browsers, including those used by digital TV companies, and, unlike Java, has a single implementation of the running environment. Java, however, is considerably more powerful, and has more options for the disabled.

With respect to the more considerable problem of accessibility for the disabled, organizations have a moral and legal responsibility to ensure the widest possible access to information. When the information is in the format of maps and images, this becomes particularly problematic, and this partly accounts for the absence of more working sites of this nature. While the present studies were prototypes and therefore made little attempt to tackle this more difficult area, the problems are not intractable. All on-line consultations should provide text-based options for those with speaking browsers. These are unlikely to have the functionality of map-based systems, but represent the best possible option at this stage. The future is considerably more promising. It would actually be possible to have attribute data read to users from the systems outlined above. However, such functionality will be a good deal easier when web browsers allow Java 2 applets to run without the present need for a plug-in. Java 2 has accessibility functionality built into the basic interface classes, and, with time, mouse-overs/keyboard navigation should be able to give the user a detailed vocal picture of the data and interface. The latest version of Netscape comes with Java 2 as standard, and the move towards a universal code virtual machine architecture by Microsoft suggests it is only a matter of time before others follow.

Accessibility recommendations

1. Target local community groups, if necessary supplying Internet enabled PCs. The sites can usually be run locally and data collected or mailed on disk by local organizations.
2. Always provide a basic text alternative, but be aware that real accessibility to Java-based systems is likely to appear over the next 3-5 years.
3. Investigate digital TV systems and the possible use of Flash-based systems, but be aware of the very real problems with accessibility and coverage.

3.4.2 GIS and IT Understandability

This could be seen to be a potential problem that needs to be addressed. An obstacle to participation could be lack of familiarity with the technology. Many people, particularly older people and people from blue-collar professions, may never have used a mouse before, indeed evidence from our case studies highlights some interesting issues relating to how people used the systems with children operating the GIS while their parents informed them on what comments to make. Recent initiatives by Age Concern, a charitable organisation for the elderly, aim to overcome this problem through hands-on training (Gaines 2001), and, as has been mentioned above, the demographics for web users are now a good deal less biased towards the managerial classes and educational groups.

In terms of understanding the material presented by the consultations, web-based systems are undoubtedly the most equitable and best adapted to the broad range of the population – provided genuine public participation is sought. Information is freely available on the Web, and can be added to by almost anyone, both within and outside the system. This provides users with a very real ability to pick information they feel is at their level, and this can be analysed to check understanding. Several major software vendors have been moving towards adaptive user interfaces and intelligent agent-based systems over the last few years, which may suggest these as mechanisms for improving individual user's experience of the consultation process. However, a number of prominent public relations disasters (most notably Amazon's alleged attempts to structure prices based on user loyalty) suggest systems that adapt to the interests and abilities of users, however helpful they may appear, would be judged morally wrong by most in this context. Instead, the studies outlined above used a structured data system to present information on the relevant issues, which had an excellent response. A broad overview was presented to the user, and they could then find out increasingly complex information on a subject if they wished. Such a system encouraged a broad range of participation and allowed the users comfort and interest.

Understandability recommendations

1. Structure information, explicitly or implicitly, as a file tree. Present a broad overview of each issue with a couple of lines of text. In the next level, give a more detailed paragraph and in the next, an entirely detailed explanation. Finally, allow users to browse detailed documentation and reports. They will pick the level and subjects they feel they can appreciate most, while getting a broad view of the totality of the issues.
2. Follow standard usability guidelines, and test with a broad range of users. See, for example, *http://www.useit.com/*.
3. Do not overload users. Attention span alters with the interface. We found that most users responded well to tick box questionnaires, but would only interact with five to seven factors on a weighting map. A potential solution for situations with more issues is to explicitly force the user to pick the issues they believe are important on a questionnaire/selection sheet before generating the appropriate styled applet by manipulating the embedding HTML.

3.4.3 Data and Copyright Issues

While many of the technical issues have been overcome there are still certain copyright issues relating to spatial data, particularly detailed map data which pose particular problems in the UK. At present, the most detailed data provided by the UK's national mapping agency (Ordnance Survey) for putting on the Internet is at a scale of 1:10,000 (Ordnance Survey 2001) and this may not necessarily be appropriate for some public participation schemes. In addition to the cost of actually purchasing the data, the actual copyright costs may also discourage public bodes from carrying out on-line PPGIS consultations.

Recommendations

1. Greater use of aerial photography could be made allowing on-line raster images to be overlayed with digitized roads or site boundaries for particular consultations. In some cases this may be a more useful 'backdrop' rather than a standard map base.
2. A case needs to be made to allow local planning authorities to make use of other Ordnance Survey products for on-line use at a much more reasonable cost.

3.4.4 Trust

Trust is essential for on-line participation to work successfully, and by 'successfully', we mean getting a good response and for the system not to backfire badly. There are two essential ingredients to public trust in systems: trust that something will be done and the system not abused, and trust in the information given and the process. If you loose either of these you will not simply have failed, you will have done positive harm to the planning process.

Recommendations

1. Belief that something is being done and the system not abused.
   - Only start the process if you are genuinely interested in a public response! Groups will concentrate on on-line presences, and you have to be able to argue your case strongly and respond flexibly.
   - Limit the on-line consultation time explicitly and give a date to return for further information – or, even better, allow users to sign up to email news bulletins.
   - Give explicit and obvious details of what will be done with the information, as recommended by the Information Commissioner, and check you are registered appropriately under the 1998 Data Protection Act. See *http://www.dataprotection.gov.uk/* for details and recommendations.
   - Allow people to opt-into giving demographic and contact details. Being explicit about the level of confidentiality given by different subsets of postcodes may help persuade people to give them.

2. Belief in the information and process.
   - Ensure that the information given is bi-partisan, and, where possible, secure the certification of data by all the groups involved. Where there is disagreement over interpretation, be explicit about this – groups will be more willing to

ratify your data if you are. In addition be sure to detail any modeling processes and areas of interpretation in what appear to be 'scientific' software solutions.

- Allow people to leave to external sites, and link to contrary viewpoints. This gives your site legitimacy by showing you have nothing to fear from debate. For example, NIREX, the people responsible for nuclear waste disposal link to their own reports, but also the sites of the major Green organizations.
- Allow users to add their own data/information – either into the site's data area or into discussion boards.
- Allow users to see other responses and to opt-into putting contact details on-line. One of the major oversights with the initial Slaithwaite project was that other responses were kept from users, in the belief that it would encourage a wider range of responses from participants. This is true, but a real advantage was gained by the off-line PfR exercise in that people could see when they were not alone in an opinion and build self-help groups dedicated to solving the problem without local government input.

3.4.5 Response Legitimacy

The flip side of trust from the public, is trust in their responses. Until a national electronic identification system becomes available, if it does, there can be no guarantees that users are responding legitimately, and this should be accepted as an inevitability, but not a malevolent one. At the present stage of technological awareness, there are actually some advantages in allowing flexibility in who enters the system. As was mentioned above, in many cases parents and grandparents allowed children to enter their demographic details at the start of our case studies, but were then observed to 'back-seat drive' their way through the actual responses. This not only allows the technically unsure to make their voice heard quasi-vicariously, but if children can see the physical responses to 'their' democratic decisions it may lead to a more engaged public over the coming years.

Recommendations

1. Joke returns are usually obvious (one user in a study on potential disposal sites for nuclear waste commented that Arsenal's football ground was perfect for the job!). Equally obvious are automated multiple returns and those initiated by on-line campaigns. Maintaining good access logs tied to the return data will allow you to assess returns from a single machine over a short period of time, and postcodes can be checked against addresses if gained.
2. Analyse demographic data, but also view the responses 'blind'. Only dismiss responses on the basis of plainly ridiculous answers, not on the age or occupation of the respondent.
3. Watch for people who start the process but drop out. If you collect demographic data early, and do not rely on a single final button push to submit information you can monitor whether drop-outs occur because of usability problems or because

particular groups are finding the system unengaging. Make sure your system does not break if people don't traverse the whole thing.

3.5 The Future Role of Public Participation and On-line Democracy

Over the past 5-10 years, the explosion of information and communication technologies has begun to transformed the way all levels of society from government to business, and increasingly the general public, function. These technologies are widely regarded as the motivation for fundamental changes to the way society operates. They have already begun to modify the nature and experience of social control, participation and cohesion. Central and local government bodies in the UK have begun to embrace these new technologies and view them as a key component of the future vitality of the country. It is inevitable that over the next decade more local authorities will make use of the medium in an attempt to re-engage the public in local democracy.

Through the application of new web-based GIS technology to real environmental decision-making problems, we have been able to identify key issues relating to how the Internet can be used to engage the public in participatory processes. We have developed applications which allow the public to:

- access information about particular decision problems;
- explore spatial and aspatial data; and
- contribute to the decision-making process.

While we have reported here on two experiments, further work is needed to test the software on different decision problems and at different spatial scales. The nature of the software that has been developed means that it is available to you to experiment with or even use in a real decision-making scenario. Further details about the software are on-line at: *http://www.ccg.leeds.ac.uk/democracy/*.

The Internet has the potential to re-engage the public in participatory processes if it is used correctly and legitimately by proving equal access to those who wish to participate and a comfortable medium for exploration of issues. The question remains as to whether the public will feel compelled to participate in line with the Government's vision for participatory democracy.

References

BBC (1999) Net catches the UK, BBC News On-line, 26 October. *http://news.bbc.co.uk/hi/english/sci/tech/newsid_486000/486413.stm.*

BBC (2001) Women overtake men on-line, BBC News On-line, 14 June. *http://news.bbc.co.uk/hi/english/business/newsid_1388000/1388408.stm.*

Cabinet Office (March, 1999) Modernising Government White Paper, Office of the E-Envoy. *http://www.e-envoy.gov.uk/publications/reports_index.htm.*

Cabinet Office (April 2000) E-government: a strategic framework for public services in the Information Age. *http://www.e-envoy.gov.uk/ukonline/strategy.htm.*

Carver, S., Blake, M., Turton, I. and Duke-Williams, O. (1997) Open spatial decision-making: evaluating the potential of the World Wide Web, Chapter 19 in Kemp, Z. (ed) *Innovations in GIS 4,* Taylor & Francis, London, pp. 267-278.

Ceccato, V. and Snickers, F. (2000) Adapting GIS technology to the needs of local planning, *Environment and Planning B: Planning and Design*, 27: 923-937.

Clark, M. J. (1998) GIS: Democracy of delusion? *Environment and Planning A*, 30: 303-16.

Craig, W. (1998) the Internet aids community participation in the planning process, *Computers, Environment and Urban Systems*, 22: 95-104.

Davoudi, S. and Healey, P. (1995) City Challenge: sustainable process or temporary gesture?, *Environment and Planning D: Society and Space*, 13: 79-95.

ESRC (2000) Virtual Society Research Programme. *http://virtualsociety.sbs.ox.ac.uk/.*

Gaines, S. (2001) Surfs up for gran, The Guardian On-line, 21 June, pp.10-11. *http://www.guardian.co.uk/internetnews/story/0,7369,510050,00.html.*

Ghose, R (2001) Use of information technology for community empowerment: transforming geographic information systems into community information systems, *Transactions in GIS*, 5(2): 141-163.

GUV (1998a) GUV's Tenth annual user survey (October 1998): Age, Graphics, Visualization and Usability Center, Georgia Tech. *http://www.cc.gatech.edu/gvu/user_surveys/survey-1998-10/graphs/general/q54.htm.*

GUV (1998b) GUV's Tenth annual user survey (October 1998): Primary Industry, Graphics, Visualization and Usability Center, Georgia Tech. *http://www.cc.gatech.edu/gvu/user_surveys/survey-1998-10/graphs/general/q30.htm.*

Healey, P. (1998) Collaborative approaches to urban planning, *Environment and Planning A*, 30: 1531-1546.

Keates, J. S. (1996) *Understanding Maps*, Second Edition. Addison-Wesley, Reading, MA.

Lake, D. (2000) The Right Tools For The Job, The Industry Standard: Intelligence for the Information Economy, 13 March. *http://www.thestandard.com/article/0,1902,12809,00.html.*

Lenk, K. (1999) Electronic support of citizen participation in planning processes, in Hague, B. N. and Loader, B. D. (eds.) *Digital Democracy: Discourse and Decision-making in the Information Age,* Routledge, London, pp. 87-95.

Liff, S., Steward, F. and Watts, P. (1999) Public access to the Internet: new approaches from internet cafes and community technology centres and their implications for libraries, *New Review of Information Networking*, 5: 27-41.

MacEachran, A. M. (1995) How Maps Work: Representation, Visualisation and Design, Guildford Press, New York.

Macgill, J. (2001) Geotools – a Java Mapping Toolkit. *http://geotools.sourceforge.net/.*

Millar, J. (2001) Government's internet project 'doomed to fail', *The Guardian*, 4 June. *http://www.guardian.co.uk/internetnews/story/0,7369,500998,00.html.*

Ordnance Survey (2001) Business Geographic Internet Licence. *http://www.ordsvy.gov.uk/downloads/ism/20967a5.pdf.*

Pickles, J. (1995) *Ground Truth: The Social Implications of Geographical Information Systems*, Guildford Press, New York.

Royal Mail (1997) *Postcode Address File,* Post Office, London.

Skeffington, A. (1969) *Report of the Committee on Public Participation in Planning: People and Planning,* HMSO, London.

Thomas, H. (1995) Public participation in planning, in Tewdwr-Jones, M. (Ed.) *British Planning Policy in Transition*, UCL Press, pp. 168-188.

Which? (2001) Annual Internet Report 2001, Which? On-line. *http://www.which.net/surveys/intro.htm.*

Web Sites

On-line democracy: *http://www.ccg.leeds.ac.uk/democracy/*
Data Protection Act 1998: *http://www.dataprotection.gov.uk/*
Standard usability guidelines: *http://useit.com/*
Society of IT Management: *http://www.socitm.gov.uk/*
Virtual Slaithwaite: *http://www.ccg.leeds.ac.uk/slaithwaite/*
Woodland Web: *http://www.ccg.leeds.ac.uk/dales/*

4 Web-based Tools and Interfaces for Participatory Planning and Design

Kheir Al-Kodmany

University of Illinois at Chicago, CUPPA Hall, 412 South Peoria Street, Chicago, IL 60607, USA

Abstract: This chapter addresses the growing interest in using screen-based maps on the Internet for collaborative planning and design. Specifically, it introduces six prototype web interfaces that invite users to read maps of their community and then make selections and type in comments in order to provide feedback into a planning process. Each of the six prototypes highlights a different tool or feature for navigating or making selections on online maps. New web technology has made it possible to create map-based surveys to receive feedback from the general public, but it is not yet clear what kind of graphic design alternatives and digital map designs and tools are the most useful for novice map-readers. This chapter describes several different interface designs with the purpose of finding which combinations of online tools and maps are most productive in soliciting feedback in a community planning process. The goal is to explore how web-based maps can be advanced beyond mere information provision to actual two-way interaction with the public. Tools such as free-hand drawing may provide new avenues for people to take a greater role in public decision-making. Examples are presented of solutions to the basic problems of creating interactive screen-based maps, which include navigating large geographic area maps and making selections on maps.

4.1 Introduction

This chapter describes research into screen-based collaborative planning and design using the World Wide Web (WWW). The potential of the Internet as a vehicle to disseminate spatial knowledge and to receive feedback on spatial issues has not yet been fully studied; especially lacking is an examination of the graphic design alternatives possible on the Internet. This chapter explores a variety of digital map designs and tools created for the dual purposes of communicating spatial information to average map users, and allowing those users to navigate and make selections on the maps in order to give feedback into a community planning process.

In the past ten years, there has been a dramatic increase in the number of people exposed to and using screen-based geographic information products for general use.

This is due to the rapid development and widespread implementation of geographic information technologies, including geographic information systems (GIS), remote sensing and image processors and global positioning systems (GPS), and the widespread utilization of the Web. The mass marketing of such products and technologies means that a growing number of lay people are interpreting and creating maps and other geographic information products using computers. Today's geographic information technologies may be used to generate paper maps, but they are just as likely to be used to display screen-based map products. However, while we suspect that there are differences in how people comprehend screen-based versus paper-based maps, cognitive research has focused only on the human ability to understand geographic information from traditional hard-copy maps. There have been no systematic attempts at studying how delivering maps on the Web differs from more traditional cartography. While our research does not explicitly measure comprehension, it is one step toward exploring the variety of graphic designs that are possible in presenting screen-based maps to a lay audience in a participatory planning setting.

In the field of participatory planning and design, particular techniques have proven useful over time. For example, a typical activity in traditional public participation planning workshops is to spread a large map on a table and have each planner or participant mark their ideas on the map. Another technique has been to provide participants with model pieces that they are free to move around. We were interested in how these kinds of activities could be simulated online using computerized tools. Publishing maps on a web site, to be viewed on a computer screen, presents unique and largely unexamined challenges (e.g. download times, browsers' compatibility and programming know-how, in addition to cartographic skills), but offers important advantages over paper maps. For example, updating digital maps is relatively easy and inexpensive; printing and duplicating costs are incurred by the user; maps are potentially accessible to an exponentially larger number of people; other types of media such as sound and photos can be linked to and accessed from the map; an on-line GIS map can be queried; *et cetera.*

This chapter seeks to address the growing interest in using screen-based maps on the Internet for collaborative planning and design. Specifically, it introduces six prototype web interfaces that invite users to read maps of their community and then make selections and type comments in order to provide feedback into a planning process. Each of the six prototypes highlights a different tool or feature for navigating or making selections on online maps. For instance, one prototype allows users to actually draw their own selections using an online 'pen'. While some of the prototype interfaces described in the chapter are presently being used in an existing community planning process, others are still being further developed and are undergoing testing.

The projects described in this chapter test the capability of present web-based GIS technology for supporting graphic tools. Three main challenges are confronted in creating internet-based participatory maps. The first is how to represent and navigate a large geographic area on a small computer screen. Unlike paper maps, computer screens limit the amount of geographic area that can be viewed at one time. Scrolling across a large map online is usually impossible because of the slow speed and large amount of computer memory that is required to reload the images. We have developed two interfaces that present alternative ways of presenting a map of a large community area that are easily navigable by novices.

The second challenge was developing a flexible, accurate way for users to select and comment on a specific neighbourhood, block or building. In the first project, we used pre-defined square units on the map (using a grid system) so that it would be possible to easily analyse the location selection data from a large group of participants. However, we subsequently found that this method could excessively limit user choices and so we developed a new interface that features freeform drawing as the method of selection. The development of web-based freehand drawing tools, where users can use an online 'pen' to select their own locations, is a particularly exciting innovation. In both of these research areas, we have experimented with the use of aerial photographs, structure base maps and GIS maps of the community as the background upon which users can select locations of concern. After presenting several options for navigation and selection, we discuss one interface where we attempted to incorporate the best features into one 'complete' or integrated interface.

The third challenge was to find a way to replicate online the traditional public participation activity of allowing participants to move three-dimensional model pieces to create their own site plans. We have developed one interface prototype that allows users to draw and move boxes on top of a background to make their own choices about siting buildings online.

This research follows the framework first developed by Lynch (1960), Nasar (1998) and Sanoff (1991). These scholars advocate users' need-driven design or a user-oriented approach. They argue that current practices do not adequately address users' needs and preferences. The research described in this chapter is propelled by the belief that public needs are underrepresented in today's planning and design practices and that web technology offers a medium of communication that can potentially address this problem. While it has become common for local governments, community planners and community agencies to utilize the Web to offer information to community members, this project takes the next step by using the Internet as a medium for two-way communication; the tool allows people to become both *receivers* and *providers* of information. This project lays out some of the potential contributions and limitations related to enabling two-way communication and participation over the Web. It will be most helpful to read this chapter in front of a computer in order to try out the interfaces.

4.2 Theoretical Background: Imageability and Public Participation in Planning

The theoretical foundations for these web interface designs can be found in the work of Lynch (1960) and Nasar (1998), who argue for the importance of discovering how city design affects citizens. By studying what kinds of evaluative images residents have of their community, planners, researchers and community leaders can derive valuable information about how to improve the physical form of their communities.

In his seminal book, *The Image of the City* (1960), Lynch uses the concept of 'imageability' as a theoretical framework for studying cognitive maps, urban form and the spatial relationships of cities. Imageability *"is that quality in a physical object which gives it a high probability of evoking a strong image in any given observer. It is that shape, colour, or arrangement that facilitates the making of vividly identified, powerfully structured, highly useful mental images of the environment. It might also be called legibility"* (Lynch 1960, 9). If most people like the imageable elements of a city, then it will probably convey a positive evaluative image. If they dislike them, the city will convey a negative evaluative image, suggesting a need for changes in the city's appearance. This aspect of city image is what Nasar calls the 'likability' of the cityscape. Likability refers to the probability that an environment will evoke a positive evaluative response among the groups of people experiencing it. Inhabitants of a city with a good evaluative image find pleasure in the appearance of its memorable and visible parts.

The city landscape may be a source of pleasure and delight to its residents and visitors, and can potentially be a restorative for the stresses of daily life. Thus, the development of a city's form should be guided by a visual plan that is concerned with visual form on the urban scale (Lynch 1960). However, to devise such a plan, we need to know how the public evaluates the cityscape and what meanings they find in it. Nasar suggests that it is possible to learn the public's preferences by empirically measuring them. Just as we weigh objects to find how light or heavy they are, Nasar says that we can measure preferences to determine the degree to which people like or dislike various areas of a city. Nasar developed and implemented a method of surveying residents (using traditional phone surveys and manual map-making) to determine which areas they like and dislike in their community, with the goal of creating a single 'evaluative image' of the community that could guide future design and development.

In recent projects at the University of Illinois at Chicago, we have adopted and advanced Nasar's original method by using the Internet and a GIS program to survey and map resident preferences. Most of the interfaces we have developed are based on the concept of surveying a group of community residents to discover their preferences and opinions about the imageability of their community. GIS is in the background, in the sense that it functions as the method for analysing the data that is collected via the Web. The Web is used as the primary platform for collecting spatial information.

4.3 Current Work on Public Participation and the Internet

Shiffer and others at MIT have researched a variety of ways of using the Web for urban planning and design. They have explored how emerging information technologies and web technologies can improve the processing and communication of planning-related information in metropolitan planning organizations. They provide case studies illustrating how to deliver spatially-referenced multimedia material for site planning and reviews using projects in Washington DC, the South Boston Seaport and Boston's waterfront development. Such multimedia interfaces, coupled with the accessibility of the Web, have the possibility of opening up a new paradigm within urban design, one which helps to communicate ideas and developments to other agencies and the general public (Shiffer 1995).

As Shiffer's work has demonstrated, one result of this move toward digital visualization of urban form and distribution of information on the Web is that there are new possibilities for involving the lay public in design decision-making. As the number of households with internet access increases, and the demographic profile of internet users diversifies, the potential for using the Web for public participation planning increases exponentially. Thus, a promising avenue for research is to explore which types of digital tools and interfaces can best engage the public in planning activities.

A survey is one important tool for public participation on the Web. Citizens can use a web-based survey to become information creators, rather than passive recipients of information. This is an important leap forward in using the Web as a communication tool. Most of these applications utilize simple feedback forms where users type in comments in response to questions and then click 'submit' to send their responses to the web server. There are fewer examples of web-based surveys that utilize graphics, maps or other kinds of visualization to inquire about the public's locational preferences. One example is the Landscape Scenic Preference Survey developed by the Macaulay Land Use Research Institute, which aims to quantify the landscape preferences of the general public. Participants are shown pairs of landscape images and asked to choose which they prefer (*http://www.mluri.sari.ac.uk:80/~mi550/landscape/*). The same survey was conducted in a traditional manner where people were shown photocopied photographs and then asked to give each landscape an evaluative score. Early indications have shown that the results from the paper-based questionnaire were not significantly different from the online version. The researchers identified areas that needed improvement but the initial reaction was that the online survey worked well. In another example, Kingston (1998) has developed several projects that use the Web to facilitate public input on several environmental problems in Britain (*http://www.ccg.leeds.ac.uk/vdmisp/*). In one project, a web-based decision-making environment was developed which allows the public to model a number of possible planting scenarios in locating areas for regeneration of native woodland (Kingston 1998).

The drawbacks of using the Web for communication between planners and designers and the public centre around the broad issue of access. First, though access to the Internet is increasing, it is still difficult to draw a random, representative sample since internet users do not yet accurately reflect the real demographic makeup of the general public. Also, users must be quite motivated to log on and find the web site in order to participate. This fact underlies a concern about creating a critical mass of users to sustain meaningful internet interactions. This prompts the question of what factors are needed to achieve a critical level of activity (Shiffer 1995). These examples suggest that web-based surveys need to be further refined but have now become quite feasible as public planning tools.

4.4 Background of the Projects

In 1993, the University of Illinois at Chicago (UIC) established the Great Cities Institute (*http://www.uic.edu/cuppa/gci/about/index.htm*) to respond to urban problems facing American cities. 'Great Cities' refers to the University's commitment to use its teaching, research and service programs to improve the quality of life in metropolitan Chicago. Under the Great Cities program, UIC has worked with Chicago communities on approximately 220 different projects and programs. One of the newest urban outreach initiatives under Great Cities is the Great Cities Urban Data Visualization (GCUDV) program, established in 1998. GCUDV's mission is to explore how advanced visualization capabilities can be used in community planning to create innovative computer applications and to test them in actual urban projects (*http://www.uic.edu/cuppa/udv/*). Researchers in the GCUDV program work to link databases, GIS and 3D visualization into a medium suited for urban and regional planning and policy. Projects focus on social and ecological data as well as urban image visualization. The program is staffed by a multidisciplinary team that includes faculty, students and recent graduates from UIC's Electronic Visualization Laboratory and from several other disciplines such as architecture, art history and information technology.

In 1998, UIC planners and designers were invited to be part of a participatory community planning process in Pilsen, which is a primarily Mexican-American community of 50,000 residents just south of the UIC campus. Like many other low-income neighbourhoods, Pilsen is faced with a host of urban challenges, including heavy traffic, dilapidated and vacant buildings and crime. Community leaders felt that the meaningful involvement of a broad range of residents in a planning process would strengthen the overall sense of community, and a cooperative effort would help present a 'unified front' when funding opportunities arose. A planning team was formed that included 25 community members and five university professionals: two architects, two planners, and a graphic artist.

At the heart of the UIC-Pilsen planning process was an introductory series of community design workshops where a variety of interactive high- and low-tech tools were used. For the next stage in the process, the research team experimented with new interactive technologies being developed on the WWW to see if the Internet could enhance existing public participation efforts. In particular, the team developed a web survey, using Nasar's basic framework, to discover Pilsen residents' spatial perceptions of their community. The web survey was employed as a supplement to, rather than a replacement for, the ongoing face-to-face workshops. The early versions of the online survey gave rise to a variety of additional interfaces. The goal was to discover the best features to facilitate online, two-way communication using visualization in a community planning endeavour.

The prototypes that we have developed have moved from simple web sites that display spatial information to more complex and interactive sites that actually allow participants to draw and comment on maps of their community. The increasing levels of sophistication can be described as follows: (i) simply displaying a map (such as disseminating the results of a spatial analysis); (ii) viewing a map with different types of data included, and some ability to navigate the map; (iii) maps that include some GIS functions so that users can retrieve attribute data along with the map; and (iv) map surveys that allow users to provide feedback directly on the maps (Kim 2001). The degree of user interaction depends largely on which tools are used, such as HTML, plug-ins, Java applets, Java servlets, and Internet Map Server technologies.

Most of the web-based GIS sites that are currently available on the Web provide one-way information delivery. Although providing proper spatial information is an important aspect of participatory planning, it is only a basic requirement. In order to become a robust tool for use in participatory planning, web-based GIS should go one step further: to become a medium for two-way information exchange. It may not be necessary in all cases; however, if it is widely utilized, the use of web-based GIS may become a critical tool in gaining important feedback in community planning.

4.5 Current Projects

The general intent of most of the interfaces described below is to discover how residents of Pilsen evaluate the appearance of their community by asking what particular places they like and dislike. The survey instructions, while not included in all of the interface prototypes, emphasize the issue of *community appearance* and encourage respondents to think of which places they find to be pleasant and which places they find to be blighted and unattractive. The results are intended to be used as guidelines for decisions on community appearance and visual form. The map tools have been developed to work on average computers and do not require the loading of GIS or Internet Map Server (IMS) applications. Local residents simply interact with web maps by clicking map locations and typing text to give us their opinions.

The interfaces we have developed are organized around the three main challenges we faced in developing web sites for screen-based collaborative planning and design: (i) navigation methods, or methods to make a large geographic area navigable online; (ii) selection methods, or methods for allowing users to select locations on an online map; and (iii) compositional methods, or methods for allowing users to develop a site plan online.

4.5.1 Navigation Methods

Two navigation methods are used: nesting maps with increasing resolution and using thumbnail photographs that help to make a large geographic area navigable online.

Nested Maps

This web survey (*http://g015.cuppa.uic.edu/gridFeedbak/final/index.html*) (Figure 4.1) attempts to evaluate participants' likes and dislikes for the four square mile community of Pilsen. In earlier versions of this web map/survey tool, we had the problem of not being able to zoom on the map. We wanted to publish the map at the highest possible resolution, yet this took up too much space on the screen; users had to scroll up and down and right and left to see the map. In this new interface, where we are trying to cover a much larger geographic area, this was a serious problem. We found a solution in the concept of a geographic hierarchy with three zooming levels. This web interface utilizes a grid and an aerial map of the community. Users are able to navigate the map by utilizing three zooming levels, each with increasing resolution.

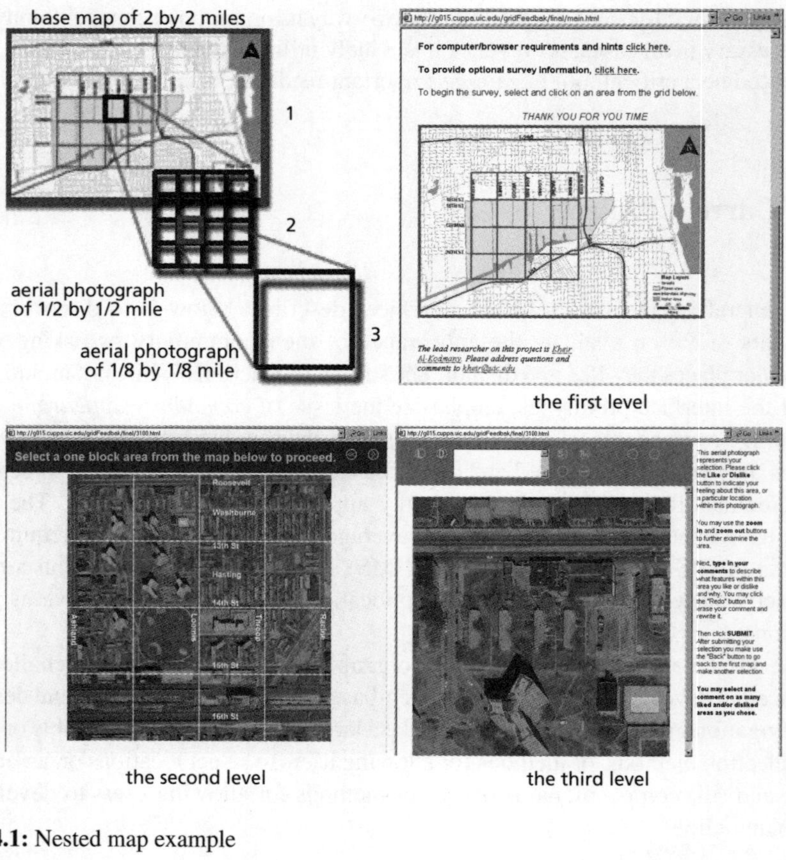

Fig 4.1: Nested map example

When the user selects one of the 16 squares on the initial map (the first level), the program zooms in, and the selected square appears as a new full-screen aerial map that is again broken down into a grid of 16 squares (the second level). When one of the 16 cells is selected from the second level, an aerial close-up of the selected area (the third level) then loads into the window. The actual geographic size of the final selection is four city blocks. The highest resolution is on this third level with six inches per pixel. At all times, the user has the ability to zoom in and out between the three levels. At the third level, users can see the fine details and distinguish buildings and streets. This square becomes the 'selection' and the user may then type in comments about the location.

Thumbnails

Another option for navigating a large community area is to show a number of very small 'thumbnail' photos, or tiles, which, upon clicking, bring up a larger map of the area. A demonstration of this technique can be found at *http://g015.cuppa.uic.edu/feng_BAK/project.html* and *http://g015.cuppa.uic.edu/ramki/image.html* (Figure 4.2). Like the other interfaces, this one is designed using an aerial map of the Pilsen community. Cells (miniature photographs) are laid out across the top of the screen with some street names visible to help orient participants. The instructions read 'Select one of the thumbnail photos by clicking on it. In the text box below the map, describe any physical changes to the community that have occurred since the map photo was taken 10 years ago'. This is only one possible use; other uses could also be envisioned. Also, more tiles could be used to cover a greater area of the community, although this will slow down the interface.

Discussion of 'Zooming' Methods

The two prototypes described above utilize different methods of 'zooming' on a web-based map, and each has pros and cons. The first method, using nested maps, uses Java. In the Java interface we write the scripts for zooming. There are four Java layers of different scales (zoom). Once you click the zooming function on the map, you call one layer after the other, simulating a 'zoom' on the photograph. When the user clicks to get the next map (zoomed version), the new map is called from the remote server. The disadvantage of this method is that it can be extremely slow, particularly for users with modem connections. The advantages are that Java provides great flexibility in creating and adding new features.

The second method (the thumbnail interface) uses Flash Shockwave software. Flash is a web design tool to create animated web pages. The strength of this tool is that it handles images in vector format. Thus, users do not lose resolution no matter how closely they zoom in on the map. This is a very fast method of zooming since the function is available within the program itself. Zooming is done via a small movie file that makes the zooming easy and in real time. Once a user enters the web site, a message appears asking the user to install Flash software. If the user clicks yes, it loads in just a few seconds. The web site contains a small movie file that automatically gets loaded on the user's hard drive and then the zooming is done from the user's own machine and it is very fast. While speed is a significant advantage of this method, the present Flash software package does not allow the incorporation of features from other programs. Thus, we cannot mesh this site with other functions that are done with the Java programming language. For instance, at this time there is no way to include a selection feature so that users can select a building within the photo.

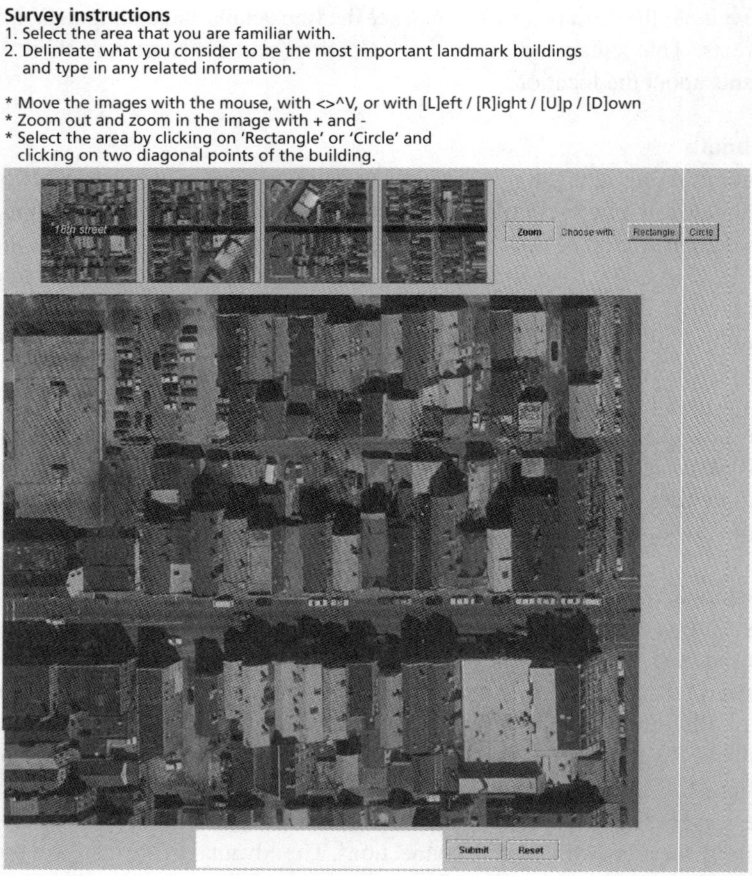

Fig 4.2: Thumbnail example

4.5.2 Selection Methods

Two methods are presented for allowing users to select and comment upon locations on an online map: a grid-based system and freehand drawing tools.

Grid-based Selection

We have created several different prototypes that utilize a grid as the underlying structure behind a given map that constrains the selection available to users and allows us to compile and analyze the spatial selections made by participants. Below we describe three different interfaces that utilize the grid-based selection method.

Our earliest project (*http://g015.cuppa.uic.edu/gridFeedbak/xarial18St/arial_18St.html*) (Figure 4.3) was a survey consisting of one exercise entitled *Urban Likability and Dislikability* (ULD) for Pilsen. Simply speaking, participants logged on to the project web site where they could view a high-resolution aerial photograph of Pilsen with a grid overlaid on top of it. We limited the geographic area to the vicinity of the 18th Street commercial district, since this area was the main focus of revitalization efforts. Participants were asked to point out areas on the map that they liked and disliked and to provide the reasons for their responses. Each square of the grid was identified by its centroid (the centre of the square). This centroid was coded as the actual longitude and latitude of the centre of the square.

Fig 4.3: Grid-based selection

Participants were asked to identify the areas of their community that they most liked and disliked by clicking on the appropriate square of the grid. The only visual assistance on the map was the name of streets. They were to use a GREEN pointer to indicate 'liked areas' and a RED pointer to indicate 'disliked areas'. By clicking on one of two radio buttons located on the side of the map, they could 'load' the pointer with either colour. When participants clicked on the square, a small window with a question mark appeared, asking them to state their reasons for liking or disliking that area of the community. When finished, the participant clicked on a button labeled 'submit' and the whole form (the user's input) was transferred to the UIC server. Each of the participants' selections could then be stored in the web access logs for analysis and feedback into the planning process. In section 4.6 we present a case study utilizing this interface.

Since the server was linked to an Oracle database and a GIS application, we had the capability of taking all the points that were selected by the participants, sorting them by longitude and latitude, and plotting them on a map automatically. The Oracle database could also group the associated comments. In this manner, a community-input database was created that contained the range of views about areas liked and disliked with the associated reasons. We then created a number of GIS choropleth maps to illustrate the intensity of likes and dislikes (urban likability and urban dislikability). We used dots to represent intensity: the number of dots in each cell of the map was proportionate to the number of times that area was selected by the residents in the survey exercise. In addition, these GIS maps were interactive; clicking on an area (or cell) of the map opened a window of text that listed the residents' stated reasons for liking or disliking that area. Since the maps provided written evaluations for each point, they were extremely useful in supplying specific directions for improvement and could easily be incorporated into the next stages of the Pilsen community planning process.

In a later prototype (*http://g015.cuppa.uic.edu/gridFeedbak/final/index.html*) (Figure 4.1) (the same project mentioned above in 4.5.1 where we used nested maps), we also used a grid but it advanced the initial project by dealing with a larger geographic scale. Not only does the user navigate using the grid, but he or she also selects a square of the grid to comment upon. As mentioned above, the primary disadvantage of this method is that users cannot select the specific buildings or combination of buildings that they wish to comment upon, since the size of the final selection square is predefined. The advantages are clearly evident in the above example where the uniform selection areas allowed us to easily create, sort and analyze the users' feedback.

In a final example of using a grid system, we created a 'real-time feedback' interactive map (*http://g015.cuppa.uic.edu/gridFeedbak/xdbgrid18St/grid_18St.html*). Participants could view a structure base map of Pilsen and choose areas of like and dislike. They could also type in comments about the selected areas. Once their selections were submitted, they could immediately receive a return map showing the accumulative evaluative image of the examined area. This worked in the following way: the update servlet on the database is contacted and the data sent to the servlet translates the visual information into numeric, string data that is stored in the database. Once the database is updated, the data servlet is called which queries the database for all the relevant information and then translates the data into visual information and sends it back to the user. The visual information includes the cumulative 'like' and 'dislike' opinions of the area, represented by increasing intensity of colour along with the user comments for each block in the grid. The data obtained from different surveys can be co-related as it is stored in the database sorted by the latitudes and the longitudes. The grid was a good structure that was easy to code and to be read by the Oracle database that created the return map.

In evaluating the grid as a selection mechanism, it has clear advantages and disadvantages. As in the above example, the grid enables very fast analysis and compilation of spatial data that can be easily compared among participants. On the other hand, we found that users did not have enough discretion in selecting the particular areas of the community and buildings that they wished to comment on. In addition, they were constrained to the square shape of the grid: even if they only wanted to comment on one building in a corner of this large square, they had to select the entire square.

Freehand Drawing and Selection Tools

This project introduces an entirely new technology for enabling user feedback using web-based maps. In this new prototype (*http://g015.cuppa.uic.edu/zhaoxia/blankcanvas/sketch.html*) (Figure 4.4), participants can go to the online survey web site and use a drawing tool to select the areas of the community that they like and dislike; their location choices are not limited by the pre-defined geographic areas of the square grid. On the initial screen, the participants view a structure base map of the community along with two buttons labelled 'Click to select area with drawing tool' and 'Click to type in comments'. When a participant clicks on the drawing button, the cursor turns into an arrow and upon pressing and holding down the left button of the mouse, it starts to draw. The user may draw any shape on the map and when the mouse is released, the lines close on themselves to form a polygon. If the user does not draw an enclosed shape, the program approximates the final piece of line that closes the shape into a polygon.

Once the participant draws a shape and releases the mouse button, the shape is immediately filled with a light colour. As the participant continues to select additional areas with the drawing tool, other shapes are filled in as well. When areas overlap, the program indicates this by increasing the density of colour in these areas. Areas that participants delineate most frequently will be the darkest, while the areas chosen the least would be lightest. This works through the placement of a very fine grid at the back of the map. When a participant delineates an area, cells underneath will be activated and will create one tone. As participants highlight additional areas, the overlapping areas would be activated twice generating 'double' tones and hence increasing the density of the colour. This shading technology is beneficial when compiling multiple users' responses, so that the darker shades would represent the most frequently chosen locations.

Once the user has drawn and selected an area with the drawing tool, the participant can then click on the text button to type in comments about the delineated area. In order to avoid clutter on the map generated by multiple annotations when a participant selects several areas, we added a check box that would allow a participant to turn the text boxes on or off. The participant may add a text box for each section of the community that is selected. It is possible to make two separate interfaces: one for drawing and writing about liked locations and one for disliked locations. This may assist in sorting positive and negative opinions in the database. Another possibility would be to simply use one interface and allow participants to record all opinions, positive or negative.

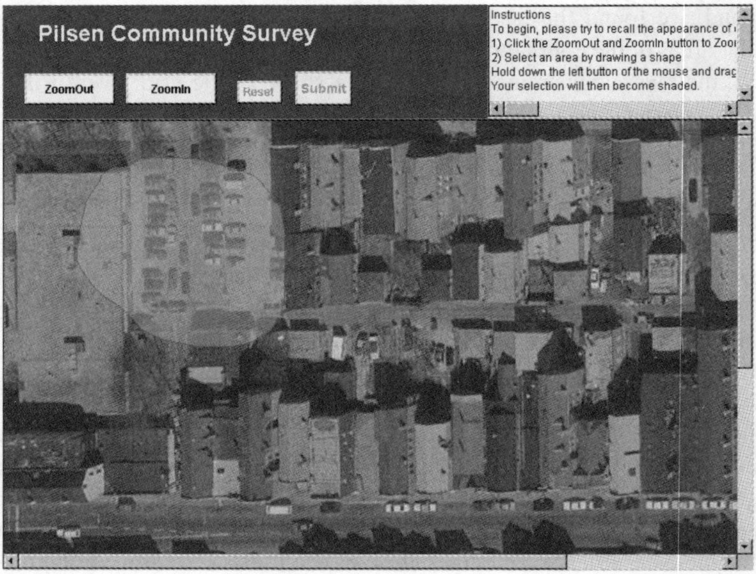

Fig 4.4: Freehand drawing example

The most important feature of this project is that we replace the Java applet grid with freehand sketching capabilities using Java to create the web interfaces so that the user is not restricted to pre-determined geographic areas within a grid. We found that Java is very good for creating all kinds of graphic functions on the Web. This new capability of doing freehand sketching on the Web may have numerous applications. People can make their own maps and then share these maps with other participants. Another possibility is that the map could be a photographic image and we could ask participants to comment on the scene by drawing directly on the picture, indicating elements that they like or dislike. Also, we hypothesize that community residents will appreciate this feature since it enhances the ability of the computer to mimic traditional public participation tools that people generally enjoy, such as drawing and writing comments on a paper map.

Toward a Complete Interactive Web Map Design

The prototypes discussed above each represent a particular feature; some are designed to sketch, others to type in comments. Some are grid-based and others are based on freeform drawing. Some have zooming, others do not. We are in the process of developing a web site (*http://g015.cuppa.uic.edu/mapTool3/mt.html*) (Figure 4.5) that will combine the best design features of the interfaces described above. Users will have options to work with or without the grid, to zoom in and out, to add layers, to type in comments, to sketch, and so on. This has apparent advantages (more choices) and disadvantages (confusion, technical difficulties to create the interface and to work out the database). This site is currently undergoing further development.

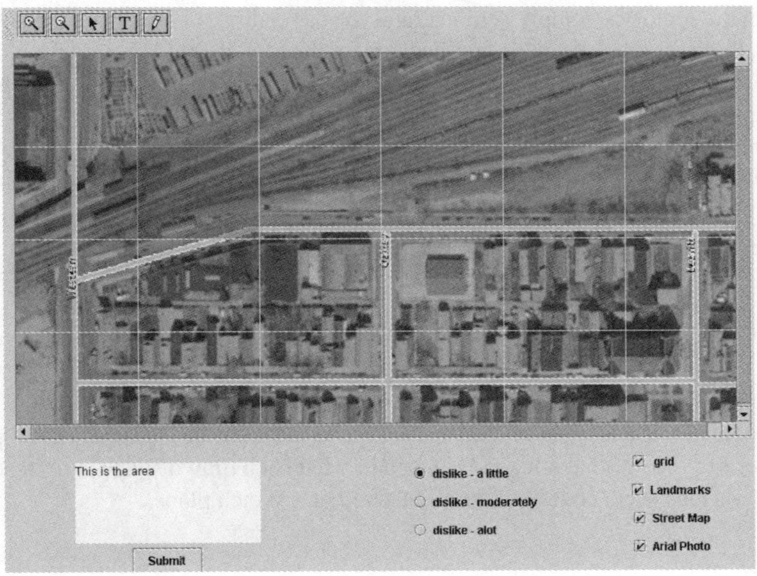

Fig 4.5: Prototype of complete interactive web map

4.5.3 Compositional Method

The final challenge we attempted to address using internet maps was how to allow participants to essentially create their own maps. This prototype (*http://g015.cuppa.uic.edu/zx/sketch2/project3.htm*) was developed for a slightly different use than the ones described above. While the others use maps of the community to help residents share their opinion and knowledge about the community by selecting locations and typing comments, this prototype aims at allowing community participants to draw alternate site plans for a specific location, such as to show a preferred arrangement of houses in a new subdivision. Called the Collaborative Decision Support System, it utilizes the freehand drawing technology described above. Users actually draw their own boxes (to represent houses, other buildings or land use zones) and then move them around into a desired configuration. The purpose of this project is to allow users to create and compose maps online.

In the current prototype, the boxes are drawn on a structure base map of Pilsen. The primary buttons available are 'draw', 'move' and 'copy'. Users may correct their work with 'delete' and 'clear' buttons. When the drawing is complete, the participant clicks 'submit' to send the drawing to the server database. The user is then able to view all the drawings previously submitted by other users. While at present the prototype simply demonstrates the capability of creating and moving the boxes, there are many possible applications for the tool. It could be used for planning work on a small scale; for instance, the survey site could open with a site map and a description of the particular site along with a set of development plans. Users would be given a set of instructions, such as asking them how they would prefer to site the structures in a development that calls for eight single family homes, 16 town homes, a community centre, and a playground. Users would then create the boxes and move them around on the screen until they developed a satisfactory plan. Some of the fixed details of the site could already be indicated on the site map, such as the direction of traffic, existing driveways, *et cetera.*

Alternatively, the proposed issue could be planning work at a larger scale, such as a land use and zoning plan. In this case, participants could help address the issue of urban sprawl by drawing shapes of different colours to indicate their preferences for placement of various types of land use zones. As the web site is further developed, it will need to incorporate a legend to explain what each drawn shape is representing so that various users could interpret one another's sketch plans.

4.6 Case Study: Using a Web Survey in Community Planning

In this section, we describe an actual community planning process in which the grid-based interface described in Section 4.5.2 and the nested map interface described in Section 4.5.1 were implemented.

4.6.1 Implementing the Web Map-Survey Interfaces

Two community-based organizations, 18th Street Development Corporation and the Resurrection Project, facilitated the survey process. They advertised the survey and sent out letters explaining the purpose of the survey and encouraged residents to conduct it online. The letters indicated the URL and the password to access the site and invited residents to come to their offices to do the survey if they did not have other access. Also, residents had the option to do the survey at the local library or a university computer lab. UIC researchers and leaders of the community organizations worked together to create as 'random' a sample as possible. Members of the community and graduate student interns did a great deal of work in designing and testing the survey, informing local residents, contacting leaders and talking to business owners.

After analyzing results from the first map-survey, which covered only the main commercial area of the community, it became clear that, in order to be more useful, the survey needed to cover the entire community area. Expanding the coverage area, however, was not a simple matter, as it posed problems in terms of image and file size, resolution and ease of navigation. The original map-survey interface took up a large amount of space on the screen, since the aerial photo was published at the highest possible resolution, six inches per pixel. Also, it required a lengthy amount of time to load onto the user's screen because the aerial photo was set as one image and was not broken into smaller pieces, or tiles, to make the loading faster. In redesigning the map-survey, we had to address the issue of scale. A smaller group of residents completed the second, redesigned survey.

In the new interface, we included the entire community area and 'nested' three different scales, or resolutions, within one map. The user must 'drill down' to reach the final map where they select the location that they like or dislike. The final map is the original digital map at its highest resolution so that users can read the fine details and distinguish buildings and streets.

4.6.2 Discussion of Survey Results

In analyzing the recorded comments, it became immediately clear that the comments from the first survey were much more specific than in the second survey. Part of the reason for this is that in the second survey, the smallest selection cell was 1/8 mile by 1/8 mile (.125 mile by .125 mile) while in the first survey the smallest cell was 1/16 mile by 1/16 mile (.06 mile by .06 mile). Also, the area covered in the first survey was a major commercial corridor that was very familiar to all residents; it was easy to relate to and to provide comments. However, in the second survey, there were some areas, particularly the industrial districts, that were rarely chosen by residents and thus we did not expect to receive very detailed information. Indeed, comments about these areas were very generic. In the first survey people said things like *"I don't like how there is always so much litter on this corner"* or *"I like how these houses look so new and clean"*. They also talked about the need for more unified landscaping, more unified signs, fixing and widening sidewalks and the need for more of a cultural touch in the design of facades. In the second survey with the larger coverage area, some respondents were not sure what to talk about and the answers were more general and imprecise, such as: *"This industrial area is really ugly"*, *"we should have these industrial areas more controlled"*, *"we should have more city regulations of these areas"*, *"the Chicago River area needs to be cleaned"* and *"the river needs to be put to use by local residents, it could be developed as tourist attraction, the present use of it is dumb for the industrial activities"*. While these answers were less helpful for pinpointing specific planning improvements in the community, they did give us insight into residents' overall evaluative image of their whole community. The composite maps that were created show that large parts of the community were fairly unfamiliar to the majority of respondents. The well-traveled areas received the highest number of responses and certain areas, such as 18th Street, the major commercial corridor, were classified as the most imageable parts of the community.

The new survey project that included the larger coverage area had both advantages and disadvantages. The major advantage of the new survey was that we could begin to understand the public's evaluative image of the whole community area, rather than of the limited area of the main commercial district. However, some of the precision of the smaller map was lost. Another issue with the larger coverage area was the cost of the aerial maps. We purchased the set of four tiles (each is one mile by one mile) for $500 from an outside vendor (Kucera International Inc., Ohio, USA), yet we were not able to get the exact coverage we wanted for our irregularly-shaped community area since the cost of the maps prohibited us from buying more than one map set.

The two evaluative GIS maps that were created in this project reveal a high degree of consensus on likability and dislikability and suggest ways to improve the appearance of Pilsen. For instance, people repeatedly commented on the high density, the lack of parking, the lack of trees and the preponderance of vacant lots. In particular, residents voiced strong dislike for spots on the north and east borders of the community, where railways, car yards, lumber yards, a garbage dump and sheds are located. Given the high visibility of the places that do appear on the maps, such as locations along the Chicago River, changes in their appearance could have a strong effect on the public image of Pilsen.

Since the maps provide written evaluations for each point, they should be useful as the planning team proposes directions for improvements. The respondents also provided some helpful comments for improving the computer interface. They wanted the familiar landmarks to be marked to give them reference points when looking at the aerial map. Others commented that the aerial map was most helpful for highlighting large, general issues in Pilsen, such as land use and traffic flow; a different kind of map interface might be better for addressing design issues such as improving the visual appearance of the retail core. The planning team hopes to incorporate this input into the next stages of the Pilsen community planning process.

4.7 Future Research

Our present work has focused on developing prototypical interfaces using the Web and GIS for two-way communication in community planning. We have explored a variety of graphic designs and web technologies that are possible in presenting screen-based maps to a lay audience in a participatory planning setting. In the future, we plan to empirically compare these interfaces to learn more about how people comprehend screen-based maps versus paper-based maps. More research is needed in order to understand the unique challenges and important advantages of web-based maps in general, and the usefulness of each graphic design and tool in particular.

The tools we have developed in these prototype interfaces have the potential to be used by a variety of agencies for multiple purposes. For example, it has been observed that transportation planning and local 'comprehensive' planning tend to occur separately, resulting in some cases in policies that work at cross purposes. A partial solution could be that comprehensive planning and transportation planning leaders could incorporate these kinds of web-based tools to both inform the public and also learn about public concerns and views. By sharing future planning ideas and learning about public concerns, costly mistakes could be avoided. This kind of communication could be used to address other issues such as urban sprawl, subdivisions, transportation planning, landscaping, and identifying advantageous options for development and environmental protection.

4.8 Conclusion

In this chapter we have described a variety of digital map designs and tools created for a dual purpose: to communicate spatial information to average map users, and to allow those users to navigate and make selections on the maps in order to give feedback into a community planning process. We have provided examples of solutions to the basic problems of creating interactive screen-based maps, which include navigating large geographic area maps and making selections on maps. New web technology has made it possible to create map-based surveys to receive feedback from the general public, but it is not yet clear what kind of graphic design alternatives and digital map designs and tools are the most useful for novice map-readers. This research attempts to explore different interface designs with the purpose of finding which combinations of online tools and maps are most productive in soliciting feedback in a community planning process.

It is increasingly important to direct research to discover the most effective methods and tools for interacting with the public using maps. As there has been a dramatic increase in the number of people exposed to and using screen-based geographic information products for general use, two-way communication of spatial information must become more efficient and more easily comprehended. Researchers in the field have estimated that *"up to 90 per cent of all business data has a geographic component"* (ArcNews 1997), while an estimated 80 per cent of all government information is spatially referenced (Huxhold 1991). Governments themselves publish geographic information online. For example, the San Diego, California Police

Department publishes data on crimes just 24 hours after the incidents occur on a public web site. Local planning agencies use the Internet as an adjunct to the traditional public meeting format. A number of U.S. Government agencies, including the U.S. Census Bureau (*www.census.gov*), the U.S. Geological Survey (*www.usgs.gov*), and the National Cancer Institute (*http://www.nci.nih.gov/atlas*), provide maps via the Internet. Without direct research into which designs, tools and methods of presenting and receiving spatial information are most easily used and comprehended by the public, we may be deluded in estimating the level of communication that is actually occurring.

Our project seeks to explore how web-based maps can be advanced beyond mere information provision to actual two-way interaction with the public. Tools such as freehand drawing may provide new avenues for people to take a greater role in public decision-making. By beginning to examine the graphic design alternatives possible on the Internet, we are taking one step further toward understanding how people comprehend and utilize screen-based maps.

References

ArcNews (1997) Visio Corporation adopts MapObjects, *ArcNews*, 19(2): 1, Environmental Systems Research Institute Inc.

Huxhold, W.E. (1991) *An Introduction to Urban Geographic Information Systems*, Oxford University Press, New York.

Kim, D. (2001) *Design and assessment of web-based Geographic Information Systems,* Masters Project, College of Urban Planning and Public Affairs, University of Illinois at Chicago.

Kingston, R. (1998) Web-based GIS for public participation decision making in the UK, A paper presented at the Empowerment, Marginalisation, and Public Participation GIS Meeting in Santa Barbara, California, 14-17 October.

Lynch, K. (1960) *The Image of the City,* MIT Press, Cambridge.

Nasar, J. (1998) *The Evaluative Image of the City*, SAGE Publications, London.

Sanoff, H. (1991) *Visual Research Methods in Design,* Van Nostrand Reinhold, New York.

Shiffer, M. (1995) Interactive multimedia planning support: moving from stand-alone Systems to the World Wide Web, *Environment and Planning B: Planning and Design* 22.

Web Sites

Macaulay Land Use Research Institute (Web survey using images): *http://www.mluri.sari.ac.uk:80/~mi550/landscape/*

Kingston's web-based decision-making systems in Britain: *http://www.ccg.leeds.ac.uk/vdmisp/*

University of Illinois at Chicago's (UIC) Great Cities Institute: *http://www.uic.edu/cuppa/gci/about/index.htm*

UIC's Great Cities Urban Data Visualization program (GCUDV): *http://www.uic.edu/cuppa/udv/*

Nested maps: *http://g015.cuppa.uic.edu/gridFeedbak/final/index.html*

Thumbnails: *http://g015.cuppa.uic.edu/feng_BAK/project.html*

Thumbnails: *http://g015.cuppa.uic.edu/ramki/image.html*

Grid-based map-survey interface (earliest project):
http://g015.cuppa.uic.edu/gridFeedbak/xarial18St/arial_18St.html

Real-time feedback interactive map-survey:
http://g015.cuppa.uic.edu/gridFeedbak/final/index.html

Grid-based map-survey interface (later prototype):
http://g015.cuppa.uic.edu/gridFeedbak/xdbgrid18St/grid_18St.html

Freehand drawing and selection tools:
http://g015.cuppa.uic.edu/zhaoxia/blankcanvas/sketch.html

Integrated interface with combined features:
http://g015.cuppa.uic.edu/mapTool3/mt.html

Collaborative Decision Support System (composing site plans online):
http://g015.cuppa.uic.edu/zx/sketch2/project3.htm)

U.S. Census Bureau: *www.census.gov*

The U.S. Geological Survey: *www.usgs.gov*

National Cancer Institute: *http://www.nci.nih.gov/atlas*

5 Community Engagement in Land Use Planning Through Web-based Technologies

Steve Ventura, Tracy Miller and Glen Barry

Land Information and Computer Graphics Facility, University of Wisconsin-Madison, B102 Steenbock, 550 N. Babcock Drive, Madison WI 53706, USA

Abstract: The Shaping Dane's Future project developed several web-based tools to support citizen-driven land use planning in a community near Madison, Wisconsin. Tools and resources included a step-by-step procedure for plan development, resource and infrastructure maps including land use-related spatial analyses, interactive mapping (WebGIS), a planning portal for additional planning-related data and information, tools for land use allocation and impact assessment, and an electronic opinion polling tool. The web-based tools and resources were developed within the context of a broader citizen-based planning process facilitated by additional information technologies.

5.1 Introduction

As part of a broader project to more effectively engage citizens in participatory land use planning, a web site has been constructed (*http://www.lic.wisc.edu/shapingdane/*) that offers the citizens of Verona in Dane County, Wisconsin, the opportunity to access a variety of web-based resources that support technology-facilitated land use decision-making. The *Shaping Dane* project web site presents a process for land use planning (the Planning Resource Center), data to provide citizens with an accurate and easy-to-understand view of their community, and tools to develop land use scenarios and understand their impacts. It includes static analyses such as a map that depicts 30-year growth patterns throughout the town of Verona and all of Dane County, showing the trend in which a growing population is being dispersed in decreasing densities. It also supports interactive mapping – WebGIS – based on ArcIMS® from ESRI and geographic information system (GIS) tools for growth impact assessment.

In addition to seeing spatial patterns, citizens needed a way to systematically evaluate growth alternatives, thereby enabling them to contribute to the land use planning dialogue. A planning toolkit known as *Planning Analyst* (Niemann *et al.* 1999) was provided as one solution. The toolkit offers software modules that support geographic data exploration and analyses, land use allocation, and impact assessment. Areas within the community can be examined in three-dimensional detail and future growth scenarios can be generated. These started as stand-alone tools implemented with ArcView® and other GIS software and scripts. Some of the capability is being web-enabled as part of on-going research and development. The *Planning Analyst* modules have been tested in public land use forums, with the Land Use Task Force in Verona, and in GIS short courses (a total of 15 groups) at the University of Wisconsin-Madison. However, these tools are still under development; they are still evolving and being refined.

The goal was that, with minimal training, citizens could create and evaluate their own alternative scenarios. Verona participants were introduced to the various components of the Shaping Dane web site and the planning toolkit. Aided by the use of these resources, citizens developed their own land use plan, which was presented to the public in draft form in July 2001. The land use plan identifies resources special to the community and allocates future land uses in ways that could not be done without GIS and web resources. As of this writing, the plan is being finalized and linked to implementation and enforcement strategies.

In this chapter, we report on the role of the *Shaping Dane's Future* web site and its role in the development of a citizen-written land use plan for the town of Verona. This was one of several information technologies that facilitated production of the town's land use plan. Other aspects are reported in more detail elsewhere (e.g. Ventura *et al.* 2001a, Miskowiak *et al.* 2000, Niemann *et al.* 1999).

5.1.1 Background

In the Summer of 1998, then Vice President Al Gore selected Dane County, Wisconsin as one of six sites across the United States to demonstrate how information technologies such as geographic information systems (GIS) and the World Wide Web can assist local land use planning efforts. This led to the community demonstration project, *Shaping Dane's Future*, a collaboration of the University of Wisconsin-Madison, Dane County, Federal Geographic Data Committee (FGDC), the Natural Resources Conservation Service (NRCS) and Environmental Systems Research Institute (ESRI Inc.).

The city and town of Verona were selected as the project focus. Like many communities, Verona is facing controversial land use issues, including protecting natural resources and watersheds, guiding urban growth, and preserving prime farmland. Rapid growth spurred by a robust economy in the nearby city of Madison had rendered Verona's 1987 land use plan obsolete. A land use task force was charged with updating the land use plan. The Verona Land Use Task Force represents a cross section of the community – from farmers to retired professors. These volunteer 'citizen planners' were charged with thoughtfully assessing Verona's urban and rural growth, with a goal of preserving their community's small-town character, its agricultural and natural resources, and the high quality of life enjoyed by most residents. Many of these citizen planners had only rudimentary knowledge of community resources, and, until recently, had never viewed the existing spatial patterns and conditions of their town nor envisioned the impacts of development decisions.

5.1.2 The *Planning Analyst* Toolkit

For years, researchers at the University of Wisconsin-Madison's Land Information and Computer Graphics Facility (LICGF) have studied how GIS and other information technologies influence the land use decision-making process. We became interested in how the Web could be used in the land use planning process as an outgrowth of the development of several software modules for understanding and using spatial data and analyses. Beginning with a series of 'hands-on' GIS classes for land use planning and a series of public Land Use Forums in Dane County focused around GIS products, LICGF developed the *Planning Analyst* land use planning toolkit. This toolkit is a place-based decision support system organized around four modules: exploration; analysis; allocation; and evaluation. These modules include software tools and educational materials that are used to automate and support the planning process.

The Shaping Dane web site (*www.lic.wisc.edu/shapingdane/*) evolved as a way to convey the concepts (and to an increasing extent, the software tools) of *Planning Analyst* within the context of the town of Verona's revision of their land use plan. The site, which provides an array of data and resources, is designed to complement, not replace, a facilitated group planning process. This site provides evidence that planning resources of varying technical sophistication can now be made available to anyone with an internet connection. The Shaping Dane web site provides a portion of the functionality of *Planning Analyst*. WebGIS provides the ability to explore existing spatial data resources, retrieving data needed for a particular site or project from a much more extensive library of available data. The web site provides graphic results of experts' analyses – for example, delineation of environmental corridors that could be used to restrict growth in environmentally sensitive areas. The ability to interactively allocate land uses via the Web is under development; it will be based on stand-alone software we call 'Place-It'. Place-It is an ArcView extension that allows users to interactively allocate land uses to parcels of land and report back statistics on the type and amount of land converted, the population changes that would be associated with such conversions, and the potential environmental, fiscal, and other impacts. Similar to analysis, the evaluation of land use scenarios on the web site is limited to pre-created experts renditions, though our goal is to include interactive tools such as water quality models and fiscal impact spread sheets.

5.2 Shaping Dane Web Site – Citizens' View

The Shaping Dane web site consists of two primary components – a 'Planning Resource Center' and 'Electronic Planning Facilitation'. The Planning Resource Center (Figure 5.1) is designed to provide quick and easy access to information on the site about planning process, data, analysis and products. It is organized under three headings:

- *On-line Atlas*
  - view on-line maps (*Map and Image Gallery* and *3D Virtual Worlds* – pre-created image files and movies for users to view and print maps based on analyses of land use planners and other experts on topics such as environmental corridors, growth management, and farmland protection); and
  - tools for interactive mapping (*Make your own map*), and downloading data and metadata (*Download Data*).
- *Existing Plans and Ordinances*
  - download existing information such as applicable statutes and existing local and county plans.
- *Land Use Planning*
  - general and local background information for planning.

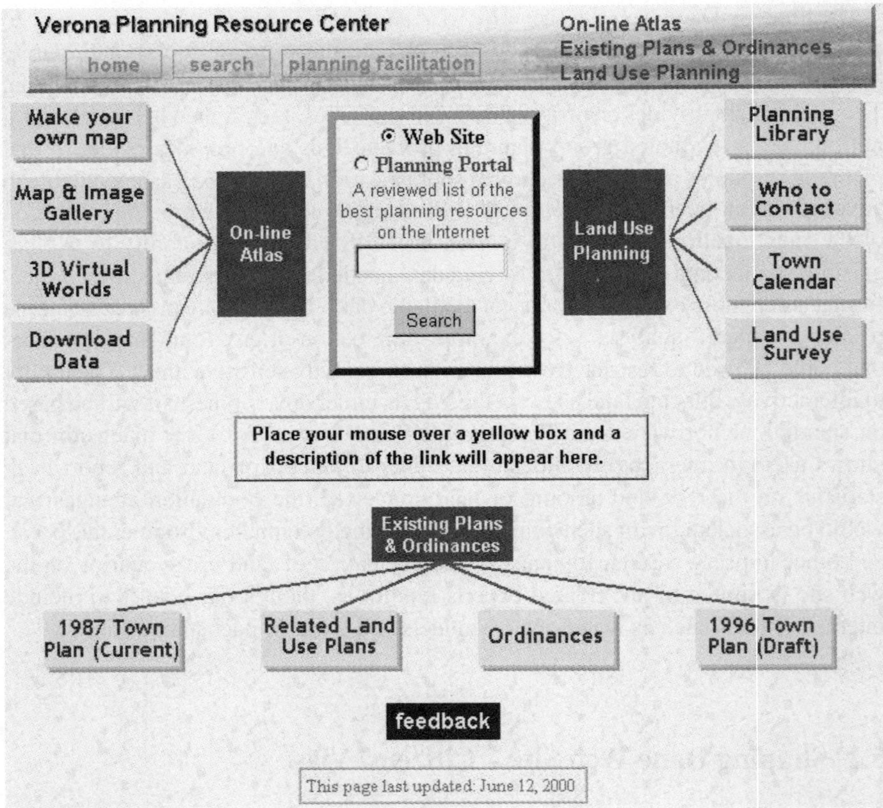

Fig 5.1: Planning Resource Center – one of two primary starting points for *the Shaping Dane's Future* web site

Using WebGIS, we made exploration of Verona land use data possible on-line. Users can make their own map from the full set of data, or view and manipulate data we used in analyses of growth management, farm priority zones and environmental corridors. Many options are available to users: zoom in on parcel data; overlay proposed farm priority zones; query for industrial zoning; locate wetlands and more. The level of GIS functionality achieved includes viewing each of the constituent layers, zooming and panning and querying the data according to attributes. The Land Use Planning section information includes an extensive Planning Library with planning guidebooks such as Ohm (1999) and Edwards (2000) that form the basis for some of the procedures and models in the planning toolkit. An additional link from the Planning Resource Center supports a spider-based 'Planning Portal' which links to hundreds of reviewed and catalogued planning information and environmental assessment resources on the Web (Figure 5.2).

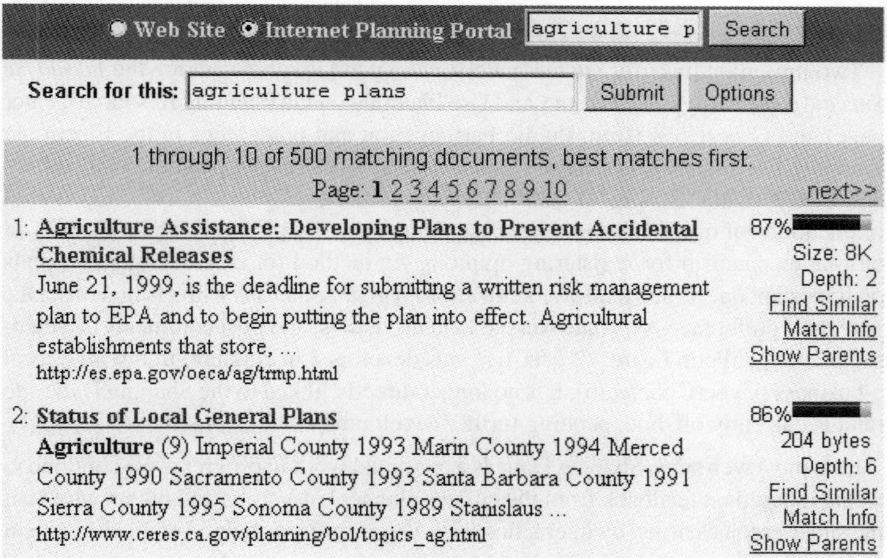

Fig 5.2: The Planning Portal – an indexed search engine for on-line planning resources

The second main option is Electronic Planning Facilitation (Figure 5.3). This avenue provides a guided trip through the planning process. It is organized around the elements of a comprehensive plan (required by the Wisconsin's 'Smart Growth' comprehensive planning law) and the stages of the planning process. For each planning step, frequently asked questions anticipate the issues and organize the provided information. Additional links are provided to various mapping and impact assessment tools as appropriate or available.

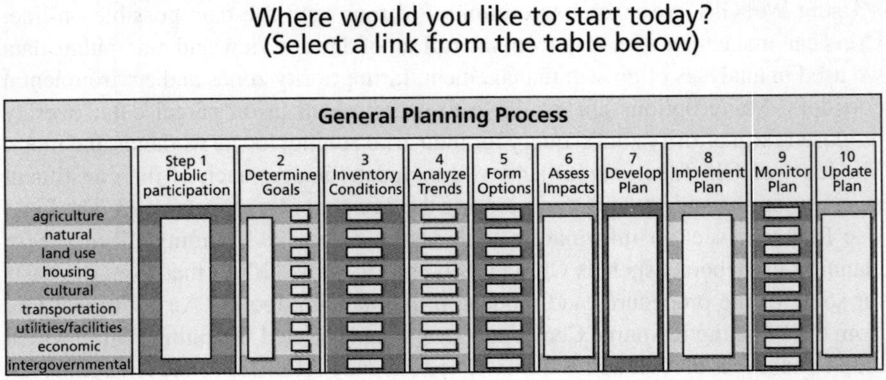

Fig 5.3: Electronic Planning Facilitation – step-by-step procedures for local land use planning

Two important links for citizen understanding and involvement are the *Land Use Survey* (most easily linked from Land Use Planning on the Planning Resource Center page) and *CyberCivic* (from Public Participation and other steps in the Electronic Planning Facilitation). The *Land Use Survey* provides results from the mail survey conducted in the Spring of 2000. Results include graphics portraying spatial implications of questions about land use attitudes and aspirations. *CyberCivic* is an on-line mechanism for registering opinions – a method for conducting straw polls on important questions. It allowed citizens to register opinions with elected officials, vote their preferences on controversial land use issues, and post comments in a chat-room and a bulletin board. *CyberCivic* was developed in collaboration with a local e-business (CyberCivic.com); it is no longer directly linked to the Shaping Dane site (and is currently off-line, pending further development and deployment).

Like most web sites, Shaping Dane is a continual work in progress. We continue to refine it based on feedback from the citizen-planners of Verona and others who visit the site. Lessons learned by interaction with Verona citizens helped guide the content and organization of the web site, both through their on-going planning activities and through specific sessions with Land Use Planning Task Force members who evaluated prototypes. Much of the analytic content of the site concentrates on agriculture, natural resources and growth management, because these are the issues we heard from citizens as important. The look and feel of the page was substantially improved when members of the Task Force tried to use early efforts. We observed and recorded their use patterns and frustrations, and attempted to rectify limitations. For example, although using early versions of the ArcIMS-based WebGIS was very easy for our spatially literate development staff, we needed to add considerable explanation and help functionality before it was useable by citizens.

5.3 Shaping Dane Web Site – Developers' View

New and developing web-based mapping and search technologies allow planning facilitators to provide the general public with interactive access to textual and map-based information and tools. The challenge is to use these tools in a manner that shares content in a clear and compelling manner to support community decision-making processes. A number of recent trends in dynamic web page generation add considerable capabilities to a general planning site. Increasingly sophisticated tools allow for more dynamic, interactive, targeted and filtered information exchange and retrieval. From the wide range of database-driven, dynamic web page creation tools that are available, we focused on two that particularly merited further development for community based planning – interactive mapping and search engines. Both technologies are relatively new, but given the continued central role of spatial patterns in land use planning and the need to find particular information within the Internet's huge search space, we are likely to see continued development for some time. We also report on some other web-based technologies that could become part of a web site for planning support – interactive surveys and image library servers.

5.3.1 Internet Map Serving – WebGIS

Internet map serving software, sometimes called WebGIS, allows basic GIS functions to be carried out over an internet connection using a standard web browser. Functions currently available with 'off-the-shelf' commercial software are limited to viewing, querying and printing of maps. Additional functionality can be programmed and will undoubtedly become part of future releases of commercial software. This will enable users to integrate local data sources with internet data sources for display, query and analysis in a web browser (e.g., Netscape Navigator®, Microsoft Internet Explorer®). Data and GIS functionality will be delivered to the desktop of anybody with a reasonably current computer and internet connection.

WebGIS is a rapidly evolving area of technology. The Shaping Dane Project sought to see how far we could get in providing WebGIS functions with commercially available software, at a time (1999) when many solutions were just arriving on the market and were not necessarily fully robust. We were fortunate to be able to test a pre-release version of internet map serving software, the ArcIMS® package from ESRI. ArcIMS – the successor to two earlier programs provided by ESRI – represented a significant leap in features and relative ease of use. Use of the software requires an internet server to be installed and configured, as well as several helper applications. Once the software is operable, the manner in which ArcIMS designs and implements an interactive map is intuitive and relatively easy, at least to people previously exposed to GIS. First, the site designer prepares a base map layout in the *Author* module. Themes (layers) are designated, along with display characteristics (Figure 5.4). This collection of available data layers and renderings are saved as a file whose characteristics can later be modified to change the map display. The next step to serve maps requires creating a 'map service' in the *Administrator* module. The process of making a map service takes saved files and prepares them to be shared on the Internet. The final step is the *Designer* module, which constructs the HTML and Java scripts that make maps interactive.

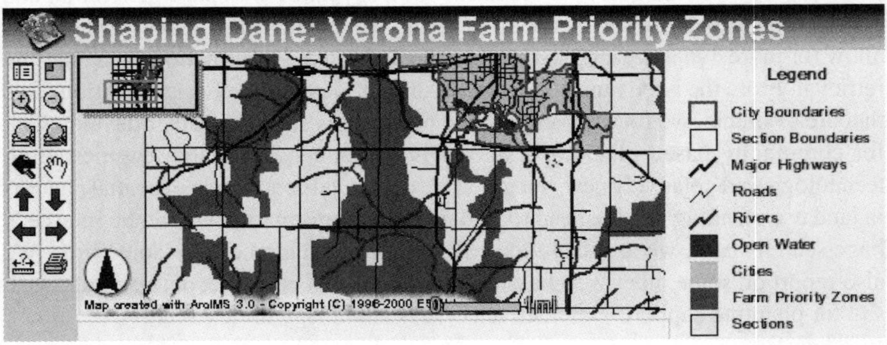

Fig 5.4: The ArcIMS Author module resembles familiar GIS map interfaces

5.3.2 Planning Portal – Reviewed Internet Search Spaces

The Shaping Dane project constructed a specialized search engine of reviewed internet land use planning information residing throughout the Internet. A set of reviewed internet resources relating to community based planning in general, and Verona's planning needs in particular, were made fully searchable in support of the planning mission. This 'Planning Portal' allows full text searching of internet content that has been carefully selected for inclusion. The limited search space ensures more relevant internet search results for the purpose of community planning, avoiding the large number of useless links resulting from a standard web search.

Search engine support software allows a technically savvy land use professional to index a handpicked set of reviewed web sites, making these fully searchable in support of the planning mission. The typical 'links' page – a list of additional resources that are hyperlinked – does this to some degree. However, increasingly sophisticated search software allows you to go the next step, and make your own 'search space', a search interface to external content that serves as a portal to the content identified and reviewed by an expert. The expert may choose, as LICGF did in our own experimental efforts, to integrate information from its locally oriented site with that from numerous other external sites into one searchable collection of information – a planning portal – in support of your planning mission. A citizen-planner thus can seamlessly search both our site and organized, reviewed external sites for information about planning and environmental assessment topics.

The sophisticated software algorithms that web search portals such as Yahoo and Lycos have used for some time to index and search web content are beginning to become more widely available, though not yet in a user friendly manner. With some effort, sites are selected using 'spider' internet technologies that 'crawl' through a site, creating search indexes of all content they encounter within established boundaries, based on content analysis. These sites are reviewed, abstracted, indexed and made fully searchable. The search interface allows users to extensively customize a search, and instantly jump to the pages that meet search criteria. This unified search capability has been implemented thus far for approximately 100 of the best planning resources and a like number of sites with information about environmental impacts of land use and population expansion, for a total of several thousand individual internet pages. Sites that were indexed for the planning portal include leading research institutions, non-governmental organizations, and public agencies at all levels of government.

At this point, the Shaping Dane developments are largely experimental. These efforts used spider indexing technologies in a significantly new manner – creating a new type of refined internet-wide search capability for land use planning purposes – but requiring some programming to do so. Portal implementation has required original database development as a front end to the command-line driven menu system of the spidering software package. Sites to be included in the portals and spider commands are stored in an organizing database that was designed at LICGF. This is used to generate scripts to periodically carry out the indexing. The spidering software constructs an index that contains every word, its frequency and its proximity to other words found in the entire content of selected external web sites. The index resides on LICGF's servers, with the content remaining in original locations. There is still potential to more fully implement and improve the search content by identifying, indexing and adding more sites.

5.3.3 Other Web Site Technologies – Future Developments

Future versions of the Shaping Dane web site, implemented in other communities may include elements of two other developments of LICGF. Techniques drawn from the Wisconsin Land Information Program survey would provide the means to electronically survey citizens and to provide tools for understanding and conveying results. The OrthoBrowser, first developed at the Massachusetts Institute of Technology (MIT), provides the basis to seamlessly serve orthophotography and other imagery that are ordinarily a challenge on limited band-width internet connections.

WLIP survey

We have developed an on-line system for conducting the annual Wisconsin Land Information Program inventory of county land information system status through a web-based interface (Ventura *et al.* 2001b). County land information offices directly update a database with details of counties' progress on fourteen 'foundational elements' and other details of the local GIS. The system supports queries of the database with tabular and graphical display of results.

The on-line survey is based on active server pages (ASPs). ASPs extend the functionality of web pages. In this case, they provide greater interaction with the user and support connections between the database, the questionnaire and the results generator. Initially, we used Microsoft's FrontPage® to build the web interface, but it has been enhanced and further developed using a variety of ASP and HTML tools, along with ActiveX Data Objects (ADO), Visual Basic scripts and Java scripts. The database is supported in Microsoft's Access2000®. Additional scripts manage and verify county passwords, build forms for adding additional questions, generate summary and completion status data, and handle database queries. The prototype survey web site (*http://www.lic.wisc.edu/wlip/nwelcome.asp*) is hosted at the University of Wisconsin-Madison. Additional technical details about how the survey was implemented are available in the technical section of this web site. Although the survey was not specifically focused on land use planning, this general approach could be used to replace paper-based surveys of beliefs and aspirations (such as the one we conducted in Verona to help shape land identification and protection strategies). It could also be used as a complement to a tool such as *CyberCivic*.

OrthoBrowser

Digital orthophotography has become a ubiquitous component of GIS, particularly for systems looking at extensive land resources such as necessitated by land use planning. Increasing resolution (as fine as 6 inches per pixel in Dane County) creates very large files. This causes excessive time delays when providing these data in a WebGIS or ftp download environment due to internet bandwidth limitations. Several groups of developers, including Ferreira (1996), have developed software to serve digital orthophotography. Their solution is called *OrthoBrowser*. The software allows a user to seamlessly select a window of any size from an extensive area and download or view data at a resolution appropriate for the size of area selected. Edge matching, re-sampling and other technical operations to create this seamless, selectable view are managed by software 'behind the scenes.'

We have adapted this software to a coastal planning context in Wisconsin to handle a very large volume of orthophotos. Historical photography was processed into orthophotos to evaluate land use change patterns, particularly 'bluffline' recession rates. Now, a user can pick a coastal county or municipality, zoom into a view of that area and select photography of varying vintage back to the 1950s (*http://www.lic.wisc.edu/mit/index.html*). When integrated with WebGIS, the modified *OrthoBrowser* can provide a powerful means to serve imagery with other data relevant to local planning, providing the small subset of data relevant to a local land use decision from an immense library.

5.4 Conclusions

At the onset of this project, we hypothesized that the Shaping Dane web site could provide useful information to the Verona land use planning task force and that this would more effectively engage citizens of the community in planning processes. We have confirmed, through post-planning interviews and unsolicited comments at public meetings, that the site was useful for the task force. It helped them understand planning processes, find and explore available data, develop land use scenarios and determine their impacts, and develop maps portraying conditions and scenarios. Perhaps most importantly, it helped them develop and hone queries for analysts using stand-alone GIS and impact assessment models.

We conducted before and after surveys of Verona citizens (Spring, 2000 and Summer 2001).The initial survey indicated that the Web was potentially a viable means to reach citizens. Over 95 per cent of households had computers and more than 70 per cent had internet access. The follow-up survey of Verona citizens in the summer of 2001 inquired about the site and how it was used by citizens within the town. It appears that access, beyond members of the Task Force and other local citizens directly involved in planning, was limited. However, citizens that did access the site found it quite useful in understanding and participating in local land use issues.

Through several extensions of standard web functions such as WebGIS, advanced query tools and data serving procedures, we have demonstrated that the Web is a technically viable means to provide data, explanations, methods and models for local land use planning. By evaluating these tools in a 'real world' land use planning effort, we were able to customize commercial software for the needs of a local community grappling with difficult land use issues. Interaction with the community provided useful feedback to developers and provided information and products used in the local planning process. Citizen decision-makers such as the Verona Land Use Planning Task Force were able to understand and use these new forms of information. Though web-based tools and data cannot replace the personal interaction necessary for successful land use planning, these can significantly facilitate the process, bringing needed insights and information.

References

Edwards, M.M. (2000) *Community Guide to Development Impact Analysis*, Wisconsin Land Use Research Project, University of Wisconsin-Madison, Madison.

Ferreira, J. (1996) *Tools to Facilitate Access to Digital Orthophotos,* Massachusetts Institute of Technology: http://ortho.mit.edu/nsdi/summary.html.

Miskowiak, D.M., Ventura, S.J. and Sutphin, T.E. (2000) Farmland Preservation and GIS, *Land Information Bulletin Technical Paper No. 3*, Land Information and Computer Graphics Facility, UW-Madison.

Niemann, B..J, Heinzel, W.M., Ventura, S.J., Sutphin, T.E. and Miskowiak, D.M. (1999) Mapping Growth Management Factors, *Land Information Bulletin Technical Paper No. 2*, Land Information and Computer Graphics Facility, UW-Madison.

Ohm, B.W. (1999) *Guide to Community Planning in Wisconsin*, University of Wisconsin-Extension, Madison.

Ventura, S.J., Niemann, B.J., Chenoweth, R.E. and Sutphin, T.P. (2001a) GIS-enhanced land use planning in Dane county, Wisconsin, in Craig, W., Harris, T. and Weiner, D. (eds.) *Empowerment and Marginalization: Public Participation GIS*, John Wiley and Sons, New York.

Ventura, S.J., Hart, D.A., Moyer, D.D. and Niemann, B.J. (2001b) Surveying the lay of the Land Information Systems, *GeoSpatial Solutions,* 11(10):34-37.

Web Sites

Shaping Dane: *www.lic.wisc.edu/shapingdane/*

Wisconsin Land Information Program Survey: *http://www.lic.wisc.edu/wlip/nwelcome.asp*

OrthoBrowser: *http://www.lic.wisc.edu/mit/index.html*

6 Tools for Community Design and Decision-making

Ken Snyder

PlaceMatters.com, 3469 W Hayward Place, Denver CO 80211, USA

Abstract: Since the fall of 1998, the US Department of Energy has organized four national meetings on Tools for Community Design and Decision Making (TCDDM). The TCDDM meetings brought together community practitioners working in land use planning and community development; tool providers offering a wide spectrum of visualization, impact analysis, predictive modelling, and collaborative decision making tools; and foundations and public agencies working to advance more effective planning tools and process techniques. In general, TCDDM participants are interested in helping communities make better decisions about land use, resource use, community development and community design. More broadly, they want to help communities become more livable and more environmentally and economically sustainable. To achieve this, TCDDM organizers believe that much of the solution relies upon the successful integration of technical tools for land and resource management and other community development objectives, with process techniques for civic participation, stakeholder involvement and consensus building. Together these elements form what might be referred to as a Planning Support System (PSS) for communities. Over the course of the last three years, TCDDM participants have identified strategies to improve decision support tools, and explored ways to make the use of these tools and process techniques more mainstream. This chapter highlights key outcomes of these meetings, including TCDDM core values, key principles of effective planning support systems, and an overview of many of the tools presented at these meetings. The tools overview considers how various tools support these principles. A more thorough analysis of the opportunities and constraints represented by these tools would be a valuable next step in helping communities better assess which are the best tools for meeting their specific needs.

6.1 The Demand for Planning Support Systems

Whether the issue is mandates for environmental impact statements, requirements to address air quality violations, critical habitat for endangered species threatened by development, or the need to balance jobs, housing and transportation goals, the need to address complex issues in an integrated fashion is overwhelming for many city and regional planning entities. Emerging tools and process techniques for community design and decision-making have the potential to dramatically improve decision-making and so change the planning profession. In particular, these tools and process techniques, collectively referred to as planning support systems (PSS), could transform decision-making in two ways: firstly, they can help communities shift land use decisions from a reactive regulatory emphasis to a more proactive performance-based approach; and secondly, they can enable more effective and efficient community involvement in decision-making processes.

An example of performance-based planning might involve changing the permitting process in a community from a more traditional approach, where a set number of building permits are issued each year on a first-come-first-serve basis, to a system that evaluates each proposal against community criteria. To make this work, local officials first meet with the community to identify the issues they care about most and help them translate those values and concerns into indicators. With impact analysis and predictive modelling tools, new development projects can be evaluated against those indicators. Projects that bring the most benefit to the community would then get approved, and projects that failed to pass a threshold of criteria would be sent back to the drawing board.

The planning process can also become much more interactive and responsive to community interests with tools that allow on-the-spot analysis. Thus, stakeholders can be present in the design phase of a project and provide feedback on how to improve the project. Using visualization and design tools, the community could rank different design elements for a specific project or envision alternative futures for the region through scenario analysis. With real-time tools, recommendations and changes can be incorporated into project proposals almost immediately and each new iteration can then be tested against community-selected indicators. The result is a planning process focused more on solutions than problems, less geared toward regulation and more toward performance, and that allows stakeholders to be more proactively engaged. Together these tools and process techniques, or PSS, have the potential to steer communities toward the long-term practice of sustainable development. In essence, this has been the primary goal of the workshops on Tools for Community Design and Decision Making (TCDDM). These types of PSS are being used successfully in a number of projects around the United States. Two particularly good examples presented at the most recent TCDDM meeting include: Dane County, Wisconsin with their use of a process called Planning Analyst to support land use allocation, impact assessment and visualization (see chapter by Ventura in this volume); and the Envision Utah project, utilizing a number of modelling/impact analysis tools coupled with a community participatory process to help the Salt Lake City region deal with growth.

6.2 Key Outcomes of TCDDM

What are the characteristics of effective PSS and what are some guiding principles to make these tools even better? In addition, are there core values that all PSS should try to respect? These are questions that participants of the first TCDDM meeting were asked to address in a large group exercise. Meeting participants Mike McKeever, Ron Thomas, Nancy Hanson, and the author developed draft values and principles prior to the meeting. These were then reviewed and refined by meeting participants through the use of 'meetingware' technology, which greatly enhanced the group's ability to provide direct feedback (see MeetingWorks® and CoVision™, section 6.3.23, as examples of meetingware technology).

6.2.1 Common Values

Seven core values emerged:

- *Citizen-based participatory democracy*: PSS should assist in developing effective, broad-based citizen participation in all aspects of public sector and civic governance, building consensus and collaboration among diverse parties and strong interrelationships between non-governmental organizations, the private sector and public institutions.
- *Place-based*: Rather than focus on large scale federal or state level programs or arbitrary jurisdictional boundaries, planning support tools should focus at a scale such as the neighbourhood, watershed or regional level that allows real consideration of the economic, social and environmental conditions of place.
- *Vision-centred*: PSS should shift the focus from rule-based to vision-centred decision-making, moving away from episodic planning to an ongoing process that creates more vibrant, effective communities.
- *Systems-based analysis*: planning tools and decision-making processes should help build an understanding of the interconnections between the natural and built economic and social systems of a community.
- *Measurable outcomes*: PSS and the use of community indicators, impact analysis and scenario analysis tools in particular, should provide citizens with a vision of alternative futures, an understanding of the trade-offs associated with different choices, and a means to establish accountability for decisions and actions based on terms that are both understandable and quantifiable.
- *Alignment of local and regional actions*: Given how economic, social and environmental issues tend to stretch beyond jurisdictional boundaries, PSS need to recognize the interrelationships between activities at the local and regional scales.
- *Alignment of short term actions with long term consequences*: activities should show tangible improvements to communities in the near term, and also lead to healthy, sustainable long-term futures.

6.2.2 Design Principles

The following list of principles also was developed at the first TCDDM meeting. The list was described as a 'wish list' for practitioners and the types of tools they would like to see made available for community planning and development. It has served as guidance for the development of more effective planning support tools. Over the years we have seen significant growth in the development of tools which more closely match these principles.

Interactive and usable at multiple levels of expertise. It should be easy to explore planning choices and their consequences in community meetings with many participants. PSS should also help communities to check the effect of proposed changes to adopted plans and produce a quantitative and graphic summary of the benefits of implementation for use in annual progress reports. Computer tools should be adaptable for a range of users, from citizens who are not computer specialists to professionals with working computer knowledge.

Example: in a community meeting a projector is attached to a laptop computer showing a land use map of a neighbourhood on a large screen. Someone wants to know what changing the land uses within one quarter mile of a transit stop from commercial to a mixture of commercial and residential would have on travel behavior, air quality and energy use. The tool should be able to quickly illustrate the land use changes and their effects to support further 'what if' thinking during the meeting.

Accessible and affordable. The hardware and software should be readily accessible and affordable.

Example: the tools should be usable on computer equipment that is widely in use and reasonably priced. The software also should be widely used, increasing the potential for and benefits of standardized approaches and cross learning between projects within a community and around the country.

Adaptable and maintainable. Computer code should be easily adaptable to support local conditions and data and maintainable through local capacity that exists or can be created.

Example: the staff of local community organizations or agencies should be able to keep the database up-to-date to ensure long-term value and usage. The system should have value to the neighbourhood, city and region, beyond the initial planning study.

Open architecture. People are suspicious of 'blackbox' computer-generated answers. Models and assumptions should be as transparent as possible.

Example: impact analysis and predictive modelling show significant reductions in vehicle miles travelled and thus related air emissions as the result of a proposed transit-oriented design project. The assumptions and algorithms behind these calculations should be easily accessible for scrutiny. Techniques should be understandable so stakeholders can validate their credibility or improve them through peer review.

High quality data. PSS should be supported by data that are local, up-to-date, and credible. Data, like language, are limited in their ability to describe reality. What and how communities measure things, inevitably contain biases – both in what it measures and the accuracy of the measurements.

Example: a welfare to work project needs to have current demographic and employment information to be effective at linking people with jobs. Cause-effect relationships based on modelled projections rather than empirical information should use the most credible information available. While the GNP tries to measure economic health, it also increases with economic activity surrounding accidents, natural disasters, and crime. A tool should help recognize these biases and wherever possible minimize them in analysis.

Local and regional scale. Local choices have regional consequences and *vice versa*, which should be highlighted by PSS.

Example: a zoning change to convert 20 per cent of the single family land in a metropolitan region to rowhouses will significantly increase land efficiency, reducing the amount of farmland that is converted to urbanization. The effect of this 20 per cent conversion needs to be understood, and the amount of acreage it would save on the urban fringe needs to be quantified and illustrated.

Comprehensive coverage and integration of issues. All of the issues that are important to citizens need to be addressed. This helps to convince a broad range of interests to participate. The tool should facilitate this cross-discipline approach. The complex cause and effect relationships between issues and disciplines should be accounted for.

Example: citizens working to revitalize an inner city neighbourhood need all kinds of information. A short list would include everything from health care programs, affordable housing, crime, environmental hazards, educational programs, techniques to attract investment in retail and other services, arts and culture, and at risk youth activities and programs. PSS must support a full range of local values and priorities. Development patterns affect impervious area, which in turn affects surface water runoff, and water quality in streams and rivers. Decisions about where to locate major employment centers affects travel times and costs, which affect disposable income.

Impact analysis including short, mid and long-term effects. Impact analysis tools that enable communities to quantify the economic, environmental and social impacts of different development choices should also be sensitive to how different groups within a community will be affected. The effects of choices over time should also be identified.

Example: a community is trying to decide whether it makes sense to build a new light rail route. The tools used should identify impacts (e.g. ridership, travel behaviour, air quality) based on current development conditions as well as probable changes to those conditions under different planning scenarios over time (e.g. 10 year increments through build out).

Engaging citizens and supporting face-to-face interactions. The tool should enhance community process, not overrun it, and be user friendly and interesting enough to engage citizens of different ages, income, and interests. High-tech tools should not be intimidating in how they are applied to local planning and decision-making. They should encourage community interaction among citizens, decision-makers and different interest groups rather than cause decision-makers to hide behind the technology for answers without engaging public participation.

Example: tools that have hands-on elements to promote and encourage everyone to be involved.

Visual imagery. A wide variety of visual images to illustrate the effects of choices should be available.

Example: photographs of local examples and alternatives from other communities, videos, three-dimensional imagery, before and after photographs, individual development scale and regional scale, and more are all very powerful tools in helping citizens to visualize and understand the impacts of various options.

Values – cause and effect. Document, track and prioritize community values and distinguish the cause and effect of community problems in order to find solutions. Cause-effect relationships based on modeled projections rather than empirical information should use the most credible information available.

Example: a tool for community process that provides a weighting mechanism for community values that helps prioritize concerns and establish a tracking system for indicators of community health and progress.

Promote identification of design options. In making development decisions, communities need tools that highlight design options through scenario analysis and provide feedback on the impact of these different development options.

Example: in considering development options for an urban infill project, tools can provide information on innovative solutions, including transit-oriented development, mixed use development and energy efficiency design practices.

Regular monitoring and reports. Tools should be designed to easily produce periodic reports of progress toward achieving community goals and if linked to a regional model, the contribution of each community toward achievement of the regional goals. Communities need tools that help document and record decision-making in process, allowing participants and others to revisit how consensus was reached and to track key decisions as the community works towards project implementation.

Example: a city council hears an annual report of progress at achieving sustainable community indicators and discovers that while new development projects have been consistent with the neighbourhood land use plan the new employers are not meeting targets for hiring employees from within the neighbourhood. The council uses this finding to initiate a dialogue with the employers and community to remedy this problem.

6.3 Overview of Tools for Community Design and Decision-making

Generally, seven types of tools are available to help communities in community planning. These key components are used below to create a taxonomy of tool functionality. The matrix represented in Table 1 helps clarify the different capacities of tools currently available.

Table 6.1: Levels of functionality for tools

	visualization tools	asset mapping	impact analysis	scenario analysis	performance-based planning	predictive modelling	community process tools
Miniature Modelling – Box City, Hackney Building Exploratory	*	*					*
Photomontages – Jeff Winston, Victor Dover, Steve Price	*			*			*
3-D Rendering – DPZ, Correa Valle Valle, HTNB,	*			*			
Environmental Simulation Center, Jeff Winston	*		*	*			
3D Visualization Room – BP Visualization Center	*						
Visual Preference Surveys – LRK, Neelson	*			*			*
Impact Analysis and Economic Forecasting – SCALDS	*		*				
InfraCycle, REMI®, Metroware™	*		*		*		
Web-based Calculators – AirHead and ICLEI Commuter Calculator			*				
Web-based Asset Mapping and Analysis – HUD EGIS	*	*	*				
GIS Analysis – CommunityViz™	*	*	*	*	*	*	*
NEWS	*		*				
INDEX®	*	*	*	*	*		*
LEM	*	*	*				
Model Builder™	*		*	*	*		
PLACE³S	*		*	*	*		*
QUEST™	*		*	*	*		*
UGrow	*	*	*	*	*	*	*
What If?™	*		*	*			
Predictive Modeling – SLUETH	*					*	
TRANSIMS	*					*	
UrbanSim	*		*		*	*	
Process Tools – FHI Sustainable Dev Toolkit	*				*		*
CoVision™, MeetingWorks®, WebCouncil™							*

6.3.1 Miniature Modelling

Box City, developed by Ginny Graves at CUBE (*www.cubekc.org*), is a simple technique of using basic art supplies (cardboard boxes, construction paper, markers) to create a small replica of a city block or street corridor. Because it is low-tech, people of all ages can easily engage in an exercise identifying the things they value the most in their community and things they would like to see changed. The Hackney Building Exploratory (*www.buildingexploratory.org.uk*) in the township of Hackney (located within the greater metropolitan area of London), has used similar techniques very effectively to educate children and adults about the built environment they live in and how historical, environmental and political issues have helped shape their community over the years. Exhibits range from high-tech computer programs developed in partnership with the Centre for Advanced Spatial Analysis (University College London) and the Housing Corporation to illuminated models, kits, maps, games and giant jigsaws made by professional artists working with local children and adults.

6.3.2 Photomontages

Several TCDDM participants, including Dover and Kohl (*www.doverkohl.com*), Duany Plater-Zyberk and Company (*www.dpz.com*), Urban Advantage (*www.steveprice.com*), and Winston Associates (*www.winstonassociates.com*), demonstrated the use of photomontage and computer imaging techniques to assist communities in visualization. A photomontage involves taking a digital image of an existing streetscape or building development and 'doctoring' the image to add design features such as decorative street lights, trees, grass medians, light rail, bike lanes, and mixed use with retail on the first floor and office space and housing above. It allows planners and interested citizens to see how an existing streetscape might be improved or give people a better sense of the type of development they would like see occur in the future. It is particularly useful for showing incremental change, such as first adding a median, then street trees, then infill with buildings, *et cetera.* This technique can be particularly useful in helping shift discussions away from density and more towards design and functionality.

6.3.3 3-D Rendering

3D rendering consists of creating a virtual model of the project (terrain, buildings, landscape) and providing a viewer with a 'walk-through', or 'fly around' experience. The model usually begins with an accurate terrain base, over which is draped either an air photo or a variety of ground covers such as grass or paving, for example. On this base are placed roads, trees, buildings and land features. Finally, add the finishing touches – shadows, cars, reflections, a diorama background from the actual site, and sky highlights – and you have a very believable representation. Models can be developed with varying degrees of 'realness' – ranging from simple massing models to very precise renderings of materials. With the use of a computer projector, models can easily be projected at very large scales greatly increasing a sense of reality. Users are finding that realistic portrayals, especially ones that can be experienced by moving through them (virtually) greatly increase the comprehension of the viewer. Greater comprehension, achieved more quickly leads to quicker decisions – whether it be in the design process with a client or in the public arena by a reviewing agency. Quicker decisions lead to less costly processes and better products. Although creating a 3D model was once a very expensive process, software and hardware improvements have reduced the cost dramatically over the last five years, making this tool increasingly accessible to both the public and private sector. Participants demonstrating 3D rendering tools include: Correa Valle Valle (*www.urbanism.com*), the Environmental Simulation Center: (*www.simcenter.org*), HNTB Technology Group (*www.hntbtech.com*), and Winston Associates (*www.winstonassociates.com*).

6.3.4 3D Visualization Room

In partnership with the University of Colorado's College of Engineering and Applied Sciences, British Petroleum (BP) has invested in the development of the BP Visualization Center (*www.colorado.edu/Research/bpVisCenter*). The Center is equipped with modeling and visualization tools to view 3D imagery on very large screens. The technology is used to display data onto three walls, and floor of a 12-foot by 12-foot by 10-foot high room set up as an immersive three-dimensional environment. In the immersion room, viewers find themselves surrounded by almost life-size 3D models, as if they were living in it. With the use of head and hand tracking devices viewers can navigate through the landscape with motions of their body and interact with or change the scene in 3D. While the BP Visualization Center was established by donations from the oil and gas industry, sponsors are interested in potential application in community visualization.

6.3.5 Visual Preference Surveys and Interactive Visioning Techniques

Interactive Visioning Surveys (*www.lrk.com/homeplans*), developed by Looney Ricks Kiss (LRK) of Princeton, New Jersey, can be disseminated throughout a community at computer terminals or on a website to obtain widespread feedback. The kiosk-type survey asks a question and presents graphic images from which users select their preferred choice. The images can focus on aspects of streetscapes or any other design topics the local government wants input on. In Hillsborough, New Jersey, this kiosk technique enabled the town to get feedback from nearly 10 per cent of its residents. The kiosk helped users visualize various street improvements (buried utility lines, wider pedestrian sidewalks, decorative lighting, *et cetera*) and what they might look like integrated into an existing streetscape. Respondents were asked to select their overall favourite and then indicate how much they would be willing to pay in taxes to implement the design they chose.

Visual Preference Surveys (*www.anavision.com*), developed by Tony Nelessen, use colour slides and simulations that enable the viewer to compare, contrast and select among images. Audience members provide feedback on each image, ranging from –10 for something they really dislike to +10 for something they would love to see in their community. Using the previously described photomontage technique, Nelessen Associates, works to eliminate biases created by things like the weather; keeping the same background sky and basic streetscape but adding those amenities they want participants to rank.

6.3.6 Impact Analysis and Economic Forecasting

Developed by the Federal Highway Administration and Parsons Brinckerhoff, Quade and Douglas Inc., the Social Cost of Alternative Land Development Scenarios (SCALDS) (*www.fhwa.dot.gov/scalds/scalds.html*) is an Excel spreadsheet model that estimates the full cost of alternative land use patterns. SCALDS estimates monetary and non-monetary costs associated with urban land development. It builds on research in a number of areas including least-cost planning, full-cost travel studies, and cost of services research. The accompanying user manual includes a literature search and an extensive bibliography documenting these studies.

InfraCycle Software Ltd. (*www.infracycle.com*) is a fiscal impact tool that calculates the life cycle cost of municipal infrastructure (fire, police, roadways, sidewalks, street light, park land, recreation facilities, storm water, sanitary sewers, garbage collection, transit, schools, school busing). The municipal revenues can be calculated from sources such as taxes, levies, development charges, application fees *et cetera*. Revenues are compared to costs to determine if revenues support costs.

Founded in 1980, REMI® (*www.REMI.com*) constructs models that predict the economic and demographic effects of policy initiatives. Using REMI® Policy Insight, you can answer 'what if' questions concerning the national, state, or local economy. A wide variety of policies that influence economic activity can be evaluated, including policies in economic development, transportation, energy, environment and taxation.

Another tool provider present at TCDDM meetings, Geostat, has developed an 'urban package' of GIS tools called Metroware™, which includes a planning model, an econometrics model, a capital improvement model, and land use model. In the past, municipal information on capital facilities such as street and water systems was separated into different and often incompatible formats while fiscal information was found in spreadsheets. Metroware™ enables municipalities to combine spatial physical and fiscal information into one information system.

6.3.7 Web-based Calculators

Airhead (*www.airhead.org*), a project of the Center for Neighborhood Technology, is designed to help people reduce the air pollution impacts of their daily activities.

The website includes an emissions calculator that helps identify and track the air pollution created by people's activities, and a product search that shows the relative air pollution impacts of a variety of consumer products.

The ICLEI Commuter Calculator (*www.iclei.org/games/comcalc.htm*) helps see the potential shifts in greenhouse gas emissions as a result of different commuter options.

6.3.8 Web-based Asset Mapping and Analysis

The US Department of Urban Housing and Development's Enterprise Geographic Information System (EGIS) (*http://hud.esri.com/egis*) provides location, type and performance of HUD-funded activities in every neighbourhood across the country; as well as a selection of EPA information on brownfields, hazardous wastes, air pollution and wastewater discharges. In addition, it is capable of conducting map analysis over an ArcIMS platform. The interactive components of EGIS will be released in phases over the next several years. As new technology and additional data sets become available, the EGIS will be upgraded and extended. At the time of this writing, the first phase of EGIS was rolled out to the public. The EGIS will contain many of the HUD data sets as well as data sets from other federal agencies, including EPA, Census and FEMA flood plain data. Users will be able to import their own data for analysis.

6.3.9 GIS Analysis

CommunityViz™ (*www.communityviz.com*) is a newly available suite of software planning tools that run on the ESRI ArcView/Spatial Analyst GIS platform. CommunityViz™ was developed by the Orton Family Foundation with technical help from ForeSite Consulting, MultiGen Paradigm, Green Mountain Consulting, and PriceWaterCooper. The program includes three tools — a Scenario Constructor which enables users to conduct impact analysis of different development options; the 3D Town Builder which allows communities to create a replica of their community and fly through it like a flight simulator; and the PolicySimulator which enables users to predict how people respond to different policy options using agent-based modeling techniques. One of the strengths of this suite of tools is that each is integrated with the other. This makes it possible for planners to add to new buildings within the 3-D TownBuilder, for example, and then immediately see the impacts of these choices in the Scenario Constructor. Similarly, the 3-D Town Builder can be used to show how development might look like as a result of population shifts due to policy changes.

INDEX® (*www.crit.com*) is a GIS-based PSS that uses indicators to measure the attributes and performance of community plans. It is marketed in both standard and custom versions to professional planners and community organizations as a support tool for three key functions in community development: creating plans through issues identification, alternatives analysis, and goal-setting; implementing adopted plans by evaluating proposed development consistency with official goals; and achieving plans by periodically measuring cumulative progress toward goals. At its heart is a set of stakeholder-selected indicators that numerically and spatially gauge conditions and proposals. The software's analytical scope encompasses land use, transportation and environmental resources, and temporally it can perform static impact or dynamic forecast analyses of up to 20 years. The spatial resolution of its indicator scoring can be set at the block, parcel or building footprint levels.

The Location Efficient Mortgage (LEM) (*www.locationefficiency.com*) is an innovative mortgage product that takes advantage of the 'hidden asset'of transit density in urban neighbourhoods to encourage homeownership and promote increased transit ridership. The program, developed by Center for Neighborhood Technology (CNT), the Natural Resources Defense Council, and the Surface Transportation Policy Project, uses GIS analysis to identify homes with easy access to transit. With a LEM mortgage, lenders are allowed to recognize the savings made by a household whose primary means of transportation is public transit rather than the private automobile. Thus lenders can 'tretch' their standard debt-to-income ratio, ensuring that more low- and moderate-income families, first time homeowners and dedicated transit users can obtain mortgages, or larger mortgages than they otherwise would quality for. Fannie Mae, the nation's largest supplier of homeownership capital is sponsoring the underwriting of LEM mortgages.

Neighborhood Early Warning System (NEWS) (*www.cnt.org/news*), also developed by the Center for Neighborhood Technology, enables community organizations, city and county agencies to monitor real estate trends and to more effectively counteract housing abandonment, commercial decline and financial disinvestment in Chicago's inner city communities. 'Problem indicators' from several city and county agencies are combined with the Department of Planning's Harris Files into a single, computerized database. The indicators are code violations, housing court cases, and fire records. Once merged, the information is made available to the community over the Internet.

Model Builder™ (*www.esri.com*), developed by Environmental Systems Research Institute (ESRI) in Redlands, California, is a suite of spatial analyst tools inside ESRI's Spatial Analyst Extension to ArcView GIS. It enables users to construct, execute, and share spatial models using process flow diagrams. The power of Model Builder™ over standard GIS platforms is that it enables users to conduct spatial analysis and modelling through a flow-chart type interface that is automated and self-documenting, thereby keeping track of the steps taken to conduct the analysis. Each node of the flowchart represents components of a spatial process. A common type of spatial model evaluates site suitability. To choose the most suitable land for a park, for example, the user is able to consider multiple factors, (such as soil type, land cost, and proximity to residents); conduct arithmetic calculations on certain layers (adding the critical habitat for different species, for example); and give priority to the most important issues through a weighting process. Creating a model in this format makes it possible to run multiple variations of the model without having to recreate the entire model and each one is saved in a separate directory for later retrieval. The models can then be easily replicated by others or easily adapted to future analysis. The user can apply the same model to different geographic areas, for example, by changing the input data, and modify the model to explore 'what if' scenarios and obtain different solutions.

PLACE^{3}S, an acronym for PLAnning for Community Energy, Economic and Environmental Sustainability (*www.sustainable.doe.gov/articles/place3s.shtml*), is a land use and urban design method created specifically to help communities understand how their growth and development decisions can contribute to improved sustainability. PLACE^{3}S has been developed in partnership with the California Energy Commission the US Department of Energy, Parsons Brinckerhoff, Fregonese/Calthorpe and Space Imaging Services. There are three main components to the PLACE^{3}S approach: public participation – a fully engaged, comprehensive group of stakeholders committed to the principles of sustainability and collaborative planning; planning and design – a clear set of principles that embody a community's values and vision of what greater resource efficiency and sustainability mean to its future; and measurement – quantitative documentation of energy, economic and environmental impacts to support informed planning choices and monitor plans as they are implemented.

QUEST™ (*www.envisiontools.com*) was developed by the Sustainable Development Research Institute at University of British Columbia and its spin-off consulting firm, Envision Sustainability Tools Inc. QUEST™ facilitates debate and discussion among a variety of stakeholders about regional sustainability. It allows users to actively explore different possible scenarios of the future for their region and evaluate the social, economic and environmental consequences of each. The tool can be custom-built for any region, and used by expert and non-expert audiences alike. QUEST™ is being used on a number of projects around the world including with the Georgia Basin Futures Project (*www.basinfutures.net*), which is looking at the coastal watershed that encompasses Vancouver, Victoria and the Strait of Georgia. The goal is to further develop and improve this interactive approach to model development, public consultation and policy development.

UGrow is a PC-based system dynamics model for urban policy design and testing. (*www.prescott.edu* site limited to 1998 year-end report). The system dynamics approach is inherently cross-disciplinary and incorporates a variety of spatial and temporal factors including those internal to a community as well as external influences. The program integrates time-based simulation modelling with a spatially oriented geographical information system and other visualization technologies. UGrow is being developed to assist community decision-makers seeking to build more sustainable communities. This model characterizes an urban area as the confluence of built, human and natural environments and currently accounts for some 300 variables grouped in major sectors. It is a large model that accepts inputs from other models. It links local sustainable development efforts to global climate change effects.

What If?™ (*www.what-if-pss.com*) is a GIS-based system that can be used to explore alternative community development scenarios and project future land use patterns and associated population, housing, and employment trends. It allows public officials and private citizens to examine the likely impacts of alternative policies for controlling urban growth, preserving agricultural land, or expanding public infrastructure in easy-to-understand maps and tables. It is designed to be used by non-technical people in public forums, allowing communities to use currently available GIS information to support community-based dialogue and collaborative decision-making. What If?™ can be used to conduct a land suitability analysis, project future land use demand, prepare a land use plan, and allocate this demand to suitable locations (see chapter by Klosterman *et al.* In this volume for more information on What If? ™).

6.3.10 Predictive Modelling

The SLEUTH model, commonly known as the Clarke Cellular Automata Urban Growth Model (*www.ncgia.ucsb.edu/projects/gig*), uses cellular automata, terrain mapping and land cover deltatron modelling to address urban growth. SLEUTH derives its name from the six types of data inputs: slope, land use, urban, exclusion, transportation and hillshading. This model simulates the transition from non-urban to urban land use using a grid of cells (cellular automaton) each of whose land-use state is dependent upon local factors (e.g. roads, existing urban areas and topography), temporal factors, and random factors (U.S. EPA 2000). It has been used in San Francisco, Chicago, Washington-Baltimore, Sioux Falls and the south coast of California. The long-term goal of the project is to develop these tools to best predict urban growth on a regional, continental and eventually global scale (see chapter by Gaunt and Jackson in this volume for more information on SLEUTH).

The TRansportation ANalysis SIMulation System (TRANSIMS) (*http://transims.tsasa.lanl.gov*) is a set of new transportation and air quality analysis and forecasting procedures developed to meet the Clean Air Act, the Intermodal Surface Transportation Efficiency Act, Transportation Equity Act for the 21st Century, and other regulations. It consists of mutually supporting simulations, models and databases that employ advanced computational and analytical techniques to create an integrated regional transportation system analysis environment. The integrated results from the detailed simulations will support transportation planners, engineers, decision makers and others who must address environmental pollution, energy consumption, traffic congestion, land use planning, traffic safety, intelligent vehicle efficiencies and the transportation infrastructure effect on the quality of life, productivity and economy.

UrbanSim is a simulation system that models the real estate development process and the locations of households and jobs based on scenarios of transportation, land use and environmental policies. It is meant to support a community visioning process by allowing communities to explore the potential consequences of alternative policy scenarios such as light rail or highway expansion, and the use of land use plans and urban growth boundaries to guide development. UrbanSim can evaluate the consequences of land use and transportation policy scenarios at a high degree of detail, subdividing a metropolitan area using a 150-metre grid to represent real estate development and prices, and household and business locations. This detail supports analysis of non-motorized transportation and neighbourhood-scale design issues, in addition to regional consequences of major land use and transportation policies. The National Science Foundation and state and metropolitan agencies have funded its development, and it is now operational in several major metropolitan areas and states including Hawaii, Oregon, Utah and Washington. The UrbanSim system is available from the Internet (*www.urbansim.org*), and has been developed as an Open Source project, meaning that the software and its source code is free and available for modification and redistribution. Current development is on data integration tools to facilitate transfer of the model to other locations, a web-based interface for community participation, and evaluation and visualization tools to assess and communicate results.

6.3.11 Process Tools

The Sustainable Development Tool Kit (*www.i4sd.org/toolkit.htm*) is a set of collaborative processes to support vision-based planning and community development that have resulted from our work in communities. They work in conjunction with GIS and place-based planning and decision support tools to aid communities in developing and implementing consensus-driven sustainable development. The toolkit is based on the philosophy that the opportunity to get the future we want depends on understanding of whole systems, and the capacity to act on that knowledge. The tools were developed to facilitate the processes of teaching learning, taking the action, and measuring progress. It is a process of continuous improvement where each new project in a community contributes to the desired future, turns NIMBY into market and creates local economic opportunity. The toolkit is designed to help provide a means for bottom-up civic participation in the planning and development process to create a blueprint for sustainable economic development.

Council™ and WebCouncil™ (*www.covision.com*) are two tools that can enhance community participation in the decision-making process. Council™ is a facilitated process supported by 'meetingware' technology including laptop computers and voter keypads to get feedback and ideas from large groups. The computers are networked together using wireless technology maximizing mobility and reducing setup time.

Similarly, MeetingWorks® for Windows (*www.meetingworks.com*) is a groupware product that includes tools for electronic brainstorming, idea organization, ranking, voting, cross impact analysis, and multiple criteria (weighted factors) analysis. Individuals can respond to questions and become aware of their whole group's perspective. Everyone in the group can express their ideas to the rest of the group either anonymously or by identifying themselves so in the end everyone feels heard. Typically used by large corporations, especially after mergers and takeovers, the tool also works well for communities.

In Washington DC, Mayor Williams has used Council™ to facilitate two large scale Citizen Summits. The first community-scale gathering took place in November 1999. A second meeting took place in October 2001. Each Citizen's Summit attracted over 2,500 participants. With 10 people at a table and one laptop per table, the mayor could discuss and ask questions about his strategic plan and receive rapid audience feedback. After discussing citywide issues, the group broke into neighbourhood constituencies to establish priorities. The first meeting produced a 300-page report outlining next steps, the second meeting focused on follow-up and implementation. WebCouncil™ is a web-based tool that enables virtual meetings, specialized discussion groups and resources to manage and track a project. It is often used to keep people involved and active in discussions and next steps in between face-to-face meetings.

6.4 Planning Support Systems in Practice

Envision Utah is an example of a community using a full range of tools at a regional scale: information resources, visualization, impact analysis and GIS tools coupled with a comprehensive public process. Established in 1997, Envision Utah represents a public/private partnership to look at growth issues and their impacts on transportation, air quality, land use, water supply/demand and infrastructure costs in the 10-county region surrounding Salt Lake City (commonly referred to as the Greater Wasatch Area). The project is a multi-year; multi-million dollar effort supported by Governor Michael Leavitt's Office of Planning and Budget, numerous foundations and private funders, as well as a number of federal and local agencies.

A key component of Envision Utah is the Quality Growth Efficiency Tools Committee (QGET) made up of representatives including leaders within state and local governments, metropolitan planning organizations, regional planning agencies and the private sector. The role of the QGET is to assist in the technical analysis of growth trends and projections, and to consider alternative growth scenarios. To date, the project representatives have met with over 6000 residents in over 175 meetings, have put in over 70,000 hours of modeling and analysis, and have received over 80,000 responses to public surveys.

The project has four phases: inventory, scenario development, quality growth strategies and implementation. A key driver behind the creation of Envision Utah has been the metropolitan region's rapid growth. Approximately 1 million additional people are expected to live in the region by 2020, a 59 per cent increase over today's 1.7 million. Interestingly, 70 per cent of that is expected to come from internal growth, giving Utah the distinction of having the highest natural growth rate in the country.

During its first year, Envision Utah held two dozen public meetings involving nearly 1300 residents to determine priority concerns of citizens. Phase II then involved the creation of four future scenarios for Salt Lake City over the next 20 years. Assuming population growth as a given, the scenarios presented a spectrum of strategies for accommodating growth. Scenario A represented the most outwardly expanding development pattern involving predominately single family housing on large lots and segregated retail zones for retail. Scenario D represented the most compact, transit oriented development, derived from significant infill development, and mixed use zoning.

Using GIS impact analysis and modeling tools, the four scenarios were then evaluated against the priority concerns or indicators of the community, which included, among other things, agricultural land consumed, air quality impacts, transit ridership, pedestrian orientation, and infrastructure costs. The scenarios and the tradeoffs associated with each development strategy were presented in a full-page newspaper insert, requesting readers to fill out a survey identifying which scenario they preferred. Figure 6.1 and Table 6.2 show how each scenario performed against certain indicators. Overall, the survey results showed a strong preference for more compact transit oriented development – referred to as 'scenario C+'.

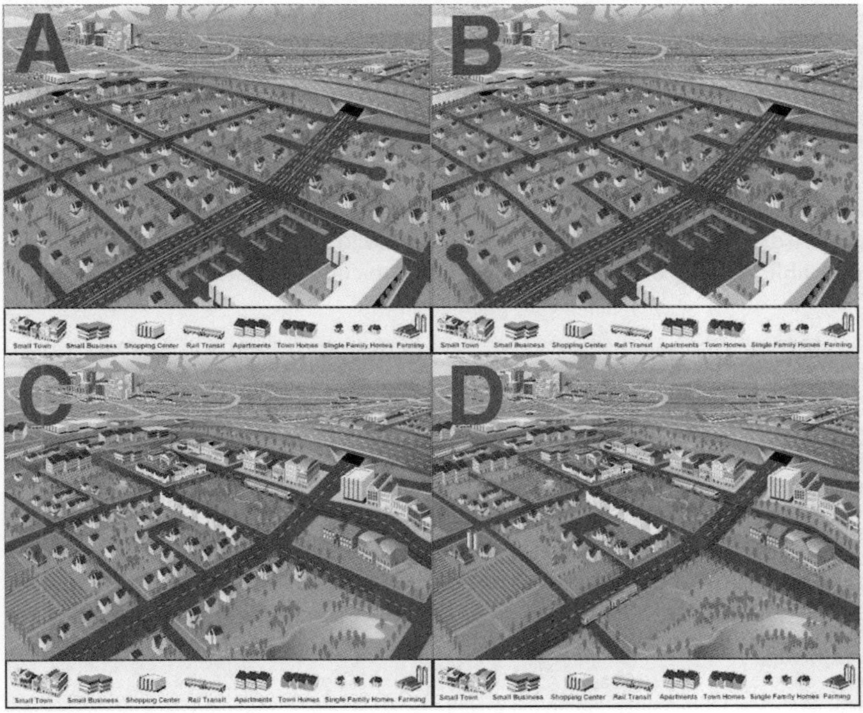

Fig 6.1: Envision Utah scenarios
Source: *www.envisionutah.org*

Unexpectedly, air quality, a top community concern, performed slightly worse in scenario D than in scenarios B and C despite higher transit ridership and walkability in D. This actually demonstrated the strength of their modelling techniques for it showed how air quality suffered due to a lack of attention given to road improvement in scenario D. Even though fewer people were driving, those who choose to drive in densely developed areas were stuck in traffic considerably more. This became an indicator to planners: if they wanted to pursue scenario-D-type transit oriented development and still enjoy air quality gains, they would have to pay adequate attention to improving the road network along with other design features.

The strength of the Envision Utah approach is that it takes advantage of community feedback to continually refine strategies for accommodating growth. The modelling and planning support tools help the public understand the inherent trade-offs associated with different development futures. The modelling tools are used to revisit and evaluate the performance of different design options and provide feedback on how to reach desired indicator outcomes.

Based on survey responses and community feedback, Envision Utah developed a Quality Growth Strategy. The strategy identified six goals: enhance air quality, preserve open space and agricultural land, conserve/maintain availability of water, promote mobility and transportation choices, maximize efficiency in public and infrastructure investments, and provide housing opportunities. Envision Utah is now in the process of implementing their vision. Phase IV has focused on specific strategies/tools for local jurisdictions. Most recently, efforts have focused on the development of toolboxes for local communities. Toolboxes have been created for water issues, model codes, and analytical tools. The analytical tools toolbox provides local jurisdictions with simple instructions on how to use GIS tools (namely ArcView and a public domain extension called X-tools) to conduct basic land use analysis.

Table 6.2: Impact analysis of scenarios

ELEMENT	CURRENT	CHOICES FOR 2020			
		scenario a	scenario b	scenario c	scenario d
average auto peak hour speeds (mph)	25.7	22.9	20.0	20.9	19.8
people who can walk to rail (1/2 mi)	0	38,755	45,557	664,991	866,765
per capita water use (gallons/day)	319	303	279	231	218
total air emissions (tons per day)	1869	2660	2511	2501	2512
average size of single-family lot	0.32 acre	0.37 acre	0.35 acre	0.29 acre	0.27 acre
land consumed	431 sq. mi	840 sq. mi	756 sq. mi	557 sq. mi	516 sq. mi
agricultural land consumed	—	174 sq. mi	143 sq. mi	65 sq. mi	43 sq. mi
cost of infrastructure 1998-2020	—	$37.6 billion	$29.8 billion	$22.1 billion	$23.0 billion

Source: Envision Utah (2000)

Some criticisms of Envision Utah include: the lack of regulatory mechanisms to implement the desired scenario; the difficulty of translating a 20 year vision into concrete immediate actions; and so much emphasis on local control that the regional context of implementing Envision Utah is sometimes lost. Overall, Envision Utah has been very well received both in the Salt Lake City Region and nationally. It has been recognized as one of the more effective regional efforts to help communities grapple with growth. The effort has been credited for mobilizing the populace behind a 0.25 cent sales tax increase, approved in November 2000, to support a region-wide public transportation system. Using GIS proximity analysis, Envision Utah estimates the proposed transportation enhancements will increase the number of residents shopping, working and living within one-half mile of rail transit from 46,000 to more than 600,000 (Envision Utah 2000).

Foundations continue to provide significant funding to help in the implementation of the project – most recently, the Hewlett foundation has committed $3 million over the next 3 years to the project. Envision Utah is an excellent example of a PSS that includes technical tools integrated into an effective community process.

6.5 Conclusion

Since 1998, the author has surveyed a number of communities to identify what characteristics commonly exist with successful projects utilizing planning support systems. Several key characteristics stood out.

The tools were simply a means to accomplishing specific tasks part of a more comprehensive, well-thought out process. Successful PSS integrate process and tools together to produce desired outcomes. The best tools in the world are not going to be effective if they are not integrated into a process that engages stakeholders in the decision making process at the right time and in the right way.

The process and implementation strategy were perceived as genuine. If the public is asked to participate in time-consuming exercises, it must feel that it will be listened to and its participation will help shape results. While Envision Utah is one of the best examples in the country of how to integrate scenario analysis, impact analysis and visualization tools into a community participatory process on the local and regional scale, it suffered from some early criticism that the process was disingenuous. Citizens were asked to fill out a survey choosing one of four futures for the Greater Salt Lake region. The Governor as well as an impressive list of public and private organizations backed the process. Unfortunately, at the same time, the Governor was lobbying for federal funding for the construction of the Legacy Highway project to be built through Salt Lake City region. This left some people with the impression that the public process around Envision Utah was superficial and that the state was pushing ahead with projects that would continue the trend towards low density, auto-dependent development.

Visualization tools were central to the process of getting public input around land use and development issues. People are highly visual in their thinking. Visualization tools using 3D rendering, photo montages, architectural drawings or even cardboard replicas of the community can be highly effective in developing a vision of the future and helping people understand options.

Impact analysis tools were used to quantify the tradeoffs associated with different decisions. Impact analysis tools help people understand the implications of their individual choices on the community as a whole. They make it possible for communities to experiment with different scenarios of development and see how they affect other issues like tax revenue, new demands on schools, impact on critical wildlife habitat and demands on natural resources.

Issues were addressed on an interdisciplinary level. Geographic mapping tools help communities recognize the links between environmental, social and economic issues. They help communities look across disciplines, and see how the natural, built, economic, and social layers relate to each other.

The process involved face-to-face meetings and a place to gather. The most effective community development projects often have meetings for people to gather and meet each other face to face. Several communities in the US are looking to establish regional resource centres as places equipped with decision-support tools and structured to promote dialogue around community issues.

The projects were set up so people enjoyed the process and, overall, had fun. If it is not fun, people are less likely to stick with the process. Participation will eventually drop off, and the project will lose a sense of group consensus.

Within the next five years, computer based PSS will most likely radically transform the field of community planning and design. Advances in 3D visualization and design programs, impact analysis and scenario analysis models, predictive modelling programs, as well as techniques in integrating these tools into effective community and stakeholder processes are drivers behind this transformation. These tools are incredibly valuable because they help planners and community citizens visualize alternative futures, understand the environmental, social and economic impacts of their choices and build consensus among diverse stakeholders about a desirable future for everyone. Economic development is about the economic stability of cities and this is driven by the land use allocation ratios that make economic development possible (Hosack 2001). Integrating these tools together with effective community decision-making processes can have a tremendous impact on the quality of outcomes of community planning efforts. Planners can see options more clearly. Stakeholders can provide input and feedback in a more effective manner. Elected Officials can be held accountable for projects, which have utilized impact analysis and visualization techniques.

In conclusion, TCDDM participant, Cindy Shea, summarized the benefits of integrating both tools and process together in article she wrote after the Stuart Florida meeting:

"As community planning becomes simultaneously more complex and more open to citizen input, computer-based tools offer an opportunity for greater involvement at a more sophisticated level. One of the many challenges is to ensure that these tools support and assist decision making, rather than overwhelm it. The most useful tools will serve users at multiple skill levels and provide immediate feedback during community meetings. At the same time, these tools need to be transparent, adaptable, and based on high quality, current data in order to earn the trust of all parties involved. Once more becomes known about the function, design, and impact of various

developments, it becomes possible to better tailor incentive structures to achieve desired outcomes. What's a few thousand dollars on landscaping, architectural detailing, and impact fees, for example, if a developer is liberated from uncertainty and lengthy delays? The clearer a community is about its goals and rulemaking processes, and the more robust its base of graphic and data support, the greater the chance of realizing its desired future" (Shea 2000).

Acknowledgements

Special thanks to Bill Becker, Elizabeth Conover, Stan Geertman, Pat Lana and John Lambie for editorial comments on this chapter. Also thanks to Elliot Allen, Juan Ayala, Larry Conrad, Geoff Dorn, Ray Essiambre, Ginny Graves, Karen Hobbs, Hoyt Johnson, Dick Klosterman, Michael Kwartler, Lenny Lind, Mike McKeever, Bill Miller, Anton Nelessen, Steve Price, Steen Rasmussen, Mike Walsh, Helen Whyte, and Jeff Winston for contributing to the individual tool descriptions.

Disclaimer

The opinions expressed in this chapter are solely those of the author and not of the US Department of Energy (the author's employer at the time of writing). The tools described here represent a sample of tools available to communities. Reference to any specific commercial product, process, or service by trade name, trademark, manufactures' name, or otherwise does not necessarily imply its endorsement, recommendation, or favoring by the United States government of any agency thereof.

References

Envision Utah (2000) *Envision Utah Quality Growth Strategy and Technical Review*, Envision Utah Publication, Salt Lake City.

Hosack M, W. (2000) *Land Development Calculations: Interactive Tools and Techniques for Site Planning, Analysis, and Design*, McGraw Hill, New York.

O'Looney, J. (2000) *Beyond Maps: GIS and Decision Making in Local Government*, ESRI Press, Redlands, California.

Shea, C. (2000) *Tools for Community Design and Decision Making: Overview of Florida Conference*, Florida Sustainable Communities Network online publications. http://Website: www.i4sd.org/toolsind.htm.

U.S. EPA (2000) *Projecting Land-Use Change: A Summary of Models for Assessing the Effects of Community Growth and Change on Land-Use Patterns*, EPA/600/R-00/098. U.S. Environmental Protection Agency, Office of Research and Development, Cincinnati

Web Sites

Airhead (CNT): *http://www.airhead.org*

BP Visualization Center: *http://www.colorado.edu/Research/bpVisCenter*

CommunityViz™ (Orton Family Foundation): *http://www.communityviz.com*

Correa Valle Valle: *http://www.urbanism.com*

Council™ and WebCouncil™ (CoVision):"*http://www.covision.com*

CUBE: *http://www.cubekc.org*

Dover and Kohl: *http://www.doverkohl.com*
Duany Plater-Zyberk and Company: *http://www.dpz.com*
EGIS (US HUD): *http://hud.esri.com/egis*
Envision Utah:*http://www.envisionutah.org*
Hackney Building Exploratory: *http://www.buildingexploratory.org.uk*
HNTB Technology Group: *http://www.hntbtech.com*
ICLEI: *http://www.iclei.org/games/comcalc.htm*
INDEX®: *http://www.crit.com*
InfraCycle™: *http://www.infracycle.com*
Location efficient Mortgages (CNT): *http://www.locationefficiency.com*
Looney Ricks Kiss: *http://www.lrk.com/homeplans*
MeetingWorks™: *http://www.meetingworks.com*
Model Builder™: *http://www.esri.com*
Neighborhood Early Warning System (CNT): *http://www.cnt.org/news*
Nelessen Associates: *http://www.anavision.com*
PLACE³S: *http://www.sustainable.doe.gov/articles/place3s.shtml*
QUEST™: *http://www.envisiontools.com*
REMI®: *http://www.REMI.com*
SCALDS (US DOT): *http://www.fhwa.dot.gov/scalds/scalds.html*
SLEUTH:—*http://www.ncgia.ucsb.edu/projects/gig*
Sustainable Development Toolkit (Florida House Institute): *http://www.i4sd.org/toolkit.htm*
The Environmental Simulation Center: *http://www.simcenter.org*
TRANSIMS: *http://transims.tsasa.lanl.gov*
UGrow: *http://www.prescott.edu* (site limited to 1998 year-end report)
Urban Advantage:*http://www.steveprice.com*
UrbanSim: *http://www.urbansim.org*
US Department of Energy Web site on Tools: *http://www.sustainable.doe.gov*
What If?™: *http://www.what-if-pss.com*
Wintson Associates: *http://www.winstonassociates.com*

PART 2

TOOLS FOR SUPPORTING THE PLANNING PROCESS

7 The Planner's TOOLBOX: A Web-based Support System for Sustainable Development

Jukka Mikkonen[1], Mika Ristimäki[2], Kari Oinonen[2] and Henning Sten Hansen[3]

[1] Regional Council of Päijät-Häme, Hämeenkatu 9 A, FIN-15110 Lahti, Finland
[2] Finnish Environment Institute, P.O. Box 140, FIN-00251 Helsinki, Finland
[3] National Environmental Research Institute, Department of Policy Analysis, Frederiksborgvej 399, DK-4000 Roskilde, Denmark

Abstract: The Planner's TOOLBOX is a web-based tool for regional planners. It offers knowledge of how to use geographical information and geographical information systems in planning. The main emphasis is on how to implement the principles and practices of sustainable development in spatial planning. The Planner's Toolbox is the end-product of a European Union INTERREG II C project to develop a 'Planning System for Sustainable Development, PSSD', involving Finnish, Danish and German partners representing a range of regional planning bodies and national research institutes. The Planner's TOOLBOX is an associative cross-reference database. It offers examples, theories, methods, indicators, tools and scientific documents concerning questions about planning and GIS. The contents can be examined by using classifications, by undertaking searches and by following hierarchical and associative links. Users can upload their own information into the Planner's TOOLBOX.

7.1 Introduction

The *Planner's TOOLBOX* is the end product of the 'Planning System for Sustainable Development' (PSSD) project that was established to promote international co-operation in the field of planning with GIS between spatial planners and researchers. It was one output of an INTERREG IIC Community Initiative relating to 'Transnational Co-operation on Spatial Planning in the Baltic Sea Region', funded by the European Regional Development Fund. The project was administered and led by the Regional Council of Päijät-Häme on behalf of the South Finland Regional Alliance and it came to an end in June, 2001. The other partners included the Finnish Environment Institute, the Geological Survey of Finland, the Technical Research Centre of Finland, the Baltic Region Healthy Cities Office (Finland), the National Environmental Research Institute (Denmark), the Danish Forest and Landscape Research Institute, the Region of North Jutland (Denmark), and the Technical University of Hamburg-Harburg. These partners were originally chosen with regard to the need for expertise in the connections between environmental and socio-economic factors and spatial planning and in the field of new information and communication techniques (ICTs) including geographical information systems (GIS).

The idea of the project originated from the Regional Council of Päijät-Häme, who did not have a computer-aided methodology or tool for analysing natural environmental and socio-economic factors relating to land use planning. Following early attempts to find a solution at regional and national levels, the European INTERREG IIC Programme provided an appropriate context in which to develop such a system since the problems confronted had been identified as international and multi-regional, precisely the criteria required by the Programme. Although the international context was beyond the initial considerations of the Regional Council, this dimension proved to be very important.

In general terms, the development of GIS-based spatial planning methods has been very fragmented in Europe because of the variation in spatial structures and problems existing in different states, regions and cities. In this context, the European Union's European Spatial Development Perspective (ESDP) is the first major attempt to create some unification of methods between the member states in their approach to spatial problems. This kind of new international framework for spatial planning requires the application of national and regional GIS-based methods in attempting to manage both common and area-specific needs at the same time.

The overall objective of the project was for the partners to produce a prototype GIS-based regional planning methodology for use within the Baltic Sea Region, and a computer-based tool to make it operational. The term 'prototype' was used to convey the idea that the project aimed to create a first version without carrying out any extensive examination of relevant potential international needs. In fact, several independent methods were produced and a computer-based system was created that could apply each of them. In practice, the key focus was on promoting sustainable spatial planning and development in the Baltic Sea Region. The GIS methods under development were mainly based on grid technology and the project aimed to exemplify the potential of the system for spatial planning and development. The results of the project were uploaded into the *Planner's TOOLBOX* on the Internet. Thus, the *TOOLBOX* enables flexible international communication as well as development of GIS-aided spatial planning methods.

The PSSD project chose the so-called DPSIR framework of the European Environment Agency for structuring its methods and tools. According to this framework, a chain of causality exists that starts with 'driving forces' (D) that lead to 'pressures' (P) and further to 'environmental states' (S) and 'impacts' (I) on human welfare, finally leading to political 'responses' (R). The data required for describing the driving forces are collected at municipal, county or national levels. Those variables relating to pressures and states, like land use and soils, are usually available for observation points or internally homogeneous polygons (Hansen 2000). A planning system for sustainable development involves collecting a wide range of different spatial data, but integrating different kinds of data may often cause problems. Many data in the social and environmental sciences are collected for polygon units, but the systems of units often differ from one variable to another. Defined environmental regions such as vegetation zones or watersheds will seldom match the boundaries of officially defined zones like counties and municipalities, for example.

Trying to reconcile two sets of data that are collected over different tessellations, such as municipalities and watersheds, is a very difficult task. First, it is necessary to decide which layer will be the 'standard' and which will be the one adjusted to match the standard. Then, the regions of one layer have to be disaggregated into building blocks that are reassembled to match the other layer. This may be a very complex and time-consuming task for a planner. Finally, having adjusted the geographical units, it is necessary to adjust the data values from one spatial system so they are consistent with the second one. However, instead of representing geographic features as vector-based polygon objects, space can be divided into discrete squares, cells, for which the values are stored. A cell is a uniform unit that represents a defined area of the earth, such as a square metre. Grid cells do not change over time in the same way that administrative units alter, for example. Therefore, a cell-based system is better suited for analyses of change over time. The cells can be encoded in raster or vector format. In the project, vector-based cells were used but it is relatively simple to convert data from vector to raster, and *vice versa*.

The interaction between humans and various analytical tools is frequently abstract in nature for both inputs and outputs (Shiffer 1995). Graphical user interfaces (GUIs) are widely used, but sometimes the user is required to remember a range of commands in order to interact with a computer-based system (e.g. ArcInfo). In addition, many of these tools produce rather abstract output in the form of numeric data. For example, numeric output describing noise levels (in decibels) may be meaningless to the layman. The Internet provides new opportunities for a standardized user interface across computing platforms (Windows, Unix, Macintosh) using web browsers. Furthermore, a modular approach with several 'helper' applications (plug-ins) is used to support a wide range of data formats. Thus, an academic paper may be published on the Internet as a pdf file, whereas a film can be viewed using, for example, Windows Media Player. Developments in multimedia offer many promising possibilities for incorporation in planning support systems. For an international project, like the PSSD, the ability of the web browsers to connect to documents and multimedia data located anywhere in the world is also of great importance.

Therefore, it was decided to develop an internet-based user interface for the presentation of the tools and methods created by the PSSD project. The main page of this so-called *Planner's TOOLBOX* is organized into a news list and two menu bars (Figure 7.1). Possible ways into the *Planner's TOOLBOX* are via 'Planning topics', 'Theories', 'Methods', 'Tools' and 'Documents'. In addition, a search engine is included. Generally, theories, methods and tools are entities referring to sustainable planning, whereas documents are full papers – often published elsewhere. Some of the methods and tools developed within the PSSD project are described in the PSSD technical report (Hansen 2001). The idea behind the entities is to obtain higher precision in searches and links. Thus, the user is guided directly to the real issue under consideration.

The Updating button plays an important role in the use of the *Planner's TOOLBOX* as well as in its further development. The idea is that registered users can upload comments concerning existing methods and tools as well as their own tools for a wider usage. Thus, the knowledge database of the *TOOLBOX* will continue to grow in proportion with its use. The *Planner's TOOLBOX* contains a link to the internet-based 'Indicator generator' developed as a part of the PSSD project by the Technical University of Hamburg-Harburg and used to define indicators of regional sustainability and to check their quality.

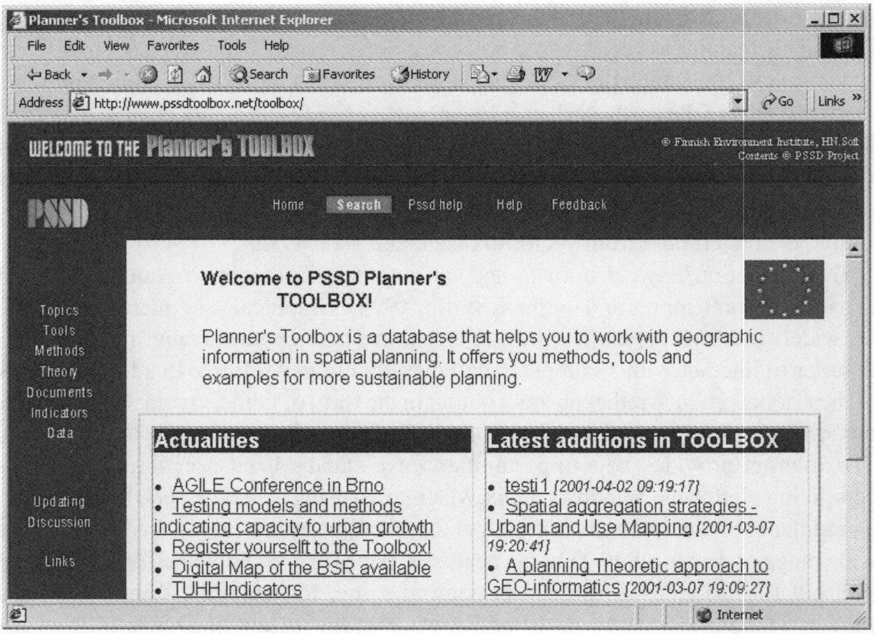

Fig 7.1: The website of the *Planner's TOOLBOX*

The remainder of this chapter describes the content and workings of the *Planner's TOOLBOX*. The structure of the system is elaborated in Section 7.2 and the functions of the system are described briefly in Section 7.3. Examples of its application are outlined in Section 7.4 and the first results of its evaluation are presented in Section 7.5. Finally, some remarks concerning the future of the PSS will be provided in Section 7.6.

7.2 Structure of the System

The basic structure of the *Planner's TOOLBOX* is straightforward. The database includes two main tables: one contains the substantive records and the other contains information from the linkages between the substantive records. Due to the database structure, the software (PHP3 scripting) generates the html pages for the user according to the demand from the database according to the user's choices or searches. Figure 7.2 describes the structure of the *Planner's TOOLBOX*.

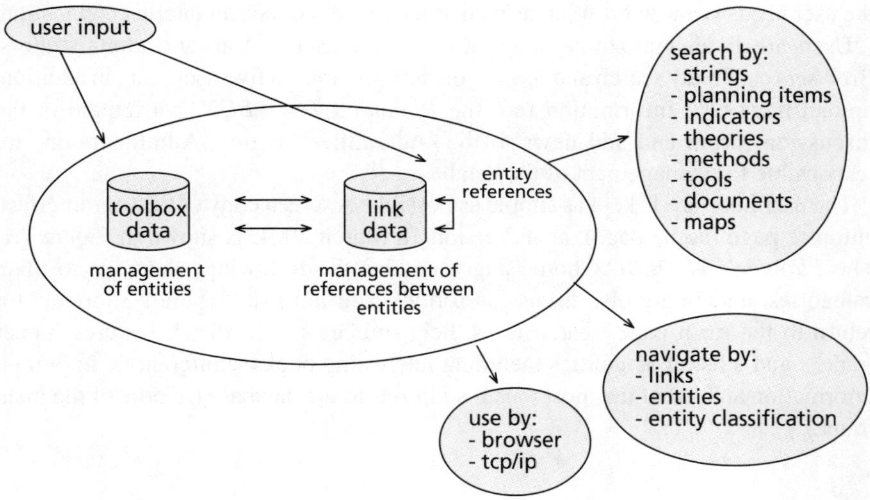

Fig 7.2: The structure of the *Planner's TOOLBOX*

The *Planner's TOOLBOX* is an associative cross-reference database. It offers examples, theories, methods, indicators, tools, and scientific documents concerning planning and GIS questions. Every issue is divided into its component parts, the so called entities. The entities of one issue are linked to each other, but also to other entities. For example, in order to take biodiversity into account in planning, it is necessary to be aware of theories about biodiversity, to understand tools like GIS macros and scripts that are used for measuring it, to be able to produce and interpret maps showing actual solutions, and to have information about data processing problems and limitations. In addition, it is also important to recognize that biodiversity is linked to other topics. Entities are linked both horizontally and vertically. The disaggregation of issues into component entities – though offering great benefits for associating information – may also be confusing. To overcome this, the *Planner's TOOLBOX* presents the main issues as unbroken documents, for example as pdf files, linked to entities. Thus, the links (references) between entities are the key elements of the *Planner's TOOLBOX*.

The *Planner's TOOLBOX* is based on free software, including a Linux operating system, Apache Web server, MySQL relational database management system and PHP3 scripting language. The main reasons for these software choices (in autumn 1999) were performance, reliability, openness, modernity and cost. The only software the user needs is a web browser, as well as having, of course, an internet connection.

Users are divided into three categories: browsers, active users and administrators. Browsers can freely search and browse the information. Active users can, in addition, upload their own information into the *Planner's TOOLBOX*, participate in the discussion forum and add news to the 'Actualities' section. Administrators are responsible for management of the database.

The user interface is kept as simple as possible. A screen copy of the user interface entrance page (home page), as in Version 18 March 2001, is shown in Figure 7.1. The *Planner's TOOLBOX* home page contains the following: shortcuts to main categories, updating tools, discussion forums, and links (left frame); shortcuts for return to the main page, search tools, help sources and feedback features (upper frame); and a list of actualities including interesting books, conferences, important information and list of the most recent additions to the database (middle of the main frame).

7.3 Functions of the *TOOLBOX*

The structure of the database in the *Planner's TOOLBOX* is simple. The main tables contain only the basic information such as the name and an abstract description of each entity. All other information is stored in other tables. This means that the basic product can easily adapted to other application areas.

7.3.1 Classifications

Each entity is structured similarly in the database. The structure of an entity consists of basic information and a set of classifications. The basic information involves:

- Author;
- Entity name;

- Abstract: an explanatory text on the entity's contents which is typically a couple of sentences. The purpose of these pieces of information is to help the user find out interesting topics by browsing filtered abstracts;
- Description: the contents of the entity. This may include, for example, a whole report text or a source code. The user may include html codes in the text for formatting purposes; and
- File/document: the file that the user wants to be involved with in the entity. Often, the file is a figure (often a map), but it may also be a word document, worksheet, PowerPoint presentation, video clip, *et cetera*.

Planning phenomena can be classified in many ways, but none of these fits all planning topics. In the PSSD project, the planning phenomena were divided as follows (an example of each topic in parentheses):

- ecological sustainability (e.g. the possibility to predict the presence of biologically valuable sites – theory);
- landscape and cultural heritage (e.g. accessibility of landscape – method);
- spatial structure (e.g. accessibility in the Baltic Sea Region – indicator);
- spatial functionality (e.g. analysis of sustainable regional policies – method);
- traffic and technical services (e.g. noise theory – theory).

To help the planner acquire knowledge, the results of the project tasks were divided into the following main categories:

- topics (the above mentioned classification);
- tools: concrete items for operations; these may include programs, macros, exact guidelines for performing geographical operations, *et cetera*;
- methods for managing sustainability or GIS tasks;
- theories of sustainability, planning, and GIS;
- indicators to estimate trends in sustainability;
- documents containing combinations of topics; the topics may include large entities. The documents are usually pdf files in order to ensure that the user sees the work as the author intended; and
- data, including datasets or links to datasets; the datasets and the *Planner's TOOLBOX* may use the same server or the datasets may be located anywhere on the Internet.

The indicator section provides the link to the Indicator Generator of the Technical University Hamburg-Harburg (TUHH). It enables the search for information about indicators of regional sustainability and helps to develop new sustainability indicators, to operationalize intra-regional changes and to visualize them for the inhabitants of the region. The Indicator Generator is a stand-alone system that can be used also without the *Planner's TOOLBOX* in the URL *http://cgi.tu-harburg.de/~sb3www/indihome.shtml.* Besides the Indicator Generator entrance, the indicator section offers the shortcut to indicators in the *Planner's TOOLBOX*.

7.3.2 Associativity of Entities

In the *Planner's TOOLBOX*, the topics (entities) are split into pieces – into their 'atomic parts'. The advantage of this division is the possibility of linking entities according to their characteristics. The entities in the database are connected both horizontally and vertically. The vertical link means that a topic like 'fragmentation' includes all theories, methods, tools, indicators, examples and documents that fall under the concept of fragmentation. The horizontal connection means linking by the factual contents of the link: for example, the method for 'Defining a Densely Populated Area' is linked to a large number of other entities, including the method of measuring the 'Efficiency of Built-up Areas' and the theory of 'Spatial Aggregation Strategies'. The links between entities are formed by the users applying their expertise. Figure 7.3 shows a sample of some links between entities.

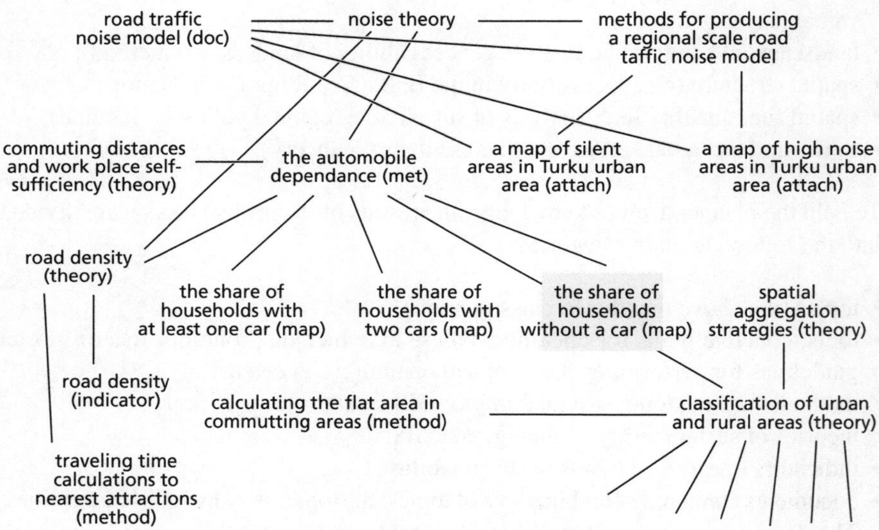

Fig 7.3: A sample of links between some entities

7.3.3 Search Capabilities

The *Planner's TOOLBOX* offers the users several ways to find information on their subjects of interest. Once the user finds an interesting topic, he/she may follow the references freely and finally end up with information new to him/her. The structure and contents of the entities are unrestricted, but some classifications exist in order to help the user. Information may be searched in several ways:

- free string search from the whole database;
- planning topic classification (see above);
- European Spatial Development Perspective (ESDP) policy aims classification:
  - polycentric and balanced spatial development;
  - dynamic, attractive, and competitive cities and urbanised regions;
  - indigenous development, diverse, and productive urban areas;
  - urban-rural partnership;
  - an integrated approach for improved transport links and access to knowledge;
  - polycentric development model: a basis for better accessibility;
  - efficient and sustainable use of infrastructure;
  - diffusion of innovation and knowledge;
  - natural and cultural heritage as a development asset;
  - preservation and development of the natural heritage;
  - water resource management;
  - creative management of cultural landscapes; and
  - creative management of the cultural heritage
- geographical level classification:
  - transnational/global;
  - national;
  - regional;
  - local; and
  - combinations of the above.
- indicator type classification: the indicator management is based on the so-called DPSIR model developed in the European Environment Agency (EEA). The integrated approach needed in spatial planning promoting sustainable development calls for a framework for structuring information on cause-and-effect relationships. Briefly, the model framework assumes cause-and-effect relationships between interacting components of social, economic and environmental systems, which are:
  - driving forces of environmental change (e.g. industrial production);
  - pressures on the environment (e.g. emissions);
  - state of the environment (e.g. poor quality water);
  - impacts on population, economy, ecosystems (e.g. diseases, lack of good quality water); and
  - response of the society (e.g. changes in the legislation);
- entity type classification (see above);
- author search (first name, last name, organization)(Figure 7.4);
- subject search (string search from subject field);
- keyword search. The keyword can be selected from a list. The users can add keywords to the list; and
- references between the entities (the most important). The search is done by clicking the links to other entities.

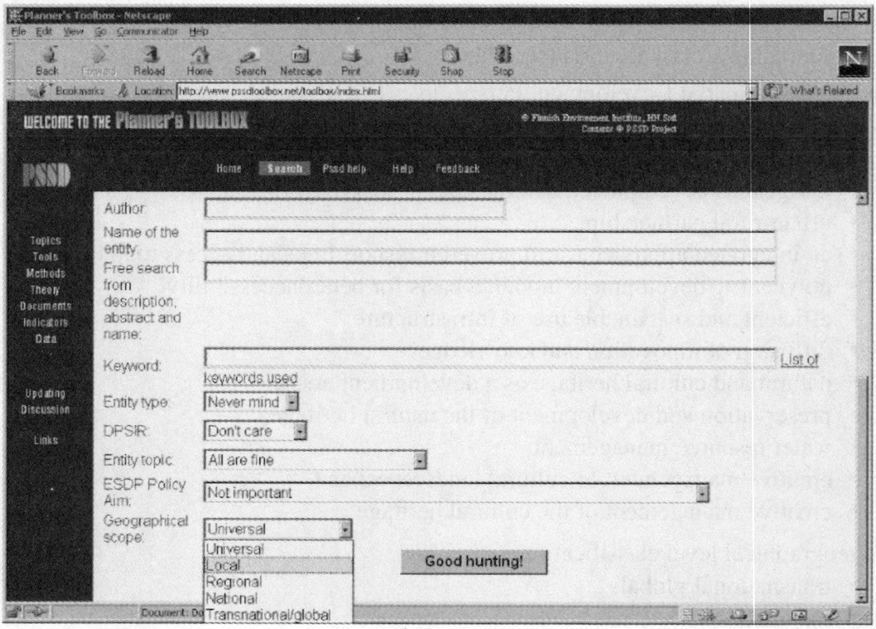

Fig 7.4: A screen shot from the search section of the *Planner's TOOLBOX*

The *Planner's TOOLBOX* relies heavily on the users' activity, because the users produce its contents. It is interactive, the contents are always up-to-date and the updates and inputs of the users are realized immediately. The *Planner's TOOLBOX* also includes a real time discussion forum intended for discussions concerning sustainability, planning and GIS.

7.4 Practical Application

The usefulness of the *TOOLBOX* in practical planning work is based on the description of GIS-based methods that are necessary for the solution of planning problems at the provincial level. Although regional planning problems and their solutions are specific and local, the descriptive methods are most often generic. The main practical purpose of the *Planner's TOOLBOX* is to serve as a databank for knowledge of GIS-based planning. It contains descriptions of best practice from real planning and descriptions of how to forecast area populations. GIS-based methods are important tools for assisting the planners in their work. It also helps to have good communication between the older and the younger planners; i.e. between those with experience and those with ideas on maximizing the potential of the new technology.

In order to select appropriate GIS-based methods for each particular planning problem in the best possible way, the methods are linked with their respective descriptive theory, a practical example of an analysis and an indicator with maps, tables or graphs, as well as the actual tool. The actual tools in the *TOOLBOX* carry out a GIS analysis; they are macros, programs or descriptions of analysis within a general GIS-based system. Ideally, these form a path that the user is able to follow from the theory to the application of the tool.

The first example of such a path in the *TOOLBOX* concerns taking biodiversity into account in land use planning. The path starts with a description of the theory of biodiversity and the imminent reduction of biodiversity. From the theory there is a link to descriptions of method, e.g. how biodiversity was investigated in a test area located in a South Finland district. The link presents e.g. the database structure, data gathered from the area for its classification, and a general method for the conversion of point data into statistical 250 x 250 metre grid data. The essential part of the method section describes the generalization of biodiversity through valuing its various elements. The method helps screen the vast set of basic descriptive data on biodiversity and render it more useful for land use planning. The valuation of biodiversity is combined with the data on the extension of densely populated areas within the last two decades, which indicates pressure for land use. Finally, an estimate is given of the distribution of areas with valuable biodiversity within densely populated areas, and how they relate to the planned land use according to the regional plan in force (Figure 7.5). As tools, the various sub-items of each theme contain detailed descriptions of the performance of the respective analyses. The sub-items are linked to other themes that utilize the same GIS method.

Another example describes a series of thematic maps produced utilising GIS methods. The maps describe several different aspects of the compatibility indicator at three different spatial scales: the provincial, national and international levels. The link from here to the planning problems presented in Figure 7.6 is the main point. Entity shows an example of a record called the 'Grid Tools' – theory that includes a fully functional set of grid operations, including the 'Accessibility of centres' shown in the figure.

The *Planner's TOOLBOX* contains methodical descriptions such as the above relating to general theories and actual planning problems. These instructions and their respective practical examples form the core of this PSS.

Naturally, it is not possible to present all GIS methods used in provincial level planning in the *TOOLBOX*. At its best, it is merely a collection of examples of GIS methods that may be used to enhance sustainable development in planning. Thus, from a practical viewpoint, instead of having to concentrate efforts on the development of technical GIS methods, the focus of attention is transferred to their utilization and application following the principles of sustainable development. Consequently, the emphasis moves from GIS to GIK – 'geographic information knowledge'. Another noteworthy aspect is that the *Planner's TOOLBOX* increases interactive working on the Internet.

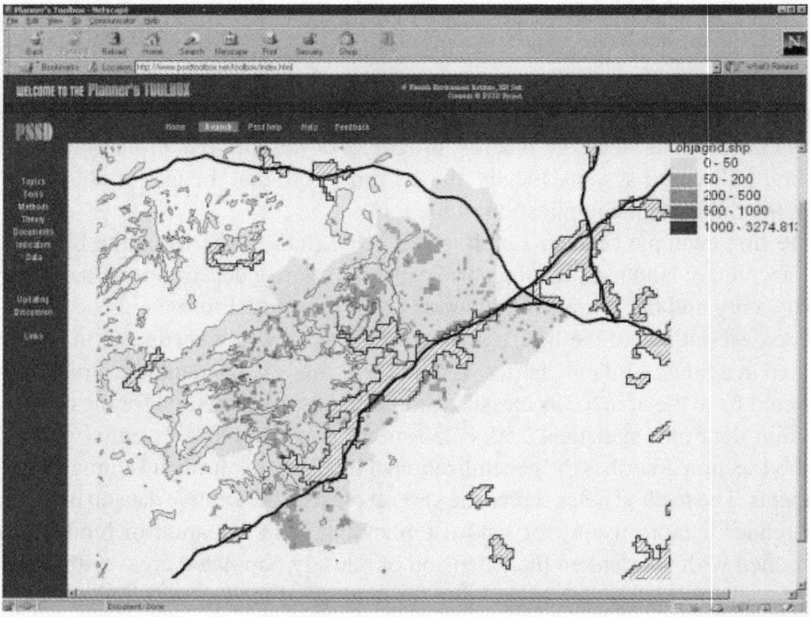

Fig 7.5: Distribution of areas with valuable biodiversity (darker shaded areas) and location of populated areas (grey areas) in Lohja municipality (southern Finland)

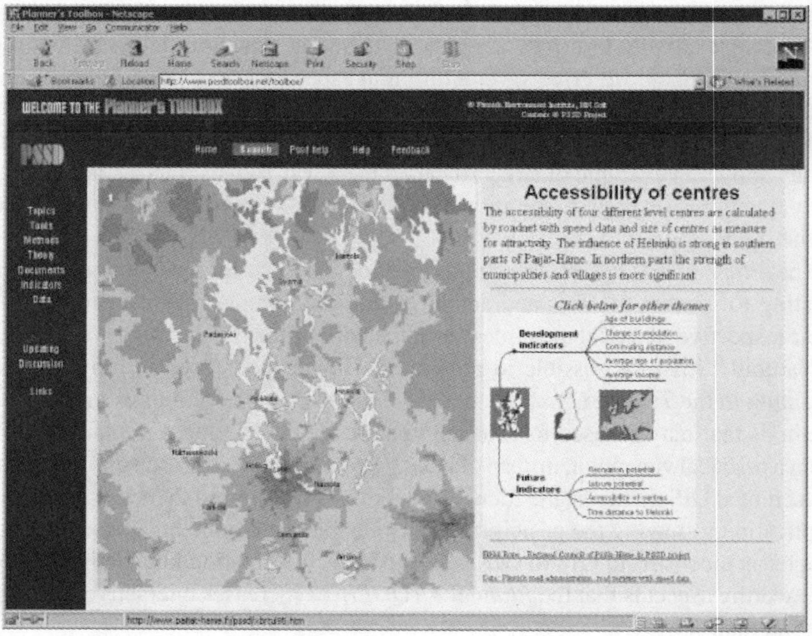

Fig 7.6: 'Accessibility of centres' in the *Planner's TOOLBOX*

7.5 Evaluation of the *Planner's Toolbox*

The purpose of the PSSD project is to offer the *Planner's TOOLBOX* to the international planning community as a prototype – or as a preliminary model, example, suggestion, incentive – for professional communication. The term 'prototype' was used to convey the idea that the project aimed to create a first version of the system without carrying out any extensive examination of relevant potential international needs, because it was felt that the discussion of such needs would need an initial incentive.

To date, two different evaluations of the *Planner's TOOLBOX* have been performed. First, some independent scientists have been asked to reflect on the PSSD project in general and on the *Planner's TOOLBOX* in particular (Möller-Jenssen 2000, Geertman 2001). This external evaluation has resulted in the following remarks. The *TOOLBOX*:

- is an operational, real-world, internet-based PSSD of which there are few, if any, other examples;
- brings together and makes explicit a diversity of knowledge for achieving sustainable planning goals;
- possesses lots of knowledge: theory, maps, tools, methods, indicators, data, models, analysis, *et cetera*;
- pays particular attention to theory and method;
- contains lots of meta-information;
- is an interesting, potentially flexible, and powerful way of disseminating the results of the subtasks in a way that facilitates their application within other areas; and
- may be improved by developing the user interface in such a way that it provides a better response to users' needs.

Besides this external evaluation, the *Planner's TOOLBOX* was tested and evaluated several times by the participating Finnish and Danish regional planners themselves during its development. Their feedback indicated that, although the system is only a prototype, it has a great deal of potential. The division of the information into smaller entities caused the most contradictory opinions. Some planners were very content with this solution, because it allows for new contexts and conceptual linkages between methods. Others criticized this aspect because of the difficulties in finding the ideal paths from theories to tools using the links between entities. One self-evident feature of the *TOOLBOX* is that, being still a prototype, it lacks a wide range of ready-to-use scripts and macros. Instead, there exist various methods to experiment with. The selection of the end-user's tools is expected to increase in relation to the growing number of people utilizing the system.

One way to measure the usefulness of the *Planner's TOOLBOX* is to estimate whether the information it contains is useful or not in real-world spatial planning. As a preliminary answer to this question, it is evident that some of the first toolbox methods have been used in planning. The counties of North Jutland (Denmark), for example, adopted the urban area noise model. The method of classification of bedrock areas was utilized in a Finnish spatial planning and groundwater protection project. The value of the *Planner's TOOLBOX* from an ordinary planner's point of view can be summed up bin the words of one user who stated that *"if I had to draw a GIS analysis map on some complex spatial planning phenomenon, I would first go and check the Planner's TOOLBOX for possible ideas"*.

7.6 Present and Future Users and Developments

The primary target area for the current contents of the *Planner's TOOLBOX* has been the Baltic Sea Region, because the project emerged from a co-operation between Finland, Denmark and Germany. However, it may be used in other areas and global level applications are not excluded. The similarity of land use planning systems within the Baltic Sea Region has been responsible for facilitating the use of common tools and methods in this region.

The *Planner's TOOLBOX* offers several significant benefits for different groups of users of which the most important ones are the spatial planners, and more specifically, the statistical region-level planners. Implementing new GIS and planning methods in practice takes time, especially for the older – and the most experienced – of the present-day planners whose experience should be used for the benefit of future generations. The international character of the *Planner's TOOLBOX* makes GIS methods already tested elsewhere accessible for the older, experienced planners. The discussion forum of the *Planner's TOOLBOX* is for discussing current problems – and solutions. Just as the future planners can learn about actual planning problems from the system, the experienced planners can also learn about practical GIS methods from the same source.

Other important users are the developers of the GIS and planning methods. The *Planner's TOOLBOX* offers continuous real-time co-operation between the developers. The system can be seen as the first version of an international virtual 'blackboard'. Moreover, it is also valuable for spatial planning and GIS education. Through using the system, planning education can be combined with international co-operation in research and actual planning. Some universities and research institutes in Finland and abroad have been very interested in using the *Planner's TOOLBOX* in education, regarding it as a concrete step towards learning on the Internet (virtual academy).

Public participation in regional planning has a long tradition in Denmark and it will become more important in Finland due to a new Building and Land Use Act. The traditional practice of public participation involved planning meetings that quite often took place in an atmosphere of confrontation (Kingston *et al.* 2000). In contrast, internet-based approaches to public participation have many advantages; the information is available at any location with web access, at any time. Furthermore, citizens are able to make their comments and express their feelings more anonymously than with the traditional method, thus encouraging ordinary people to participate in the regional planning process. In Northern Jutland County, this approach has been used recently with a high degree of success. One of the future development goals of the *Planner's TOOLBOX* is to make scenarios, enabling planners to evaluate alternative approaches to sustainable planning. In addition, the public could interact using simplified scenarios, in order to achieve a better understanding of the pros and cons of each alternative scenario.

The *Planner's TOOLBOX* has been available for unrestricted use since spring 2001. However, it is still only a prototype and the starting point for further development. Maintenance and further development are the responsibilities of the Finnish Environment Institute, the National Environmental Research Institute of Denmark, and the Technical University of Hamburg-Harburg. The practical maintenance will be undertaken by the Finnish Environment Institute, who will also be responsible for granting the userids. The shortcomings of the current system will be amended by the Finnish Environment Institute in cooperation with all users. The database structure presented in a previous section allows for further development of the TOOLBOX in a flexible way. It is now freely available to everyone for browsing and the users who wish to join the *Planner's TOOLBOX* 'Society' may register at *http://www.pssdtoolbox.net/plannertool.htm.*

References

Geertman, S. (2001) PSSD – Some Reflections, Unpublished presentation in PSSD Seminar, Lahti, 12.3.2001.

Hansen, H.S. (2000) Integration of environmental and socio-economic data, *Proceedings ESRI European, Middle Eastern and African User Conference*, Istanbul, 18-20 October, pp. 1-9 (CR-rom).

Hansen, H.S. (ed.)(2001) PSSD – Planning System for Sustainable Development. The Methodological Report, *NERI Technical Report No. 351*, National Environmental Research Institute, Denmark.

Kingston, R., Carver, S., Evans, A. and Turton, I. (2000) Web-based public participation geographical information systems: an aid to local environmental decision-making, *Computers, Environment, and Urban Systems,* 24: 109-125.

Møller-Jensen, L.(2000) Evaluation of the PSSD project, Unpublished presentation at PSSD Seminar Espoo, 12 December.

Shiffer, M. (1995) Interactive multimedia planning support: moving from stand-alone systems to the World Wide Web, *Environment and Planning B*, 22: 649-664.

Web Sites

Indicator Generator: *http://cgi.tu-harburg.de/~sb3www/indihome.shtml.*

Planners Toolbox: *http://www.pssdtoolbox.net/plannertool.htm.*

8 STEPP: A Strategic Tool for Integrating Environmental Aspects into Planning Procedures

Gerrit J. Carsjens[1], Ron J.A. van Lammeren[2] and Arend Ligtenberg[3]

[1] Land-Use Planning Group, Wageningen University and Research Centre, P.O. Box 9101, 6700 HB Wageningen, The Netherlands
[2] Landscape Architecture Group and Centre for Geo-Information, Wageningen University and Research Centre, P.O. Box 9101, 6700 HB Wageningen, The Netherlands
[3] Centre for Geo-Information, Wageningen University and Research Centre, P.O. Box 9101, 6700 HB Wageningen, The Netherlands

Abstract: This chapter describes a research project that aims to improve the quality of physical plans on a local level and to create a basis for communication between disciplines and stakeholders that are dealing with environmental and spatial planning. A GIS tool for integrating environmental aspects into planning procedures has been developed. Although the research shows the potential for the use of the tool in practice, we still are very far away from supporting participatory planning. The research clearly indicates that the current planning practice is not well equipped to deal with real-time participation.

8.1 Introduction

Local authorities have a responsibility for creating a sustainable environment. Physical planning at a local level has direct impacts on the urban and rural environment. In practice this means finding a sustainable balance between the influences of present and proposed human activities and the sensitivities of the urban and rural environment. However, simple and uniform tools that analyse and visualise the environmental aspects of local physical plans are not available.

A research project of Wageningen University and Research Centre intends to develop a GIS-based *Strategic Tool for integrating Environmental aspects in Planning Procedures* (*STEPP*). The basic aims of the project are to improve the quality of physical plans on a local level and to create a basis for communication between disciplines and stakeholders that are dealing with environmental and spatial planning. *STEPP* is intended to support the work of local authorities throughout the spatial planning process, especially with regard to identifying options in the early phases of the process. It can also be used as a tool for communication with stakeholders or to serve educational purposes. Section 8.2 gives a brief history of the Dutch interest in the environmental aspects of spatial planning. Section 8.3 presents the principles of *STEPP*, while Section 8.4 describes the professional experiences with *STEPP*. The conclusion in Section 8.5 focuses on the communicative perspectives of this type of PSS.

8.2 Background

The Dutch interest in the environmental aspects of spatial planning has been strongly influenced by the concept of the 'compact city'. This concept, from the early nineteen-eighties, was adopted as the Dutch spatial planning answer to the problem of urban sprawl. Taking into consideration the small size of the country, the high population density, and the impressive economic growth, the compact city approach was intended to stop any unnecessary waste of open space (rural area). Furthermore, by concentrating urban activities, the benefits could be less commuter traffic, less pollution and possibly less energy use. Despite these good intentions, however, a new problem arose. This was how to preserve the environmental quality of the 'grey space' (urban and semi-urban areas) as well as the 'green space' (rural and natural areas). By concentrating the rural activities, the environmental problems related to the 'grey space,' like noise, odour, dust, toxic pollutants and calamities, became more prominent. This conflict between the spatially oriented compact city concept and the environmental drawbacks has been named 'the compact city paradox' (de Roo 1999).

The concern about the environmental quality initiated some new environmental policies. The year 1989 was a pivotal one. It was the time during which these policies were developed. It was also in 1989 that the Dutch Ministry of Housing, Spatial Planning and Environment (VROM is the Dutch abbreviation) introduced a system that allowed for the integration of sector environmental standards and their related spatial zones into one wholly integrated environmental zone. The Integrated Environmental Zoning (IEZ) system was intended for areas around industrial sites (VROM 1990). The IEZ approach is based, for the most part, on the calculation of influence zones. These zones are the combination of environmentally intrusive activities and their sensitive surroundings. The sensitivity of the surroundings is derived from the sensitive (land use) functions in the surrounding areas.

The IEZ approach offered an impressive and challenging field for information system developments. The Ministry stimulated the efforts to develop computer systems, preferably based on geo-information that could support the IEZ system. On the other hand, the Ministry carefully scrutinised the effect of this top-down approach on the local and regional authorities. The local and regional authorities could submit project proposals to scale down the IEZ system. The municipality of Zwolle (approximately 100,000 habitants) worked on the municipality version of an IEZ system (Streefkerk 1992). This particular system, so called BM (initials for the Dutch words for Environmental Quality Assessment) focused on intrusive environmental influences like local air pollution, noise, local terrain contamination and hazards. From the earliest beginnings of the development of BM, Wageningen University was involved and had the task to develop a dedicated GIS application (van Slagmaat *et al.* 1992, van Lammeren *et al.* 1994). A comparative study undertaken by Groningen University (Humblet and de Roo 1995) qualified the BM as the most promising of all comparable systems.

In subsequent years, the Dutch Government intended to adapt the IEZ system to a more consensus-based approach and to make it more convenient to future oriented objectives of the different planning actors. There were important reasons to do this. The local political impact of the IEZ was one reason. Because of the top-down approach, the results of the IEZ looked more like missives than politics. A procedure to change this 'parachuting' was based on the consensus approach and called ROM, a Dutch acronym for Spatial Policy and Environment (de Roo 1999).

The ROM procedure dealt with 'grey' as well as 'green' environments. In the rural cases, new 'intrusive' components like acidification, drying out, fragmentation and manure-based soil contamination were taken into account. The activities of Wageningen University continued by adapting the original BM application into a system that could support the ROM procedure. Acidification especially had been the main focus (van den Brand 1992) which finally resulted in ADIS (van Lammeren 1994), a geo-information system that supports planning by information input about agricultural activities in relation to the output of ammonia. Despite much interest by local authorities and consultants, the BM application has hardly ever been used in practice. The main reasons seem to be a lack of promotional activities and insufficient computer tools and facilities. In addition, the possibilities to use the BM outside of local governments were limited.

In 1998, a follow-up project at Wageningen University began (Carsjens *et al.* 1999, Groenveld 2000, Kluskens 2000). The main objective of this project was to develop a GIS-based *Strategic Tool for integrating Environmental aspects in Planning Procedures* (*STEPP*). *STEPP* is based on the previous experiences with BM. The basic aims of the project are to:

- improve the quality of physical plans (green as well as grey) on a local political level;
- create a basis for communication between spatial planning and environmental disciplines; and
- update and extend BM to a more uniform tool.

STEPP aims to support the work of local authorities throughout a spatial planning process, while also offering possibilities for communication with stakeholders. The first edition of *STEPP* was implemented in the GIS software ArcView and its programming language Avenue.

8.3 Challenging Approach

In comparison to the IEZ system and according to the ideas of the ROM approach, *STEPP* offers an analytical as well as a design component (Figure 8.1). The analytical component (A) of *STEPP* includes a definition of the sensitivity of delineated receptive areas in a certain geographical domain, and a definition of land use areas (LU) with a specific degree of intrusion of pollutants or contamination. A spatially based confrontation of sensitivity and intrusion gives a certain environmental quality (EQ). The location and the nature of human activities in an area can be adapted to generate modified land use (LU′). The design component (B) offers an opportunity to apply different scenarios, based on new spatial arrangements of human activities.

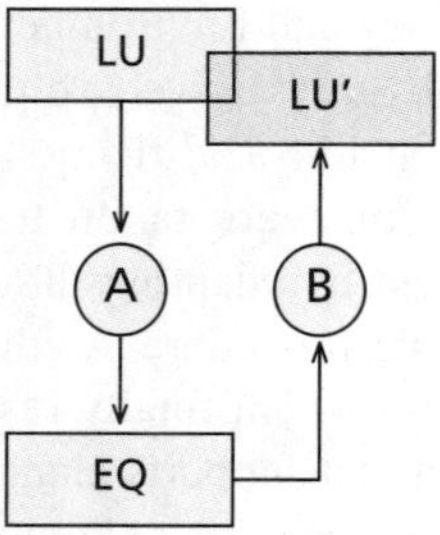

Fig 8.1: *STEPP* components

The analytical component consist of five different steps (Figure 8.2):

1. Determining the human activities and types of receptive environment in a study area, based on land use data.
2. Determining the areas where environmental impacts occur (zones of influences).
3. Combining the zones of influence and the sensitivities of receptive areas to produce an environmental quality map for each category of environmental impact.
4. Aggregating the zones of influence into an integrated environmental impact map.
5. Combining the individual quality maps into an integrated environmental quality map.

Information on land use is a very valuable source to determine spatially located human activities (step 1). Most human activities are related to physical objects like industrial and company buildings, farmland parcels, roads and railways. Some typical environmental impacts of these activities are smell, dust, nuisance, fire and explosion hazards, and local air and soil pollution. The data on activities can be derived from (mostly analogue) information available through the Dutch municipalities.

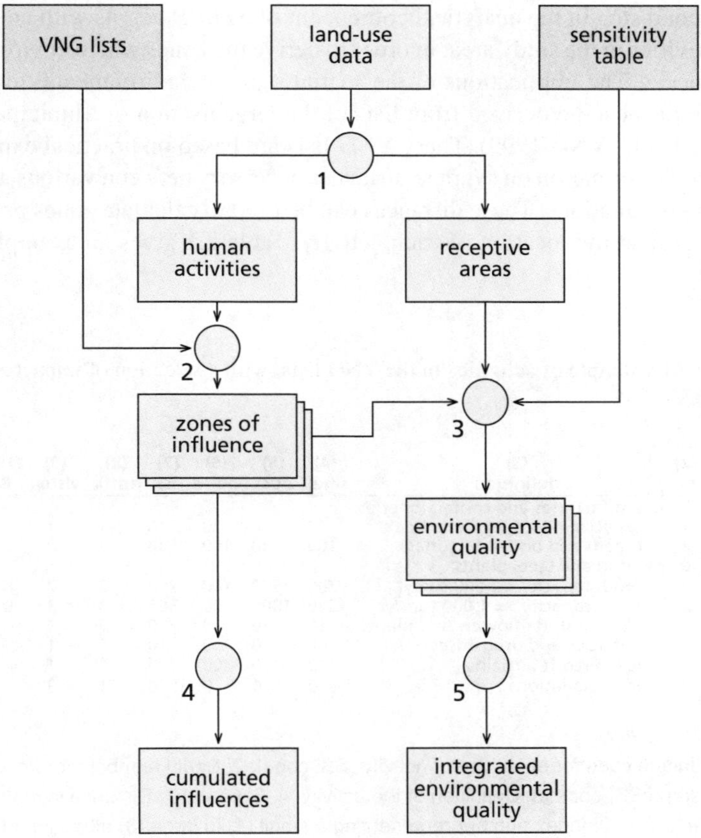

Fig 8.2: The basic steps of the analytical component of *STEPP*

Information on land use can also be used to define receptive areas (step 1). Two basic geographical information databases, the Digital Topographical Map 1:10,000 (published by the Dutch Topographical Service) and the Digital Land-Use Statistics (published by the Dutch Central Bureau of Statistics) provide the necessary information. *STEPP* distinguishes three categories of receptive environment:

- areas with human inhabitants, subdivided in areas with large numbers of humans (such as shopping malls), areas where humans have permanent residence (such as private residences) and areas that humans occasionally visit (such as recreation areas);
- areas with (unique) ecological values (such as parks and nature areas); and
- areas without (unique) ecological values, subdivided in soil and water areas (such as gardens and canals) and paved areas.

The second step in the analytical component of *STEPP* begins with an analysis of the activities in the study area, in order to derive the zones where environmental impacts occur. The implications of the spatial scale of environmental impacts of most activities can be derived from lists of the Organisation of Municipalities in the Netherlands (VNG 1999). These VNG lists are based on practical experience, and contain information on average distances necessary between various activities and sensitive functions. These distances can be used to calculate zones per impact category around the location of each activity. Table 8.1 gives an example of the VNG lists.

Table 8.1: An example of activities in the VNG lists, with a selection of impact zones per category

(1) SBI	(2) nr	(3) definition	(4) smell	(5) dust	(6) noise	(7) risk	(8) traffic	(9) visual	(10) B	(11) D	(12) L
1581	0	bread factories and confectioneries:									
1581	1	capacity < 2500 kg flour/week	30	10	30	10	1	1			
1581	2	bread- and biscuit factories	100	30	100	30	2	2			
271	0	pig iron and steel plants:									
271	1	prod. capacity < 1,000 ton/yr	700	500	700	200	2	2	B		
271	2	prod. capacity >= 1,000 tons/yr	1500	1000	1500	300	3	3	B		L
5122		wholesale trade flowers and plants	10	10	30	0	2	1			
5231, 5232		pharmacies and drugstores	0	0	0	10	1	1			
9272.2		model aircraft airfields	10	0	300	100	1	1			
I 0037		radar installations	0	0	0	1500	1	3		D	

Source: (VNG 1999)

Notes: 1: Unique code for each type of activity (SBI code); 2: Serial numbers for activities within the same SBI-code; 3: Definition of the activity; 4, 5, 6, and 7: Distance in metres per impact category; 8, 9: Index number representing a small (1) to large (3) attraction of traffic or visual disturbance; 10, 11: Codes indicating soil (B) and air (L) pollution; 12: Code indicating a diversity of activities within the SBI code.

The concept of influence distances of the VNG lists has been adopted in *STEPP*. The additional information on environmental impacts (the codes and index values) is converted to distance measures in order to make them suitable for *STEPP*. Three additional impact categories with influence distances are added to *STEPP*: (i) visual disturbance; (ii) traffic disturbance; and (iii) local soil pollution. Table 8.2 describes the conversion that takes place.

In addition to the individual activities listed in the VNG lists, *STEPP* also includes a tool to introduce four different types of industrial zones. Each type of zone is characterised by average influence distances of representative activities in the VNG lists. The four types of industrial zones are: (i) service industries; (ii) wholesale trade; (iii) manufacturing industry; and (iv) heavy industry.

The third step starts with an analysis of the sensitivity of functions for each impact category. Table 8.3 gives an example of how these sensitivities can be classified.

Table 8.2: Conversion from index values and codes in the VNG lists to distances (in metres) in *STEPP*

	impact category in the VNG lists						
	traffic			visual			soil pollution
impact category in STEPP	1	2	3	1	2	3	B
traffic disturbance	30	100	300				
visual disturbance				30	100	300	
soil pollution							10

Note: In case an index value or code is not applicable the impact distance is zero

Table 8.3: Sensitivity of functions for each impact category

sensitive function	code	level of sensitivity per impact category						
		smell	dust	noise	hazard	air poll.	soil poll.	visual
human beings								
concentrations	MEC	3	3	2	3	2	3	2
permanent	MEP	2	2	2	2	2	3	2
occasional	MEN	2	2	1	2	2	2	1
ecological values								
unique ecological values	EC	2	2	2	1	3	3	1
no ecological values								
soil and water	BW	1	1	1	1	1	3	1
materials	MA	1	1	1	1	1	1	1

Note: 3 = very sensitive; 2 = sensitive; 1 = not sensitive

STEPP compares the impact zones for each impact category with the sensitivity of the functions present in each zone. This gives a quality indicator map for each impact category. The general idea is that:

- areas without influences have good quality (indicator value 0);
- areas with functions that are not sensitive for a specific impact have a reasonable quality (indicator value 1);
- areas with functions that are sensitive to a specific impact have a moderate quality (indicator value 2); and
- areas with functions that are very sensitive to a specific impact have a poor quality (indicator value 3).

The fourth step incorporates the different impact maps into an integrated environmental impact map. It calculates the different impacts at any one location. The fifth step incorporates the different quality indicator maps into an overall integrated environmental quality map. This process entails summarising the values of the different quality indicator maps. The importance of an individual quality indicator map depends on the situation and/or the sensitivities involved. Thus it is then possible to rank one map higher than another. Figure 8.3 gives a schematic example of the computations in the fourth and fifth step.

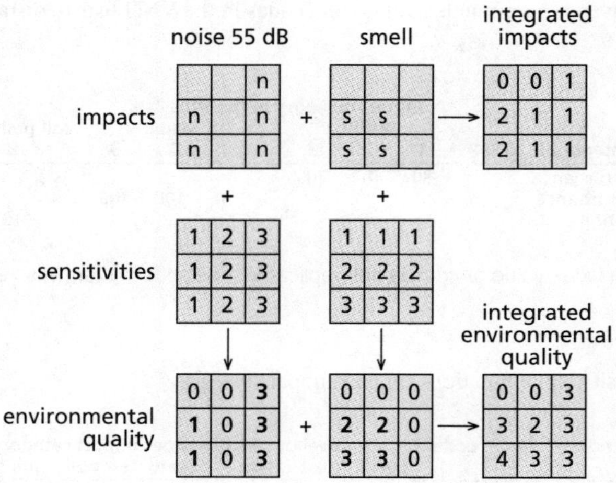

Fig 8.3: Example of the computations in the fourth and fifth steps (without ranking the individual maps)

The design component (B in Figure 8.1) offers an opportunity to work with scenario studies based on new spatial arrangements of human activities. Most input data and parameters that are used by *STEPP* can be modified, e.g.

- the locations of (human) activities (see Figure 8.4a);
- the shape and size of specific areas;
- the different types of receptive areas (function of an area);
- the sensitivity of receptive areas (Figure 8.4b, see also Table 8.3); and
- the activity-based environmental impact parameters (the influence distances of the VNG lists as well as impact distances of individual activities) (Figure 8.4c, see also Tables 8.1 and 8.2).

Intended measures to improve the environmental quality (e.g. filter installations or noise barriers) have to be 'converted' to a reduction in impact distance before they can be introduced in *STEPP*.

8.4 Professional Experiences

STEPP has been used for educational purposes as well as in actual planning practice. Two cases using the application will be described: firstly, the application of STEPP in two planning courses at Wageningen University and secondly, the results of a workshop where *STEPP* was used by planning consultants for a housing project in the town of Goor in the Netherlands.

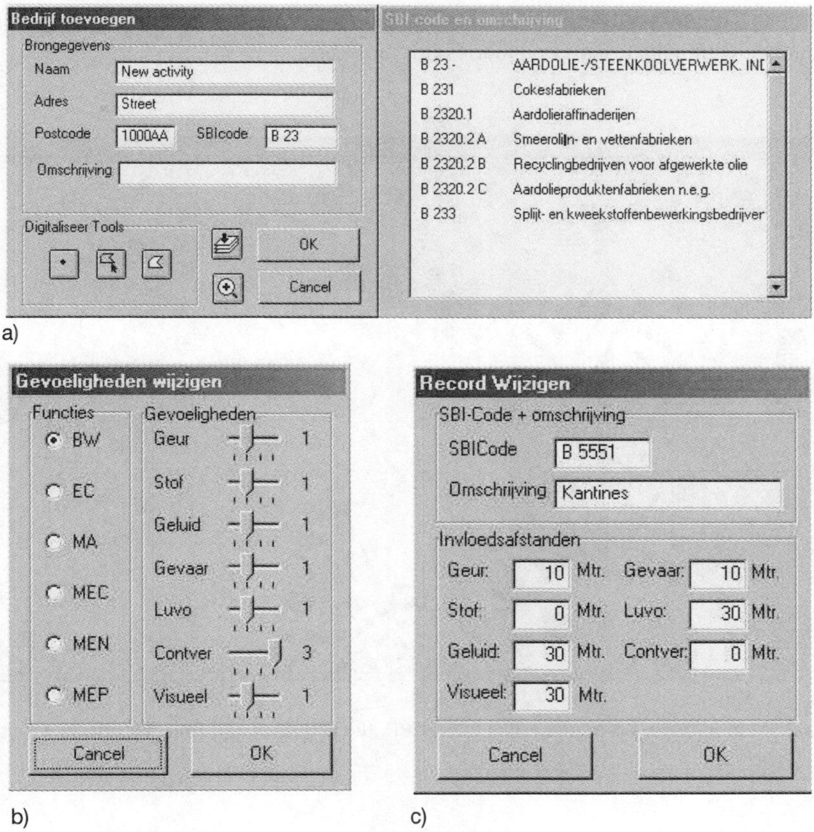

Fig 8.4: Screens to (a) introduce new activities, (b) modify the sensitivities of receptive areas and (c) modify the influence distances of the VNG lists

Planning courses: Since 1998, *STEPP* has been used in two spatial planning courses at Wageningen University. One course focuses on the relationship between planning and environmental aspects (Carsjens 2001) and a second course focuses on physical planning at local scale level (Jaarsma and van der Knaap 2000). Both courses contain practical assignments, where students themselves are working with the application. The experiences with both courses helped to test and improve the user-interface of *STEPP*. The results of both courses indicate that *STEPP* stimulates the discussion on the environmental aspects of physical plans. *STEPP* was also shown to be easily accessible.

Workshop Goor: In March 2001, a one-day workshop was organised for a dedicated group of planners and environmental specialists of the SAB office for urban planning (Arnhem). The main objective of the workshop was to evaluate the applicability of *STEPP* in practice. The workshop aimed at a development project of a new residential area in the municipality of Goor. The proposed location of the housing project is a vacant site within the town's boundaries (Figure 8.5).

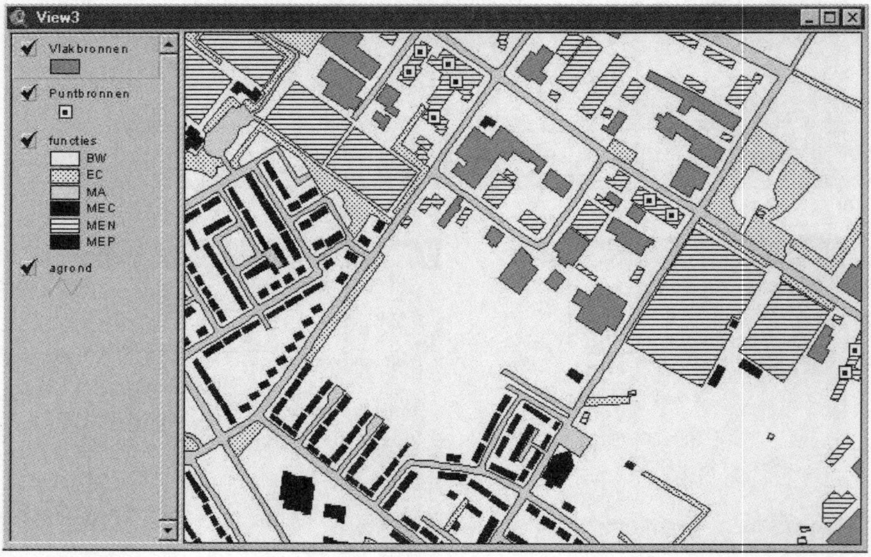

Fig 8.5: Map with the basic input data of STEPP: the activities and functions of the Goor study area

The location is surrounded by residential areas to the south and the west, but also borders an industrial park in the northeast. The activities are defined as point or area objects (depending on the size of the activity). For an explanation of the function codes, see Table 8.3. The 'empty' area in the middle of the map is the proposed site of the new residential area. The main questions were which potential impacts could occur at the proposed site and how to mitigate these impacts. *STEPP* was used to analyse this problem. The morning programme of the workshop included an introduction to *STEPP* and some basic exercises. In the afternoon the housing project in Goor was analysed. One of the resulting maps is shown in Figure 8.6.

The large map shows the location of the residential area (code MEP) with a transition zone on the north side, containing a mixture of private houses and company premises (code MEN). The zones on the north side represent the different environmental quality zones. The remaining area, outside these zones, is not influenced and has the best quality. The transition zone prevents the influence zones reaching the residential area. The only activities that are allowed in the transition zone are those without significant environmental impacts (e.g. offices) and that are less sensitive to the impacts from the industrial park. The five small maps show the environmental quality zones for each of the different impact types.

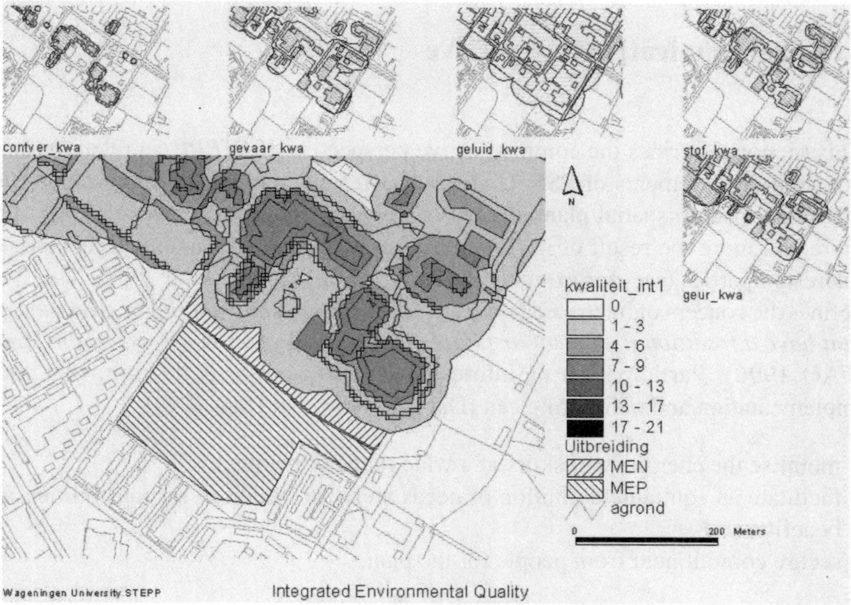

Fig 8.6: An integrated environmental quality map for the Goor study area

An extensive questionnaire was used to record the experiences with the different aspects of *STEPP*. The results were also discussed. The main conclusions from the questionnaire and the discussion are that:

- *STEPP* is a fast, efficient and representative tool to analyse the environmental aspects;
- *STEPP* is user-friendly and easily accessible, especially given the fact that only one of the consultants had some basic knowledge of ArcView;
- *STEPP* at present is not equipped to analyse the environmental impacts of road traffic;
- the defined categories of receptive environment (the six function classes) should be more detailed, especially regarding the distinct types of residential areas;
- additional information can improve the applicability of *STEPP*, e.g. information on traffic intensity and qualitative aspects of residential houses;
- *STEPP* stimulates discussion between professional disciplines; and

- the application will be purchased by the participants of the workshop (SAB) and will be used for at least two years with different types of projects. This will, of course, give an even better perspective on the applicability of *STEPP* in practice, and will help to improve and develop the application further.

8.5 Communicative Perspective

This section describes the communicative perspective of *STEPP*, in relation to the potential developments of PSS. Given the increasing number of interested parties that participate in spatial planning, the outcomes of planning processes tend to be more and more the result of negotiations, based on the best mutual acceptance of interested parties (see, for example, de Jong 1986, van Woerkum 2000). The FAO defines the concept of interested parties as *"individuals, communities or governments that have a traditional, current or future right to co-decide on the use of the land"* (FAO 1996). Participation of interested parties in data collection, planning, implementation and monitoring can (Dalal-Clayton and Dent 1993):

- mobilise the energies and skills of a wide range of people;
- facilitate an equitable definition of needs and distribution of responsibilities and benefits; and
- secure commitment from people for the plan.

The increasing level of participation has led to a shift in the perception of planning, from a centralised view towards a more systemic and process oriented approach. By encouraging an increasing number of participants to join the spatial planning process, the demands on communication with interested parties has moved beyond the traditional simple (formal) hearings. Interested parties want to have an active role from the initial stages (formulation of goals, exploration of alternatives) right up to the final decision-making.

Information and communication technology (ICT) can have a major influence in supporting participatory planning. ICT tools can support participatory planning at five levels of participation, also called the ICAPD levels (after NRLO 1998). Furthermore, we also distinguish four levels of time and place of use of these tools by participants. Both types of classification are presented in Table 8.4.

The current trend in participatory planning is towards advice and co-producing of spatial plans. This is catalysed by the advances made in ICT like the mass acceptance of email and internet, allowing interested parties to access information more easily and enable them to respond in a more direct way.

Applications that facilitate planning at the same time in the same place can be found in the field of groupware solutions. These include software that facilitates co-operative brainstorming and decision-making, for instance, among interested parties. There are examples of approaches within the domain of land use planning that can be found in the GEVIS project (Bongers *et al.* 2000) and OntwerpGis (Geertman *et al.* 2000). These approaches, however, are restricted in a way that only invited parties are able to join the planning process. The current *STEPP* application is also an example of a same time, same place application. It can facilitate discussions and the formulation of alternatives at 'runtime'. The intuitive interface, the clearly defined steps in the program, the qualitative and cartographic representation of the model results, and the reasonable requirements for hardware and software enable an quick and efficient use, even by people who are not an expert in GIS (see the conclusions in Section 8.4).

Table 8.4: The different levels of participation and levels of time and place of use of ICT tools

levels of participation		time and place of use of ICT tools	
I	inform	SS	same time, same place
C	consult	DS	different time, same place
A	advise	SD	same time, different place
P	co-produce	DD	different time, different place
D	co-decide		

The second approach can be found in examples such as information pillars installed in town halls, that are a result of the Dutch national program, *Public Counter 2000* (in Dutch: Overheidsloket 2000) (*http://www.ol2000.nl*). This program aims at creating a nationwide network of desks, to be consulted by citizens, trade and industry for information and public products and services, based on the use of ICT. Information pillars are devices that enable citizens to consult digital information and communicate with the 'ones behind', using email and other electronic means of communication. The pillars can also be used to present plans to citizens allowing them to then submit their opinion. Co-producing spatial plans is not really possible using these devices due to their limited interactivity. However, these devices have proven to be eminently suitable for informing, consulting and asking advice about spatial plans.

The third and fourth approaches can be established through the Internet (and perhaps in the future, by mobile phones and personal digital assistants as well). The Internet allows interested parties to participate regardless of time and/or place. Many provinces and some larger cities in the Netherlands are already presenting their spatial plans and related spatial information on the Internet. Examples are the websites of the province of Gelderland (*http://geodata.prv.gelderland.nl/km/kaartenbak/*) and the city of Groningen. Groningen created a website (*http://www.groningen.nl/forum*) where citizens can discuss and give their opinion on reconstruction plans for each district of the city. In 2000, the Dutch Ministry of Housing, Spatial Planning and Environment started the project *Digitally Exchangeable Spatial Plans* in order to make all spatial plans in the Netherlands available through the Internet by the year 2010. A new standard (IMRO) has been developed (Ravi 2000) that enables an exchange of information between the different types of spatial plans to occur.

A part of the research project of *STEPP* at Wageningen-UR focused on the aspect of communication and participation using the Internet. A pilot internet application has been designed called LODERWEB (*http://www.geo-informatie.nl/loderweb*) (Kluskens 2000) that extends the *STEPP* method to a system that allows interested parties to express their concerns, opinions or alternatives. A map-server was used to present the *STEPP* outcomes on the Internet. This gave additional means for users to zoom, pan and query the outcomes. Users can comment by clicking on a location in the map and enter their opinions, alternatives, *et cetera*. In addition, the system enables users to view and comment on the points made by others.

As previously mentioned, modern ways of spatial planning require simple and clear methodologies and tools. *STEPP* has proven to be able to deliver valuable tools that enable participants to quickly explore and interact with various scenarios (Section 8.4). The qualitative results of these interactions also enable non-experts to interpret the outcomes. The clearly defined steps within *STEPP* make it easier to comprehend the model thus preventing it from being a black box.

8.6 Conclusion

Although experiments clearly show the potential for the use of distributed applications in spatial planning, we still are very far away from supporting a new way of planning. Figure 8.7 indicates the current status of PSS in the Netherlands, with respect to the levels of participation and the autonomy of the participants. The grey colour indicates the current levels of use in the Netherlands.

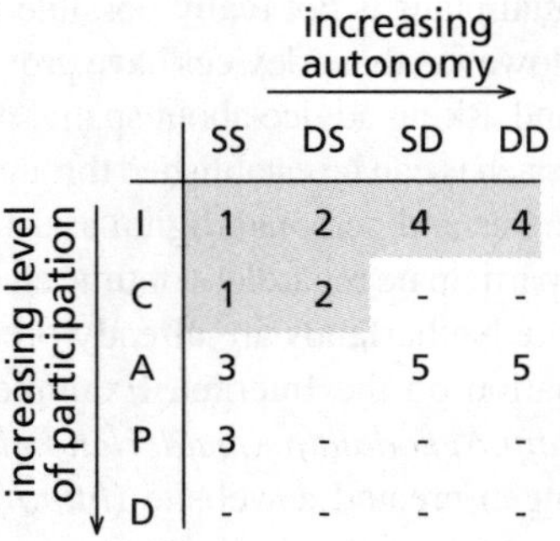

Fig 8.7: The status of planning support systems
Note: 1 = *STEPP*, 2 = information pillars, 3 = GEVIS/OntwerpGIS, 4 = the province of Gelderland, 5 = LODERWEB

Co-producing and co-deciding spatial plans is still not practically applicable. The research on *STEPP*/LODERWEB clearly indicates that, besides a number of technical issues, current planning practice is not well equipped to deal with real-time participation. Important questions that remain are: how to process the often conflicting opinions; how to give feedback to the contributions given; how to deal with the increasing time constraints imposed by the direct-react-direct-response expectations of the new media; *et cetera.* Currently, we consider participation through these technologies limited to only the first two levels of ICAPD. The search for a sustainable balance between the influences of human activities and the sensitivities of the urban and rural environment is not merely a quest for a suitable tool, but also means a quest for a suitable participative communication.

Acknowledgements

Funding for this study was provided by the Delta research program of Wageningen University and Research Centre. The authors also wish to thank Ries Kluskens and Gerard Groenveld for their contributions.

References

Bongers, F., Hollan, C., Heusden, M. and Hidding, M. (2000) *Een theoretisch model voor een GEVIS (concept),* Landbouw Economisch Instituut, Den Haag.

Carsjens, G.J. (2001) *Relaties tussen landgebruik en milieu. College en practicum. Syllabus onderwijselement K100-251*, Leerstoelgroep Landgebruiksplanning, Wageningen Univeriteit, Wageningen.

Carsjens, G.J., Kluskens, R. and van Lammeren, R.J.A. (1999) Methodology to integrate environmental aspects into physical planning, in Musy, A., Santos Pereira, L. and Fritsch, M. (eds.) *Emerging Technologies for Sustainable Land Use and Water Management*. Presses Polytechniques et Universitaires Romandes, Lausanne, abstract p 32, CD-ROM 7 p.

Dalal-Clayton, B. and Dent, D. (1993) Surveys, plans and people: review of land resource information and its use in developing countries, *Environmental Planning*, Issue no 2. IIED, London.

de Jong, M.J. (1986) *Idee- en consensusvorming in de ruimtelijke ordening: een sociaal-wetenschappelijk onderzoek naar de totstandkoming van het Streekplan Friesland 1982*, Proefschrift Wageningen Universiteit LU-1104, Wageningen.

de Roo, G. (1999) *Planning per se, Planning per saldo: Over Conflicten, Complexiteit en Besluitvorming in de Milieuplanning*, Sdu Uitgevers, Den Haag.

FAO (1996) Planning for sustainable use of land resources – towards a new approach, *Land and Water Bulletin 2*, Food and Agriculture Organization of the United Nations (FAO), Rome.

Geertman, S., van Deursen, W. and Vos, E. (2000) OntwerpGis: ondersteuning voor participatieve planvorming, in Fendel, M.F. (ed.) *Geo-Informatiedag Nederland 2000: Presentaties en Demonstraties over Onderzoek, Ontwikkeling en Praktijktoepassing op het Gebied van Geo-informatie,* VGVI, TU Delft.

Groenveld, G. (2000) *Uitbouw van de BM methode met een implementatie in ArcView*. Centrum voor Geo-Informatie, Wageningen Universiteit, Wageningen.

Humblet, A.G.M. and de Roo, G. (1995) *Afstemming door inzicht; een analyse van gebiedsgerichte milieubeoordelingsmethoden ten behoeve van planologische keuzes*, Geo Pers, Groningen.

Jaarsma, C.F. and van der Knaap, W.G.M. (2000) *Practicum Planning 1. Syllabus onderwijselement K100-250*, Wageningen Universiteit, Leerstoelgroep Landgebruiksplanning.

Kluskens, R. (2000) *An Internet based environment to facilitate multi-actor informing and consulting for environmental planning based on the BM methodology*, Centre for Geo-Information, Wageningen University, Wageningen.

NRLO (1998) *ICT: mogelijkheden voor sturing en ontwerp in landelijke gebieden*, Ontwikkelingen in wetenschap en technologie, Nationale Raad voor Landbouwkundig Onderzoek (NRLO), Den Haag.

Ravi (2000) *Informatiemodel Ruimtelijke Ordening, Geactualiseerde versie publicatie 97-2*, Ravi overlegorgaan voor vastgoedinformatie (RAVI), Amersfoort.

Streefkerk, N. (1992) *Handboek Beoordelingsmethode Milieu; Inhoudelijke Achtergronden en Handleiding*, VNG, Den Haag.

van Lammeren, R.J.A. (1994) *Ammoniak Depositie Informatie Systeem*, Geografendagen, Den Burg.

van Lammeren, R.J.A., Algra, R. and van den Bosch, G. (1994) Evaluation of the method for assessing environmental quality (BM), in Molenaar, M. (ed.) *Remote Sensing and Geographical Information Processing: Concepts and Applications for Land-use Monitoring and Agriculture,* Wageningen Agricultural University, Wageningen.

van Slagmaat, M., van der Meer, A.M. and van Lammeren, R.J.A. (1992) *Prototype Beoordeling Milieukwaliteit: Handleiding*, Landbouwuniversiteit Wageningen, Wageningen.

van den Brand, P. (1992) *Een Regionale Beoordelingsmethode Milieukwaliteit voor het Thema Verzuring*, Scriptie Landbouwuniversiteit Wageningen, Wageningen.

van Woerkum, C. (2000) *Communicatie en interactieve beleidsvorming*, Samsom, Alphen aan den Rijn.

VNG (1999) *Milieureeks (kgm) Nr 9: Bedrijven en Milieuzonering,* VNG Uitgeverij, Den Haag.

VROM (1990) *Ministeriële Handreiking voor een Voorlopige Systematiek voor de Integrale Milieuzonering, deel 6*, Integrale Milieuzonering. Directie Geluid en Omgeving, Directoraat-Generaal Milieubeheer, Ministerie van Volkshuisvesting, Ruimtelijke Ordening en Milieubeheer, Den Haag.

Web Sites

Public Counter: *2000: http://www.ol2000.nl*

Gelderland: *http://geodata.prv.gelderland.nl/km/kaartenbak/*

Groningen: *http://www.groningen.nl/forum*

LODERWEB: *http://www.geo-informatie.nl/loderweb*

9 Flowmap: A Support Tool for Strategic Network Analysis

Stan Geertman[1], Tom de Jong[2] and Coen Wessels[3]

[1] URU/Nexpri, Utrecht University, Faculty of Geographical Sciences, P.O. Box 80.115, 3508 TC Utrecht, The Netherlands
[2] URU, Utrecht University, Faculty of Geographical Sciences, P.O. Box 80.115, 3508 TC Utrecht, The Netherlands
[3] Nexpri, Utrecht University, P.O. Box 80.115, 3508 TC Utrecht, The Netherlands

Abstract: This chapter explains the characteristics of Flowmap, a planning support system instrument dedicated to performing network analysis. Initially, discussion focuses on the history, primary purpose, development, capabilities and users of the Flowmap software. Special attention is given to the current and potential functionality of the system. Thereafter, three very different case studies are presented that illustrate how the capabilities of Flowmap have been used to solve real world planning problems. Some final remarks are made about the experience of using this planning support tool and about intentions for further development.

9.1 Introduction

Since the early 1990s, interest in information systems that can be used to store, manage, process, analyse and present geographical data has been booming. In tandem with the increase in the number of users of this technology, new avenues for application have been opened up. The expansion started in sectors like defence, forestry and mining and extended into geo-marketing, insurance, car navigation and telecommunications. Moreover, geographical information systems (GIS) expanded in size, functionality, and diversity (Tomlinson 1987; Burrough 1986; Ottens 1993; Steinitz 1993). More and more commercialised information systems were put on the market and new systems or updated versions of old systems have been introduced, offering novel capabilities (e.g. connection to the Internet) and well-known functions in a way hitherto unknown (e.g. 3D mapping/animation).

Despite the worldwide booming interest in GIS, its rise as an instrument to support policy in the field of physical planning has been rather limited (Geertman 1999). The use made of these instruments has been largely restricted to so-called *lower-order* (operation-oriented) planning activities, such as those performed by public utilities, where the technology is primarily used for the management of data relating to spatial phenomena (e.g. maintenance of sewage waste infrastructure) and the cartographic visualisation of these data (see, for example, Couclelis 1989; Webster 1993; 1994; Klosterman 1994; Worrall 1994). Although these activities are indispensable, they make only limited use of the potential of these systems. So-called *higher-order* (strategy-oriented) planning activities in the field of strategic physical planning primarily still lack a large-scale application of GIS. In particular, physical design and research activities within the phase of policy preparation are good examples of higher-order strategic physical planning activities where the technology still remains under-used as a policy support instrument. More analytical and modelling applications of the technology for the analysis and simulation of geographical phenomena are still under-used. The need for support of these kinds of strategic physical planning activity is substantial, given the growing complexity of society in general and of planning in particular. However, the application of GIS therein has been primarily restricted to pilot studies instead of fully-fledged real world applications (Geertman 1996).

This chapter describes a GIS-based planning support system (PSS) known as *Flowmap* and illustrates ways in which this instrument can support the process of finding solutions for so-called higher-order planning activities. Initially, we discuss the support instrument itself, the objectives for which it has been constructed, its development, its capabilities and its users. Special attention is paid to the current and potential functionality of this instrument. Thereafter, we take a closer look at its application and illustrate its potential by using three very different real-world planning case studies. These case studies illustrate the capabilities of *Flowmap* PSS. Some final remarks based on our experience of using *Flowmap* are made in conclusion.

9.2 *Flowmap*: The Tool

Flowmap is a stand-alone software package dedicated to the analysis of spatial interaction data. In this context, spatial interaction data is considered as all data that describe the movement of people, goods, energy or information through space on a regular or irregular basis usually via some form of transport network. A typical example of frequently and regularly occurring interaction is commuting whereas an example of intermittent and irregular interaction is population migration. *Flowmap* has the capabilities to select and display these kinds of data and uses various methods and models to analyse actual interaction data and predict potential interaction flows. These latter capabilities make *Flowmap* a very suitable tool to incorporate in PSS, especially in the fields of facility or service location and accessibility analysis. Although *Flowmap* contains many GIS functions, it is not designed as a full GIS package, but rather as an extension for specific spatial analyses that are difficult or impossible to realise within mainstream GIS.

Often the question is posed 'why go to all the trouble of developing and maintaining your own GIS tools when there are many readily available commercial GIS packages that offer more and/or better functionality?' The reason for this is simple. *Flowmap* is not a proprietary GIS, but a GIS-based instrument that focuses on functional distance (flows through space of people or goods) and related analysis. This topic is outside the main interest of most GIS users and therefore only gets limited attention from commercial vendors. Admittedly, most commercial packages now have some functions for finding the shortest path through a transport network and most of them are likely to operate faster than *Flowmap*. But finding the shortest path is usually considered the end of the analysis in proprietary GIS, whereas for typical *Flowmap* applications, it is just the beginning, enabling the user to go on to carry out accessibility analyses, gravity modelling and flow assignment without the need to write or buy any additional scripts or extensions. Moreover, algorithms dealing with spatial clustering based on functional distance, like Intramax or the Hitchcock redistribution of catchment areas, are not simple scripts. Thus, the additional efforts to keep developing and maintaining *Flowmap* as an independent system are considered to be justified.

Flowmap is designed, implemented and supported by the staff of the Faculty of Geographical Sciences of the Utrecht University (*http://flowmap.geog.uu.nl*). The actual development began in 1990, when there was a need for the automated generation of so-called 'desire line' maps: lines that indicate where people desire to go. In the period after the inception of *Flowmap 2.0*, various supporting functions were added and gradually the product developed into version 4.4, the final DOS version, in 1997. At that time, further development work was performed in close collaboration with an American company called Deltasystems (later Genasys II Ltd.) and an two-track approach was followed. Firstly, all algorithms were implemented directly in the commercial GIS package Deltamap (later Genamap). Secondly, the same functions were also made available on a PC-platform (*Flowmap*) to insure the possibility of application in all environments. With the demise of Genasys, it was decided to make *Flowmap* the only platform for further development. With the support of the CSIR (Council for Scientific and Industrial Research, Pretoria, South Africa), all transport network-related functionality was duplicated from Genamap to *Flowmap* and the first full MS-windows version was born in 1997 (*Flowmap 6.1*). Using *Flowmap* for background processing, the CSIR added more GIS functionality and a sketch planning tool to create its own PSS called ACCESSMAP, of which various versions are now being marketed in South Africa. The current *Flowmap 6.3* is available in two versions: the 'Educational' and the 'Professional' releases. The main feature of the Professional version is the possibility to run the software in batch mode instead of in an interactive mouse-driven fashion. For external communication purposes, *Flowmap* makes use of the well-known BNA geometrical data format, but has been extended with functions to convert both ArcView shapefiles and MapInfo MID/MIF files to this format. For the exchange of attribute data, *Flowmap* makes use of the widely used dBASE file format.

When looking at the capabilities and functions of *Flowmap* we can distinguish five fields as follows.

- Functional distance analysis – mainly applications of the Intramax procedure for spatial clustering of areas by maximization of intrazonal interaction. This option was discontinued after release 4.4, but will be reinstated in release 6.4. Figure 9.1 shows how a set of municipalities in the Zeeland province in the South West Delta region of the Netherlands are aggregated together to minimize the volume of commuting between different numbers of zones.

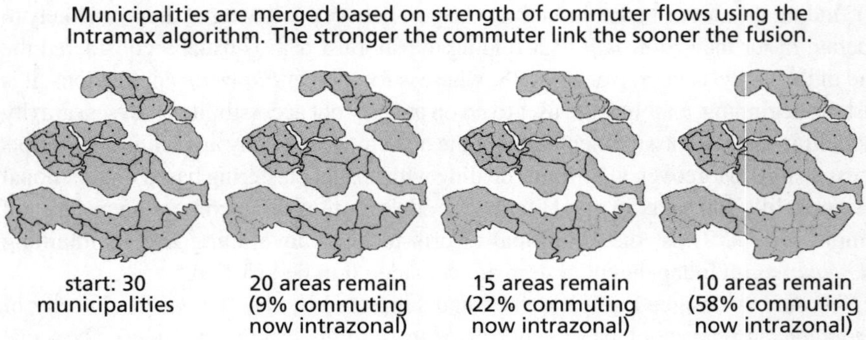

Fig 9.1: Functional distance analysis

- Display of interaction data, like commuting and migration flows, in the form of desire lines, for which diverse options for proportionality and shape are available. Four of these options are shown in Figure 9.2.

- Transport network analysis, including the creation of equal distance zones around centre locations ('zoning'), the allocation of network segments to nearest centres ('districting'), calculation of shortest (fastest/cheapest) paths between sets of origins (e.g. living areas) and destinations (e.g. working areas) and the creation of origin-destination distance tables based on shortest paths via air or via a transport network (Figure 9.3).

- Accessibility analysis (Figure 9.4), including various methods based on distance between origins and (nearest) destination. The first group of methods are catchment area analyses; in its basic form, all origins are simply allocated to the nearest destination. A typical example is the allocation of pupils to schools. In case destinations have only a limited capacity, average travel time may be minimized by applying a post-processing technique called 'Hitchcock relocation'. The second group of methods is the construction of 'location profiles' and 'catchment profiles'. The location profile involves the creation of a graph for each destination (e.g. a specific school) showing, in a cumulative fashion, how demand (e.g. pupils per origin location) grows as distance gradually increases. The catchment profile is a similar graph, but for the whole study area instead of each individual destination. From a location profile, one can read how many pupils live within say 15 minutes travel from a particular school; from a catchment profile one can read, for instance, how many pupils in the whole study area have access to schools within a given amount of travel time. Finally, there is the calculation of so-called potential accessibility measures. The simplest of these measures is the 'proximity count' that represents the number of origins (e.g. pupils) within a user-defined distance of each destination (e.g. schools). Very similar is the 'threshold distance' measure that indicates what the distance is to the nearest threshold amount of customers (not shown in Figure 9.4). Other accessibility measures include the classic 'potential value' in which pupils that live near the school get a greater weight than pupils that live further away and the 'gravity surface' in which competition between existing and proposed destinations is taken into account using gravity modelling (not shown in Figure 9.4).

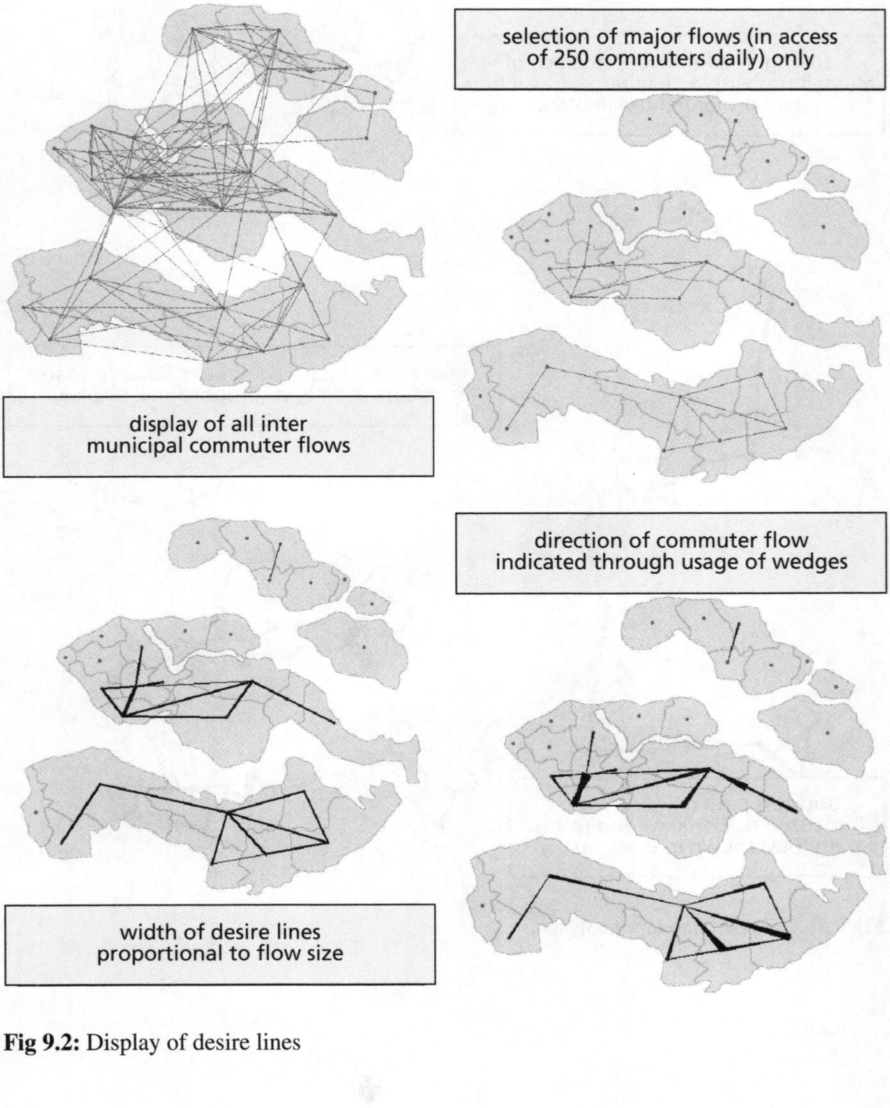

Fig 9.2: Display of desire lines

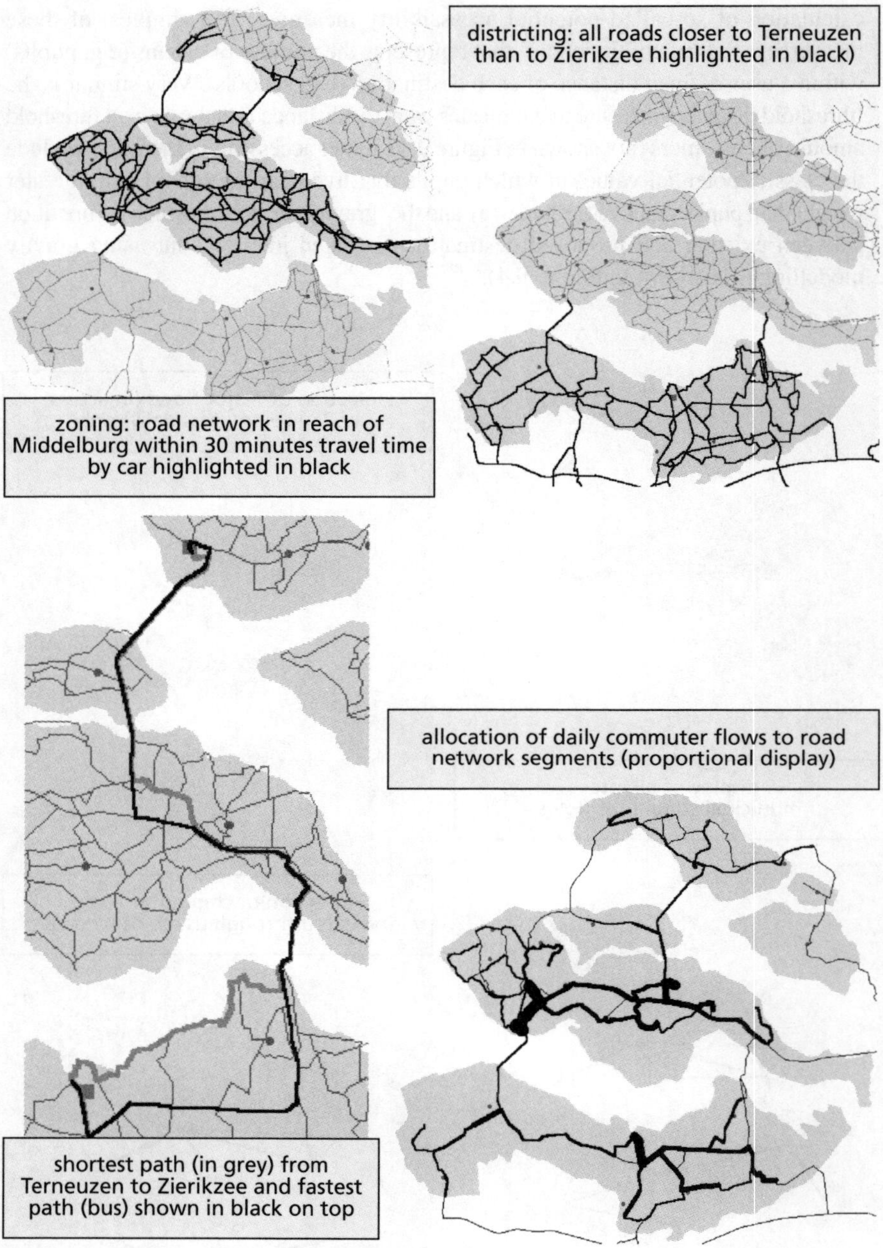

Fig 9.3: Transport network analysis

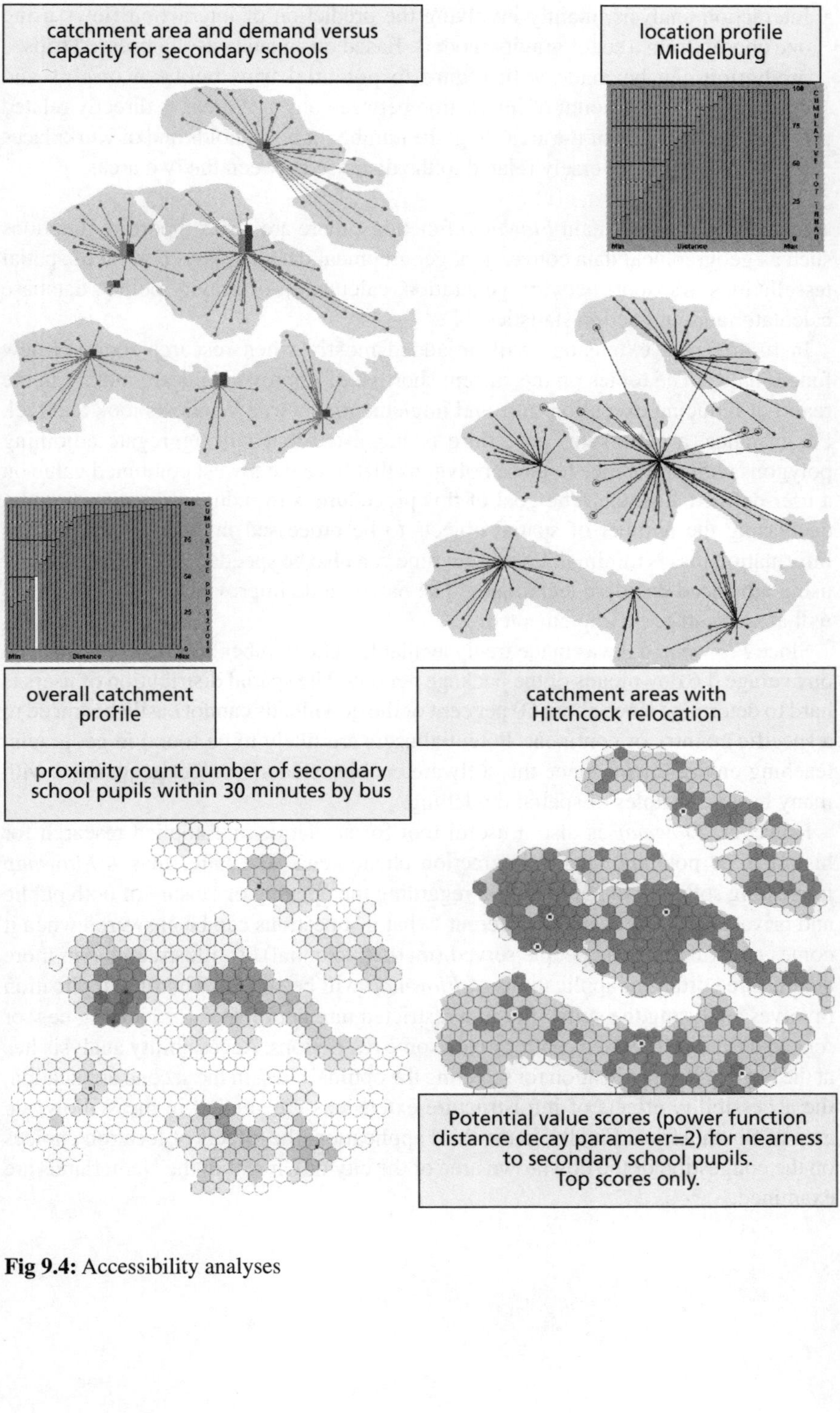

Fig 9.4: Accessibility analyses

- Interaction analysis, mainly involving the prediction of interaction flows using one chosen from a set of gravity models. Based on an analysis of observed trips, a prediction can be made with regard to potential trips between origins and destinations. The amount of interaction between any two areas is directly related to the attractiveness of the areas (e.g. the number of households and of workplaces respectively) and inversely related to the distance between the two areas.

In addition to these main *Flowmap* functions, there are also supporting functions such as geographical data conversion, geographical data display, creation of spatial tessellations, transport network generation, calculation of gravity points, database calculator and interaction statistics.

In future, new extensions will be added mostly when research requires new functionality. The topics on the current shortlist of improvements are mainly in the realm of reducing execution time and implementing a true Windows look and feel. For instance, a 'Minisum' procedure is suggested that will aggregate adjoining polygons with preference to those polygons that have the lowest combined value on a user-defined attribute. The goal of this procedure is to reduce execution time by decreasing the number of spatial objects to be processed in such a way that the information loss is minimal. Execution time can also be speeded up considerably by using advanced database techniques. The batch mode improved scripting facilities will at some stage be implemented.

Since *Flowmap 6.3* was made freely available in September 2000, there have been on average 3.6 downloads of the package per day. The spatial distribution of users is hard to determine since about 50 per cent of the downloads cannot easily be traced to a specific country or continent. Potential users are likely to be found in geography teaching environments since the software enables students to get acquainted with many basic principles of spatial modelling.

However, *Flowmap* is also a useful tool for academic and applied research for historical or potential spatial interaction phenomena. The latter makes *Flowmap* interesting software to apply in PSS regarding the opening or closure of both public and private services, as many different 'what if' questions can be answered when it comes to estimating the people served (market potential). To illustrate this in more detail, three different applications of *Flowmap* will be shown. The first application involves analysing the suitability of a restricted number of sites for locating new or renewed financial service centres given some restrictions. Accessibility analysis lies at the heart of this application for selecting the optimal sites. In the second application, the accessibility effects of infrastructure extensions are studied in the Willowvale area of South Africa. Finally, in the third application, the effects of location choices on the congestion of the downtown area of the city of Utrecht in the Netherlands are examined.

9.3 Case Study 1: Improving the Efficiency of Financial Service Centres

With the increased use of information and communication technologies (ICTs) – in particular, the Internet – more and more organisations in the private and public sectors are redefining the ways and levels of accessibility of their services for their customers. Especially within the private sector, this redefinition often results in a strategy to improve efficiency through the concentration of services and personnel into fewer but larger service centres, and the closure of small service centres.

In this real-world case study, an international financial organisation with its head quarters in the Netherlands requested such an efficiency-improving operation. The organisation served its Dutch customers through 16 financial service centres (FSCs) located all over the Netherlands. Due to the increasing diversity of requests for services (and the impossibility for small FSCs to handle these in an appropriate way) and the increasing possibilities of using the Internet to provide efficient, 24-hour communication with customers, the organisation started an efficiency operation for the period 2001-2004. Advisers from ETIN (Spatial and Socio-Economical Consultancy) and NEXPRI-Utrecht University (Scientific Geo-ICT Consultancy) were asked to assist in this operation by providing sufficient background data and information to perform this efficiency operation in an appropriate way.

In fact, the operation to improve efficiency had four main prerequisites:

- the maximum number of FSCs is 16 (current situation) and the minimum is two;
- each FSC should have a minimum of 100 employers;
- the regional labour market should be taken into account in proposals for closure or extension of existing FSCs; and
- for all 1,521 FSC employees in the Netherlands, 'social regulations' are applicable; the main implications of this is that the travel times from home to work for the present FSC employees below a certain salary level are not allowed to exceed a fixed maximum.

The research consisted of two main parts, the outcomes of which were put together in a multi-criteria analysis (MCA) before presenting them to the board of directors of the financial organisation. In the first part of the work, the ETIN advisors practiced a so-called 'demand approach' in which economic considerations and the current and potential labour markets were taken as a starting point. For each of the 16 existing FSCs and the eight potentially new FSCs, their actual/potential function as FSCs with a minimum of 100 employees was investigated. In particular, the present and potential ability of the labour market to provide the FSCs with new and skilled personnel in the case of an extension, on the one hand, and its present/potential ability to adopt superfluous personnel in case of closure of the FSCs, on the other hand, were examined. This part of the research resulted in an overview of pros and cons of the labour market of each of the actual and potential FSCs.

In the second part of the research, NEXPRI-Utrecht University carried out a so-called 'supply approach' in which the social regulations and the present labour force of the FSCs were taken as a starting point. Therefore, several network analyses were performed with the help of *Flowmap* and ArcView. Initially, the employees of the FSCs were assigned to four digit postcode areas within the Netherlands, based on their home addresses. Next, the locations of the present FSC were digitised. In addition, these locations and the centroids of the postcode areas were linked to both a road network and a public transport network. These networks possess an attribute that indicates, for each segment, the road distance or the public transport travel times respectively. After a visual check of the correctness of these shortest on-the-fly links (sometimes this appears to result in very strange links) and their subsequent correction, the network analyses were performed. The accessibility was determined from each of the postcodes to the nearest FSC with a maximum value of 75 minutes for public transport and 40 kilometres for private transport. This was undertaken for all existing 16 FSCs together, and also for a situation in which just four upgraded FSCs would remain. The analysis shows how many employees from each origin can or cannot reach the nearest FSC within the given distance/travel time maximums (Figures 9.5 and 9.6).

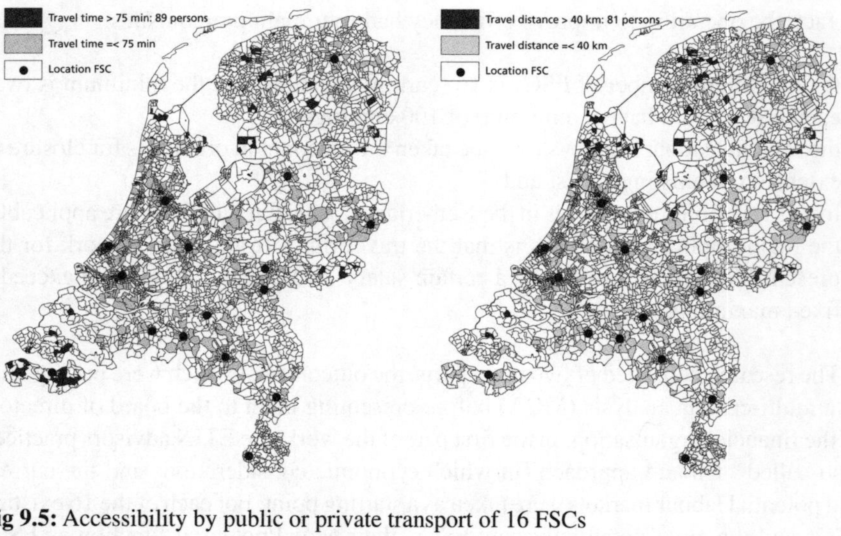

Fig 9.5: Accessibility by public or private transport of 16 FSCs

After putting together the outcomes of both parts of the research within a MCA framework, just four of the original 16 FSCs were suggested on the basis of economic and accessibility considerations to be prime centres in which to concentrate the financial service activities of the organisation. These four centres were Amsterdam,

Rotterdam, Eindhoven and Groningen. Moreover, social considerations (poor labour market) indicate that two additional FSCs (Enschede and Terneuzen) were required. In addition, for location reasons, the region Arnhem/Nijmegen was recommended to the board of directors of the financial organisation. In total, these seven FSCs accommodate about 91 per cent of the present market. In the end, the board of directors accepted these recommendations – with the exception of the latter – and reduced the number of FSCs from 16 to six.

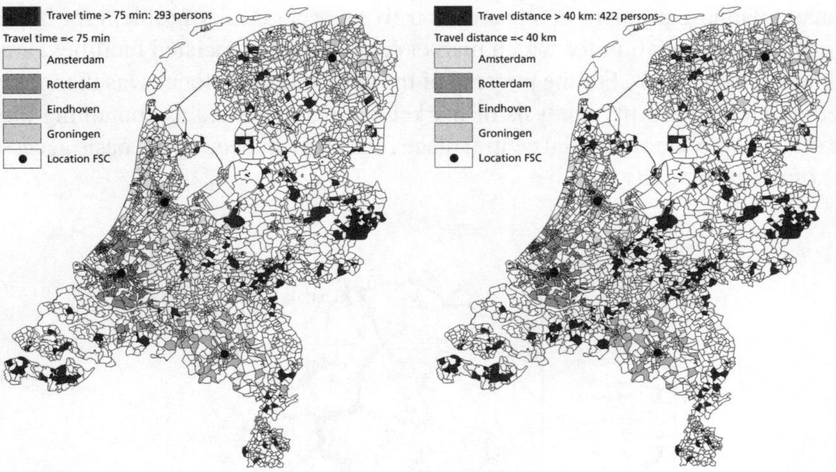

Fig 9.6: Accessibility by public or private transport of 4 FSCs

9.4 Case Study 2: Analysing the Accessibility Effects of a Proposed New Road

This case study involved using *Flowmap* in a typical rural development planning context. Figure 9.7 shows the general study area of the Wild Coast in South Africa that has a rural population of approximately 475,000. Most of the existing district towns indicated on the map have a population of around 1,000 or less. Only Butterworth and Idutywa have higher numbers of residents. Given that there are about 900 villages, and facilities such as primary and junior secondary schools, clinics and even hospitals seem to be fairly arbitrarily scattered, this signifies a poorly developed service centre hierarchy.

One of the reasons for the present alignment of the N2 highway from East London to Umtata via Butterworth is that the intervening area on the coastal side is dissected by a relatively large number of rivers that run into the spectacular coastline. In this area, there are only a limited number of river crossings, mostly located approximately 20 km inland. The roads are generally in a poor condition which, together with the absence of a continuous route close to the coast, means that tourists tend to visit only one, or at most two places during a vacation or tour of the area. As part of the commitment to rebuilding South Africa and achieving sustainable economic growth a Wild Coast Spatial Development Initiative (SDI) has been initiated which aims to stimulate the development of coastal tourism, forestry, vegetable and fruit production through establishment of market centres and a coastal tourist/market access route. The main planning problem in this area is truly an accessibility problem for large sections of the population for which market centres (and associated facilities) are out of reasonable reach. For the purpose of the case study, the focus was therefore placed on an accessibility analysis of market centres, regarded as containing the basic set of economic and social central place facilities for meeting the basic access needs of the community.

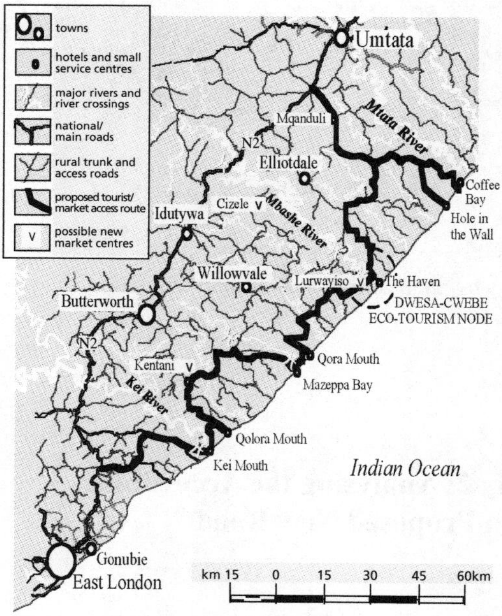

Fig 9.7: The Wild Coast study area

Accessibility of existing market centres

The populations of 188 out of 900 villages in the Wild Coast area have to travel more than one hour to reach the nearest available existing market centre. To facilitate decision support, this situation can be translated into an accessibility map. Figure 9.8 gives an overview of the accessibility to the four current market centres in the area: Elliotdale, Idulywa, Willowvale and Butterworth. Komga and Umtata are major towns just outside the study area but are included in the analysis. Based on travel times measured from the existing road network, a catchment area analysis was undertaken to establish which villages and what proportion of the population could be reached within different access time ranges.

In Figure 9.8, isochrone lines (enclosing areas within the same travel time interval from the nearest centre) are drawn at 30-minute intervals around the four centres. The influence of the northwest to southeast flowing rivers on the accessibility surface can be deduced from the lack of any clear-cut circular patterns around the centres. If we assume that these centres are the only places where it is possible to obtain a basic set of central place goods and services, it is instructive to note that only 58 per cent of the population are within the desired access time of 45 minutes. The existing network means that 77.4 per cent are within one hour of a centre and 91.5 per cent are within 75 minutes.

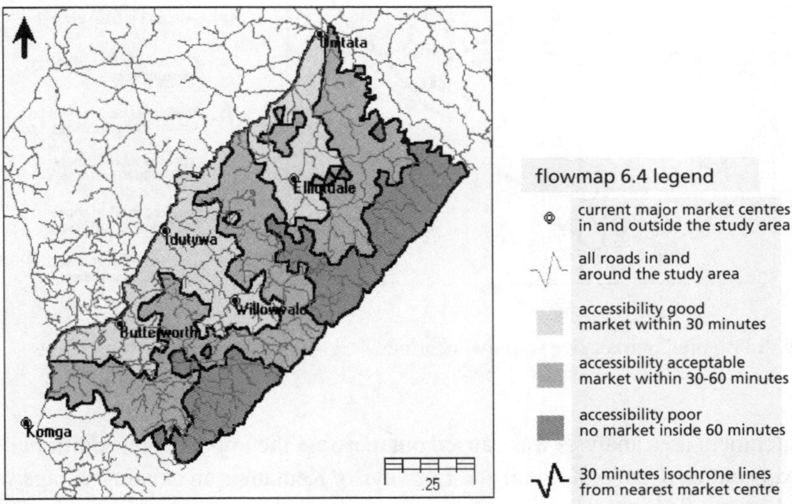

Fig 9.8: 30-minute isochrone lines around the markets

Scenario 1: Increasing the number of market centres

By adding new market centres, obviously the overall travel time for all villagers will decline, as will the travel time for villagers living in disadvantaged areas. If the planning of the location of these new market centres was to take place *de nouveau,* decisions could be further supported by using *Flowmap* through the generation of trend surface maps, indicating the most suitable locations for new development.

Figure 9.9 shows as background information the current market centres, all officially proposed sites and the area outside a 60-minute reach from current centres. In addition, the trend surface for new development is displayed; a darker colour indicates a higher suitability for a new market centre development, based on a proximity count analysis for inaccessible villages only. Note that most of the best locations are found, in fact, within the 60-minute reach of current centres, in places where roads connect different clusters of inaccessible villages. The planning reality, however, is that new potential locations for market centres have already been selected based on criteria other than geographic accessibility. Adjacent to the existing centres, a small number of villages are to be 'upgraded' and given the role of market centres.

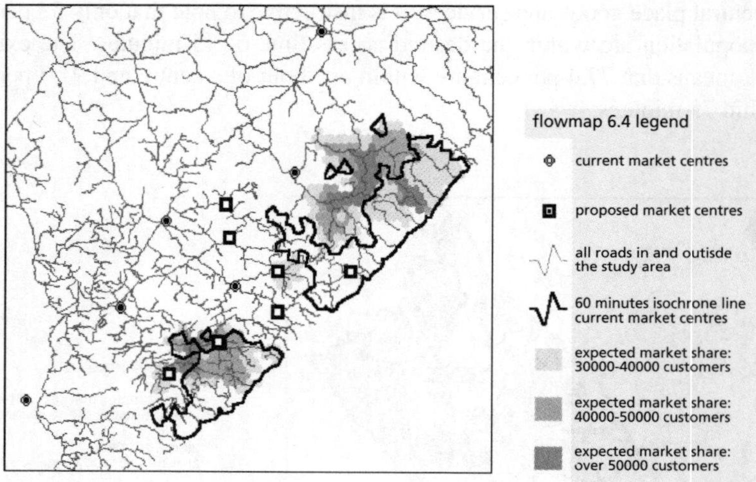

Fig 9.9: Potential market size for new locations

Catchment area analysis was carried out to assess the impact of an additional three market centres: Cizele, Kentani and Lurwayiso. Kentani is an existing village which may be provided with all the facilities of a fully-fledged, permanent market centre. In the case of Lurwayiso, the centre could play an important role as a market where handicrafts and fruit are sold to tourists from the Dwesa-Cwebe eco-tourism node, and local produce is exchanged among the local population. Cizele was selected as market centre with a mainly and social/educational role in an area which appeared to be fairly isolated. Figure 9.10 shows that a significant reduction of distance to the nearest market centre is apparent for each village in the central and southern part of the study area (e.g. around Willowvale/Kentani) where two of the three new centres are located. In the northeastern coastal zone (i.e. towards Coffee Bay), the impact of adding the three new centres is, understandably, quite restricted. The percentage of people within an hour access time only increases from 77.4 per cent to 85.9 per cent.

Of course, a far better overall result might be achieved by choosing a village more to the northeast of the study area instead of Cizele or Lurwayiso. Alternatively, Lurwayiso could be better connected to the northeast by construction of the coastal road, the effects of which we will show in the next scenario.

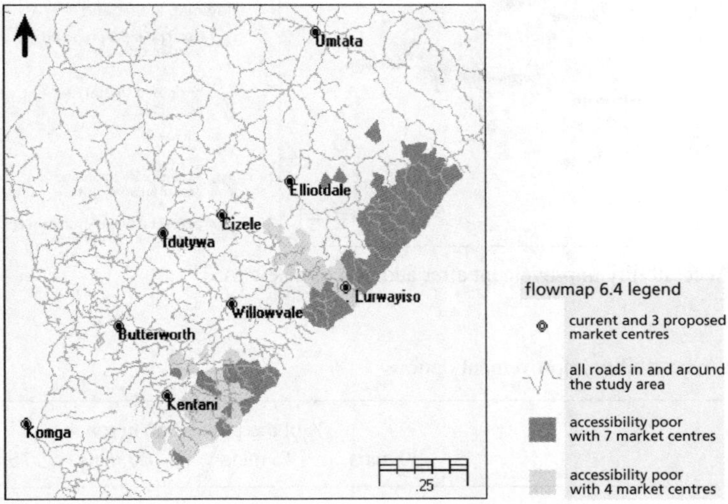

Fig 9.10: Accessibility improvement after adding new markets

Scenario 2: Addition of a new road to facilitate development

An interesting question that the population-based analysis cannot alone address concerns the effect of an improvement of the infrastructure in the area on accessibility. In this context, the local authorities in the region are very interested in the development of a coastal road which could provide all weather, yet still slow-moving, access to a series of tourist destinations and, at the same time, provide a channel for both the local marketing and export of garden produce. The effects of such a new road turn out to be hardly noticeable in terms of accessibility improvement, unless a market centre scenario is considered where the village of Lurwayiso is a proposed site. In that case, the village would experience a significant increase in market share. But in comparison with the original situation, Figure 9.11 shows that still only the nearest villages to the northeast of Lurwayiso would benefit from the coastal road. The number of villages outside the above-discussed one-hour limit only decreases from 114 to 95 and Table 9.1 shows that overall effects are also very limited. So improving the accessibility of a new market centre is hardly an argument to justify the high cost of building a new coast road.

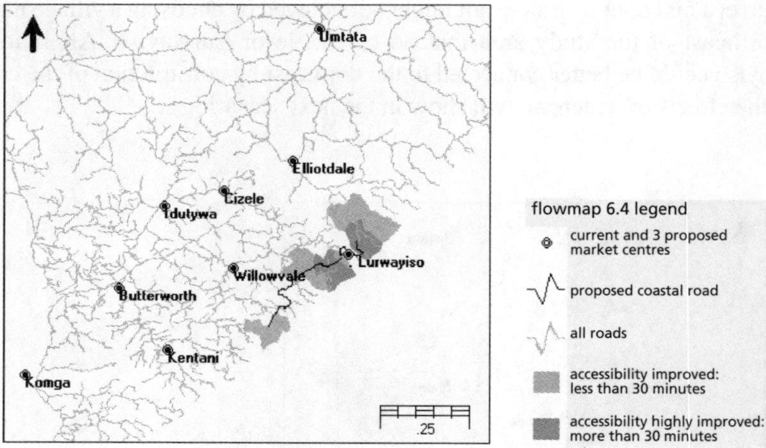

Fig 9.11: Accessibility improvement after addition of a new road

Table 9.1: Accessibility improvement options

Scenario	% of the population in reach			
	30 mins	45 mins	60 mins	75 mins
Present situation (4 centres)	34.2	58.0	77.4	91.5
7 centres (3 extra centres)	46.1	68.3	85.9	95.3
7 centres plus coastal road	47.7	72.7	89.2	96.8

9.5 Case Study 3: Analysing the Effects of a Location Decision on Transport Network Usage

Since the foundation of the first Islamic primary school in the Netherlands ten years ago, there has been a rapid growth in the number of these schools and their pupils. There are now 29 Islamic primary schools with approximately 10,000 pupils and the demand is still growing. Recent research in Rotterdam has shown that 10 per cent of the households wanted to send their children to an Islamic primary school. That would mean 8,500 pupils, which is ten times the number of pupils in these schools in Rotterdam. One out of thirteen households stated that they also wanted to send their children to an Islamic secondary school. The constitutional freedom of education in the Netherlands enables all religions or philosophies of life to found a school that will be paid for by the Government. This holds true also for the city of Utrecht that has a high percentage (13.1 per cent) of Muslims in the population. Now the question arises whether it is also possible to found Islamic secondary schools.

In this example (Floor *et al.* 2001), we use *Flowmap* to study the effects on traffic loads resulting from the location of a new Islamic secondary school. The analysis will be based upon the spatial distribution of the two largest groups of Muslims in Utrecht, Moroccans (8.5 per cent of the population) and Turks (4.6 per cent of the population). The population of Moroccan and Turkish origin resides mainly in the social housing sector in the city of Utrecht. The Moroccans (19,685 in 2000) show a concentration in the southwest of Utrecht and the northwest. The percentage of Moroccans in the population of the neighbourhoods differs greatly with the highest value of 40.3 per cent (Bolt 2001). The Turks (10,670 in 2000) in general have the same spatial distribution as the Moroccans with a greater emphasis on the west and northwest of the city. The neighbourhood with the highest value (18.6 per cent) is the same as in the case of the Moroccans. The spatial distribution of the two population combined groups as shown in Figure 9.12 is in absolute numbers and based upon a 100 x 100 metre grid system for the city of Utrecht.

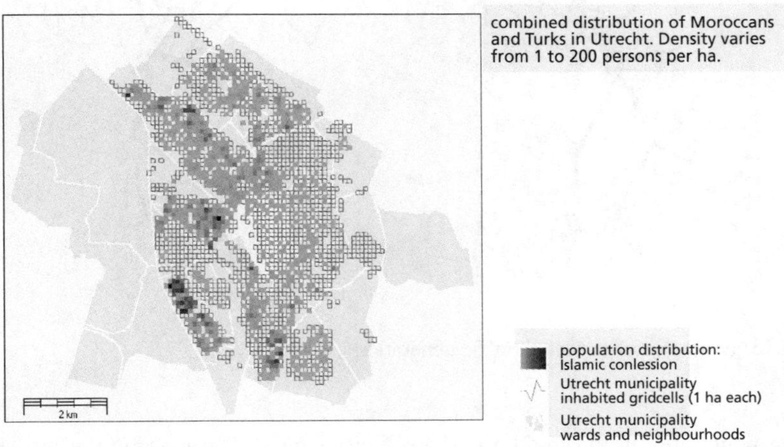

Fig 9.12: Spatial distribution of Morrocans and Turks in Utrecht

From this spatial distribution, it is not directly clear where to put the new school and planning support methodology can be applied to support the decision-making process. Apart from the more usual criteria like minimising overall travel time, a special criterion has also been addressed. In Utrecht there is a heavy congestion in the old inner city centre. The general policy is therefore that new services be located in such a way that any additional traffic generated should avoid the city centre as much as possible. In this example, we will focus on the application of *Flowmap* to address this special criterion.

There is no school busing system in the Netherlands and schoolchildren of this age will predominantly travel by bicycle. Therefore *Flowmap* has been used to analyse inner city traffic loads generated by the various potential school locations. A digital network of bicycle pathways is available for the whole of the Netherlands from the National Traffic Bureau. First, the whole build-up area of the city was divided in 1 hectare cells, and all cells that were potentially suitable for building a school were selected. A set of 3,600 regularly-spaced potential sites remained. Analysing these alternative sites one by one, we assume that schoolchildren will take the shortest (bicycle-)path from their home to the school at a site. After calculation of all shortest paths, we can allocate expected traffic to actual road segments. Figure 9.13 shows the results for two of the alternative sites.

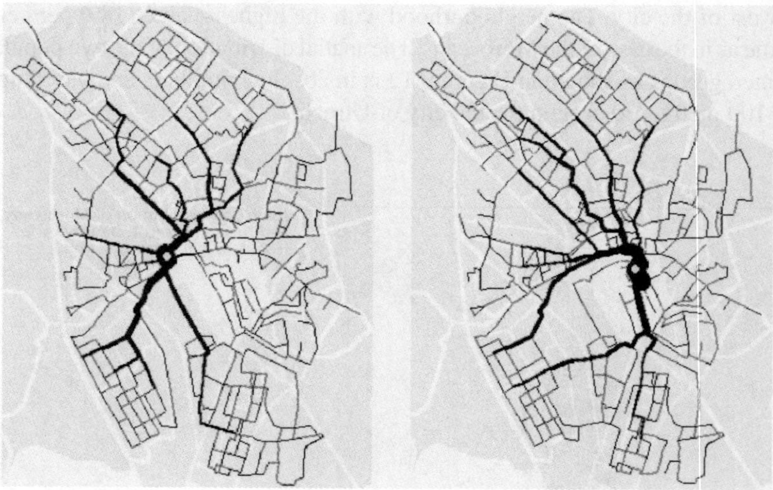

Fig 9.13: Examples of expected traffic generation

The white area in the middle is the old city centre. The map on the left shows traffic loads where the school is situated to the west of the centre, the right hand map shows the same for a central school location. The thickness with which road-segments are displayed is proportional to the traffic load. Once traffic loads have been allocated, we can distinguish between road segments inside and outside the city centre. When we multiply the traffic load with the length of the road-segment for all roads inside the city centre and divide by the total number of schoolchildren, we get for each alternative location the average road-length travelled through the city centre. This score is used to compare the various potential sites. The resulting numbers vary from 80 to 1,760 metres downtown travel per school child. Figure 9.14 shows the spatial variation over all potential locations.

The lowest (best) scoring location (80 metres) is to be found in a scarcely populated area to the west of the top of the downtown area whereas the highest scores are concentrated in the eastern half of the downtown area. In general, the whole southeast section of town generates relatively high scores and therefore is disqualified as a potential site. Of course, reducing the inner city traffic load is only one of the criteria involved in searching for a suitable site. The results from this analysis can be combined (for instance by using MCA) with similar results, addressing criteria to minimize overall travel time for schoolchildren and to minimize the worst case travel time. In the end this will generate an overall picture that indicates the suitability of new school locations from both accessibility and traffic perspectives.

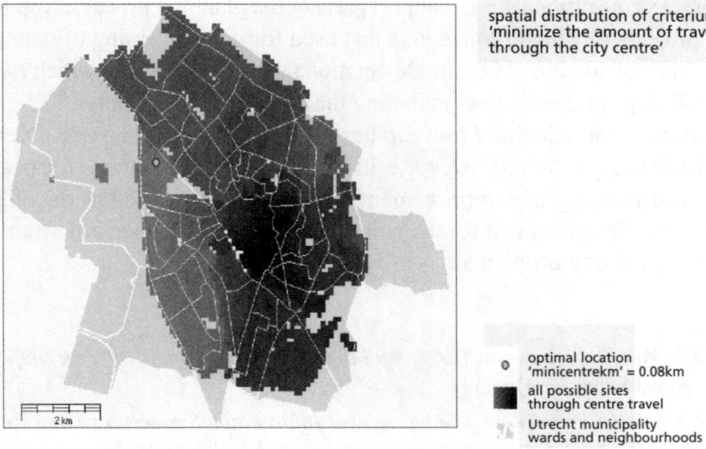

Fig 9.14: Spatial variation in generated downtown traffic

9.6 Concluding Remarks

In the introduction it was stated that GIS still lacks large-scale application in the context of so-called higher-order planning activities in the field of strategic physical planning. This is despite the fact that the need for support of these kinds of strategic physical planning activities is substantial, given the growing complexity of society in general and of planning in particular. In this chapter, we have indicated and illustrated in three case studies the value of the combined use of GIS and a GIS-based PSS named *Flowmap*. From this, several conclusions can be drawn.

Whilst *Flowmap* is a PSS instrument that is capable of producing very valuable results, some important reservations should be noted. Firstly, the quality of the output (analysis and modelling results) is primarily dependent of the quality of the input (data), so when the data are poor, the results will also be limited. Secondly, *Flowmap* adds very useful network-based capabilities and functions to proprietary GIS software and can be considered in that respect as a very functional and valuable (part of a) PSS, dedicated to support planning practice but is not embedded within proprietary GIS which must be acquired independently. Thirdly, although *Flowmap* is Windows-based, it is primarily designed for expert use. As a consequence, new looks and feels to improve user friendliness are added all the time, mostly in answer of requests from educational applications. Fourthly, *Flowmap* is not a finished product, but will continue to grow within its scope by adding and expanding functionality on demand of planning practice and geographical research. Finally, *Flowmap* is not aimed at full automation of the planning process; its rationale is to support parts of the planning process. Applications vary from projects in which *Flowmap* is just used for pre-processing of data like for tessellation and calculation of Origin-Destination tables to projects in which *Flowmap* is used for all steps of the analysis including the display of end results.

In general, we conclude that *Flowmap* has proved its worth in a wide diversity of real world planning contexts. Despite its somewhat rudimentary form and the continuous restructuring and improvement that have characterised its development, it is a very valuable instrument for the solution of so-called higher-order (analytical and strategic) planning problems.

References

Bolt, G. (2001) *Wooncarrières van Turken en Marokkanen in ruimtelijk perspectief*, Faculty of Geographical Sciences, Utrecht.

Burrough, P.A. (1986) *Principles of Geographical Information Systems for Land Resources Assessment,* Monographs on soil and resources survey 12, Clarendon Press, Oxford.

Couclelis, H. (1989) Geographically informed planning: Requirements for planning relevant GIS, Paper presented at the 36th North American Meeting of the Regional Science Association. RSA, Santa Barbara.

Floor, H., Jong, T. de and Ritsema, van Eck, J. (2001) The adaptation of optimization techniques for usage in combination with Planning Support System, including a case study of siting an Islamic secondary school in the city of Utrecht, the Netherlands, in Proceedings CUPUM 7, Honolulu.

Geertman, S. (1996) Methodological issues of the Geo-IT application in planning, in Rumor, M., McMillan, R. and Ottens, H. (eds.) *Geographical Information; From Research to Application through cooperation*, IOS Press, Amsterdam/Oxford/Tokyo/Washington-DC, pp. 579-583.

Geertman, S. (1999) Geographical information technology and physical planning, in Stillwell, J., Geertman, S. and Openshaw, S. (eds.) *Geographical Information and Planning*, Springer Verlag, Berlin, pp. 69-86.

Jong, T. de, Ritsema, van Eck J. and Floor, H. (2000) *Flowmap 6.3. Manual and web site: http://flowmap.geog.uu.nl,* Faculty of Geographical Sciences, Utrecht.

Klosterman, R.E. (1994) International support for computers in planning, in *Environment and Planning B*, 21: 387-391.

Naude, A., Jong, T. de and Teeffelen, P. van (1999) Measuring accessibility with GIS-tools: a case study of the Wild Coast, South Africa, *Transactions in GIS*, 3 (4).

Ottens, H. (1993) The application of GIS technology in Europe, in Harts, J., Ottens, H. and Scholten, H. (eds.) *EGIS '93 Proceedings*, EGIS Foundation, Utrecht, pp. 1-8.

Steinitz, C. (1993) Geographical information systems: a personal historical perspective, the framework for a recent project, and some questions for the future, in Harts, J., Ottens, H. and Scholten, H, (eds.) *EGIS '93 Proceedings*, EGIS Foundation, Utrecht.

Tomlinson, R.F. (1987) Current and potential uses of geographical information systems: the North American experience, in *International Journal of Geographical Information Systems*, 1: 203-218.

Webster, C.J. (1993) GIS and the scientific inputs to planning. Part 1: description, in *Environment and Planning B*, 20: 615-748.

Webster, C.J. (1994) GIS and the scientific inputs to planning. Part 2: prediction and prescription, in *Environment and Planning B,* 21: 145-157.

Worrall, L. (1994) The role of GIS-based spatial analysis in strategic management in local government, in *Computers, Environment and Urban Systems*, 18(5): 323-332.

Web Site

The current version is Flowmap 6.3. The distributed version is the so-called Educational Release. After sending in a registration form this can be upgraded to the Professional release for a small fee: *http://flowmap.geog.uu.nl.*

10 Geovisualisation for Planning Support Systems

Bin Jiang[1], Bo Huang[2] and Vit Vasek[3]

[1] Division of Geomatics, Institutionen för Teknik, University of Gävle, SE-801 76 Gävle, Sweden
[2] Department of Computer Science, Keele University, Staffordshire, ST5 5BG, UK
[3] Laboratory on Geoinformatics and Cartography, Masaryk University, Brno, Czech Republic

Abstract: Planning support systems can be defined as spatial decision support systems consisting of three important components, namely: data, models and geovisualisation (Klosterman 1999). In this chapter, we focus on geovisualisation in planning decision-making, in which it has two roles: firstly, collaboration, and secondly, exploration, particularly when it is integrated with the other two components. We report several prototypes of geovisualisation for exploration and collaboration in the context of urban and environmental planning. These prototypes integrate recent developments in internet GIS, geospatial virtual environments and multi-agent simulations. The techniques reported here have been used to develop a working planning support system based on the Internet.

10.1 Introduction

Planning Support Systems (PSS) can be characterised as spatial decision-making systems with particular application for planning, which involves a wide range of professionals with diverse backgrounds and the general public concerned. Current developments in GIS towards 'Group Decision-Making Systems' and 'Public Participation GIS' (PPGIS) seem to have the same target. Due to the diverse nature of people involved in the planning process, specialist software and platforms certainly do not meet the various requirements from different people. However, all professionals and the general public as human beings are used to visual approaches, just as planners use maps and sketches to communicate and exchange ideas about planning scenarios. We concentrate on geovisualisation, which provides efficient and effective visual methods and tools for understanding complex geographic phenomena in planning decision-making activities.

Researchers in urban planning are among those who use geovisualisation extensively to promote their professional activities. Various experiments have also been conducted towards exploratory visual tools for urban planning, mainly using multimedia integrated with geospatial databases in GIS (e.g. Shiffer 1995, Batty *et al.* 1999, Gouveia and Camara 1999, Laurini 2001). Several studies have been made on using maps and visual tools for collaboration. Florence *et al.* (1996) proposed the metaphor of the 'wallboard' for group decision support systems. Rinner (1999) adopted an argumentation map in a web-based planning support system. More recently, MacEachren (2000, 2001) has presented a comprehensive overview towards collaborative geovisualisation, focusing on using maps and GIS as a mediator for collaboration. Thanks to recent developments in VRML and various other 3D standards, nowadays large-scale geographic databases can be easily visualised and disseminated over the web. GeoVRML, as a new extension of VRML97, implements some more nodes which are particularly suitable for geographic applications, and overcomes some limitations of VRML97 (Reddy *et al.* 1999).

The research reported in this chapter is oriented towards two important functions of geovisualisation in planning processes: exploration and collaboration. There are various ways of undertaking visual exploration through interactive techniques such as dynamic geographic representation and multimedia with nonlinear hyperstructure. The exploratory geovisualisation we are going to elaborate here adopts different techniques. The first technique is the adoption of geospatial virtual environments, through which its counterpart real world environment can be explored extensively, and planning scenarios can be previewed with different levels of detail. The second technique is to integrate analytical models for visual interrogation. Thus, modelling processes can be explored visually and intuitively. To the best of our knowledge, this kind of exploratory visualisation has not yet received much attention, with a few exceptions (e.g. Bishop and Karadaglis 1996). Furthermore, these two kinds of exploratory geovisualisation can be realised through the Internet for collaboration in planning processes. To illustrate this, the chapter introduces several experimental prototypes in the context of urban and environmental planning.

The remainder of this chapter is organised as follows. Section 10.2 focuses on two important functions of geovisualisation in planning, namely exploration and collaboration. Section 10.3 presents several geospatial virtual environments (GeoVE), which can be used for exploration of real-world environments. Section 10.4 introduces exploratory analysis environments for conducting what-if experiments and Section 10.5 deals with geovisualisation servers on the Internet for planning tasks. Finally, in Section 10.6, we draw some conclusions.

10.2 Geovisualisation for Exploration and Collaboration in Planning

Geovisualisation in this context is simply referred to as geographic or cartographic visualisation. But it integrates knowledge and expertise from modern cartography, GIS, scientific visualisation (McCormick *et al.* 1987), information visualisation and virtual environments (Chen 1999). Highly interactive and dynamic exploratory tools can facilitate planning decision-making. Through these tools, end users can thoroughly explore real-world environments, and new design scenarios can be previewed through well-constructed virtual environments. More importantly, such effective and efficient geovisualisations can be implemented in a decentralised context, which supports collaborative decision-making. While geovisualisation is integrated with analysis and simulation models, what-if modelling can be carried out for planning tasks.

Geovisualisation, or visualisation in a more general sense, sounds like an interface to geospatial database, or a means to show modelling results. However, it has several different roles in data management and scientific activities. Inspired by exploratory data analysis, DiBiase (1990) identifies four roles in two distinct domains for geovisualisation. The first is the private domain, in which professionals, such as scientists or engineers, use geovisualisation for visual thinking. Two relevant roles are exploration and confirmation. Then what is discovered through visual thinking is synthesised and presented to the lay professionals or the general public in the public domain – the second domain. This geovisualisation conception has been extended by MacEachren (1994) by adding one more interactivity dimension. Obviously planning activities involve both professionals and the general public. In this context, we focus on two important roles of geovisualisation in planning processes, i.e. exploration and collaboration. It can be used for exploring planning sites, previewing and communicating planning scenarios.

Conventional maps provide fewer capabilities for exploration, as they are static. The map reading process can be thought of as *reactive*, i.e. maps are passive and readers have to make an effort to understand the map symbols and what they represent. This is very similar to the way human beings react in the real world. Another reason why conventional maps have fewer exploratory capabilities is that they serve as geographic databases. In other words, visual representations and geographic information are not separate. This situation has changed rapidly since the advent of interactive graphic techniques. Particularly in the GIS environment, maps no longer serve as databases. Maps displayed on the computer screen are supposed to be *interactive*, although maps are still considered to be static to some extent. Such interactivity can occur in various ways, such as map display layer by layer, clickable map, and graphic manipulation like zoom and pan, or alternatively changing colours or symbols of maps. In this context, visual displays or geovisualisations in a more general sense are considered to be visual interactive tools to facilitate planning processes, rather than as geographic databases as discussed for instance in Craig and Elwood (1998).

The advent of multimedia has brought far more opportunities for the exploration of geospatial data, as an important complement to maps. The technology has been further pushed forward by the Internet, WWW in particular, which integrates all kinds of multimedia formats. With respect to the limitation of interactive graphics in that the interaction between system and user is constrained by a limited number of commands pre-defined by system developers, Buttenfield (1993) proposed the notion of *proactive graphics*, to refer to these exploratory capabilities not expected by system designers. She argues that with proactive graphics like multimedia with nonlinear hyperstructure, data should be explored in a manner consistent with the associative power of the human intellect. She and her colleague implemented a working prototype, focusing on spatial query in a form of visual exploration of biogeographical data (Buttenfield and Weber 1994).

The second role of geovisualisation in planning is collaboration, which has much to do with visual communication. Collaboration occurs between various groups at different stages of planning, for example, between various professionals and between professionals and lay professionals or the general public. On the one hand, various professionals have to understand each other (professionals likely to be involved include planners, cartographers, environmentalists, politicians and investors). On the other hand, the general public should be consulted about environmental issues. In this case, it seems more appropriate to refer to collaboration as participation. Geovisualisation facilitates collaboration through providing interactive and dynamic visual tools. For instance, 3D photorealistic representations are used to show urban redevelopment; and dynamic computer simulations are used to show possible pollution diffusion over the next few years. These various techniques can dramatically increase the degree of understandability of PPGIS, should they be implemented (Kingston *et al.* 2000).

Collaborative planning can be conducted in both centralised and decentralised manner. Centralised collaboration usually takes place in a committee room or a computer facilities room. GIS can be adopted for such tasks, but it needs redevelopment towards group decision support systems (Armstrong 1994). As far as public participation is concerned, collaboration has to be done in a decentralised manner. Both centralised and decentralised collaborations can occur at the same time or at different times. Thus, it leads to four different scenarios of collaboration: same time-same place, same time-different place, different time-same place, and different time-different place. Corresponding to these scenarios, various new technologies have been discussed as to how they facilitate public participation in an urban planning context (Shiffer 1998). The rapid development of internet technology provides a good platform to realise all these scenarios. The Internet integrates various interactive and proactive techniques that can be used for planning processes. Participants can be involved in discussion and exchange of ideas through internetworked GeoVE. More importantly, the Internet can overcome time constraint, leading to an asynchronous and long time span debate for some controversial planning tasks.

10.3 Exploratory Geospatial Virtual Environments

Following the line of proactive graphics, geospatial virtual environments (GeoVE) opens far more possibilities for proactive exploration. Firstly, VE provide realistic or photorealistic representations, which are hard to achieve with both conventional maps and multimedia. Bill Jepson, a pioneer in urban simulation, once posed an interesting question – *"if a picture is worth one thousand words, what's a 3D model worth?"* (Potel 2000). This question presents a great challenge for planning professionals and developers of PSS. Secondly, the interactivity of VE is very high. Using a head-coupled viewing device like a Head Mounted Display (HMD), participants can be fully immersed in VE with the highest sense of presence and it allows thorough exploration of complex datasets or phenomena. Most often, low cost VE is a non-immersive one based on desktop PC through the Internet browser, but it can be implemented in a more distributed manner. Our experiments are more targeted to the latter non-immersive platform.

Realistic geovisualisation can involve important tools for communicating with lay professionals and the general public in PSS. Assuming in an urban planning task, one sketches a 2D plan and converts it automatically into a 3D scene. With the 3D scene, planning scenarios can be investigated in various exploratory ways, such as zoom, pan, fly over, walking though, rotate or change perspective. Should the 3D scene be published over the Internet, it could attract wide public debate and participation. Several years ago, in a small town in California, it took several months to paint a historical building to meet the needs of community stakeholders (Potel 2000). However, such tasks can be done in a virtual environment within a few weeks. Equally, to locate mobile phone transmitters in a city, a 3D scene can help to examine the best sites in terms of the largest coverage. It will dramatically improve the original planning proposals.

We have developed various visual representations of real-world environments using affordable cutting-edge technologies such as internet GIS and virtual reality. GIS is used because these representations are supposed to be georeferenced and maps can be used as texture for 3D representations. These representations can be disseminated through the Internet and viewed through an internet browser with an appropriate plugin. In order to introduce these representations, we herewith propose a classification scheme which involves two axes, as shown in Table 10.1. The first axis shows dimension ranging from 2D to 3D, while the second axis shows the material these representations are based on. As these representations are in contrast to the real world, they can be said to be virtual or semi-virtual, corresponding to screen-based and material-based respectively.

Table 10.1: A classification of geovisualisation

	2D	2.5D	3D
screen-based (/virtual)	maps, plans, orthogonal photos	perspective photos, sketches	3D models (CAD, or VRML)
material-based (/semi-virtual)	printed maps, plans, orthogonal photos	printed perspective photos and sketches	printed 3D polyester models

Note: 2.5D meanly refers to perspective photos and sketches, but it is sometimes very confusing. For example, screen shots shown in figure 1 and 3 are 2.5D, but they are initially 3D representation using VRML. In this situation, we still refer to them as 3D representation rather than 2.5D.

Although we focus on 3D representation, this type has a close link to 2D and 2.5D representations. The 3D representation is derived from the 2.5D representation using appropriate tools with 2D maps and photos for textures. The VRML models are interactive and dynamic, allowing users to fly over or walk through. They also have hotlinks to other hyperlinked information or scenes. More interestingly, the 3D representation can be printed out using durable polyester through 3D printers. There are various kinds of 3D printers on the market, and they are basically used in mechanical engineering for output from CAD models. The size of hardcopy is scalable, so the hard copy is also portable and rather convenient for dissemination. Despite the increasing spread of the Internet, there is still a certain percentage of people who have difficulty getting access to the 3D virtual environments on the Internet, not to mention the people with visual impediments. Thanks to the affordable cost, the hard copy of 3D representations provides an excellent complement to the virtual environments in the computer.

Our first set of examples is at building scale (see *http://www.hig.se/t-inst/virtualrc/virtual_campus/ for a live demo*). The examples were built in a reasonable size, so it is easy to navigate and explore through any internet browser. The models were derived from perspective photos of the built environment. Several editions of the models were created. Figure 10.1 demonstrates a campus model shown in Cosmo player through Netscape communicator (left), and a hardcopy version with polyester material (right, where a pen gives a sense of size scale). These two versions of the model are in bird's-eye view. In contrast, Figure 10.2 presents another photorealistic version of the campus model, assuming one is 'immersed' in the virtual environment. Please note that the trees in the model were planted virtually and do not exist in reality. This shows some potential for previewing planning scenarios.

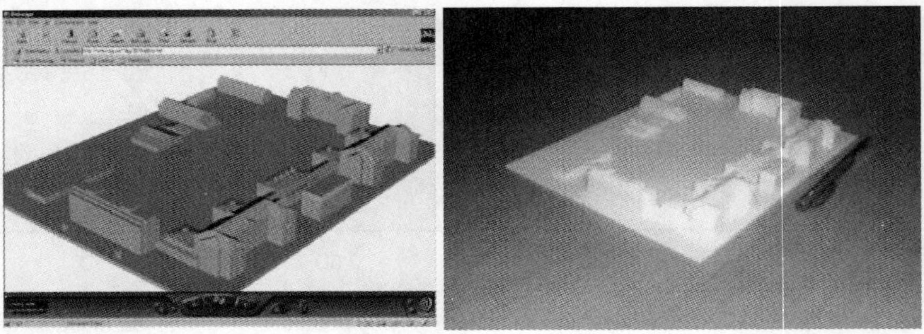

a) b)

Fig 10.1: A campus model shown (a) in browser and (b) in hardcopy

Our second set of examples is at a larger urban scale, and they were built with ArcView 3D Analyst. Figure 10.3 illustrates a city model of Gävle, a small town located north of Stockholm. For a more meaningful visualisation, different colours can be used to show different building functions; for example, red colour for residential houses, black for industrial, and grey for public buildings. Not only physical environment, but also socio-economic phenomena can be visualised in 3D. It is particularly of interest for the general public to comprehend and perceive social aspects of their environments. In this connection, Figure 10.4 demonstrates population density both in 2D and 3D formats.

Fig 10.2: An exploratory version (a) in browser and (b) in quicktime player

Fig 10.3: A virtual city model of Gävle

For the example shown in Figure 10.4, the 2D representation to the left is supposed to be for professionals while the 3D representation to the right is for the general public. One day we showed the 2D representation of population density to one of our friends, who has lived in the town for two decades. She first seemed puzzled about the meaning of the different colours. However, as she was shown the 3D scene, she immediately realised that the height of the rectanglular pillars represented some value. When we had explained to her how the two scenes showed the local population distribution, she showed preference for the 3D representation, which is more intuitive and understandable.

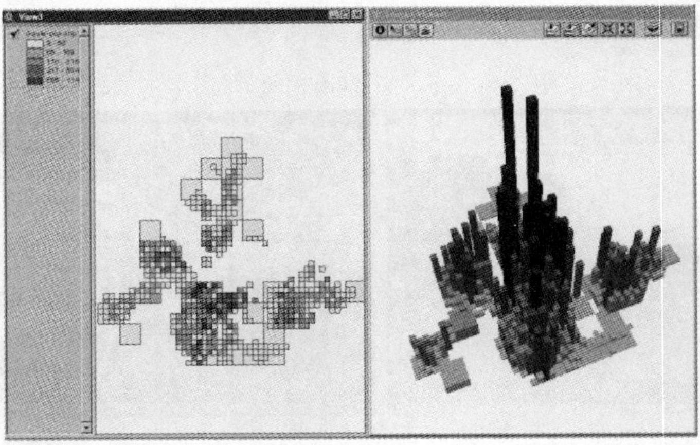

Fig 10.4: Demographic distribution of Gävle in 2D and 3D representations

10.4 Exploratory Analysis Environments

Instead of data display, geovisalisation can be useful tools for data analysis, that is, using geovisualisation as an interface for exploratory data analysis or for modelling experiments. Such highly dynamic and interactive environments can be thought of as a virtual laboratory, with which planners can post questions to conduct what-if modelling. To illustrate, let us consider a prototype for street accessibility analysis based on space syntax (Hillier and Hanson 1984). According to how each street segment (which is the longest visibility line) links to all the other segments, it derives a range of parameters to show the level of street accessibility. These parameters can be used to predict pedestrian or vehicle flows in urban systems based on well-established empirical studies. With this tool, planners can foresee how pedestrian flows move around in a proposed street network. Thus, it provides a nice model for analysing person movements. Obviously intuitive geovisualisation provides not only an interface to show the analysis results, but also an exploratory environment for conducting what-if modelling. For example, how does traffic flow change if a street is built up across streets A and B? Using central Gävle as an example, Figure 10.5 shows a typical interface from our prototype implemented as a plugin to ArcView GIS (Jiang *et al.* 1999). The environment consists of several visual components such as 2D visibility lines, 3D accessibility scene, an attribute table and a statistical chart. These components are interconnected, thus facilitating exploratory data analysis. In both 2D and 3D graphic windows, the degree of street accessibility is represented by a spectrum colour legend, i.e. red represents the most accessible street, and blue represents the most inaccessible street, and other colours represent some value between the two extremes. These are actually pedestrian flow distributions if we adopt one of these parameters – so called local integration. Within the environment, we can easily add one assumed street across streets A and B, re-calculate these parameters, and observe the changes of distribution of street accessibility.

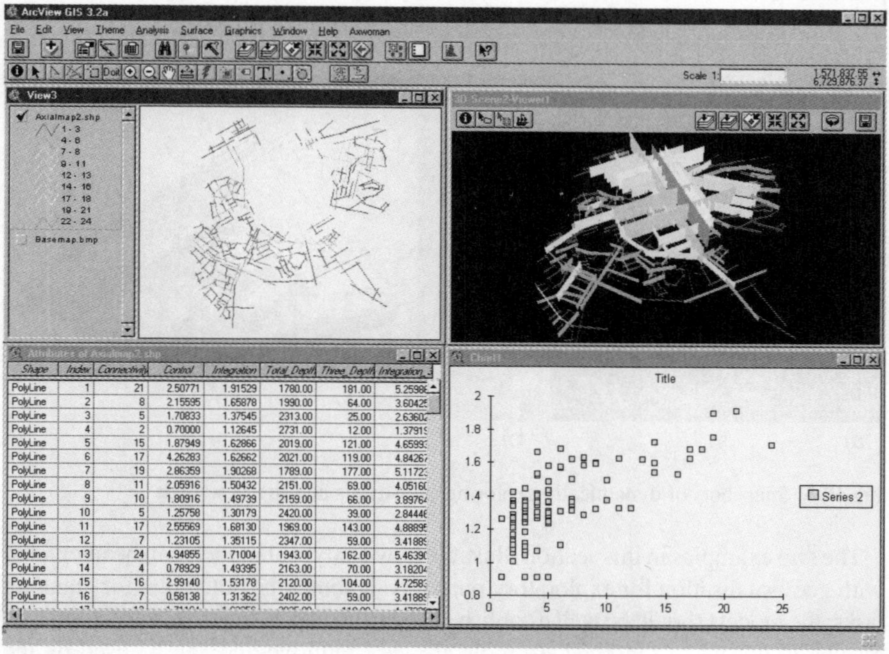

Fig 10.5: An example of exploratory analysis environments

To have an analytical model implemented with user-friendly geovisualisation environments is not a straightforward task, as it needs highly efficient algorithms for analysis. If a question is posed, a system should have an instant response to the question; a long time delay is not allowed for interactive visual thinking. In this connection, the parallel processing method provides a good technique for such tasks. Computation of an 'isovist' – a visual field from a standing point of view in an urban environment, has been a difficult task in conventional computational geometry, as it involves the calculation of line of sight. Instead, we adopt a multi-agent approach for the task and the algorithm is described as follows. Fill in space with agents and let them explore space freely. Basically agents move in every direction to explore how far they can go, and the explored results are stored as the property of each location of space (Jiang 2000). Based on the computation, isovists can be explored dynamically, i.e. whenever a mouse pointer passes, it shows the pattern of the visual field (Figure 10.6a). Such an exploratory capability can stimulate planning scenarios. For example, one can observe how visual fields will change by positioning a new building (represented as a polygon) in an area. In the same line, we implemented an exploratory environment based on viewshed analysis for Digital Elevation Models (DEM). Depending on the resolution of TIN structure from DEM, we extract a large set of discrete points, and calculate viewsheds from individual points. With interactive technique, all these viewsheds can be explored dynamically by clicking individual points (Figure 10.6b).

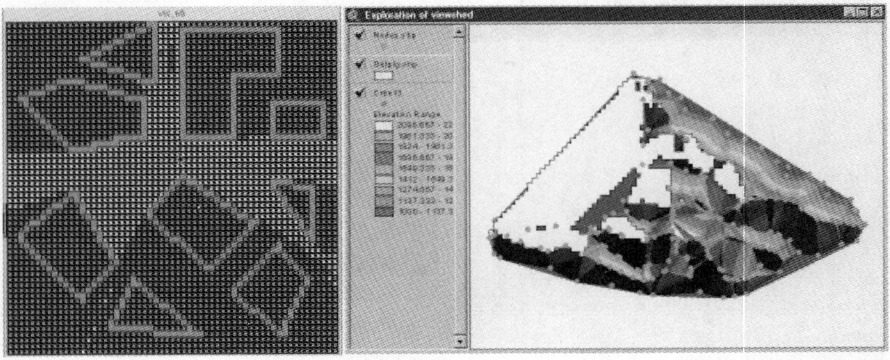

a) b)

Fig 10.6: Snapshots of dynamically exploring (a) isovists and (b) viewsheds

The two examples in this section illustrate how analytical models can be integrated with geovisualisation for exploratory purposes. It should be noted that for planning tasks, the models should be well established and validated. Otherwise, the exploratory results do not make sense. These tools are very intuitive and can be used by the general public for planning tasks.

10.5 Towards a Geovisualisation Server

The examples discussed so far are still considered for centralised environments, although they can be converted into VRML files, disseminated across the Internet and viewed on the Web. However, for a practical PSS, its three components (data, analysis model and geovisualisation) need to be considered coherently to support decentralised collaborative planning. One of the important challenges in this direction is a well-designed server-client architecture to meet the requirements of geovisualisation for dynamic and interactive exploration and collaboration over the

Internet. The notion of a geovisualisation server we have used here refers to a special kind of server. Such a server overcomes the disadvantages of current client-server architecture, and has the special capability of dealing with data transferral, analysis and interactive display in order to meet real-time geovisualisation purposes (Huang *et al.* 2001).

Using the server-client architecture of the Internet technique, existing GIS and geovisualisation software can be mounted over the Internet, which provides an interoperated environment for distributed planning processes. Databases, GIS analysis models (or modelling tools) and 3D representations can be made available for remote clients for collaborative planning purposes. It provides a fast and economical solution to developing practical PSS using existing software resources.

We have implemented a geovisualisation server using ArcView GIS and its extensions. The prototype is based on an architecture that consists of three parts: a Java-based client, a web server, and a geovisualisation server (Figure 10.7). The Java-based client resides with the web server like Apache. This serves also an interface in which end users interact with the system by sending requests and receiving results. The web server is integrated with an Internet Mapping Server (IMS). Existing GIS and geovisualisation software together constitute the geovisualisation server. For this implementation, we have used ArcView IMS combined with a web server, ArcView and its extension 3D Analyst on the geovisualisation server. Therefore the prototype is an interoperated environment with a range of interactive visualisation and analysis functions.

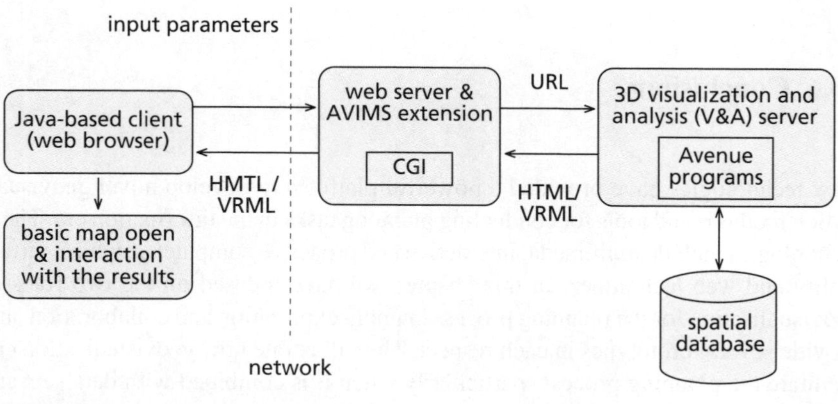

Fig 10.7: A geovisualisation server built with internet server-client architecture

The communication between server and client is bi-directional, and it is achieved through various web and GIS programming such as Java, HTML and AVENUE. For example, the client side of the system provides several pages embedded with Java applets, which allow users to select operations, define properties of 3D scenes or input parameters for 3D analysis. End users then submit the request in the form of

Unique Resource Locator (URL) to the geovisualisation server via a CGI program within ArcView IMS running on the web server, or conduct some operations (e.g. 2D map browsing) locally. The correctness of parameters is also examined by the Java-based client before a command is submitted. The main mechanism for Java-based clients to communicate with the geovisualisation server is through the encoding of parameters in URL. After receiving a URL request from the Java-based client, the geovisualisation server extracts all the necessary parameters, processes them, and then delivers the results in the HTML-compatible format to the client for display. The results could be a 2D map or a 3D VRML model. In both cases, end users can still interact with them via zooming, panning, and querying on the client side.

To illustrate, let us introduce two screenshots of the prototype. Figure 10.8a shows an interface of 3D scene parameter setting in order to create 3D perspective displays by extruding spatial features in 2D maps. Attributes of spatial features (e.g. stories of buildings) are usually taken for height information (Z-value). Such an extrusion changes the form of a feature: points into vertical lines, lines into vertical walls, and polygons into 3D blocks. After parameter inputs, clicking the 'Create 3D-VRML' button will generate a 3D scene and its VRML model on the fly (Figure 10.8b). In addition, a range of 3D analysis can be conducted such as viewshed analysis, and profile graph creation. It should be stressed that all these geovisualisations and analyses are conducted in a decentralised manner, that is, users send requests in normal web pages which go through different servers and finally receive the results in their screen. These parameters are still too technical to be understood by the general public, so it is more oriented to professional users.

10.6 Conclusions

New technologies have provided a powerful platform to develop novel geovisualisation methods and tools for conducting planning tasks in the information era. These technologies include multimeda, interactive and proactive computer graphics, virtual reality and web techniques. In this chapter, we have focused on the two roles of geovisualisation for the planning process, namely exploration and collaboration, and provide several prototypes in each respect. They illustrate how geovisualisation can facilitate the planning process, particularly when it is combined with data sets and analytical models.

Planning activities involve a large amount of data, which includes topographical, socio-economic and demographic data. Thus, massive data geovisualisation in realistic ways like VRML is a pre-requisite for a successful PSS. Due to the continuous development of computer performance, massive photorealistic geovisualisation can be made possible in desktop environments through the Internet. What-if modelling is very important for planning tasks, as it helps to explore planning scenarios. For more ambitious and difficult implementation, it would be necessary to develop

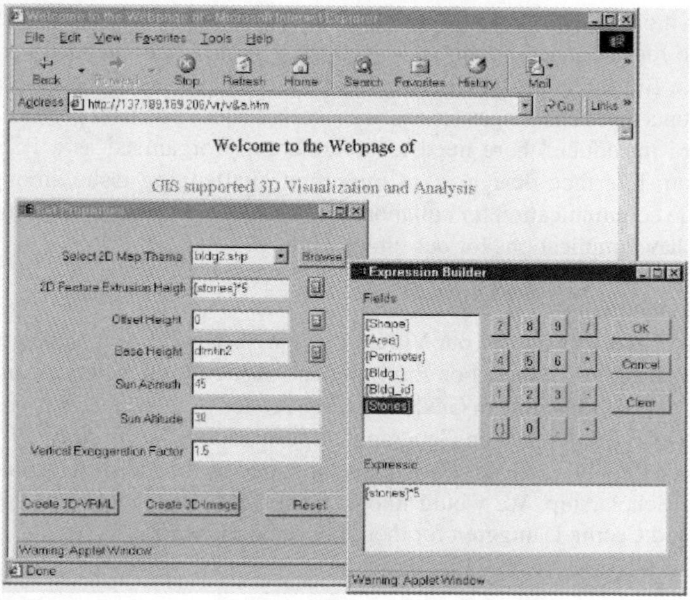

a)

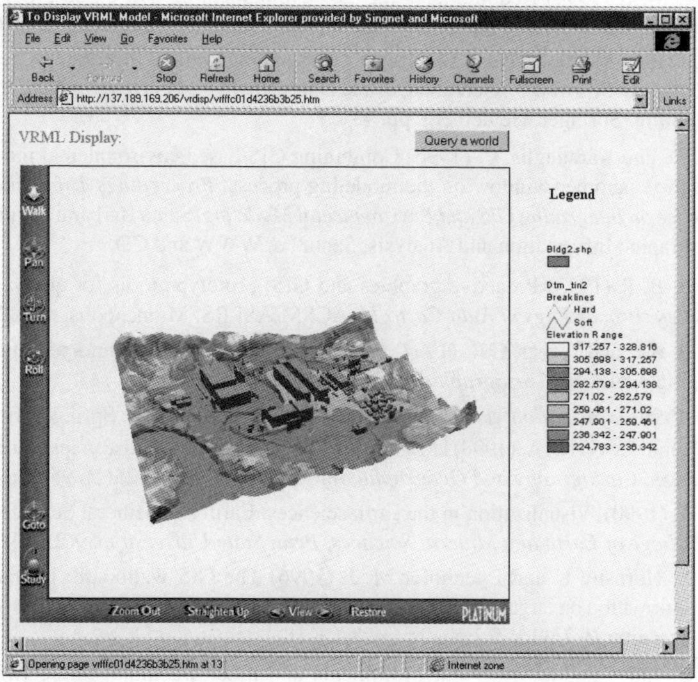

b)

Fig 10.8: Parameters input (a) for creating a 3D scene in VRML (b) specialist models with real time calculation. This kind of modelling may not be applicable for the general public to use as these models are intended for planning specialists. But there is a way to simplify parameter inputs for the general public. This presents more challenges for research work. From a practical point of view, the prototypes introduced here need to be coherently organised as a PSS. In this connection, interface design is an important challenging issue among others. Multimode communication for collaboration is likely to be one of the requirements. All these have implications for our future work.

Acknowledgements

This chapter presents part of our Virtual Planning Studio project, supported by the Swedish Research Council For Environment, Agricultural Sciences and Spatial Planning, and GIS Institute in Gävle. An early version of this paper was presented at the 4th AGILE Conference on Geographic Information Science, April 19-21, 2001, Brno, Czech Republic. Vit Vasek's work is supported by the Swedish Institue through the Guest Scholarship. We would like to thank Prof. Bengt Rystedt, Prof. David Douglas and Cecilia Ljunggren for their advice on the work.

References

Armstrong, M. P. (1994) Requirements for the development of GIS-based group decision-support systems, *Journal of the American Society for Information Science*, 1(1): 669 – 677.

Batty M., Dodge M., Jiang B. and Smith A. (1999) Geographical information systems and urban design, in Stillwell, J., Geertman, S. and Openshaw, S. (eds.) *Geographical Information and Planning,* Springer, Heidelberg, pp. 43-65.

Bishop, I. D. and Karadaglis, C. (1996) Combining GIS based environmental modeling and visualisation: another window on the modeling process, *Proceedings Third International Conference on Integrating GIS and Environmental Modeling*, Santa Barbara: National Center for Geographic Information and Analysis, Santa Fe, WWW and CD.

Buttenfield, B. P. (1993) Proactive graphics and GIS: prototype tools for query, modelling and display, *Proceedings of Auto Carto 11,* ACSM/ASPRS, Minneapolis, pp. 377-385.

Buttenfield, B.P. and Weber, C.R. (1994) Proactive graphics for exploratory visualization of biogeographical data, *Cartographic Perspectives,* 19(3): 8-18.

Chen, C. (1999) *Information Visualisation and Virtual Environments*, Springer, Berlin.

Craig W. J. and Elwood S. A. (1998) How and why community groups use maps and geographic information, *Cartography and Geographic Information Systems*, 25(2): 95 - 104.

DiBiase, D. (1990), Visualization in the earth sciences. Earth and Mineral Sciences, *Bulletin of the College of Earth and Mineral Sciences, Penn State University*, 59(2): 13-18.

Florence J., Hornsby K and Egenhofer M. J. (1996) The GIS wallboard: interaction with spatial information on large-scale displays, in Kraak, M. J. and Molenaar, M. (eds.) *Advances in GIS Research II,* Taylor & Francis, pp. 8A.1-18.

Gouveia C. and Camara A. (1999), Multimedia and urban planning, in Geertman, S., S. Openshaw, and J. Stillwell (eds.) *Geographical Information and Planning,* Springer, Heidelberg, pp. 391-402.

Hillier B. and Hanson J. (1984) *The Social Logic of Space*, Cambridge University Press.

Huang B. Jiang B. and Lin H. (2001), An integration of GIS, internet and virtual reality for the visualisation, analysis, and exploration of spatial data, *International Journal of Geographical Information Science*, 15: 439-456.

Jiang, B., Claramunt, C. and Batty, M. (1999) Geometric accessibility and geographic information: extending desktop GIS to space syntax, *Computers Environment and Urban Systems*, 23: 127-146.

Jiang, B. (2000) Agent-based approach to modelling urban and environmental systems within GIS, in Forer P., Yeh A. and He, J. (eds.) *Proceedings of the 9th International Symposium on Spatial Data Handling*, Beijing, 10-12 August, pp. 3b.17-3b.29.

Kingston, R., Carver, S., Evans, A. and Turton, I. (2000) Web-based public participation geographical information systems: an aid to local environmental decision-making, *Computers, Environments and Urban Systems*, 24: 109-125.

Klosterman, R.E. (1999) New perspective on planning support systems, guest editorial, *Environment and Planning B*, 26: 317-320.

Laurini, R. (2001) *Information Systems for Urban Planning: A Hypermedia Co-operative Approach*, Taylor & Francis, London.

MacEachren, A. M. (1994) Visualization in modern cartography: setting the agenda, in MacEachren A.M. and Taylor D.R.F. (eds.) *Visualization in Modern Cartography*, Pergamon, pp. 1-12.

MacEachren, A. M. (2000) Cartography and GIS: facilitating collaboration, *Progress in Human Geography*, 24(3): 445-456.

MacEachren, A. M. (2001) Cartography and GIS: extending collaborative tools to support virtual teams, *Progress in Human Geography*, also available at: http://www.geovista.psu.edu/publications/amm/ammP00.pdf.

McCormick, B.H., DeFanti, T.A. and Brown M.D. (1987) Visualization in scientific computing, *Computer Graphics*, 21(6).

Potel M. J. (2000) Visualisation in urban planning: they did not build LA in a day, *IEEE Computer Graphics and Applications*, May/June, pp. 10-16.

Reddy, M., Leclerc, Y., Iverson, L and Bletter, N. (1999) TerraVision II: visualizing maasive terrain databases in VRML, *IEEE Computer Graphics and Applications*, March/April, pp. 30-38.

Rinner, C. (1999) Argumaps for spatial planning, Proceedings of TeleGeo99, *First International Workshop on Telegeoprocessing*, Lyon, France, May 6-7, pp. 95-102, also available at: *http://set.gmd.de/~crinner/pub/telegeo.pdf.*

Shiffer, M. (1995) Interactive multimedia planning support: moving from standalone systems to the World Wide Web, *Environment and Planning B*, 22: 649-664.

Shiffer, M. J. (1998) Multimedia GIS for planning support and public discourse, *Cartography and Geographic Information Systems*, 25(2): 89-94.

Web Site

Virtual campus: *http://www.hig.se/t-inst/virtualrc/virtual_campus/*

11 Key to Virtual Insight: A 3D GIS and Virtual Reality System

Gert van Maren

Asset Information Systems, PO Box 102 032 NSMC, Auckland, New Zealand

Abstract: Virtual Reality (VR) is a new way of visualising and manipulating GIS data. Two-dimensional (2D) geographical information systems (GIS) data can be used, together with three-dimensional (3D) Computer-Aided Design (CAD) data, to generate a virtual world in which the user can navigate, query and manipulate the GIS and 3D CAD data. Over the past three years, we have developed the Key to Virtual Insight or K2vI system which is a VR interface on several standard GIS databases that supports visualisation, manipulation and editing of the GIS and CAD data in a virtual environment. The system uses a multi-view mode approach where each view mode has a preferred VR display.

The system is used by the School of Architecture, Auckland University, New Zealand, to generate a 3D city model of Auckland. After completion of the model, urban planners will be able to employ K2vI as a design and analysis tool in their planning process. The system supports manipulation of the GIS and CAD data and will offer functionality to perform analyses such as visual impact analysis and traffic flow analysis.

11.1 Background

Virtual Reality (VR) techniques are increasingly used for the large-scale depiction of the built environment in the process of design planning for urban settings. By creating a (make-believe) reality, designers, politicians and citizens are able to evaluate the prospective situation in a very direct way. This kind of presentation is often only the last step in the whole process of problem orientation (plan study), plan development (model study) and decision-making. To be better than that, VR has to contribute more to the planning process than just a real-time renderer, which is nowadays also possible with existing CAD systems.

The aim of the development of the *K2vI* system was to integrate GIS, CAD and VR in such way that the whole process of design planning could be supported by one purpose-specific GIS and VR system. The system was originally developed and built by the Departments of Geodesy and Technical Informatics of the Delft University of Technology (DUT) in the Netherlands, in a co-operation of several engineering, visualisation companies and the ITC. The research was supported by the Dutch Land Water Environment Information Technology Foundation. This support program concluded at the end of 1999 and *K2vI* is now being further developed for urban, military, sports and forestry applications in a joined effort of Asset Information Systems in New Zealand and the DUT.

In this chapter, we first describe the design and functionality of the basic *K2vI* system. Thereafter, an overview is presented of existing and new application areas including urban planning, military mission control, flight corridor visualization, factory design, virtual sports simulation and forest harvest planning.

11.2 *K2vI*

K2vI is a VR interface on spatial data that supports visualisation, manipulation and editing of the spatial data from within a VR environment. The software uses the multi-view mode approach where each view mode has a preferred VR display system.

11.2.1 Multiple Views

Research by Maren *et al.* (1998) showed that potential users of a GIS and VR system would prefer multiple views on the data to be able to interact with that data in a virtual environment. We therefore decided to implement three views on the data.

- *Plan view*: where geographic data is visualised in a conventional cartographic map format (Figure 11.1a). Spatial objects are represented by 2D points, lines, and polygons. The plan view is most efficient on a monitor with window and mouse interaction.
- *Model view*: where visualisation is kept simple and the user looks down on the model from the bird's eye viewpoint, as if it is an ordinary 3D-scale model (Figure 11.1b). For this view, a virtual workbench (Krüger *et al.* 1995) is an appropriate display system.
- *World view*: where the visualisation is the first person view of the area of interest. The purpose of this view is to give a realistic impression of the landscape, using both visual and audio output. The user can 'walk through' the geographic data, which is visualised using detailed 3D CAD models and textures (Figure 11.1c). The world view is best viewed with a truly immersive display, such as a Head Mounted Display (HMD) or a surround projection display, for example, the CAVE (EVL 2000).

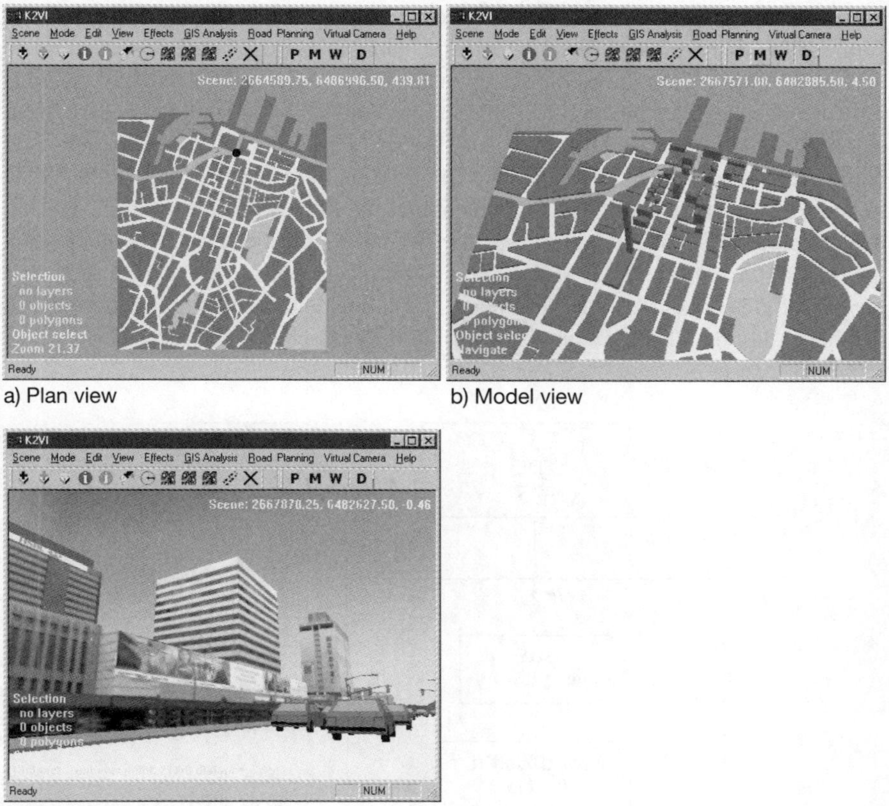

a) Plan view

b) Model view

c) World view

Fig 11.1: Centre of the city of Auckland in New Zealand
Note: The dot in the plan view (centre, bottom) represents the position of the viewpoint for the world view.

11.2.2 System Components

The major components of the system are the GIS databases and the VR system. We selected the WorldToolKit (WTK) from Sense8 (SENSE8 2000) as the VR system. As for the GIS databases, *K2vI* supports three different GIS databases at the moment:

- the Spatial Database Engine (SDE) from ESRI (ESRI 2000);
- ESRI shapefiles (ESRI 2000); and
- ASCII text files.

We are working on a direct port to Oracle Spatial 8i. This means that in the near future, *K2vI* will be able to directly load and manipulate spatial data stored in Oracle Spatial 9i. Figure 11.2 shows the set-up of *K2vI* with SDE as GIS database. *K2vI* is built as a client application on SDE. This way, existing GIS applications (like ArcView or ArcInfo) can make use of the same data. The shape co-ordinates, which are retrieved from SDE, ArcView shapefiles or ASCII textfiles, are used to define the WTK geometries for the plan, model and (part of the) world view. Shape attribute data are used to correctly generate and/or display 2D and 3D geometries. CAD model data are imported from external CAD systems. At this stage, editing of the 3D CAD is not part of the *K2vI* system. In the future, we might decide to add extra functionality to be able to edit the 3D CAD data from within the system.

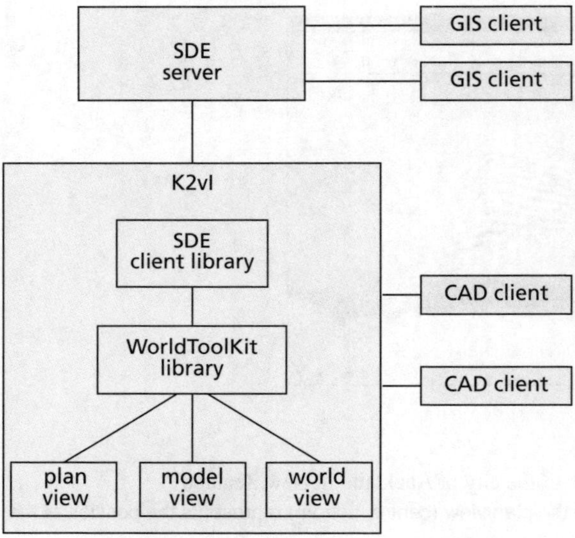

Fig 11.2: The *K2vI* system set up with SDE as underlying GIS database

11.2.3 Functionality

One of the main objectives of the research project at the DUT was to build a client/server application, that would allow real-time manipulation and editing of GIS data in a VR environment. Furthermore, the user would have to be able to perform basic GIS analyses to support the manipulation of the data.

Manipulation: Within *K2vI*, several functions have been developed to be able to manipulate the GIS and 3D CAD data directly via the different VR interfaces. Basic translation and rotation functions can be used to manipulate individual or groups of objects (Figure 11.3).

a) One object selected

b) Translation and rotation

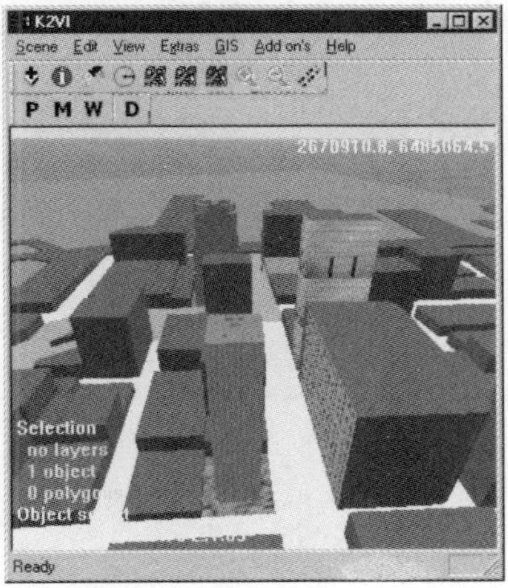

c) Scaling

Fig 11.3: Manipulation of a 3D CAD object

GIS: Basic GIS functionality is available in all views to support the manipulation of the scene. At this stage of the project, the following GIS functionality is implemented:

- *Identify*: GIS objects can be identified from a plan, model or world view by selecting an object. The record containing the object's thematic data is then retrieved from the appropriate GIS database and displayed in 3D (Figure 11.4a). It is also possible to play videos in the 3D world displaying additional information about the object.

- *2D buffering*: Performing a 2D spatial query or buffer is a typical GIS operation for which *K2vI* provides a front-end. A user selects one or more GIS objects and sets buffer parameters like radius and the GIS layer in which the spatial query is to be performed. The actual buffer analysis is performed on the GIS server and the result is sent back to *K2vI* (Figure 11.4b).
- *3D measuring*: Measuring distances is a simple GIS analysis, which can easily be integrated in a VR environment, when accuracy is not an issue. With *K2vI*, users can interactively 'measure' distances between 2D or 3D GIS objects by pointing and dragging a 3D line. This allows a quick analysis of object-to-object distances, CAD model detail dimensions and object heights (Figure 11.4c).

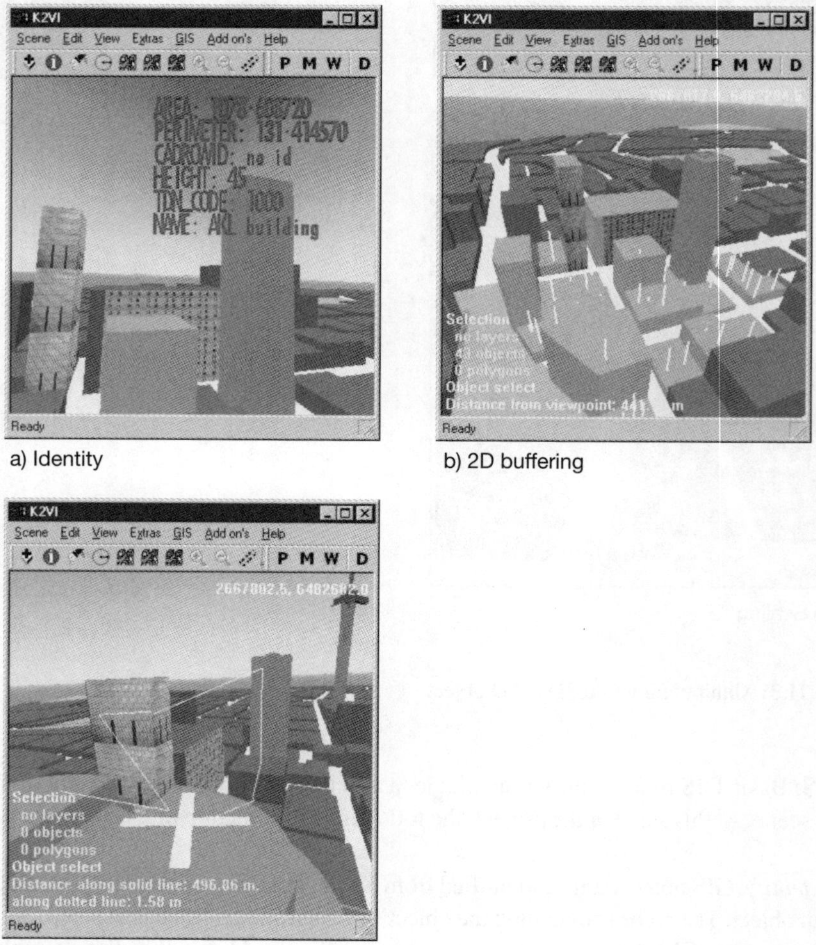

a) Identity

b) 2D buffering

c) Measurement

Fig 11.4: GIS functionality

An extended overview of the multi-view approach, system components and functionality of the software can be found on *http://www.k2vi.com* and in Maren *et al.* (1998), Maren and Germs (1999) and Verbree *et al.* (1999).

11.3 *K2vI* in Practice

The School of Architecture at Auckland University is currently using *K2vI* to generate a 3D city model of Auckland (Figure 11.5). Jules Moloney, Senior Lecturer, School of Architecture, University of Auckland stated *"Software developed for building science is used to measure and test building performance and while there may be graphic representation, the focus tends to be on data rather than visualization of the architecture. By contrast, software used in conceptual and developed design concentrates on visual appearance with minimal performance data. The aim of K2vI was to integrate GIS, CAD and VR in such a way that the whole process of design planning could be supported by one data representation system"*. The School preferred *K2vI* over VRML or other VR software because of this ability to combine a range of file types in one responsive database.

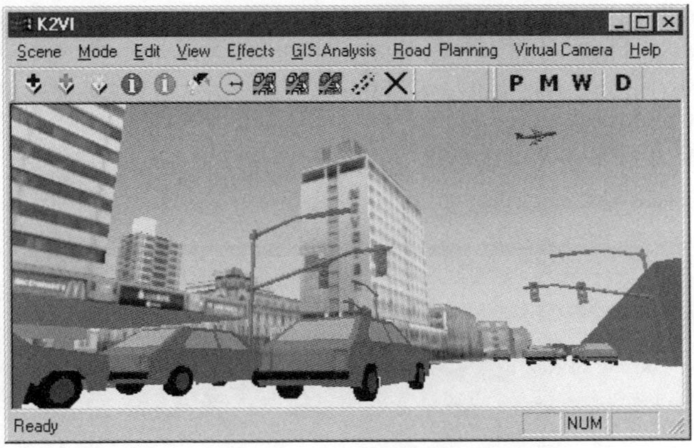

Fig 11.5: Downtown Auckland with animated cars and planes

With this aim the software was installed on several PCs at the University and students have started to build up a 3D city model of Auckland. *K2vI* is used to build up the model and explore it. Users can view the model in stereo with 'crystal eyes' hardware, walk, fly or drive through their 3D scene avoiding obstacles (collision detection), all the while tracking their position on a 2D map. Furthermore they add, move, rotate, scale or delete objects (such as buildings) from within this virtual world.

A host of GIS analyses can be performed including SQL query, measurements between GIS and CAD in a 3D space and line of sight analysis. The database information is displayed as 3D 'billboarded' text always facing the viewer. Recently added functionality allows the students to animate objects such as cars and attach stereo sound to the cars to simulate traffic flow. Special effects such as fog increase the realism of the scene. After completion of the model, urban planners will be able to employ *K2vI* as a design and analysis tool in their planning process.

11.4 New Application Areas

We are exploring a range of new application areas for *K2vI* such as military mission control, flight corridor visualization, factory design, virtual sports simulation and forest harvest planning.

11.4.1 Battlefield Simulation

The New Zealand armed forces are evaluating the software for battlefield simulation. *K2vI* allows them to animate objects over or on the terrain, for example airplanes, bombs or tanks (Figure 11.6). A mission controller can follow the objects using the navigation devices or attach the viewpoint to the object and, for example, 'be' the pilot. *K2vI* can be configured to read streaming data from a server over the Internet which allows for real time tracking of airplanes, bombs and tanks.

Fig 11.6: Battlefield simulation

11.4.2 Airspace Visualization

A recent marketing trip to the USA showed us that there is a huge demand for visualization of airspace around airports. Similar to building a 3D model of a city environment, *K2vI* can be used to build up models of airports including 3D airspace, all buildings and surrounding terrain. Using the new animation functionality, it is possible to visualize the movement of objects like airplanes in the 3D scene. Figure 11.7 show the 3D airspace of the Memphis airport in the USA.

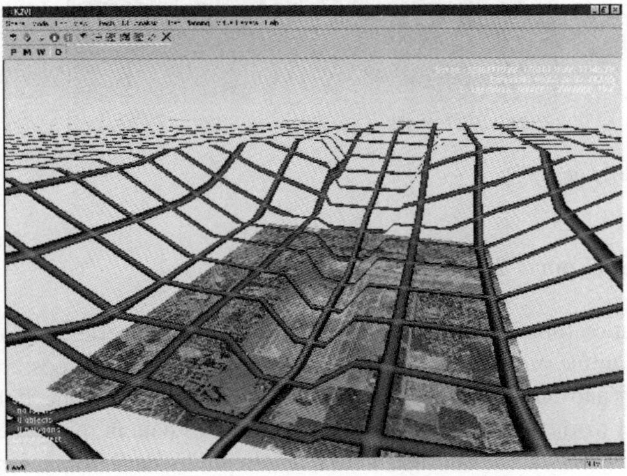

Fig 11.7: Airspace visualization

11.4.3 Factory Design

Designing a large factory is a spatial problem similar to urban design. All objects (buildings, machines) have a geographic location so *K2vI* can be used to create an optimal factory layout in realtime. Figure 11.8 shows a factory building with several machines placed inside. *K2vI* allows you to navigate around the building, enter the building and get a 'feel' for the layout of the factory. Once inside, you can change the representation for every machine or add a new machine to the production line. Next, *K2vI* allows you to place the new or changed machine anywhere in the 3D scene to line up with the existing production line. 3D measure functionality allows for correct distance measurements in 3D space. The user is present in the 3D scene at all times during the design process, getting instant feedback on how the newly designed environment looks.

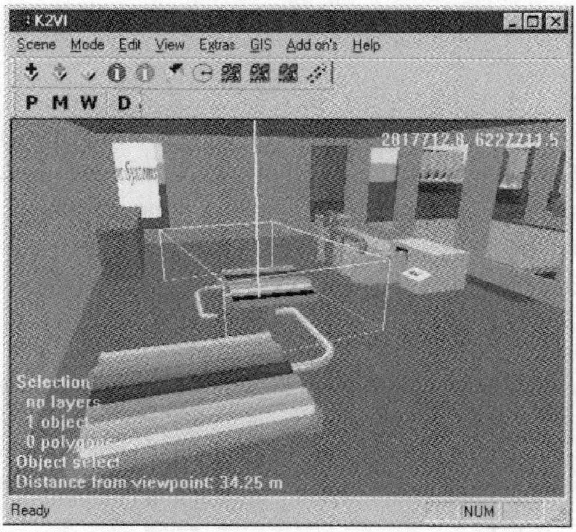

Fig 11.8: Factory design

11.4.4 Virtual Sports Simulation

K2vI's animation functionality opens up the possiblity of experiencing a sports game in VR, for example rugby (Figure 11.9). *K2vI* can read streaming position /time data for the players and animate the players in the virtual world in realtime. If the positional data is in real world coordinates, integration with GIS data in the scene is possible. Furthermore, *K2v1* can be configured to read streaming data from a server over the Internet. Other sports games like tennis, sailing, soccer can be experienced with *K2vI* as well.

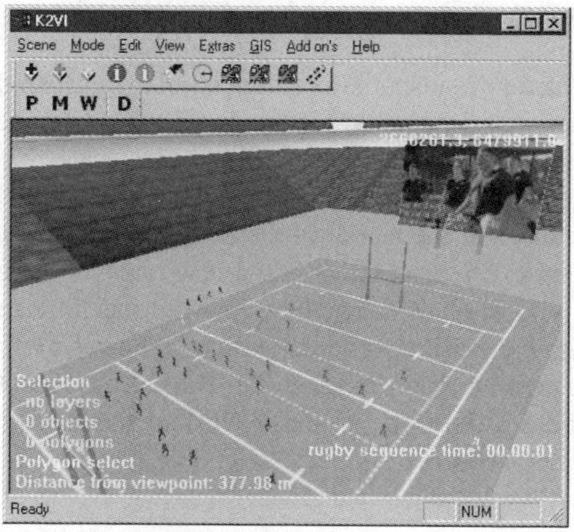

Fig 11.9: Simulate a rugby match

11.4.5 Forest Harvest Planning

A virtual harvest planning system can be extremely helpful during different stages of the harvest planning process (Hammond 1995). First, planners need to get a feel for the way the country is put together. Traditionally, this is done with the help of 2D contour maps and a considerable amount of field checking which can prove to be quite difficult due to poor ground visibility. The amount of time in the field can be greatly reduced by a system such as *K2vI*. Planners can get a feel for the country by flying over it in virtual reality and switching the vegetation on and off (Figure 11.10a and b). In addition, analyses such as the planning of roads (Figure 11.10b and c), the locating of harvesting landings and analysis of different harvesting options can be done faster using *K2vI*.

a) Radiata pine forest on textured terrain model

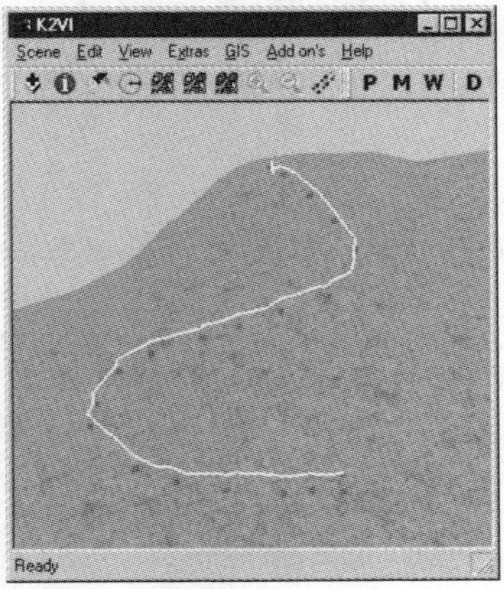

b) Forest layer switched off for interactive road planning

c) Calculated road showing slope and curve radius

Fig 11.10: Virtual harvest planning system

11.5 Future Developments

At present, the *K2vI* software is operational on PC, Virtual Workbench and the CAVE. We will concentrate on the existing application area (urban planning) as well as the new application areas. Furthermore, we will develop new functionality like interactive terrain manipulation and 3D volume calculations. Furthermore, the School of Architecture at the University of Auckland is using *K2vI* as a tool for urban design. A model of the Auckland central business district that will incorporate GIS data, CAD models and video is under construction. The model will be used to undertake a user survey of responses from designers and planning professionals as well as the public. The research is expected to generate user guidelines for the future development of functionality specific to urban design. It is anticipated that the city model will also be used as part of the evaluation process for an international competition on the development of the Auckland waterfront.

References

ESRI (2000) Homepage of Environmental Systems Research Institute, Inc., Accessed February 2000. http://www.esri.com/base/products/sde/sde.html.

EVL (2000) Homepage of Electronic Visualization Laboratory, Accessed March 2000. http://www.evl.uic.edu/EVL/VR/systems.shtml.

Hammond, (1995) *Forestry Handbook 1995*, New Zealand Institute of Forestry.

Krüger, W., Bohn, C. A., Fröhlich, B., Schüth, H., Strauss, W. and Wesche, G. (1995) The Responsive Workbench: A Virtual Work Environment, IEEE Computer Society, *IEEE Computer*, 28(7): 42-48.

Maren, G.J.M. van, Germs, R. and Jansen, F. (1998) Integrated 3D-GIS and VR: Design and implementation of the Karma-vI system, in *Proceedings of the 10th Annual Colloquium of the Spatial Information Research Center (SIRC 98)*, University of Otago, Dunedin, New Zealand.

Maren, G.J.M. van, Germs, R. (1999) Karma-vI: A Virtual Reality Interface for the Spatial Database Engine, in *Proceedings of the 1999 ESRI International User Conference,* San Diego, USA.

SENSE8 (2000) Homepage of the Sense8 Corporation, Accessed February 2000. http://www.sense8.com.

Verbree, E., van Maren, G., Germs, R., Jansen, F. and Kraak, M. (1999) Interaction in virtual world views – linking 3DGIS with VR, *International Journal of Geographic Information Science*, 13(4): 385-396.

12 Cellular Automata and Multi-agent Systems as Planning Support Tools

Paul M.Torrens

Department of Geography and Centre for Advanced Spatial Analysis, University College London, 1-19 Torrington Place, London WC1E 6BT, United Kingdom

Abstract: 'Traditional' urban simulation models have a number of weaknesses that limit their suitability as planning support tools. However, a new wave of models is currently under development in academic circles, and it is beginning to find application in practical contexts. Based around two simulation techniques that have origins in artificial life and artificial intelligence – cellular automata and multi-agent systems – it offers great potential for planning support tools, with the capacity to simulate individual households and units of the built environment in a truly dynamic, realistic and highly flexible manner. This chapter presents an overview of traditional land use and transport models as planning support tools and examines their fragilities before reviewing a new wave of urban models. Additionally, it considers the challenges facing the use of new techniques in operational models.

12.1 Introduction

In the early 1970s, the field of urban simulation was all but written-off as a failure. Douglas Lee's article entitled 'Requiem for large-scale models', published in the *Journal of the American Institute of Planners* (Lee 1973), served as a harbinger of doom for a generation of urban simulation models applied in planning contexts. (If you really want to make a name for yourself in academic circles, publish a paper declaring the demise of your discipline!) Lee provided a broad justification for retiring large-scale urban models as planning support systems (PSS) and many of his arguments were appropriate at the time. To some extent, the challenges posed in his paper were muted by subsequent developments in computer hardware, computer software, dataware, new simulation techniques and important breakthroughs in our understanding of cities. But several important weaknesses remained with urban models used as planning support tools, and without much in the way of a viable empirical alternative for supporting decisions in cities, urban planning and management agencies continued to use those technologies to assist them in their duties despite the flaws associated with the models.

In recent years, however, a confluence of several related and distinct developments across disciplinary boundaries has provided the foundation for a new generation of urban simulation models with the capacity to revolutionize our capabilities for simulating cities. Models have been developed in academic circles that enable users to simulate the complex dynamics of urban systems at the level of individual households and buildings, in some cases approximating real time representations. Moreover, these simulations have begun to migrate from the confines of the laboratory and into real world applications.

This chapter presents an overview of traditional urban simulation models and discusses their weaknesses before reviewing a new wave of urban models. Additionally, it considers the challenges facing the use of new techniques – cellular automata and multi-agent systems – as operational planning support tools.

12.2 'Traditional' Urban Models

The conceptual framework for a traditional urban model is outlined in Figure 12.1, illustrating the general land use and transport model that was (and in many cases still is) used as a planning support tool by city planning and management agencies. Generally, these models are comprised of several sub-components. Cities are usually simulated from two distinct standpoints: land use (with sub-models for land supply, land demand, and mechanisms for reconciling the two) and transport (with sub-models for potential demand and trip generation, trip distribution, modal split and trip assignment). In recognition of the fact that several relationships exist between land use and transport in the real world, the simulations of land use and transport are commonly linked via some connecting mechanisms. Traditional simulations are most commonly devised as combinations of spatial interaction models, spatial (or discrete) choice models and simple functional statements.

Spatial interaction models (also known as gravity models) allow us to predict the size and direction of flows through urban spaces using independent variables that measure some structural properties of the area being modelled. The models are commonly used to assign activities to locations in land use simulations and to model trip generation and assign trips to routes in transport simulations. For example, the geography of journey-to-work flows might be modelled using structural variables such as the distribution of workers, the distribution of employment and the costs of traveling to work. Spatial interaction models are formulated based on Newton's laws concerning gravitational attraction. Newton asserted that the force of attraction between two bodies (F_{ij}) is the product of their masses (m_i, m_j) divided by the square of the distance between them (d_{ij}^2):

$$F_{ij} = G \frac{m_i m_j}{d_{ij}^2} \tag{1}$$

where G is a constant: gravity. In urban models this translates to:

$$T_{ij} = k \frac{W_i W_j}{d_{ij}^{\alpha}} \tag{2}$$

Put simply, the above equation represents flows as the result of push and pull factors. Specifically, the flow between two places is considered to be a function of the ability of an origin i (e.g. a residential neighborhood) to generate flows (e.g. trips), the capacity of a destination j to attract these flows (e.g. through employment), the distance over which the flow must pass (d_{ij}) and some weighting mechanism that discourages flows over long distances: a distance-decay effect (α). In the above equation, k is a scaling constant; it needs to be included because the independent variables W_i and W_j are not measured in units of flow (Thomas and Huggett 1980).

Significant variations on this basic description of the gravity model include the production-constrained model, the attraction-constrained model, the production-attraction-constrained model, and the entropy-maximizing model (Torrens 2000b). The motivation behind applying these enhancements to the basic framework is to provide some form of balancing or accounting in the predictions that the model makes. To put it another way, constraints straightjacket a model into compliance with known data.

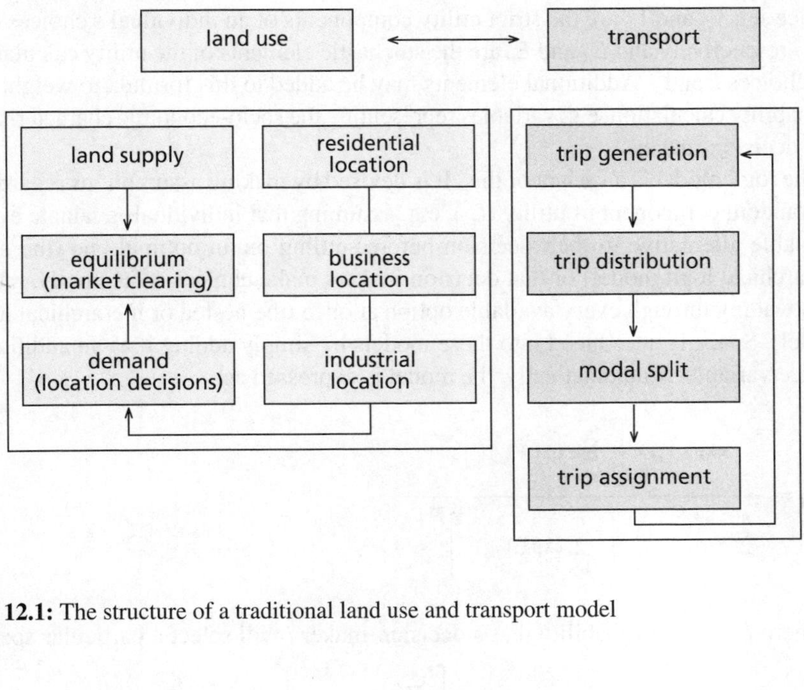

Fig 12.1: The structure of a traditional land use and transport model

Spatial choice models represent a behavioural approach to urban simulation. They are also used to simulate elements of the scheme outlined in Figure 12.1. They are commonly applied to the simulation of location decisions on the land use side and to modal choice in transport simulation. In some instances, the technique is also applied to simulating development decisions on the supply side. Spatial choice models use various assumptions to simulate decision-making or spatial choice in urban contexts, commonly on an aggregate level. First, decision-making is assumed to take place from a discrete set of choice alternatives. Second, choices are assumed to be made in such a way that the most utility, or satisfaction, is yielded. In an urban sense, this might represent a household making a location decision amongst a set of given locations that a city has to offer so that a combination of utilities is maximized (e.g. cost, amenities, quality of the school system). Third, it is assumed that choices are made in a probabilistic fashion – decision-makers have a likelihood of making certain choices. Fourth, it is assumed that the utility of a decision can be divided into two components: one measuring 'strict utility', the fixed and measurable attributes of utility, and the other dealing with 'stochastic utility', an error or disturbance term that reflects the unobserved attributes of a given decision (De la Barra 1989).

There are several variants on this general theme. The most general, the discrete choice model, is expressed mathematically as:

$$P_{ik} = \Pr(V_{ik} + E_{ik} > V_{ij} + E_{ij}) \qquad \forall k \neq j,\ j = 1,k,n \tag{3}$$

where P_{ik} is the probability of an individual i choosing an option k from a given choice set. V_{ik} and V_{ij} are the strict utility components of an individual's choices of k and j respectively and E_{ik} and E_{ij} are the stochastic elements of the utility calculation for choices k and j. Additional elements may be added to this formula to weight the probability calculation, e.g. variables representing the socio-economic characteristics of a decision-maker.

The logit model is a variant of this. It is devised by making assumptions regarding the random component of utility (E_{ij}), e.g. assuming that individuals evaluate every available alternative to their decision before settling on an optimal one (the non-hierarchical logit model) or that decision-makers make choices sequentially, rather than wading through every available option at once (the nested or hierarchical logit model). Space is introduced into these models by simply adding it as an additional choice variable. Mathematically, the model is expressed as:

$$P_{is} = \frac{\exp(V_{is})\left[\sum_{k\varepsilon s}^{n} \exp(V_{ik})\right]^{\sigma}}{\sum_{s}^{n} \exp(\mathrm{V}_{\mathrm{is}})\left[\sum_{k\varepsilon s}^{n} \exp(V_{ik})\right]^{\sigma}} \tag{4}$$

where P_{is} is the probability that a decision-maker i will select a particular spatial cluster s to focus its decision in and $\left[\sum_{k\varepsilon s}^{n} \exp(V_{ik})\right]$ is termed an inclusive value describing the attractiveness of a cluster as a function of the individual alternatives available within that cluster (Fotheringham and O'Kelly 1989). σ represents the extent to which decision-makers process their information hierarchically and ranges in value from zero to one, with $\sigma=1$ denoting decision-makers who do not process their information hierarchically at all.

Once a decision-maker has selected a given spatial cluster, *s*, to narrow the choice set, all that remains is for an option (or alternative), *k*, to be settled upon. The likelihood of a decision-maker selecting a particular alternative *k*, within the selected spatial cluster *s*, is then calculated as:

$$P_{ik\varepsilon s} = \frac{\exp(V_{ik})}{\sum_{k\varepsilon s}^{n} \exp(V_{ik})} \quad \forall k \,\varepsilon\, s \tag{5}$$

In addition, the probability of a decision-maker selecting *k* from the set of all alternatives is:

$$P_{ik} = P_{is} P_{ik\varepsilon s} \tag{6}$$

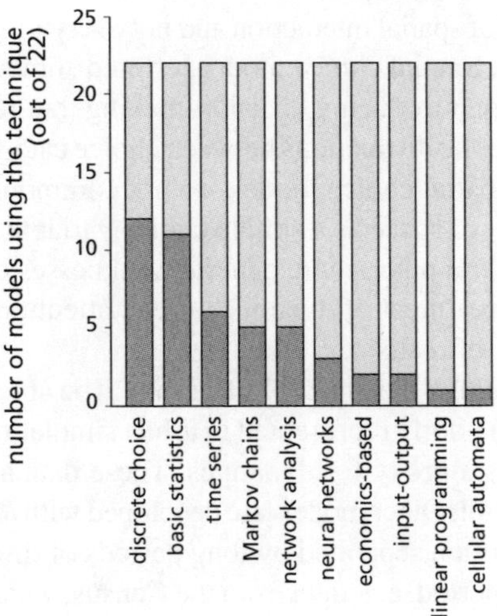

Fig 12.2: Common simulation techniques underpinning operational land use and transport models
Source: Data from US EPA (2000 Appendix A)

Between them, spatial interaction and spatial choice models comprise the bulk of material underpinning land use and transport models as planning support tools. In his 1994 review of operational modeling, Wegener identified 12 models that he considered to be state-of-the-art. Of those, 10 were formulated predominantly as discrete choice models (Wegener 1994). More recently, a similar review by the US Environmental Protection Agency (2000) identified 22 operational models and, once again, discrete choice models featured most prominently (Figure 12.2). Despite their popularity, however, these techniques have several weaknesses.

12.2.1 Criticisms of the Traditional Approach

As mentioned in the introduction, traditional approaches to urban simulation have come under heavy criticism over the years. Lee (1973) provided one of the most well known articles documenting their failures; another influential article was written by Sayer (1979). Several of the complaints forwarded in those papers were resolved over time, simply through innovation in computer hardware, software, data capture techniques and the development of new methodologies (see Batty 1994; Harris 1994; Klosterman 1994). While the models may be regarded as reasonably successful for providing a very general simulation framework for planning support, they have some important weaknesses and any success achieved with their application is tempered by the lack of an available alternative. The assumption that flows of people in an urban system – whether for travel or migration – take place in manners akin to those described by Newton is a simplification. Furthermore, spatial interaction models convey only one form of spatial interaction and not everything in a city passes through urban space as a flow. Spatial choice models offered somewhat of an improvement over gravity models by introducing decision-making, but retained weaknesses. It is widely understood that the distinctions between choice categories may often be fuzzy rather than discrete; spatial choice models do not commonly accommodate this. In addition, the treatment of space as a simple choice variable in a utility calculation is severely limiting. Several other, more generic weaknesses of traditional models can be identified: a poor treatment of dynamics, weak attention to detail, shortcomings in usability and a lack of realism.

Urban systems are highly dynamic. Models should be able to capture that property, but dynamics are often poorly represented in urban simulations. *Cross-sectional* data are commonly used as a proxy for dynamics. These data are collected for a single period in time: a snapshot. Other models are developed with *longitudinal* data, offering a series of snapshots, often separated by long periods of time with little information about the intervening period, e.g. data from the Census, which is commonly reported on a ten-year basis. While longitudinal data are much richer in the information they convey, they still constitute a weak proxy for dynamics – a lot can happen in a city in ten years! Consider the chart represented in Figure 12.3.

In this instance, were we to use two snapshots as the basis for explaining dynamics in a model of the stock market – say, September 1998 to September 2001 – we would be looking to explain a flat market; the values at the start and end of the time period are essentially identical at around $1,500. The processes that yield market stability are much different from those that generate the sort of volatility that a more fine-scaled examination of the period would reveal. Any results derived from a model designed to explain stability would be wholly inappropriate in this case. The same may well be true for urban models calibrated with longitudinal data.

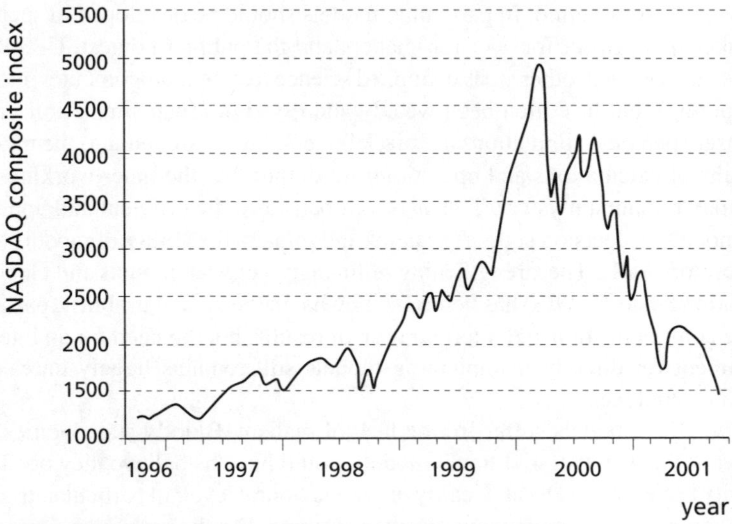

Fig 12.3: The NASDAQ composite index for a five-year period
Copyright 2001 Bloomberg LP. Reprinted with permission. All Rights reserved.

Traditional models are often weak in handling detail. For the most part, this is due to a lack of data available at fine-scale spatial resolutions. Traditional models generally adopt the Traffic Analysis Zone (TAZ) as a minimum level of spatial resolution. TAZs are aggregate levels of geography: a medium-size city would be divided into just a few hundred TAZs, for example. From this level of geography, one can only infer information at the level of individuals or entity-level geographies of urban space and to do so invokes issues of ecological fallacy and modifiable areal unit problems (MAUP) (Openshaw 1983). At the heart of the MAUP is the issue that there are almost an infinite number of spatial objects that can be defined and modified for any given area of inquiry but few, if any, modifiable entities. Census data, for example, are collected in many cases for spatially non-modifiable entities (e.g. people, households) and are reported across spatially modifiable units (e.g. counties, ZIP code geographies, census tracts and block groups). Ecological fallacy is a closely related problem. An ecological fallacy occurs when it is inferred that results based on aggregate data can be applied to the individuals who form the aggregated group. Traditional models also represent discrete socio-economic groupings in a city in a relatively aggregated manner. Household and employment categories are often divided into a handful of classifications. Microsimulation models have gone some way toward disaggregating the groupings represented in simulations, but they are still a long way from an accurate representation of the variety and diversity of people and activities in an urban system. This lack of detail can, in some cases, be regarded as a serious limitation of traditional models because many of the processes that take place in urban systems operate at finer resolutions.

It is vitally important that urban models intended for use as PSS are developed with the end-user in mind. In particular, models should be developed in such a way that makes them easier for decision-makers and the public to digest. Usability has long been a concern in other areas of applied science (e.g. human-computer interaction in computing), but has often been weakly addressed in urban simulation. In many cases, users perceive simulations as 'black boxes'; inputs are fed into the model and the results of calculations and operations are output, but the inner workings of the model may remain a mystery. This acts as a barrier to the efficient and appropriate use of models as decision support systems and impairs the ability of models to serve as exploratory tools. The strengthening of linkages between models and Geographic Information Systems (GIS) has helped somewhat in the area of usability, particularly with the communication and interpretation of results, but the need for an interactive environment for directly manipulating models still remains largely unrealized in operational contexts.

Finally, urban models suffer from a lack of realism. Bluntly stated, cities do not really work the way that traditional models would have us believe they do. There is a disparity between models and reality on a behavioural level. In particular, traditional models adopt a reductionist view of urban systems. For the most part, assumptions are made that portray cities as operating from the top down. This implies dissecting cities into constituent local components from aggregate conditions in order to understand them. In many cases, this is appropriate (when studying planning constraints, large-scale infrastructure improvements, for example). However, in other instances, it is inappropriate (when studying housing demand, commuting, for example). Many components of urban systems do not work in a top-down manner; on the contrary, aggregate conditions emerge from the bottom-up, from the interaction of large numbers of elements and entities at a local scale.

12.3 A 'New Wave' of Urban Models

In recent decades, social scientists have begun to work with a new class of simulation techniques that we might term complexity models, geocomputation models, or simply geosimulation models. This 'new wave' of simulation opens up exciting possibilities for simulating systems of all descriptions, and in particular, the simulation of behavioural processes and the structures that they generate. While the use of computers and computation in urban simulation is by no means new, the geosimulation approach – modelling systems at the scale of individuals and entity level units of the built environment – is particularly innovative from an urban simulation standpoint and offers some significant advantages for the use of urban simulation as a planning support tool.

A number of key developments across disciplines have supported the introduction of these new techniques. Within the geographical sciences, geosimulation models have been supported by a flood of detailed geographic information that has become easily attainable in recent years. These data have been made available in a variety of media and covering phenomena that would not have been possible a relatively short time ago, e.g. multi-spectral and fine-scale resolution remotely sensed data on land use and land cover change in urban areas. The provision of these data has been directly responsible for addressing some of the weaknesses we have just explored; a lack of detail in traditional models, for example. Also, it has had indirect impacts on urban simulation by supplying new insights into how urban systems operate, thereby allowing us to develop better-informed simulations. Furthermore, GIS have been developed to store, manipulate and display spatial data. There is now a rich tradition of use of these systems in PSS contexts.

Object-oriented (OO) programming languages such as Java and C++ have also contributed to the development of geosimulation-style models. OO approaches offer obvious advantages for the treatment of discrete entities of urban systems, e.g. land parcels, buildings, administrative zones, households and individuals. The basic unit in OO programming is the object. Objects are associated with data and their behaviour is mediated through methods (conditional statements and calculations that determine how objects should interact and evolve over the lifetime of a program run). The conceptualization of pieces of inanimate code as objects with related data and methods mimics the way that we think of real world objects ourselves: as discrete units with associated attributes and behaviours. This has several benefits for making models more flexible to build, as well as making them easier to convey and understand.

Ideas stemming from complexity studies have also been instrumental in the development of new generations of urban models. The main idea in complexity is that of emergence. In emergent systems, a small number of rules or laws, applied at a local level and among many interacting entities are capable of generating surprising complexity and often ordered patterns in aggregate form. Additionally, these systems are dynamic and change over time without the direction of a centralized executive. Complex patterns often manifest themselves in such a way that the actions of the parts do not simply sum to the activity of the whole. Essentially, this means that there is more going on in the dynamics of the system than simply aggregating little pieces into larger units. Examples of emergent systems abound. Stock markets are a good example: markets such as the New York Stock Exchange (NYSE) are comprised of millions of traders buying and selling in a bid to maximize their own individual profits. In the eighteenth century, the Scottish economist Adam Smith postulated the idea of an 'invisible hand' that set the level of equilibrium between supply and demand in the market place. Individual investors in stock markets act without any centralized control, yet their activities often lead to aggregate outcomes that are relatively efficient, as efficient as if they *were* controlled. Many urban systems are also complex in this sense. From the local-scale interactive behaviour (commuting, moving) of many individual objects (vehicles, people), structured and ordered patterns may emerge in the aggregate, such as peak-hour traffic congestion (Nagel *et al.* 1996) and the large-scale spatial clustering of socioeconomic groups by residence (Benenson 1998). In urban economics, large-scale economies of agglomeration and disagglomeration have long been understood to operate from local-scale interactive dynamics (Krugman 1996). In addition, cities exhibit several of the signature characteristics of complexity, including phase transitions, fractal dimensionality and self-similarity across scales, self-organization and emergence (Batty and Longley 1994; Allen 1997; Portugali 2000).

Complexity studies have shed new light on our thoughts regarding the inner workings of cities and have had profound impacts on our approach to urban simulation. They suggest a detailed, decentralized and dynamic view of urban systems. They also offer a framework for considering answers to questions of the form, 'How do cities work?' in terms of the myriad and evolving interactions of individuals and the urban spaces that they inhabit. This is a much more generative approach than the reductionist view that is traditionally adopted in urban studies. Simply dissecting cities may not provide all the answers; on the contrary, there may be a need to build them up from the bottom and in doing so, we may learn a lot about how they work. As will now be explored, this may also have some direct analogies in the way we simulate cities.

12.3.1 The Geosimulation Approach

The geosimulation approach offers a unique perspective that traditional simulation has commonly lacked: a view of urban phenomena and the spatial processes that shape them as a result of the collective dynamics of multiple urban animate and inanimate objects. Geosimulation can be considered as an extension of traditional urban modelling in several ways.

First, and as we have already seen, the intellectual roots of geosimulation differs from that of previous generations of models. Ideas from economics, physics and engineering are still adapted to spatial contexts in urban models, but the range of inspiration is now much wider. Geosimulation models borrow from developments in several fields that were not widely considered, particularly computer science (artificial life and artificial intelligence) and natural science.

Second, the geosimulation approach differs in its depiction of spatial units. While traditional urban models have focused on aggregate partitions of urban space – essentially modifiable spatial units – geosimulation works with discrete and spatially non-modifiable individual objects, such as houses, lots, householders and landowners.

The third feature relates to the portrayal of spatial interactions. As we saw, traditional models have focused on describing flows of matter and information between aggregate spatial units. Geosimulation models contrast by concentrating on the interactive behaviour of elementary geographic objects, and that interaction may take many forms: flows, distance-decay, diffusion, dispersal, action-at-a-distance, centripetal and centrifugal activity, linear and non-linear relationships, *et cetera.* If interactions are modelled at higher-level units of urban space, they are simulated in geosimulation as the outcomes of collective interactions at micro-scales. The choice sets used in spatial choice models also appear in geosimulation models, but in this case those rules are mediated spatially in a variety of ways, above and beyond the use of a simple choice set to represent space and geography.

The fourth characteristic is concerned with the treatment of time. As we saw, traditional models are essentially static. Geosimulation models, by comparison, are very dynamic and in some cases, they can approximate real time interaction within a simulation environment.

Finally, traditional and geosimulation models differ in their attitude to urban simulation. Geosimulation could be considered as a reconsideration of the goals of simulation, with a new emphasis on building scenario-generating games – tools to think with – rather than predictive models. Various simulation scenarios can be designed, each based on different suggestions regarding factors of urban dynamics, and played through to likely conclusions before being tried in the real world. The intuitive and transparent nature of the models should facilitate discussion by model users, rather than providing prescriptive remedies based on simplified assumptions. This has obvious advantages for the application of simulation tools to PSS.

The two main techniques used to build geosimulation models for application in urban studies are cellular automata (CA) and multi-agent systems (MAS). Both CA and MAS share the traits of geosimulation models discussed above. In terms of urban simulation, CA are perhaps best used to represent the dispersal of activity and characteristics between discrete spatial units of urban infrastructure and land. MAS may be more suited to simulating urban population as collectives of individuals with associated behaviours and traits and the capacity for spatial mobility and communication.

12.3.2 Cellular Automata

Cellular automata were originally pioneered in computing (Sipper 1997) but have since seen uses in a wide variety of fields, including urban studies (see Batty *et al.* 1997; Torrens 2000a). A cellular automaton is a finite state machine (an engine of sorts) that exists in some form of tessellated cell-space (Figure 12.4). The term automaton refers to a self-operating machine, but one of a very distinct nature: *"An automaton is a machine that processes information, proceeding logically, inexorably performing its next action after applying data received from outside itself in light of instructions programmed within itself."* (Levy 1992, p.15)

Additionally, CA are *parallel* automata: more than one automaton is active at any given instance. CA are comprised of five components. The *lattice* of CA is the space in which they exist. This might be considered equivalent in an urban context to a city, an environment, a landscape, or a territory. The lattice can also be generalized to represent urban spatial structures, networks of accessibility, the physical structure of the city. CA *cells* represent the discrete confines of individual automata. They are the elemental building blocks of a CA, just like individual land parcels or buildings in a city. CA cells are, at any time, in a particular *state*. The cell state offers a flexible framework for encoding attributes of a city into an urban simulation model, e.g. land use, density or land cover.

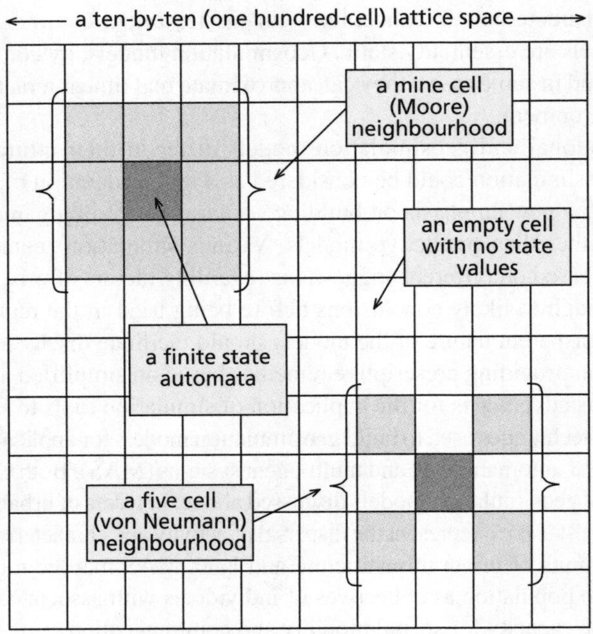

Fig 12.4: The characteristics of a basic cellular automata

Neighbourhoods are the localized regions of a CA lattice (collections of cells), from which automata draw input. Neighbourhoods in an urban CA might represent spheres of influence or activity, e.g. market catchment areas or commuting watersheds. The real driving force behind CA are *transition rules*. These are simply a set of conditional statements that specify the behaviour of cells as CA evolve over time. The future conditions of cells are decided based on a set of fixed rules that are evaluated on input from neighbourhoods. CA rules can be devised to mirror how phenomena in real cities operate. Additionally, we might discern another component – *time* – that is generally discrete and proceeds in iterative steps. In CA, the dynamic processes of change are represented through local actions (governed by transition rules) that are applied in the immediate proximity of the various objects that comprise the system of interest (Batty *et al.* 1999b). Because CA are dispersive in their action through space, this can generate structures on a macro-scale.

CA offer a range of advantages for urban simulation and in several ways they remedy particular deficiencies of traditional models. Whereas space is represented in a very simplistic manner in traditional models (commonly as a single variable or a distance calculation), CA provide for a much richer representation. Urban systems can be represented at a variety of spatial scales simultaneously, directional elements can be introduced to mimic the way that cities develop in sectors or wedges, different distance functions can be added to characterize distance-decay, and neighbourhoods can be set up hierarchically to represent action-at-a-distance. Additionally, the structure of space can be modified in a variety of ways from irregular to regular representations.

CA transition rules have been devised that mimic those of spatial interaction and spatial choice models. However, CA are highly flexible and transition rules can be devised to represent just about any process or variable you can imagine: geographical inertia, hierarchy, accessibility, suitability, potential for development, utility functions, constraints, exogenous growth, feedback effects, and deterministic and probabilistic processes. Importantly, they offer the potential for weaving theory and planners' knowledge and expertise directly into the cogs of model dynamics. The framework for CA (and for MAS) is not as constraining as those of traditional models might be considered to be.

Additionally, CA can be designed with attention to detail. They are inherently decentralized. They are dynamic, as well as being intuitively useful and behaviorally realistic. Additionally, they have a 'natural affinity' with raster data and GIS (Couclelis 1997), as well as with OO programming. CA also provide a mechanism for linking micro- and macro-approaches and for connecting patterns with the processes that produce them.

CA have been used quite widely in urban studies in recent years. The most common applications of the models are to land development, urban growth, and land-use transition. In terms of land development, CA have been used to investigate the role of density constraints in development and the spatial distribution of growth activity (Batty *et al.* 1999b). Other models have characterized the development process as a profit calculation, mediated through space with the use of decision-making regimes from game theory (Wu and Webster 1998). CA have also been used to simulate development as a function of demand, supply, and potential (Batty 1998).

CA models have also been applied to simulating urban growth processes, as well as specific forms of urban growth such as polycentricity (Wu 1998). CA have been used to represent the evolution of urban form through growth cycles (Batty *et al.* 1999b). Growth has also been modelled proceeding from historically identified 'seed' cells, using self-modifying transition rules that mimic the adaptability of cities over time (Clarke *et al.* 1997), and using predator-prey algorithms (Batty *et al.* 1999a). White and Engelen (2000) have developed CA models that rely on exogenously defined growth engines that reflect the position of cities in larger economic regions and economies.

CA have been quite widely applied to simulating land-use transition. Land-use dynamics have been modeled as a hierarchical process (White and Engelen 1997) and as a function of the rent-bidding power of individual sites and the externalities that they might produce (Webster *et al.* 1998). The role of inertia in land use transition has also been widely investigated, as has the role of constraints such as accessibility (White 1998), density, and topological factors (Clarke *et al.* 1997).

12.3.3 Multi-agent Systems

While CA are most suitable, in urban simulation contexts, for representing infrastructure and land, MAS are better used to model population dynamics. MAS also have origins in computer science, although their development post-dates that of CA by some years. Most commonly, MAS are used in computing as artificial intelligence systems or artificial life forms (Kurzweil 1990; Levy 1992). Additionally, there are 'species' of agents that serve as 'network bots', 'web crawlers', and 'spiders' (Leonard 1997). Network agents are used to navigate computer information networks, to 'mine' data, retrieve it, and return it to human users. There is also a tradition of using software agents to explore entomological behaviour (Bonabeau *et al.* 1999) and the actions of agents in economic systems and markets (Luna and Stefansson 2000).

Agents are quite similar to automata in their formulation but have less well-defined characteristics. They constitute pieces of software code with certain attributes (states) and behaviours (rules) (see Ferber, 1999 for a general introduction to intelligent software agents). They differ from CA mainly in their spatial mobility: agents can be designed to navigate (virtual) spaces with movement patterns that mimic those of humans, while CA are only capable of exchanging data spatially within the confines of their neighbourhoods. In some senses, agents could be considered as mobile CA. Their states (e.g. occupation, income, age) describe their characteristics in the same way as in CA and they exist in some space (although this need not necessarily be a cellular space). Agents may have a neighbourhood of influence in which they operate, but the criteria for defining that neighbourhood are much more flexible than those applying to CA. Like CA, MAS are driven by transition rules that govern the behavior of the agents. These rules may be characterized as 'goals' that agents seek to satisfy (e.g. minimizing travel distance) or even 'preferences' that agents may possess (e.g. likes and dislikes for certain locations in space), or may be derived from their state variables.

MAS are excellent tools for representing mobile entities in urban environments, e.g. people, households or vehicles. However, their application to urban studies has not been as widespread as that of CA, although there seems to be no particular reason why this should be the case. MAS hold all of the advantages of CA, but with the additional capacity for representing a wider array of spatial processes. Like CA, MAS are easily programmed in OO environments, as well as offering advantages in terms of detail, flexibility, dynamics, usability and behavioural realism. They have been used in urban contexts to simulate pedestrian movement in dense urban environments (Batty 2001). Here, agents are designed 'choreographically' with the capacity for realistic movement through space. Additionally, agents may be equipped with realistic behaviors (such as shopping habits) derived from life-like socio-economic profiles.

Another widespread application of MAS is in simulating residential location patterns. Whereas with spatial interaction and spatial choice models, residential location processes were modelled as simple flows between destinations or as aggregate-level decisions, MAS can model the process in a more life-like manner. Individual households can be represented with realistic profiles and preferences, then sent out to interact in virtual housing markets (Benenson 1999). In one example, residential location is modelled using both CA and MAS in a hybrid fashion (Torrens 2001a). Home-buying and home-owning agents negotiate the sale of properties through the help of MAS, while neighbourhood effects that influence the attractiveness of certain areas of the city are simulated using a CA.

12.4 Epilogue

Although CA and MAS models are developed, for the most part, in academic contexts, they are beginning to find their way into PSS. The TRANSIMS model – a CA simulation of traffic dynamics, capable of simulating interactions among hundreds of thousands of vehicles in real time – is one of the most ambitious urban simulation projects ever developed (Nagel *et al.* 1999). That model is scheduled for practical application in a range of cities in the United States. The models developed by White, Engelen and colleagues have also been used in practical contexts. In addition, Clarke's SLEUTH model has also seen considerable application in practical contexts. However, it is unlikely that CA and MAS will be used in isolation as planning support tools. They have yet to be widely tested in planning contexts, simply because the field of research is quite new. More traditional approaches, on the other hand, have enjoyed decades of use by planners, policy-makers and urban managers. It is more likely that CA and MAS will supplement existing systems rather than supplanting them. It is here that CA and MAS can make their immediate contribution, by filling in the gaps that traditional models leave, particularly those left open at the micro-scale.

The discussion thus far has been quite optimistic about the potential of new techniques to revitalize operational simulation. The techniques themselves do certainly represent the possibility for a 'revolution' in the way we simulate urban systems. However, there are some imposing barriers to putting those techniques into practical use in the real world (Torrens and O'Sullivan 2001). Ironically, computing power poses one of the most pressing limitations. CA and MAS models have been developed and tested for several cities, but most are quite modest in the number of entities that they simulate. Scaling those models up to represent an entire metropolitan area, populated with individual agents numbering millions would require daunting levels of computing power. However, even here advances are being made, particularly in the use of distributed computing to provide the computational engines for detailed models. As mentioned, the TRANSIMS model at Los Alamos National Laboratories is one such example (Nagel *et al.* 1999).

In addition, there are data limitations on the development of these models for practical uses. Conceptually, the idea of simulating individuals and the buildings that they inhabit is quite appealing. However, data are not widely available at the scale of the individual householder or building. In addition, there are several moral issues that arise from the use of individual-level – and often private – data in operational simulations.

Furthermore, micro-scale models, particularly dynamic and process-driven simulations, are quite difficult to calibrate, even if data are available. Organizing the models as a hybrid that interfaces with traditional aggregate-level models allows the possibility of scaling up the simulation to meso-scales for validation purposes (Torrens 2001b). This is a reasonable solution, but ideally, micro-models would be calibrated at the scale of the entity or the individual. The emphasis thus far in geosimulation modelling has been on pattern-based validation techniques: pattern recognition, and measures of match such as the chi-squared and kappa-statistics. Generally, this sort of calibration focuses on validating the patterns that CA or MAS models have generated and comparing those results with historical maps of urbanization or land-use patterns, using that as a justification for simulating ahead in time. The weaknesses of these approaches have been well documented (O'Sullivan and Torrens 2000).

Some significant advances have been made in using GIS to analyze the patterns that geosimulation models generate, including some 'fuzzy' recognition techniques. However, much of this effort is still bogged down in pattern-based approaches, ignoring the fact that geosimulation models comprise pattern and process, form and function. Future research may therefore have to look to new process-related validation measures such as Monte Carlo averaging, spatial information statistics and measures of complexity.

Additionally, working at the micro-scale, in some cases, reveals inadequacies in the theory of how cities work. The micro-approach betrays some theoretical gaps in our understanding of the dynamic interactions that shape our urban systems. Indeed, one of the important contributions of these models in academic terms has been in providing some justification for a 'new urban geography' of the micro-scale.

The point that this chapter is intended to convey, however, is that – at least methodologically and increasingly in practice – the techniques discussed here represent a move towards more theoretically sound, behaviorally realistic, and ultimately more useful simulation environments as planning support tools. Certainly, these simulations can be developed as proof-of-concept tools and the methodologies can be refined in academic contexts in preparation for a day in which these tools can be used to plan and manage better cities. It is also likely that several of their features will begin to see application in the real world, alongside existing traditional models (Torrens 2001b); in fact, this is already occurring. In the meantime, even as abstract tools, these simulations can do a lot for our understanding of how cities work and perhaps provide new insights into how we might construct a more sustainable urban future.

References

Allen, P.M. (1997) *Cities and Regions as Self-Organizing Systems: Models of Complexity,* Gordon and Breach Science Publishers, Amsterdam.

Batty, M. (1994) A chronicle of scientific planning: the Anglo-American modeling experience, *Journal of the American Planning Association,* 60 (1): 7-16.

Batty, M. (1998) Urban evolution on the desktop: simulation with the use of extended cellular automata, *Environment and Planning A,* 30: 1943-1967.

Batty, M. (2001) Theme issue: pedestrian modeling, *Environment and Planning B*, 28 (3).

Batty, M., Couclelis, H. and Eichen, M. (1997) Special issue: urban systems as cellular automata, *Environment and Planning B,* 24 (2).

Batty, M. and Longley, P. (1994) *Fractal Cities.* Academic Press, London.

Batty, M., Xie, Y. and Sun, Z. (1999a) The dynamics of urban sprawl. CASA Working Paper 15. University College London, Centre for Advanced Spatial Analysis (CASA). London. http://www.casa.ucl.ac.uk/working_papers.htm.

Batty, M., Xie, Y. and Sun, Z. (1999b) Modeling urban dynamics through GIS-based cellular automata, *Computers, Environment and Urban Systems,* 23 (3): 205-233.

Benenson, I. (1998) Multi-agent simulations of residential dynamics in the city, *Computers, Environment and Urban Systems,* 22 (1): 25-42.

Benenson, I. (1999) Modelling population dynamics in the city: from a regional to a multi-agent approach, *Discrete Dynamics in Nature and Society,* 3: 149-170.

Bonabeau, E,, Dorigo, M. and Theraulaz, G. (1999) *Swarm Intelligence: From Natural to Artificial Systems,* Oxford University Press, New York.

Clarke, K.C., Hoppen, S. and Gaydos, L. (1997) A self-modifying cellular automaton model of historical urbanization in the San Francisco Bay area, *Environment and Planning B,* 24: 247-261.

Couclelis, H. (1997) From cellular automata to urban models: New principles for model development and implementation, *Environment and Planning B,* 24: 165-174.

De la Barra, T. (1989) *Integrated Land Use and Transport Modelling: Decision Chains and Hierarchies.* Cambridge University Press, Cambridge.

Ferber, J. (1999) *Multi-Agent Systems: An Introduction to Distributed Artificial Intelligence.* Addison-Wesley, Harlow (UK).

Fotheringham, A.S. and O'Kelly, M.E. (1989) *Spatial Interaction Models: Formulations and Applications,* Kluwer Academic Publishers, Dordrecht.

Harris, B. (1994) The real issues concerning Lee's "Requiem", *Journal of the American Planning Association,* 60 (1): 31-34.

Klosterman, R.E. (1994) Large-scale urban models: retrospect and prospect, *Journal of the American Planning Association,* 60 (1): 3-6.

Krugman, P. (1996) *The Self-Organizing Economy,* Blackwell, Malden, MA.

Kurzweil, R. (1990) *The Age of Intelligent Machines,* MIT Press, Cambridge, MA.

Lee, D.B. (1973) Requiem for large-scale models, *Journal of the American Institute of Planners,* 39: 163-178.

Leonard, A. (1997) *Bots: The Origin of a New Species,* Hardwired, San Francisco.

Levy, S. (1992) *Artificial Life: The Quest for a New Creation,* Penguin Books, London.

Luna, F. and Stefansson, B. (eds). (2000) *Economic Simulation in Swarm: Agent-based Modelling and Object Oriented Programming*, Kluwer, Dordrecht.

Nagel, K., Beckman, R.J. and Barrett, C.L. (1999) TRANSIMS for urban planning, LA-UR 98-4389, Los Alamos National Laboratory. Los Alamos, NM. http://transims.tsasa.lanl.gov

Nagel, K., Rasmussen, S. and Barrett, C.L. (1996) Network traffic as self-organized critical phenomena. TSA-DO/SA MS-M997 and CNLS MS-B258, Los Alamos National Laboratory. Los Alamos, NM. http://transims.tsasa.lanl.gov/.

Openshaw, S. (1983) *The Modifiable Areal Unit Problem,* GeoBooks, Norwich.

O'Sullivan, D. and Torrens, P.M. (2000) Cellular models of urban systems, in Bandini, S. and Worsch, T. (eds) *Theoretical and Practical Issues on Cellular Automata*, Springer Verlag, London, pp. 108-117. http://www.casa.ucl.ac.uk/working_papers.htm.

Portugali, J. (2000) *Self-Organization and the City.* Springer Verlag, Berlin.

Sayer, R.A. (1979) Understanding urban models versus understanding cities, *Environment and Planning A,* 11: 853-862.

Sipper, M. (1997) *Evolution of Parallel Cellular Machines: The Cellular Programming Approach.* Springer, Berlin.

Thomas, R.M. and Huggett, R.J. (1980) *Modelling in Geography: A Mathematical Approach.* Barnes & Noble Books, Totowa, N.J.

Torrens, P.M. (2000a) How cellular models of urban systems work, *CASA Working Paper 28*, University College London, Centre for Advanced Spatial Analysis. http://www.casa.ucl.ac.uk/working_papers.htm.

Torrens, P.M. (2000b) How Land-Use and Transportation Models Work, *CASA Working Paper 29*, University College London, Centre for Advanced Spatial Analysis. http://www.casa.ucl.ac.uk/working_papers.htm.

Torrens, P.M. (2001a) Can geocomputation save urban simulation? Throw some agents into the mixture, simmer, and wait.... *CASA Working Paper 32,* University College London, Centre for Advanced Spatial Analysis. London. http://www.casa.ucl.ac.uk/working_papers.htm.

Torrens, P.M. (2001b) A hybrid geocomputation model for operational land-use and transport simulation. 97th Annual Meeting of the Association of American Geographers, New York. http://www.casa.ucl.ac.uk/geosimulation/publications/aag_2001.pdf.

Torrens, P.M. and O'Sullivan, D. (2001) Cellular automata and urban simulation: where do we go from here?, *Environment and Planning B*, 28 (2): 163-168.

U.S. Environmental Protection Agency. (2000) Projecting land-use change: A summary of models for assessing the effects of community growth and change on land-use patterns. EPA/600/R-00/098. U.S. EPA. Washington, D.C.

Webster, C., Wu, F. and Zhou, S. (1998) An object-based urban simulation model for interactive visualisation. Third International Conference on GeoComputation, University of Bristol, http://www.geog.port.ac.uk/geo98/.

Wegener, M. (1994) Operational urban models: state of the art, *Journal of the American Planning Association,* 60: 17-29.

White, R. (1998) Cities and cellular automata, *Discrete Dynamics in Nature and Society*, 2: 111-125.

White, R. and Engelen, G. (1997) Cellular automata as the basis of integrated dynamic regional modelling, *Environment and Planning B,* 24: 235-246.

White, R. and Engelen, G. (2000) High-resolution integrated modelling of the spatial dynamics of urban and regional systems, *Computers, Environment and Urban Systems*, 24: 383-400.

Wu, F. (1998) An experiment on the generic polycentricity of urban growth in a cellular automatic city, *Environment and Planning B,* 25: 731-752.

Wu, F. and Webster, C.J. (1998) Simulation of land development through the integration of cellular automata and multicriteria evaluation, *Environment and Planning B,* 25: 103-126.

13 The Application of Case-based Reasoning in Development Control

Anthony Yeh[1] and Xun Shi[2]

[1] Centre of Urban Planning and Environmental Management, The University of Hong Kong, Pokfulam Road, Hong Kong SAR, China
[2] Department of Geography, University of Wisconsin, WI 53706, Madison, USA

Abstract: Case-based reasoning (CBR) is a type of knowledge-based system (KBS) whose philosophy is to use previous cases to interpret or solve a new problem. There are many advantages of CBR over rule-based reasoning, the method that is adopted by most current KBS for development control. The knowledge representation scheme of CBR is better than rule-based reasoning in recording and representing knowledge that is hard to express with explicit rules or is too case-specific. It can overcome the black box inference process of rule-based reasoning. CBR can simulate the present working style of a planner in dealing with development applications that are based on his/her knowledge of past application records. The results derived directly from real cases in CBR are more convincing and acceptable to planners. A system that integrates a CBR shell (ESTEEM) and GIS package (ArcView) was developed to demonstrate the process in building a case-based system (CBS) to assist planners in the Planning Department of Hong Kong in handling planning applications in development control. Because it is case-based and not rule-based, the system is also applicable to other cities in dealing with development applications.

13.1 Introduction

Urban planning often involves a huge amount of data, regulations and guidelines from various sources. It also requires the experiences and expertise of planners and the assistance of planning models and analytical methods in its complicated decision-making process. But the capability of people in handling large amounts of information and the availability of experienced planners are often very limited. Therefore, computerized knowledge-based systems (KBS) or expert systems (ES) which have the capability of easy storage and retrieval of the required information and knowledge and *"can reach a level of performance comparable to – or even exceeding that of – a human expert"* (Turban 1995 pp. 16), would be, at least theoretically, very helpful to planners in their work. Garcia and Chien (1991 pp. 2-3) and Stefik (1995, pp. 296-297) discuss the terms *knowledge-based systems* and *expert systems.* This chapter will use the more general one, knowledge-based systems, although it does not totally include the concept of *expert system* according to Garcia and Chien (1991).

KBS and ES have aroused the interest of planners for many years. Many potential applications have been explored by researchers and a few prototypes and even operational systems have been built (Kim *et al.* 1990). However, there are many limitations in using KBS in urban planning (Barbanente *et al.* 1995, Borri *et al.* 1994, Han and Kim 1990b, Marchand 1993, Ortolano and Perman 1990). The two main obstacles in successfully applying KBS to urban planning are difficulties in building a practical KBS for planning and difficulties encountered to *"inspiring confidence in the use of it"* (Marchand 1993). Some of these problems can be overcome by the use of case-based reasoning (CBR) which uses previous similar cases to suggest solutions to new cases (Yeh and Shi 1999). In contrast to rule-based reasoning or model-based reasoning, CBR directly uses concrete knowledge and its inference is basically the processes of retrieval and adaptation. These features make CBR able to avoid some of the problems in building KBS, such as knowledge elicitation bottlenecks, and can gain the confidence of users. This chapter will examine the use of CBS in development control and its advantages over conventional KBS.

13.2 Case-based Reasoning and Urban Planning

13.2.1 An Introduction to CBR

Originating from the work of Schank and Abelson (1977), CBR has developed into a mature and important field of AI (Aamodt and Plaza 1994, Kolodner 1993, Schank and Leake 1989, Watson 1997). Increasingly, many domains are using CBR, such as architecture design, law, medicine and customer services (help-desk) (Watson 1997). The homepages of the Navy Center of Research for Artificial Intelligence (*http://www.aic.nrl.navy.mil/~aha/research/case-based-reasoning.html*) and AI-CBR Case-Based Reasoning web site (*http://www.ai-cbr.org/*) contain useful information, examples and references of CBR. Generally speaking, CBR refers to a concept and its corresponding technique in the field of KBS. Its philosophy is to use previous similar case(s) to help to solve, evaluate or interpret a current new problem (Kolodner 1993). It simulates human behaviour in solving new problems. When confronted with a new and difficult problem, it is natural for a human problem-solver to look into his/her memory to find previous similar instances for help. An important reason why a person can become an expert is that he/she can remember and properly use suitable previous cases in solving new ones.

In simulating the human brain in problem solving, a computerized case-based system (CBS) using CBR sees knowledge as encapsulated memories and its knowledge base is a *case library* storing these memories in the form of concrete stories. After the user inputs the descriptions of a new problem into the computer, the computer will look for similar old cases in its case library according to the predefined matching algorithms. Cases which meet certain criteria will be retrieved and their solutions or any other parts of the stories required by the user will be either directly proposed to the user or, if necessary, be adapted to meet the new situation before being proposed. The user can evaluate or test whether the proposed solutions work well or not and modify them. When a satisfactory solution is obtained, the newly solved problem can be stored in the case library as a new case and the knowledge of the system increases (for details of the techniques of CBR, see Aamodt and Plaza 1994, Kolodner 1993, Watson 1995, 1997). The CBR process can be typically represented by a schematic cycle comprising the four REs (Figure 13.1):

- RETRIEVE the most similar case(s);
- REUSE the case(s) to attempt to solve the problem;
- REVISE the proposed solution if necessary, and
- RETAIN the new solution as a part of a new case.

A new problem is matched against cases in the case library and one or more similar cases are *retrieved.* The solution (transformational reuse) and the past method that constructed the solution (derivational reuse) of the retrieved cases is then *reused* and tested for success in application. Unless the retrieved case is a close match to the new problem, it is most likely that the solution has to be *revised* in producing a new case that can be *retained.*

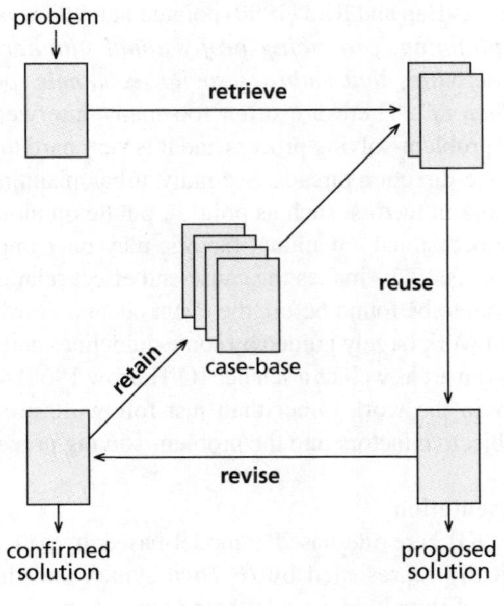

Fig 13.1: The CBR cycle (adapted from Aamodt and Plaza 1994 by Watson 1995)

13.2.2 Advantages of CBR Over Rule-based Reasoning to Urban Planning

In building a practical KBS for urban planning, the main difficulties exist in the process of *knowledge acquisition* and *knowledge representation.* The successful construction of a practical KBS is largely based on the assumption that the domain knowledge for problem solving can be properly extracted and represented. Many efforts have been made to refine these processes. However, both knowledge acquisition and knowledge representation have always been the major difficulties in knowledge engineering. They are generally referred to as *knowledge acquisition problem* (Musen 1993) and *knowledge elicitation bottleneck* (Han and Kim 1990, Kolodner 1993, Musen 1993, Watson 1995). Unfortunately, in urban planning this bottleneck seems to be even more difficult to overcome than other domains.

Knowledge Acquisition

Many authors asserted that KBS (rule-based or model-based) is based on the assumptions that (adapted from Han and Kim 1990):

- the problem can be clearly specified and is well-bounded;
- the relations between the factors or elements of the problem are known and can be expressed explicitly;
- problem-solving methods can be articulated; and
- experts agree on solutions, that is, the cause-effect relationships are clearly defined.

However, most of the problems in urban planning seem to deviate far from these assumptions. Firstly, as Han and Kim (1990) pointed out, "*Because of the distinctive nature of urban planning, practicing professional planners must deal with multidisciplinary activities that embrace social, economic, political, and even anthropological factors*". There are often too many interweaving factors and relationships in the problem-solving process and it is very hard to analyze them into clear, definite and one-direction threads. Secondly, urban planning is also a domain in which many uncertain factors, such as politics, public opinions, personalities of planners and other occasional situational factors, may play important roles in its problem-solving process. This makes the cause and effect relationships difficult to be predefined or even to be found before the event occurs. Thirdly, although urban planning is a kind of work largely guided by some guidelines and regulations, many planners regard it as an art as well as a science (O'Harrow 1968). They would like to use their creativity in the work rather than just following prototypes. This will introduce many subjective factors into the problem-solving process.

Knowledge Representation

Most of the existing KBS are rule-based or model-based. In a rule-based system, the knowledge is typically represented by *IF-Then* syntax and the control scheme constructed by rules-of-thumb. In a model-based system, more *in-depth knowledge* is required and the knowledge is represented with causal models. To represent knowledge in these ways requires two prerequisites: 1) the relationships between a situation and its consequent situation can be explicitly predefined; and 2) the knowledge should be generalized and can be directly applied to routine situations. There are two main difficulties in representing some planning knowledge by *rules* and *models*. Firstly, if the knowledge is difficult to articulate in natural language, as mentioned above, it will be more difficult to represent in the strict form as *rules* and *models*. Furthermore, there is often much *informal information* involved in planning, for example, personal judgements, hunches, intuition, hearsay and personal experiences (Han and Kim 1990), which is obviously not suitable to be recorded as rules or defined as causes and effects. Secondly, some knowledge, especially informal information, is usually very case-specific and hard to abstract into a general rule. The abstract knowledge used by conventional KBS usually abandons these individual characteristics in order to reach a general rule.

Acceptance of KBS to Planners

User acceptance is a key problem in deploying successful AI systems (Leake 1996). This problem seems to be more apparent in planning which distrusts the black-box approach in decision-making. User acceptance has been regarded as a major obstacle to successfully applying KBS to urban planning (Marchand 1993) and has drawn the attention of researchers ever since the application of KBS in urban planning (Ortolano and Perman 1990).

Advantages of CBR over Rule-based Reasoning

Watson (1995) and Kolodner (1993) summarized the general advantages of CBR from the view of both sides of a system builder and system user. Here, we would just like to elaborate some advantages of CBR when it is applied to urban planning. From the perspective of a KBS builder, since CBS tends to store knowledge in the form of concrete instances rather than abstract and general rules or models, the system builder can largely avoid meeting the *knowledge elicitation bottleneck*. Firstly, extracting knowledge *"becomes a simple task of acquiring past cases"* but not grasping the underlying reasoning in solving the problems (Watson 1995). Secondly, the knowledge-storing scheme of CBS provides a good way to record the *informal information*. This information can be easily recorded in the narration of a 'story' and because this narration is on the level of a concrete case and has no effect on other cases, the system builder does not need to worry about the problems related to generalization. Thirdly, different solutions to different cases of a similar problem, which quite common in planning, can be explicitly and conveniently represented. Finally, when the new problem is solved and the effects of the solution in the real world are known, the user can conveniently save this new case in the case-base and increase the knowledge of the system. This learning method is very simple and can be performed by the users themselves in their daily work without the help of knowledge engineers. Even when conflicting cases are found, the experts can check what caused this conflict at case level, while in a conventional KBS, the conflicts may cause a large modification to the whole knowledge base.

From the perspective of the users, the enlightening style of CBR may be more welcome to planners. There are five main reasons for this.

- *The reasoning process is more visible to the users:* From the retrieved cases, the user can directly obtain solutions to previous similar problems and, more importantly, these results are based on actual past cases that can be presented to the user to provide more concrete support for the system's conclusions and recommendations (Leake 1996).
- *The user can take part in the inference process:* Rule-based reasoning tries to abstract the experts' problem solving methods into some fixed routines and its inference is largely a close iterative process that usually will reach a final result. But, the reasoning procedure of CBR can be open and the user can take part in or even control the problem-solving process. For example, when a planner has retrieved several similar cases for a new problem, it is easy for him/her to understand how the solution was generated, why it was successful or failed, and what might be the consequence of a solution.
- *The way CBR provides help is not compelling:* CBR will not just tell the planner what he/she should do. Instead, it presents cases related to a previously handled similar problem and suggests what has been done under a similar situation, what are the consequences of the applied solutions and what tips the precedents can provide for dealing with the current problem. This process is similar to what often happens in our brains, and is the way through which many experts teach novices or handle a difficult problem.
- *The cases can provide knowledge about exceptions:* As discussed before, knowledge for handling exceptions is often what the user mostly required, particularly in a domain such as urban planning that is characterized by uncertainty and subjectivity. While many conventional methods of knowledge representation try to omit the exceptions, but a case record is all the knowledge about each concrete instance, including the exceptions (Kolodner 1993).
- *A real case is more inspiring than abstract knowledge:* In their practical work, planners often refer to previous real examples but not abstract guidelines or regulations to inspire their creativity. A computerized system that facilitates the retrieval of relevant cases and provides required information and tools of visualization, analysis, evaluation and adaptation will relieve the burden of the planner in trying to remember everything. It will help him/her to have more time to understand the cases.

In summary, a CBS leaves much room for the intelligence of planners to recognize, understand and generate creative solution to a new problem while it provides raw but comprehensive and original information, and tools to facilitate the decision-making process. By studying previous cases, a new planner may broaden his/her experience in handling similar problems, get tips and lessons, while an experienced planner, through reviewing the old cases, may recall his/her experience and knowledge, and even generate inspirations in solving new problems.

13.3 Why Use CBS in Development Control ?

Development control is an important function of urban planning. Processing planning applications is a daily work of government planning agencies and there is a statutory regulation on the length of time in processing an application. Planning applications are especially important in the UK's planning system because any development has to obtain planning permission from the planning authority (Bruton and Nicholson 1987). It is a relatively well structured function of urban planning and, as a result, many attempts have been made to apply KBS to it (Leary and Rodriguez-Bachiller 1989, Borri *et al.* 1994). However, a rule-based KBS may still encounter difficulties similar to those in other fields of urban planning. One of the problems is knowledge acquisition. Some developers of KBS for development control found that *"it is extremely difficult to find an expert who is able to formalise his/her decision processes and to communicate his/her knowledge in a satisfactory way, as a domain experience is often based on automatic physical and mental processes, and the expert is frequently led to mix facts and factors and to judge by his/her intuition and imagination"* (Barbanente *et al.* 1995, p. 573). Although Leary and Rodriguez-Bachiller (1989) argued that this expertise was more structured and able to be articulated than expected, their decision tree method seems to have the following problems:

- often in development control, it is not easy to subdivide a goal into several definite sub-goals;
- the decision tree may over-simplify the relationships among the sub-goals; and
- the method tends to tell the user what he/she should do or should not do, rather than how to do it.

Some interesting conclusions can be inferred from the practical experience of some research (Barbanente *et al.* 1995, Borri *et al.* 1994). Human experts often resort to similar cases to streamline their problem solving. Memory is very important to an expert. One of the main reasons for a person to be an expert is that he/she can remember suitable precedents in a certain situation and use such experience in solving new problems. Another problem is how to represent case-specific knowledge. Although the problems in development control and their solutions are largely well-structured compared with problems in other fields of urban planning, special considerations and characteristics of a planning application case are often also important in making a decision. Borri *et al.* (1994, p. 36) found that *"different context stimuli or pragmatic suggestions arising from the analysis of relevant context elements may produce microchanges in the problem-solving scheme"*.

Finally, although development control is relatively well defined, it is far from being without any uncertainty. Leary and Rodriguez-Bachiller (1989, p. 20) pointed out that *"officers and committees have discretion to interpret plans and policies"*, also, *"aesthetic and political judgements introduce an element of arbitrariness in the decision-making process"*, thus *"layers of complexity and uncertainty"* are added to *"what at first glance is a rigid rule-based system"*. There are always grey areas in development control and often these are the areas where the expertise and experience of the experts are most required. When dealing with such kinds of problems, the experts usually draw upon 'previous experiences' or *"the memory of how similar cases had been solved in the past"* (Borri *et al.* 1994, p. 37). This is the type of problem that is most suitable for CBR.

The advantages of using CBS in development control are as follows.

- It is a retrieving tool that can help planners find relevant cases quickly.
- It is an intelligent library that can relieve the burden of experts' memory and also increase novices' memory. It is a corporate memory system that will retain knowledge of experienced experts even when they are absent or retired.
- It will provide knowledge directly at case level. At this level, the operations of the experts to a specific problem and their results, as well as the comments from the experts are just descriptive records. There is no need to extract and generalise knowledge.
- It will use concrete cases, instead of control branching, to demonstrate the templates of different routines or methods.
- To those problems with uncertainty, it will help planners to make further considerations. From similar cases, the planner can get tips, hints, instructions, templates or precedents, depending on what he/she needs.
- From the old cases the user can learn HOW to do to a problem, but not just what to do or what not to do.

13.4 System Architecture and Building

As a case study of building a CBS planning support system, we developed a system that integrates a CBR shell, ESTEEM and a GIS package, ArcView, to support planners in the Planning Department of Hong Kong in handling planning applications. We went through the whole process of system building in developing this system – how to define a case of development control; how to define the index of such cases; how to build retrieval and adaptation methods; and how to apply such system to the practical work of development control (Shi and Yeh 1999).

13.4.1 Planning Application in Hong Kong

In Hong Kong, Outline Zoning Plans (OZPs) are mainly used to control land development. These plans indicate areas that are zoned for various uses, for example, residential, commercial and industrial. Attached to each plan is a Schedule of Notes, which shows two columns of land uses that are permitted under different land use zones. *Column 1* contains land uses that are always permitted without the need to seek planning permission. *Column 2* shows land uses that may be permitted with or without conditions on application to the Town Planning Board (TPB), the statutory body for making decision on a planning application (Table 13.1). The Board may either refuse the application or grant permission, with or without any conditions. If the proposed development is neither a *Column 1* nor a *Column 2* use, then it is not a permissible use under the zone.

Table 13.1: Example of an explanatory statement of an Outline Zoning Plan (OZP) in Hong Kong

residential (group a)

column 1 uses always permitted	**column 2** uses that may be permitted with or without conditions on application to Town Planning Board
ambulance depot	broadcasting, television and or/ film studio
ancillary car park	commercial bathhouse
canteen	educational institution
clinic/polyclinic	exhibition or convention hall
government refuse collection point (in public housing estate only)	government refuse collection point (other than in public housing estate)
flat	hospital
cooked food centre	hotel
government staff quarters	massage establishment
...	...

Before a planning application is submitted to the Town Planning Board, it will first be inspected by the planners in the Planning Department. They will examine the new application according to the planning ordinances, regulations and plans, collect opinions on this new application from relevant government departments and prepare the evaluating documents for this new application. The application, together with the evaluating documents, will be submitted to the Town Planning Board that is composed of mainly non-government members to decide whether the application will be approved or not.

The aim of our system is to help the planners in the Planning Department to prepare reasonable, comprehensive and consistent recommendations to the Town Planning Board for making decisions on planning applications. After analyzing more than 100 cases of planning applications submitted in the past several years in details, we have the following observations:

- The planners in the Planning Department often encountered similar cases.
- When handling a new application, the planners usually referred to previous similar cases if they existed. They usually retrieved old cases that are located close to the new application to see what are the processing results of these cases. These similar cases were usually marked on the location map of the new application that is one of the basic documents of an application.

- Both the Town Planning Board and the applicants often cited the decision of previous similar cases to support their arguments.

These findings are encouraging for developing our system because their decision making process and logic are similar to that of CBR and case-related knowledge is the main strength of CBS over other KBS.

13.4.2 System Architecture

CBS is a promising methodology for building and applying more practical KBS for planners (Yeh and Shi 1999). Based on the spatial nature of planning, GIS has inevitably become a widely used tool in planning (Webster 1993, 1994, Yeh 1999), especially as a planning support system (Batty 1995, Harris and Batty 1993). Hence, a system that integrates CBS with GIS will be very helpful to planners. In such an integrated system, CBS will provide the method for decision support in providing relevant experiences, tips, lessons and possible solutions in handling new problems based on past cases to the planners. The GIS of the system will provide the functions of storing, retrieving and displaying spatial data. In the handling of spatial data, GIS can be a data generator, a DBMS and a visualization tool. It can store, display and generate spatial information related to a planning case. In spatial retrieval, GIS can locate the cases by their absolute positions, for example, to locate a case by its coordinates or its address. It can also retrieve an old case that has a similar spatial relationship with the new problem.

ESTEEM and ArcView were chosen to be the software for developing the integrated system because both of them are relatively inexpensive software, and also are MS Windows-based and thus could be easily integrated through DDE of MS Windows. Figure 13.2 shows the framework of the integrated system. From the functional perspective, the system has four basic parts – the user interface, the GIS module (ArcView), the CBR module (ESTEEM) and the case library. With relatively high flexibility in building input/output interface, ArcView was chosen to be the platform for the integration. Except for some special operations, such as creating a new template of retrieval method, all the user inputs will enter the system through ArcView and all the retrieval or adaptation results from the system will be displayed by ArcView. The system provides two kinds of retrieval methods – spatial and nonspatial. For spatial retrieval, the system will let the user click within the outline of a specific case on an index map to retrieve that case. When the user clicks on that case, an identifying code will be passed to the system and ArcView will directly retrieve all the components of that case according to the identification code from the case library and display them to the user. As for nonspatial retrieval, ArcView first receives from the user the formatted text or numeric descriptions of a new problem and then passes these descriptions to ESTEEM. ESTEEM looks for the similar cases through matching the descriptions of the new problem and the index values of the old cases. Then, ESTEEM returns the identifying codes of those matched cases to ArcView. With these codes, ArcView brings the components of these cases to the user. All the user-interfaces and data exchange were built by programming with Avenue scripts.

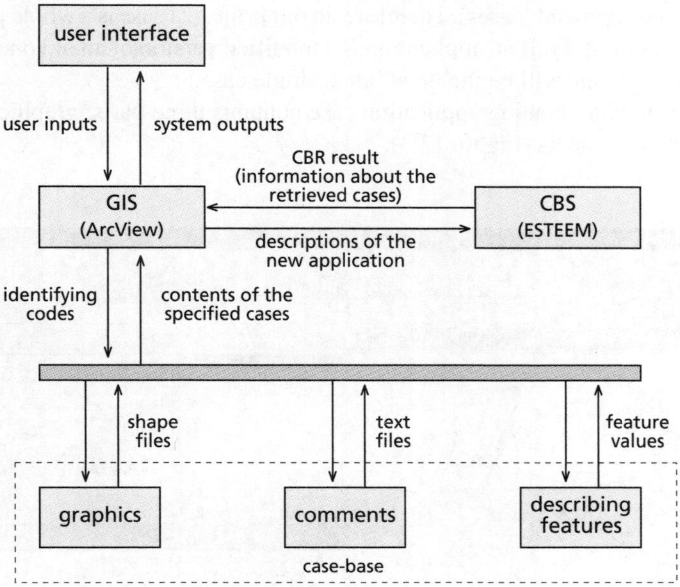

Fig 13.2: The framework of the integrated CBS and GIS system

13.4.3 Building the CBS

The building of a CBS includes building the case library, defining an index for the cases, and building retrieval and adaptation methods (Kolodner 1993). For our system, we also developed a simple method for the user to do quantitative evaluation with the retrieved cases.

Case Library

Case library is the knowledge-base of a case-based system (CBS). It is the fundamental core of a CBS. Hence, the construction of case library is the first and most important step in building a CBS. Two questions must be answered before building a case library: What is a case in this domain? And what should be used to describe a case or what is the content of a case? In the domain of handling planning applications, it is natural to think that each submitted application is a case. But, if a planning application had been rejected, its applicant might submit a modified version again, each with some improvements in design or transportation. If we treated each of these applications as a single case, the planner will not get the whole history of the application.

Furthermore, the system could not correctly evaluate the new problem. For example, if the application had been rejected for several times but finally got the permission, when evaluating a new similar application, the system would draw the conclusion that the new application would tend to be rejected because the rejected 'cases' are more than the approved 'cases'. Therefore, in our project, a case is a whole procedure of an application story. If an application is a modified version of an old one, both the old and the new one will be included into a single case.

In our system, a planning application case contains three parts: graphics, tabular features and comments (Figure 13.3).

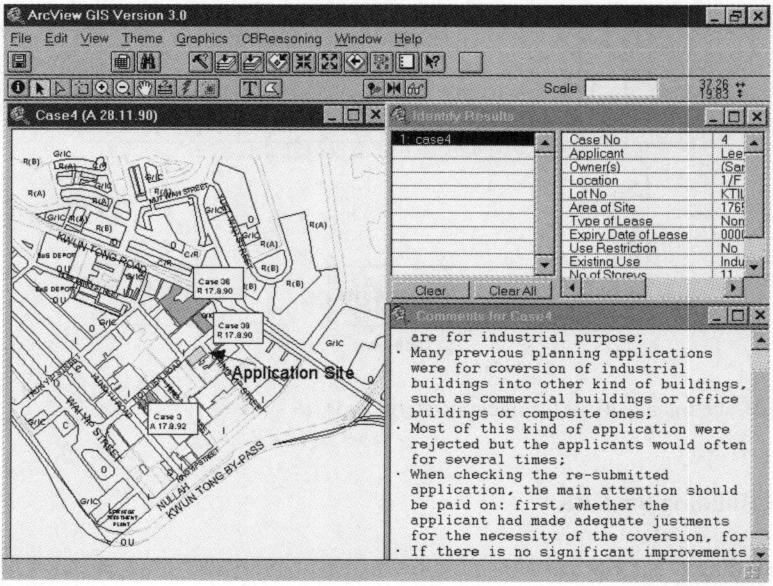

Fig 13.3: Representation of a planning application case

Graphics

The documents of a planning application currently stored in the Planning Department usually include a map that shows the position and coverage of the land use referred to by the application. Often on this map, the nearby previous cases are also located and outlined with their results and dates labelled besides their sites. In our system, this basic map will be an ArcView shapefile and through the query and display functions of ArcView, this electronic map can provide or link to much more information than the original paper map (Figure 13.3). Also, the user can click on a specific old case outlined on the map to retrieve the materials of that case. Besides this basic map, the documents of a planning application case sometimes also contain charts, photos or other graphics. In our system, these graphics will be handled by ArcView.

Tabular Features

Besides graphics, inevitably the case should include features (attributes) represented with numeric values or texts in tabular format. These features are listed in Table 13.2. In our system, a planning application case includes four kinds of tabular features, although sometimes the boundaries between them are not very clear.

- *Features for organizing information:* These features are used to manage the case library, that is, to organize the whole case library, to identify cases and to retrieve related documents. With these features, the case library can be well organized and the efficiency of the system can be improved. The examples of this kind of features are 'Case No.' and 'Applicant'.
- *Features containing information about the problem itself:* These features describe the application itself. Most of the items in Table 13.2 belong to these kinds of features, for example the applied land use, the planned land use and the existing land use.
- *Features containing solution information:* This is one of the objectives of the user. The main solution of the problem of processing planning applications is the Planning Department's View, that is, the recommendations that should be submitted to the Town Planning Board.
- *Features containing outcome information:* This is another objective of the user. The final result of the application, that is to be approved or rejected by the Town Planning Board, can be regarded as part of the outcome of the solution. Although the decision of the Town Planning Board does not totally depend on the opinions of the Planning Department, it is largely affected by these opinions.

The next question is how to record each feature in a computer-based case library. This is a question related to symbolization. The symbolization of a problem is an important work in building many AI systems (including CBS) because the reasoning performed by the computer has to be based on symbols (Stefik 1995). However, for the time being, our system is primarily an aiding and advisory system. The main objective is to retrieve useful cases for the user and would not carry out much automatic inference work such as automatic adaptation (Kolodner 1993, Leake 1996). We do not want the representation of our cases to be over-symbolic because the descriptions of the cases should be understandable as much as possible to the user so that he/she can quickly and directly get lessons from the cases or detect the differences between current situation and the old cases, without much interpretation of the symbols. In addition, the cases should involve details, especially case-specific details as much as possible, because details are very important for distinguishing one case from another and will directly affect the final decision.

All these feature values are stored in a table file in dbf format. The table file is independent from both ArcView and ESTEEM. It is convenient for the user to modify the case-library. The description of each case is a record in this table and can be identified by its identifying number (Case No) when ArcView accesses the table.

Table 13.2: Example of representation of the tabular features of a planning application case

feature name	feature value
general information:	
applications Included	A/XYZ/01, A/XYZ/02 and A/XYZ/03
beginning date	6/12/1989
end date	18/1/1991
applicant(s)	ABC Investment Company Ltd.
location	1/F DDD building, 64 EEE Rd, FFF Area, KL
use applied for	office/industrial composite use
zoning on statutory plan	industrial
existing use	industrial and associated offices
result of the application	C
approval score	0.8
details of the existing use	
lot no	YYYY nn
area of site	1,765 sq. m.
expiry date of lease	30 June 2047
type of lease	non-Renewable
use restriction (if any)	no
lease clause (design, disposition and height)	no
other restrictions (traffic implications)	rate of space for parking, loading/unloading: >=1 vehicle/1000sq ft of GFA or >=1 vehicle/5000sq ft of the area of the lot
storey no	11
building type	industrial
approx. total floor area	15,500 sq. m.
no. of occupants (residential)	N/A
details of the proposed development :	
zoning Plan No	S/K14/3
new building storey no (if applicable)	N/A
total site coverage (if applicable)	N/A
proposed use of floors: basement	N/A
proposed use of floors: ground floor	carparking, loading/unloading, lobby
proposed use of floors: first floor	industrial/office
proposed use of floors: second floor	industrial/office
proposed use of floors: third floor and above	industrial/office
number of houses	N/A
total domestic ground floor area	15,500 sq. m.
number of car parking spaces	N/I
vehicular access	yes
justifications from the applicant:	1 Hong Kong's industry was undergoing a structural change that had resulted in demand for such building to accommodate industrial back-up services. 2. The site, being in close proximity to the FFF MTR Station and FFF Town Centre, was suitable for a composite industrial-office building. 3. The proposed conversion is technically feasible (a set of building plans was attached).
the site and its surroundings:	The site is in close proximity to the FFF MTR Station and FFF Town Centre.
planning intention:	Industrial development, but the offices that need to be co-located with industrial operations tend to be permitted.
traffic aspect:	1. The site is close to the FFF MTR Station. This is a commuting convenience for the people working in the building. 2. The site is at the junction of YYY Rd. and XXX St., thus the provision of parking and loading/unloading space is required.

C: approved with conditions
N/A: not applicable
N/I: no information

Table 13.2: (continued)

feature name	feature value
lands aspect:	The site is suitable for a composite industrial-office building.
design aspect:	The initial building plan was not up to the design requirements of the Board: 1. Separate entrances and lift lobbies for goods and passengers should be provided. 2. The prescribed window area and light angles were insufficient, on some floors by more than 65%. The applicant changed the building design: 1. The use of ground floor was changed from "storage, carparking, loading/unloading" to "carparking, loading/unloading, lobby". 2. The design of window area and light angle is modified.
drainage aspect:	N/A
consultation:	N/I
processing:	Two fresh applications were submitted, in which building design was modified according to the suggestions of the Board.
Planning Department's view	1. The application was in line with the Board's guidelines to permit offices that need to be co-located with industrial operations to be established in industrial buildings. 2. The site is suitable for such a composite industrial-office building because of its good commuting situation. 3. Every unit within a composite industrial-office building must be designed, constructed and made suitable for both industrial and office uses. Where building design requirements for industrial and office buildings differ, the more stringent requirements under the Buildings Ordinance should apply. Board should pay attention on: - the window area and light angle design; - the provision and layout of parking and loading/unloading facilities.

C: approved with conditions
N/A: not applicable
N/I: no information

Comments

The content of *Comments* is the advice on a case. It has two parts. One part contains any valuable tips about the case, including anything special to this application which cannot be processed in a routine way or should pay more attention to, anything difficult to deal with, anything uncertain, any mistakes made, any tricks used, and any other things in this case that deserve special attention. The other part of the *Comments* is a list of cases that are relevant to this case which the user may be interested to examine. The *Comments* of a case is stored in an independent text file and can be identified by its file name.

Case Indexing

As discussed above, the description of the case is not much symbolized so that the user can easily learn lessons from it. But such description is not suitable for the computer to do matching and retrieving. Furthermore, using the whole description to do matching and retrieving is usually not efficient and practical. Thus, we need to develop an index for the case. The index of a case is the symbolized and simplified or partial description of the case. With the index, cases can be more efficiently and appropriately retrieved (Kolodner 1993). Two types of indexes are used in our system. The first index is a spatial index built with ArcView. With ArcView, the user can directly retrieve cases through the index maps. The second index is a conventional index of CBS by using ESTEEM. The vocabulary for indexing cases includes two parts – the dimensions describing the retrieving situation and the symbols describing each dimension (Kolodner 1993). The dimensions of the index for retrieving the cases with similar land uses to a new application and the data types of these dimensions are listed in Table 13.3.

These dimensions can be classified into two categories. The first-level dimensions are for what is called 'partial matching', that is, to find all the cases that are potentially applicable to the new application (Kolodner 1993, Leake 1996). These cases are at least similar in some aspects to the new problem, but not necessarily all of them will make real contributions to handling the new problem. The most useful one(s), if it exists, should be among these partial-matching cases and will be picked up by further retrieving. In our system, the first-level dimensions include:

- *Time of End*: the time that the final decision to the application is made;
- *Planned Use*: the land use type of the proposed site on the zoning plan;
- *Existing Use*: the current use of the proposed site;
- *Applied Use*: the proposed development for which the permission is applied by the applicant;
- *Relationship (Applied-Existing)*: the relationship between *Applied Use* and *Existing Use*, can be *new*, *conversion*, *regularize*, *redevelop*, and so on;
- *Relationship (Applied-Planned)*: the relevance of the *Applied Use* to the *Planned Use*, represented by numbers, where 1 represents the most relevant uses.

These six dimensions were chosen as the first-level dimensions because they can basically classify cases and can be used to retrieve all the potentially applicable cases with similar land use types. They are those that first come to the minds of planners when they want to look up the precedent cases with similar land use types.

When all the potential applicable cases are found by using first-level dimensions, the user may want to further find out what is (are) the most useful one(s). Then the other dimensions (second-level) will be used. The second-level dimensions include four kinds of features:

- Justifications of the applicant:
  Justification: whether the applicant made justifications for his/her/its application or not.

Table 13.3: The dimensions of the index for retrieving cases with similar land uses

dimension	data type	value range
time of the end	number*	1980 - 2000
planned use	one of a list (an element from a defined list)	I (industrial) C (commercial) Ra (residential group A) Rb (residential group B) Rc (residential group C) G/IC (government/institution/community) CDA (comprehensive development areas) V (village type development) OS (open space) GB (green belt) OST (open storage) SSSI (site of special scientific interest) UU (unspecified use)
existing use	text	(Simple words describing the existing land use, such as industrial, office, oil depot, residential, natural protection, vacant, warehouse, agriculture, and so on. If the information is not available, N/I will be assigned to this dimension.)
applied use	text	(Simple words describing the proposed development, such as industrial, warehouse, office, low-density residential, recreational, kindergarten, comprehensive development, PFS(petrol filling station), oyster cultivation, and so on.)
relationship (applied – existing)	one of a list	regularize conversion new redevelop relaxation of restriction
relationship (applied – planned)	number	1. (the applied use is very relevant or conformable to the planned use) 2. (the applied use is somewhat relevant to the planned use) 3. (the applied use is not relevant or inconformable to the planned use)
justification	yes or no	yes, no
review	number	0, 1, 2, 3 ...
fresh application	number	0, 1, 2, 3 ...
petition	number	0, 1, 2, 3 ...
result	one of a list	R (rejected) C (approved with conditions) A (approved without conditions)
reason for rejection	one of a list	U (because of concern on land use) T (because of concern on traffic situation) D (because of concern on design of the proposed development)
tip	text	(Simple words describing the handling, especially why the application was approved or why the Board's decision changed. The words may include: change design, compatible with planning intention, more justification, temporary permitted, compatible with environment, and so on.)
approval score	number	0 - 1

*ESTEEM does not provide date data type, so the date is stored as numeric data type.

- The features that describe the handling process:
  Review: how many reviews were made during the process of the application. According to the Town Planning Ordinance of Hong Kong, if the applicant is aggrieved by the decision of the Town Planning Board, he/she/it can apply to the Board for a review of the Board's decision.
  Fresh: how many fresh applications are there in a case after the initial one. If the application had been refused in the initial application, the applicant may applied for the same use on same site more than one time.
  Petition: whether the applicant made petition to the Appeal Board.
- The features that describe the application result:
  Result: the final decision made to the proposed development, i.e. *approved* or *rejected*
- The features that describe the evaluation to the application:
  Reasons for Rejection: the main reasons for the Board's rejection to some or all applications included in the case.
  Tips: the reasons for the Board changed opinions, especially for the cases in which the proposed developments were initially rejected but finally approved.
  Approval Score: the probability of approval which ranges from 0 (absolute rejection) to 1 (absolute approval).

These second-level dimensions will be able to help a user to find the most useful case(s) from the partial-matching set. In our system, we have the computer (ESTEEM) to make retrievals just based on the most basic features (first-level dimensions) to find all the potential cases, while leaving the work to find the most useful case(s) based on second-level dimensions to the planning professionals. Firstly, this is because the most similar case is not necessary the most useful one (Leake 1996). Letting the user decide which case(s) is(are) the most useful one(s) may be the most efficient and effective way because only the user exactly knows what he/she really wants to get from the cases. This is also a way more acceptable to the user. Secondly, too many dimensions in the automatic retrieval may slow down the system's performance and the system may even lose some useful cases when making retrieval since the weights were dispersed and the importance of the most basic features were reduced. Finally, to require the user to input too much information at the beginning is neither a natural way nor a feasible way. What first comes to a user's mind must be those basic or important features of the problem and the user would not like to pay much attention to those less important ones at the beginning. Also, some information is not available for a new application, such as that about the handling process, the final result and the evaluation.

Retrieval Methods

For performing the retrieval, the system should know how to match each feature (dimension) of a new application with an old case, and how to evaluate the similarity between the new application and the old case. Figure 13.4 shows a similarity-assessment template to define the feature matching method in finding the similarities of cases. This template was built with the *Similarity Definition Editor* of ESTEEM. In this template the similarity-assessment methods for evaluating each feature and the whole case are defined and the value of each parameter of the algorithm is set. The user can modify this template according to his/her experience. He/she can create more than one such template for different situations or tasks.

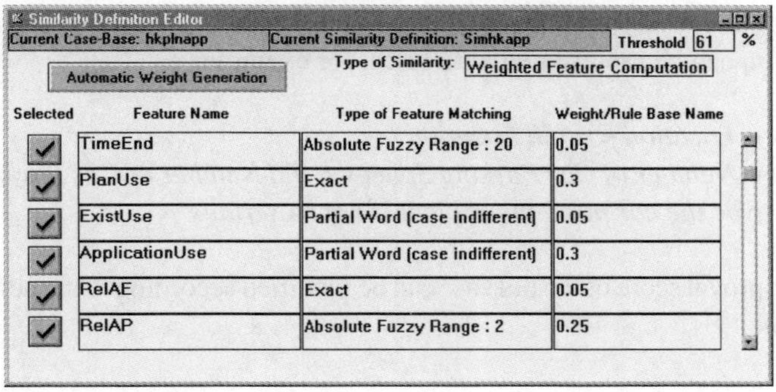

Fig 13.4: A similarity-assessment template for retrieving cases with similar land uses

The Nearest Neighbour algorithm is used by the system to find the similarity between a new problem and an old case. It is one of the most popular algorithms for comparing cases in CBS (Kolodner 1993, Watson 1995). The Nearest Neighbour algorithm is:

$$S = \frac{\sum_{i=1}^{n} W_i * sim(f_i^I, f_i^R)}{\sum_{i=1}^{n} W_i} \qquad (1)$$

where *s* (similarity score) is the sum of weighted similarities of each feature and only those cases whose *s* is larger than a predefined threshold (for example, 60 per cent) will be retrieved as the similar cases; w_i is the importance of feature *i*, *sim* is the similarity assessing function for comparing feature f_i, and f_i^I and f_i^R are the values for feature f_i in the input new case and the retrieved old case respectively (Kolodner 1993). The weights (w_i) used in the study are assigned by the authors on the advice from the planners in the Planning Department. They can be adjusted, modified and refined as the system develops.

Adaptation Methods

Apparently, a new planning application has little chance of being totally the same as an old case. Although the planners could learn a lot directly from similar old cases, they have to adapt this knowledge according to the particular context of the new problem. Automatic adaptation is one of the most challenging topics in CBR research

(Leake 1996). In our system, the Planning Department's View of old similar cases can be modified to some extent to fit the new situation. Usually the Planning Department's View is composed of several opinions on a new planning application. Thus, the results of adaptation could be suggestions for modifying the old opinions, such as, under the new condition, a new opinion may need to be added, an old opinion may not be needed any more, or an old opinion should be modified. The algorithms of adaptation can be represented with rules, for example:

IF *new Location <> old Location*
and *new Number of Car Parking Spaces > old Number of Car Parking Spaces*
THEN *maybe the car parking problem is not so serious*

Also, the approval score of the old case can be modified according to the new situation. For example:

IF *new Location = a*
THEN *old Approval Score = old Approval Score – 0.1*

Evaluation

The experimental system performs a very simple evaluation operation for the user. It will calculate the approval score of a new planning application based on the previous similar cases. It can be used for considering the consistence of the decisions. The new approval score (NA) is calculated by the following formula:

$$NA = \frac{(((a_i\text{-}0.5)s_i\text{+}0.5)s_i^2)}{s_i^2} \qquad (2)$$

where NA is the calculated new approval score of the new application, a_i is the approval score of each retrieved old case, and s_i is the similarity score of each retrieved old case (similarity score is automatically calculated by ESTEEM). The difference between a_i and 0.5 represents whether the application tends to be approved or to be rejected. For example, if the difference is positive, the application tends to be approved. The larger the difference, the more significant this tendency is. Because the old case is not totally the same as the new application and the similarity is represented by s_i, so the difference, $(a_i - 0.5)$, should be discounted by multiplying s_i. The square of s_i is used as the weight of each case if more than one similar cases are found. The more similar the case is, the more important it is to the calculation of the new approval score.

Learning Process of the System

There is no special learning scheme in our system. The planner can add or delete cases at any time when he/she thinks it is necessary, and then the knowledge of the system is modified. A new planning application can be stored as a new case when the planners in the Planning Department finish handling it, or when the result (approved or rejected) has been obtained from the Town Planning Board, or when any further processing has been performed (Figure 13.5). The planner needs to make a judgement on what approval score which ranges from 0 to 1 to be input to the new case before it is stored in the case library. It is based on his judgement on whether the decision by the Town Planning Board is unanimous or not and whether it encounters any problems in making before approval or rejection. If it is unanimously approved or rejected, the approval score of the new case can be 1 or 0 respectively. An application that was approved quickly should have a higher score than one which was approved after reviews and appeals. If an applicant applied more than once for a development, the planner can combine these applications to form a single case after the final decision is made. After the case is stored in the case library, the planners can still modify it later with new knowledge about it, such as the consequences of the approval or rejection and any new similar cases stored in the case library. The planners can also delete obsolete cases when the situation changes. Because the cases are independent from each other, adding or deleting a case has no effect on the other cases. Thus, the planners can easily make the system learn from the practical work. When put into daily work, this CBS can be improved by the planners themselves even without the assistance of knowledge engineers.

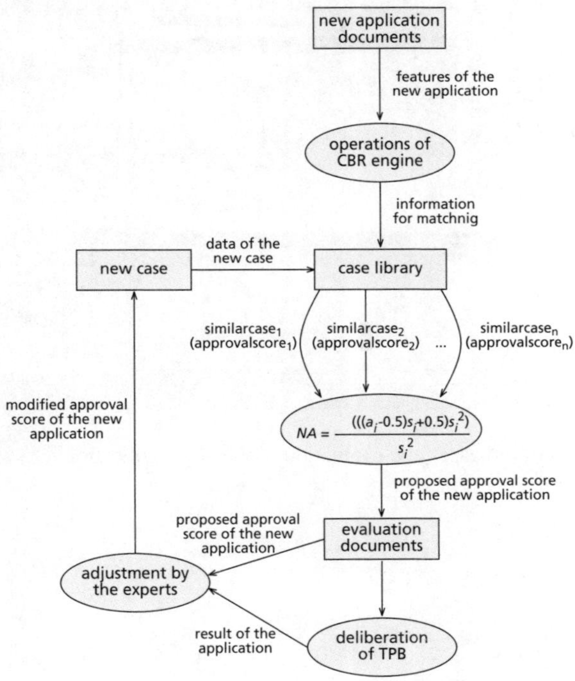

Fig 13.5: Flow diagram of the system

13.5 Implementation of the System

The operation of the system conforms with the four REs cycle of a typical CBR system. When the descriptions of a new application are input, the system will retrieve similar cases in the case library and adapt them according to the defined algorithms (Figure 13.5). Detail information of similar cases and the suggested solutions from adaptation will be displayed to the user (Figure 13.6). When retrieving cases, the system calculates the similarities of the old cases in the case library. Only those cases whose similarity values are larger than or equal to a threshold set by the user in the Similarity Definition Editor will be retrieved. If more than one similar cases are found, the general information of them will be listed in a table and the user can choose from this table the cases which he/she is interested in looking up for further details. The user can also examine the retrieved cases to find hints and information and make his/her own judgement in addition to the suggestion of the system.

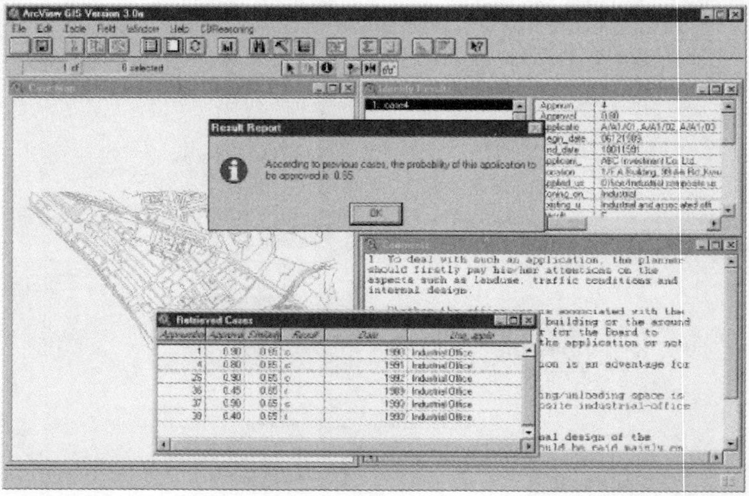

Fig 13.6: Retrieval of old cases and evaluation of the new application

13.6 Conclusions

This chapter shows how CBR can be used to handle planning applications in development control. The system simulates the way of handling planning applications in which the planner has to recall and make reference to similar planning application cases. Instead of doing it manually and having to rely on the experience of the planner who dealt with the case, the CBR system will help the planner to re-use previous similar cases in making a decision on the new applications, regardless whether the planner is new or experienced. As all the relevant past cases are revealed to the planners to help him/her in making a decision using the recommended score of the system as a guide, this overcomes the black-box approach of a rule-based system. As the decision rests with the planner, he/she can still be creative in making the final decision but he/she has to justify why the decision deviates from the past cases that have been revealed by the system.

The system will be an effective and efficient PSS in helping planners to make consistent recommendations to the Town Planning Board and meeting the statutory time period in processing planning applications which is an important daily work of a planning department. As it is using past cases to give guidance to new applications, it can help in making consistent decisions for similar applications. This is especially useful in developing countries where development control is mainly done by planners. A planner has to justify why his/her decision on the application is different from previous similar cases as found by the system. This can avoid inconsistency and corruption in making decisions. Because it is case-based and not rule-based, once the system is developed, it can be applied to other cities in dealing with development applications by inputting the real cases of a particular city into the case library of the system.

The system is a corporate memory system (Smith *et al.* 2000, Stein 1995) which frees the planners from retaining a mental record of all past decisions that can be easily lost through oversight of the person involved or frequent change of staff. In addition to providing planning decision support to planners, it can also be used as an office automation system for easy retrieval of relevant past cases. It can be used to enhance the present database system that is being used for storing and retrieving past cases. Our system that integrates the functions of CBS and GIS can provide the ability to easily retrieve the required relevant cases spatially or non-spatially. By integrating with GIS, it is also a visualization tool to bring the relevant graphics, data and texts to the planner.

The Planning Department of the Hong Kong SAR Government has implemented a GIS-based planning application information which is quite similar to this system, except that it is not using case-based reasoning to retrieve the cases and give recommendations based on past cases. It is an office automation system, which enable user to do a spatial or textual search to find and display cases that belong to the same type of applications. However, the system will not tell the user how similar is one case from another. Their system can be easily improved by adding the present CBS into it.

This system is at its initial stage of development. Its capability of automatically adapting cases needs further exploration and development. For example, the system should automatically recognize the differences in the critical aspects between the new application and its precedent cases and make an evaluation according to these differences and then proposed modifications to the old solution. The capability of matching and adapting cases based on spatial similarities and relationship is another important topic that needs further research. For example, a pollution source has the spatial relationship of distance and direction to a residential building. The pollution source and residential building constitute a kind of spatial relationship. How to use GIS to automatically recognize this relationship and automatically retrieve cases with a similar pattern needs further research.

The use of case-based reasoning is not without limitations. One of the obvious limitations is the lack of cases for new problems and changes in the planning environment that makes previous cases be irrelevant. For the absence of similar cases, the system will alert the user so that he/she has to make judgment without making the assistance of previous cases as a general guideline. For a change in legislation, the system should alert the user to such changes or delete related cases from the system's case library.

There is also a need to explore the integration of case-based reasoning with rule-based reasoning. Different reasoning engines can be complementary to each other (Bartsch-Sporl 1995, Smith *et al.* 2000, Kolodner 1993, Koton 1993). The integration of case-based reasoning with other KBS will make more powerful and efficient system by taking advantages of the strengths of the other techniques (Medsker 1995, p. 215). For example, case-based reasoning may be better at enlightening user's own creativity while rule-based reasoning may be more efficient in dealing with routine processes. Rule-based system can be used when there is no case in the CBS which is similar to the new application or when new regulations are introduced which make old cases in the case library to be irrelevant. The integration of these two reasoning engines may lead to a more powerful system for development control in future.

References

Aamodt, A. and Plaza, E. (1994) Case-based reasoning: foundational issues, methodological variations, and system approaches, *AI Communications,* 7(1): 39-52.

Barbanente, A., Borri, D., Maiellaro, N. and Selicato, F. (1995) Expert systems for development control: generalizing and communicating knowledge and procedures, in Wyatt, R. and Hossain, H. (eds.) *Proceedings of 4th International Conference on Computer in Urban Planning and Urban Management*. Melbourne, Australia, pp. 571-586.

Bartsch-Sport, B. (1995) Towards the integration of case-based, schema-based and model-based reasoning for supporting complex design tasks, in Weloso, M. and Aamodt, A. (eds.) *Case-Based Reasoning Research and Development*, Springer Verlag, Berlin, pp. 145-156.

Batty, M. (1995) Urban planning and planning support systems, *Regional Development Dialogue*, 16(1): v-viii.

Borri, D., Conte, E., Pace, F. and Selicato, F. (1994) Norm: an expert system for development control in underdeveloped operational contexts, *Environment and Planning B,* 21:35-52.

Bruton, M. and Nicholson, D. (1987) *Local Planning in Practice*, Hutchinson, London.

Garcia, O.N. and Chien, Y.T. (eds.) (1991) *Knowledge-based Systems: Fundamentals and Tools*, IEEE Computer Society Press, Los Alamitos, CA.

Han, S.Y. and Kim, T.J. (1990) Intelligent urban information systems: review and prospects, in Kim, T.J., Wiggins, L.L. and Wright, J.R. (eds.) *Expert Systems: Applications to Urban Planning*, Springer Verlag, New York, pp. 241-261.

Harris, B. and Batty, M. (1993) Locational models, geographic information and planning support systems, *Journal of Planning Education and Research* 12: 184-198.

Kim, T.L., Wiggins, L.L. and Wright, J.R. (eds.) (1990) *Expert Systems: Applications to Urban Planning*, Springer Verlag, New York.

Kolodner, J. (1993) *Case-Based Reasoning*, Morgan Kaufmann Publishers, Inc., San Mateo CA.

Koton, P. (1993) Combining causal models and case-based reasoning, in David, J.M., Krivine, J.P. and Simmons, R. (eds.) *Second Generation Expert Systems*, Springer Verlag, Berlin Heidelberg, pp. 69-78.

Leake, D.B. (1996) CBR in context: the present and future, in Leake, D.B. (ed.) *Case-Based Reasoning: Experiences, Lessons, and Future Directions*, AAAI Press/The MIT Press, Menlo Park, CA, pp. 3-30.

Leary, M. and Rodriguez-Bachiller, A. (1989) Expertise, domain-structure and expert system design: a case study in development control, *Expert Systems,* 6(1): 18-23.

Marchand, D. (1993) Expert system in urban planning: new tools or new toys?, in Pumain, D. (ed.) *Systemes D'information Geographique et Systemes Experts,* GIP RECLUS, Montpellier, pp. 88-91.

Medsker, L.R. (1995) *Hybrid Intelligent Systems*, Kluwer Academic Publishers, Norwell, Mass.

Musen, M. (1993) An overview of knowledge acquisition, in David, J.M., Krivine, J.P. and Simmons, R. (eds). *Second Generation Expert Systems*, Springer Verlag, Berlin, Heidelberg, pp. 405-427.

O'Harrow, D. (1968) Preface, in Goodman, W.I. and Freund, E.C. (eds.) *Principles and Practice of Urban Planning*. International City Managers' Association, Washington, D.C, pp. ii.

Ortolano, L. and Perman, C.D. (1990) Applications to urban planning: an overview, in Kim, T.J., Wiggins, L.L. and Wright, J.R. (eds.) *Expert Systems: Applications to Urban Planning*, Springer Verlag, New York, pp. 3-13.

Schank, R. and Abelson, R. (eds.) (1977) *Scripts, Plans, Goals and Understanding,* Lawrence Erlbaum Associates, Hillsdale, NJ.

Schank, R. and Leake, D. (1989) Creativity and learning in a case-based explainer, *Artificial Intelligence*, 40(1-3): 353-385.

Shi, X. and Yeh, A.G.O. (1999) The integration of case-based systems and GIS in development control, *Environment and Planning B,* 26(3): 345-364.

Smith, H.G., Burstein, F.V., Sharma, R. and Sowunmi, A. (2000) Organisational memory information systems: a case-based approach to decision, in Kersten, G.E., Mikolajuk, Z. and Yeh, A.G. (eds.) *Decision Support Systems for Sustainable Development*, Kluwer, Boston, pp. 277-290.

Stefik, M. (1995) *Introduction to Knowledge Systems*, Morgan Kaufmann Publishers Inc., San Francisco, CA.

Stein, E.W. (1995) Organisational memory: review of concepts and recommendations for management, *International Journal of Information Management,* 15(1): 17-32.

Turban, E. (1995) *Decision Support Systems and Expert Systems*, Prentice-Hall, Inc., Englewood Cliffs, NJ.

Watson, I. (1995) An introduction to case-based reasoning, in Watson, I.D. (ed.) *Progress in Case-Based Reasoning: First United Kingdom Workshop*, Salford, UK, pp. 3-16.

Watson, I. (1997) *Applying Case-Based Reasoning: Techniques for Enterprise Systems*, San Francisco, Morgan Kaufmann.

Webster, C.J. (1993) GIS and the scientific inputs to urban planning, Part 1: description, *Environment and Planning B*, 20(4): 709-728.

Webster, C.J. (1994) GIS and the scientific inputs to urban planning, Part 2: prediction and prescription, *Environment and Planning B*, 21(2): 145-157.

Yeh, A.G.O. and Shi, X. (1999) Applying case-based reasoning to urban planning – a new PSS tool, *Environment and Planning B*, 26(1): 101-116.

Yeh, A.G.O. (1999) Urban planning and GIS, in Longley, P., Goodchild, M., Maguire, D. and Rhind, D. (eds.) *Geographical Information Systems: Principles, Techniques, Management and Applications*, John Wiley, New York, pp. 877-888.

Web Sites

Navy Center of Research for Artificial Intelligence:
http://www.aic.nrl.navy.mil/~aha/research/case-based-reasoning.html

AI-CBR Case-Based Reasoning: *http://www.ai-cbr.org/*

14 Fuzzy Algorithms to Support Spatial Planning

Dmitry Kurtener[1] and Vladimir Badenko[2]

[1] Agrophysical Research Institute, St. Petersburg, 195220 Grazhdansky pr. 14, St. Petersburg, Russia
[2] St. Petersburg State Technical University, 195251 Polytechnical 29, St. Petersburg, Russia

Abstract: It is argued that the process of application of fuzzy set theory is very useful in supporting the process of decision-making in spatial planning. Combining a Geographical Information System (GIS) with applications of fuzzy set theory is an appropriate methodology to support location choice and land suitability assessment. In this chapter, GIS Fuzzy Modelling (GISFM) is described and some models are defined. Some examples of the application of GISFM as a planning support tool for the analysis of environmental situations are presented.

14.1 Introduction

Geographical information systems (GIS) are important tools for spatial planning (Dueker and Barton 1990, Scholten and Stillwell 1990) that encompasses three key activities: location choice, land suitability assessment and collaborative decision-making (Joerin *et al.* 2001). Location choice is defined as the identification of the best location for a particular activity or investment. For example, Carver (1991) has used GIS to evaluate various alternative sites for nuclear waste disposal, and Yen and Hong (1996) have applied GIS and location-allocation modelling for public facility planning. In the sense that it is searching for an optimum, land suitability assessment is similar to choosing a location, but its goal is to map a suitability index for an entire study territory (Hall *et al.* 1992, Senes and Toccolini 1998, Joerin *et al.* 2001). Collaborative decision support systems (CDSS) are user-friendly computer systems combining GIS, decision-making techniques and environmental modelling (Geertman and Toppen 1990, Jankowski *et al.* 1997, 2001).

Spatial planning involves decision-making techniques that are associated with techniques such as multi-decision criteria analysis (MCDA) and multi-criteria evaluation (MCE). Combining GIS with MCDA methods creates a powerful tool for spatial planning (Eastman *et al.* 1993, Pereka and Duckstain 1993, Jankowski 1995, Joerin *et al.* 2001). MCE is perhaps the most fundamental of decision support operations in GIS (Jansen and Rietveld 1990, Burrough *et al.* 1992, Jiang and Eastman 2000). Another category of decision-making techniques utilised in spatial planning is based on the application of fuzzy set theory (Banai 1983, Bogardi *et al.* 1996, Burrough 1989, Eastman and Jiang 1995, Mays *et al.* 1997, Smith 1992, Wang *et al.* 1990, Xiang *et al.* 1992).

Fuzzy set theory was introduced by Zadeh (1965) although the underlying concepts predate this. Overviews have been provided by Zimmermann (1985), Dubois *et al.* (1993) and Jager (1995). Nowadays, fuzzy set theory is a hot topic and is used in many different fields and technical arenas to address a variety of questions and problems, both mundane and abstract (Ross 1995). Combining GIS with the application of fuzzy set theory presents a relatively new approach to location choice and land suitability assessment. Joerin *et al.* (2001) have used this approach to evaluate the suitability of land for housing with respect to noise impact.

The combination of GIS with fuzzy set theory and deterministic models is known as GIS Fuzzy Modelling (GISFM). During the last few years, several applications of this approach for spatial planning have been attempted. In particular, GISFM was used for the multi-dimensional assessment of urban areas after flooding (Kurtener and Badenko 1999), for the assessment of polluted agricultural fields in order to a design of strategy for territorial prophylactic actions (Kurtener *et al.* 1999a), for the assessment of burned forest areas with the aim of planning land restoration (Kurtener *et al.* 2000), for land suitability assessment in the process of agricultural experimentation (Kurtener and Badenko 2000a), for assessment of agricultural lands to plan site-specific residue management (Kurtener and Badenko 2000b), and for the multi-dimensional evaluation of areas on the land market (Kurtener and Badenko 2000c, Yakushev *et al.* 2000).

The aim of this chapter is to describe the GISFM approach as a method for supporting planning and to present several examples of its application in the analysis of environmental planning. In Section 14.2, the GISFM approach is outlined. Section 14.3 describes a case study of its use for the analysis of the contamination of agricultural areas by heavy metals (Kurtener and Badenko 2000d). Section 14.4 considers the application of GISFM in the assessment of land drainage plots to help design a strategy of renovation. Finally in Section 14.5, the methodology is used in a case study of land use sustainability.

14.2 The GIS Fuzzy Modelling Approach

The central concept of fuzzy set theory is the membership function, which represents the relationship of an element to a set. The membership function of a fuzzy set is expressed on a continuous scale from one (full membership) to zero (full non-membership). One of the principal benefits of the application of fuzzy set theory in real world planning situations is that most indices of the planning process can be estimated by using membership functions. In particular, by setting the membership function to one, we symbolise a maximum value of the indicator (or criterion) of the planning process, and setting the membership function to zero, we represent a minimum of value of the indicator.

The application of a fuzzy set membership function can be very useful for standardisation where there are several criteria. It provides a very strong rationale for the process of standardisation (Jiang and Eastman 2001). Another important advantage of fuzzy set theory is its ability to deal with linguistic variables (Zadeh 1994). Jager (1995) states that *"the main motivation of fuzzy set theory is apparently the desire to build up a formal, quantitative framework that captures the vagueness of human knowledge as it is expressed via natural languages"*. Parameters of the planning process can be evaluated using linguistic values (or variables). The approach is based on the observation that experts most often use linguistic constructs for the evaluation of environmental or social situations. The construction of a linguistic scale is based on a perception of the situation under consideration.

We suggest that the term *fuzzy model* is used as an umbrella term to describe an end product of fuzzy modelling. In the same way that various applications of fuzzy set theory can be developed, combinations of fuzzy applications with crisp models could be the end products of fuzzy modelling. GISFM interfaces GIS with fuzzy models. According to Hartkamp *et al.* (1999), the terms *interface* and *interfacing* may be used to indicate the simultaneous use of GIS and modelling tools, since they do not imply a specific level of interaction between them. Hartkamp *et al.* suggest the terms *linking*, *combining*, and *integrating* indicate the degree of interfacing. In this context, the term *linking* is defined as merely the exchange of input and output data between a GIS and a fuzzy model. The term *integrating* implies incorporating one system into the other. The term *combining* can be defined as a special case when one part of the fuzzy model is linked with GIS and second part of the fuzzy model is integrated into GIS.

The four main steps used to realise a GISFM approach are as follows:

- *Structuring phase*: perception of problem, identification of input and output data, obtaining missing data using crisp models, definition of alternatives and criteria;
- *Fuzzy modelling phase*: building of membership functions, selection of fuzzy algorithm for integration in GIS environment;
- *Programming phase:* selection of suitable existing software or designing of a new one; and
- *Evaluation phase:* creation of thematic maps, perception of results obtained.

The distinguishing feature of the GISFM approach is the fact that indicators or criteria are defined using membership functions. The structuring phase includes the identification of criteria as membership functions. A membership function is selected for each indicator or criterion. The choice of a membership function is somewhat arbitrary and should mirror the subjective expert opinion. Usually, with this aim in mind, two types of membership function can be identified (Jager 1995). Firstly, there is the *linear* approach, which is represented by a 'straight' membership function like that shown in Figure 14.1 which has a triangularly shape. Secondly there is the *non-linear* approach that is illustrated by a Gaussian or bell-shaped membership function.

By using a triangular-shaped membership function, (*x*), we can present an indicator of the decision-making process as follows:

(x) = 0 if x < a (1)
0 < (x) 1 if a x b (2)
1 > (x) > 0 if b < x c (3)
(x)= 0 if x > c (4)

where *a, b* and *c* are reference points of *x*. The membership function in graphical form is shown in Figure 14.1 and one can see that, if *x* is below *a* or upper *c*, so *(x)* = *0*, but, if *x* changes from *a* to *b* then *(x)* increases from 0 to 1 and if *x* changes from *b* to *c,* then *(x)* decreases from 1 to 0.

Let us consider a particular example case of study of soil contamination caused by the failure of a landfill (Bogardi *et al.* 1996). In this study, one of criterion is the groundwater table. This criterion was identified by the triangular-shaped membership function with the three reference points on the abscissa: *a* = 0.61m, *b* = 1.53m, and *c* = 1.83m. In this case study, the groundwater table was measured in metres.

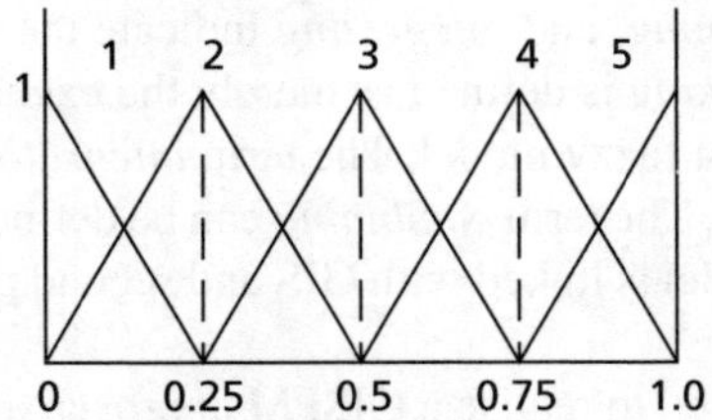

Fig 14.1: Fuzzy triangular-shaped membership function

Thus, each indicator (or criterion) is defined by a specific fuzzy sub-model. The GISFM approach is an adequate way to integrate an expert's subjective perception as a statement of the meaning of a criterion. The approach can handle simultaneously quantitative and qualitative variables. Examples are found in Section 14.4 where we consider the application of GISFM in the assessment of land drainage plots to help design a strategy of renovation. Here, ecological and social criteria are evaluated using linguistic values with a five-point scale (Table 14.1). In this case study also, the economic and technological criteria are described numerically by a stepwise linear function.

GISFM software is built up in accordance with a special kind of interfacing. In the case of the integration of a fuzzy model into a GIS environment, the same programming language should be used both in the GIS software and for programming of the fuzzy model. For example, if it is assumed that the GISFM approach will be carried out on the basis of MapInfo® software, then MapBasic® should be used as the language for programming of the fuzzy model. Each of the examples of the approach reported in this chapter uses MapBasic®. Joerin *et al.* (2001) have noted that it is quite simple to program the ELECTRE-TRI algorithm in MapBasic®, allowing multi criteria analysis to be run within MapInfo®.

14.3 The Application of GISFM in the Study of Soil Contamination

Agricultural land use efficiency, environmental stability and sustained productivity are key goals in environmental management. This example involves the use of GISFM to assess the impact of heavy metal contamination of agricultural land (Kurtener and Badenko 2000d).

14.3.1 Structuring Phase

In the process of planning amelioration actions following heavy metal contamination, it may be necessary to estimate the current state of the agricultural areas concerned in order to determine the seriousness of the pollution problem. The objectives of the exercise can therefore be established. These are as follows:

- to evaluate current state of contamination of the agricultural area by specific heavy metals; and
- to estimate composite effect of the contamination from several heavy metals.

The GIS database used in this study involved an area of land located in the suburbs of Saint Petersburg and all the data are stored in MapInfo® tables. Agricultural lands polluted by heavy metals are introduced as alternatives. Indicators were required to measure the soil contamination by each specific heavy metal, and a composite measure was required for all the pollutants together.

14.3.2 Fuzzy Modelling Phase

The starting position for building a membership function in this case study is as follows:

- the concentration of heavy metals can be very small and soil contamination is negligible;

- fixed criteria for acceptable levels of soil contaminants (threshold values) are being defined in Russia by law;
- the concentration of heavy metals can be equal to or greater than the upper threshold value and soil contamination is considered as impermissible; and
- the concentration of heavy metals is neither very small nor very large and it is not quite clear whether the soil contamination is serious or not.

Mathematically these all situations can be described by an appropriate membership function (x) as follows:

$$(x) = \begin{cases} 0; & x < th_{low} \\ f(x); & th_{low} \quad x \quad th_{up} \\ 1 & x > th_{up} \end{cases} \qquad (5)$$

where th_{low} and th_{up} are lower and upper threshold values of the contamination under consideration, $f(x)$ is function, and x is the current value of the heavy metal contamination. It is easy to see that (x) has grades in the interval [0,1]. $f(x)$ can be described by different mathematical expressions. In this study, we use a three-section linear function as follows:

$$f(x) = (th_{up} - x)/(th_{up} - th_{low}) \qquad (6)$$

where x is the current concentration of a pollutant (mg/kg) and th_{up} is equal to threshold value of the pollutant $(th_{low} = 1/3th_{up})$. The membership function in graphical form is shown in Figure 14.2 and one can see that if the concentration is below the lower threshold, the soil contamination is negligible, but if the concentration is above the upper threshold the soil contamination is fully classified as not permissible.

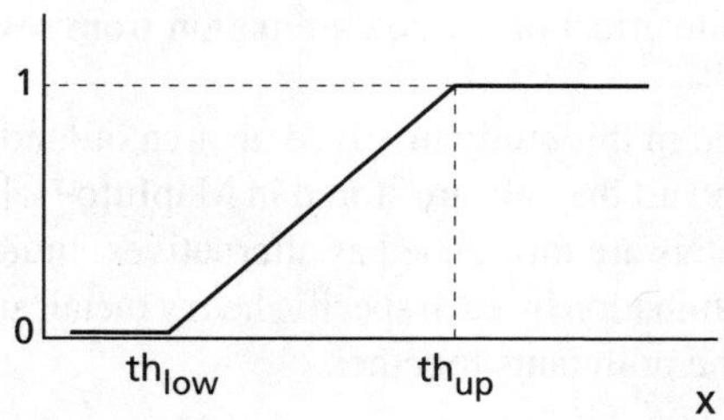

Fig 14.2: Fuzzy linear membership function

Equations (1) and (2) are used in order to define the indicators of soil contamination by each heavy metal. The composite indicator is described by summing the individual indicators using the algebraic sum for fuzzy sets as follows (Zimmermann 1985):

$$_{com} = 1 - (1 - \ _{1}) * (1 - \ _{2}) * \ldots *(1 - \ _{j}) \quad (7)$$

where $_{j}$ is the index of indicator of soil contamination by i – pollutant, defined by (6).

14.3.3 Evaluation Phase

The study evaluated the contamination of 106 agricultural areas polluted by heavy metals. The thematic maps in Figure 14.3 illustrate the spatial variations in soil contamination by copper, ftorum, arsenic and the composite index. In particular, Figures 14.3a – 3c show the current state of contamination of the agricultural area by these pollutants. The application of fuzzy set membership functions as specific indicators permits an end user to (a) see immediately the more polluted areas, (b) estimate straight away the dimensions of contamination, and (c) compare maps of soil contamination by different pollutants because all indicators are standardised.

The use of the algebraic sum of fuzzy sets as the composite indicator allows the creation of an integrated description of the current state of contamination of the areas (Figure 14.3d). We emphasise that there are other ways for defining the composite indicator, including, for example, the *weighted linear combination* (Jiang and Eastman 2001).

14.4 The Application of GISFM to Assess Land Drainage

An allocation of investment is needed for the renovation of drainage systems. Following an analysis of the current state of a set of land drainage plots, an application using GISFM was developed by Kurtener and Badenko (2001).

14.4.1 Structuring Phase

In the process of planning to renovate a drainage system, it is necessary to assess the current state of each land drainage plot. In connection with this, the problem can be defined as requiring the following:

- an evaluation of the efficient use of funds for the renovation;
- an evaluation of labour-intensiveness of the renovation;
- an examination of the possible ecological effects of the renovation;
- an evaluation of social reactions to this activity; and
- the calculation of a composite index of the effects of technological, economic, ecological and social factors.

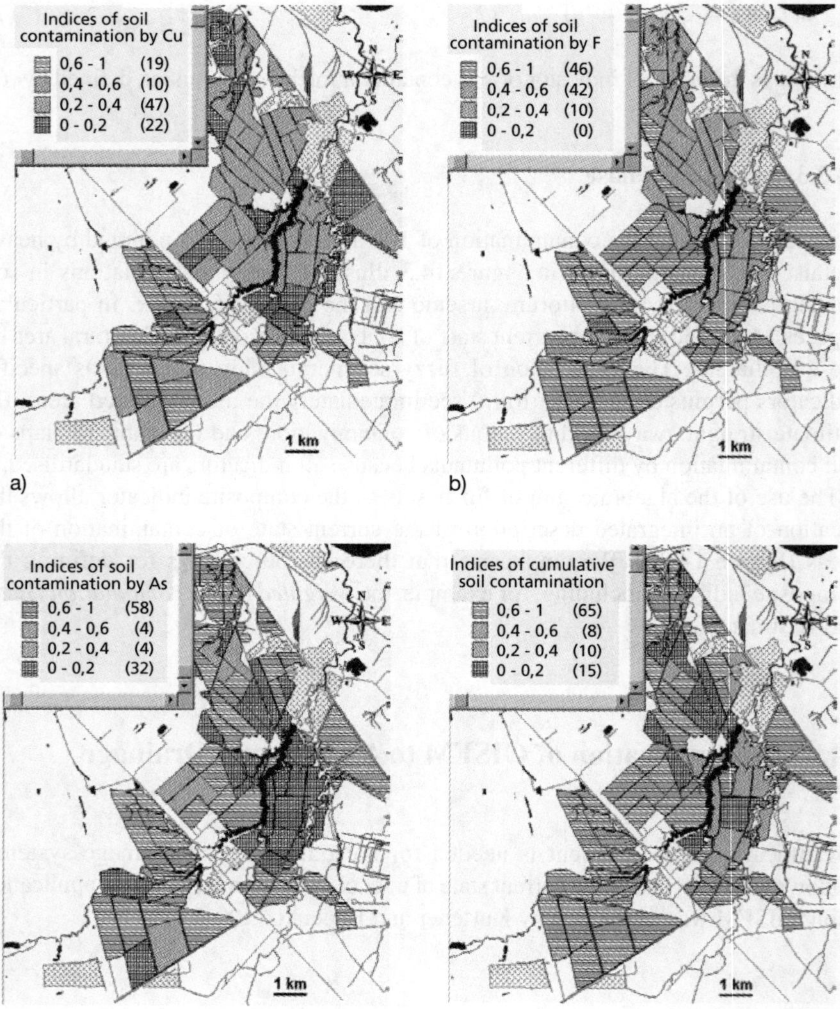

Fig 14.3: Thematic maps of indices of soil contamination by (a) copper (b) ftorum (c) arsenic and (d) the composite soil contamination index

As in the previous example, the GIS database used in this study involved agricultural lands located in the suburbs of Saint Petersburg. Land drainage plots are introduced as alternatives and the five criteria for evaluation of the plots are defined as follows:

- *Economic criterion (EC)*: defines the economic efficiency of the use of funds for renovation. Thus, EC =1 if the economic efficiency is high; EC = 0 if it is not.
- *Technological criterion (TC)*: defines the labour-intensiveness of the renovation. TC =1 if labour-intensiveness is normal; TC = 0 if labour-intensiveness is greater.

- *Ecological criterion (ECC)*: defines the ecological effect of renovation. ECC =1 if the ecological situation has improved greatly; ECC = 0 if it has not improved very much.
- *Social criterion (SC)*: defined as the human reaction. SC =1 if the human reaction is very strong; and SC = 0 if it is not.
- *Index of multiple evaluation (ME)*: defines the composite effect of the economic, technological, ecological and social factors. ME =1 if land drainage plot is the best for investment; ME = 0 if it is the worst.

14.4.2 Fuzzy Modelling Phase

In this case study, the economic and technological criteria are described by a stepwise linear function as shown in Figure 14.2. In the case of the economic criterion, variable x represents the effectiveness of an investment in renovation (th_{up} is equal to the normal level of effectiveness; $th_{low} = 1/3\ th_{up}$). With the technological criterion, variable x represents labour-intensiveness associated with the renovation (th_{up} is equal to the normal labour-intensiveness; $th_{low} = 1/5\ th_{up}$.). Hence equations (1) and (2) can be used in order to define indicators for the economic and technological criteria.

Ecological and social criteria are evaluated using linguistic values. The approach is based on the observation that experts usually use linguistic constructs for the evaluation of ecological and social situations. A five-point membership scale ranging from *Absence* to *Very High* values has been assigned (Table 14.1). The expert's job is to evaluate each land drainage plot and then choose the most suitable linguistic value to describe the evaluation of the ecological or social effects of the renovation.

Table 14.1: Linguistic construction with five-points for membership function scale

Linguistic values	Quantitative equivalent
Absence	0
Low	0.25
Medium	0.5
High	0.75
Very high	1

The index of multiple evaluation (ME) is defined using a method of multi-criteria assessment (Borisov *et al.* 1990). It is based on the section of fuzzy set theory that is devoted to multi-criteria assessment and choice of alternatives. In particular, if there are several criteria $C_1, C_2, \ldots, C_n$ and if there are coefficients of relative significance of these criteria, $a_1, a_2, \ldots, a_n$, the rule for the selection of the best land drainage plot taking into account a_i will be written as an intersection of $C_i a^i$. Coefficients of relative significance, a_i, are determined by the procedure of comparison of a pair of criteria.

14.4.3 Evaluation Phase

The criteria for the evaluation of land drainage plots are mapped in the MapInfo® environment (Figure 14.4). In particular, the application of a fuzzy set membership functions as specific indicators permits the categorisation of land drainage plots according to technological, economic, ecological and social criteria. An end user can readily compare these maps because all indicators are standardised. It is easy to see that the priority according to technological and ecological criteria should be given to the third land plot (Figure 14.4a and 14.4b), but according to the social criterion, the preference should be given to the fourth land plot (Figure 14.4c), and according to the economic criterion, the privilege should be given to the first land plot (Figure 14.4d).

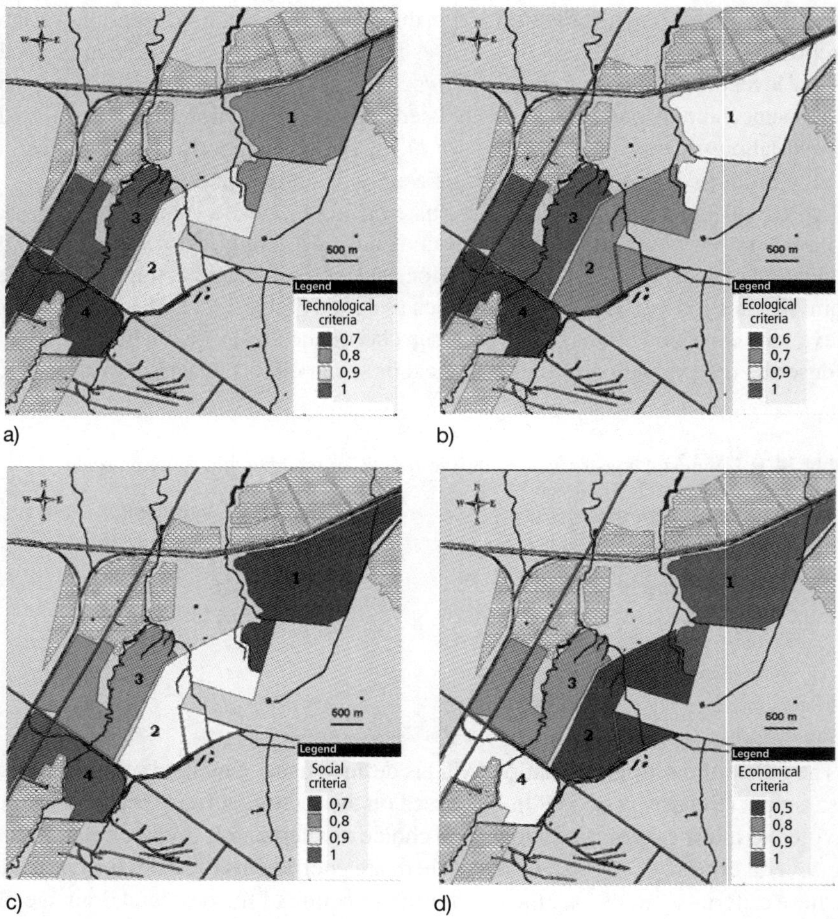

Fig 14.4: Thematic maps of indices of (a) technological efficiency, (b) ecological improvements, (c) social reaction and (d) economic efficiency

The use of the algorithm of multi-criteria assessment (Borisov *et al.* 1990) for definition the index of multiple evaluation allows the creation of an integrated description of the current state of the drainage areas (Figure 14.5). The study was carried out for two special cases: task 1, where the ecological factor is more important (Figure 14.5a), and task 2, where the economic factor is more significant (Figure 14.5b). It is clear that in the first case the land drainage plots are ranked as follows: plot 3, plot 1, plot 2, and plot 4. In the second variant, the priority sequence becomes: plot 1, plot 4, plot 3 and plot 2.

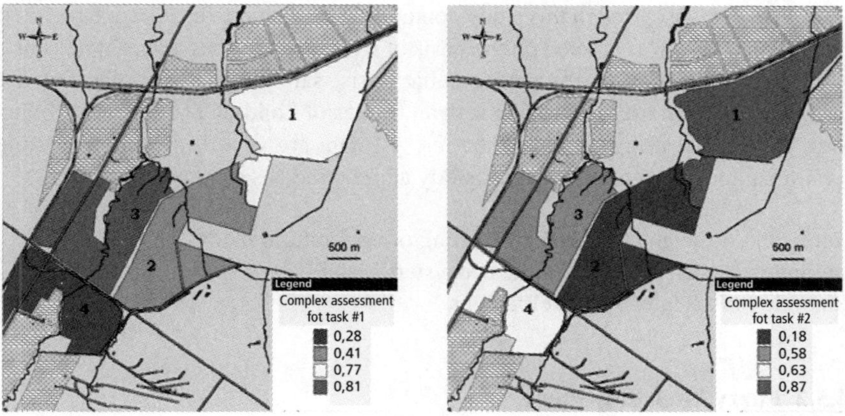

a) b)

Fig 14.5: Thematic maps of index of multiple evaluation of land drainage plots when (a) the ecological factor is more important and (b) the economic factor is more important

We emphasise that there are other ways for multi-criteria assessment, including, for example, the *MAX integrated operation* (Jager 1995, Jiang and Eastman 2001).

14.5 The Application of GISFM to Estimate Land Sustainability

This application recognises that sometimes it may be necessary to estimate land sustainability. It is evident that the sustainable development of land is dependent on an antropogeneous loading (contamination) and the resistance of its ecosystems (tolerance to the contamination). Bogardi *et al.* (1996) suggested a theoretical approach to resolve this problem using fuzzy set theory. Kurtener *et al.* (2000e) applied the methodology in the context of the development of agricultural land.

14.5.1 Structuring Phase

The problem of evaluating the opportunity for sustainable development of land can be set up as three tasks as follows:

- an evaluation of an antropogeneous loading of the agricultural land;
- an assessment of the resistance of the ecosystems; and
- an examination of the possibility for sustainable development.

The GIS database used in this study contains data on land in the Volga River delta. The GIS database was created on the basis of field observations and remote sensing data. For the creation of the cartographic basis, satellite imagery distributed by Eurimage has been used, including several frames of Landsat TM 5 satellite images using pixels measuring 30 by 30 metres. All data are stored in MapInfo® tables. Certain land areas were selected for study and the indices used were as follows:

- an index of the antropogeneous loading of agricultural lands;
- an index of the resistance of the ecosystems; and
- an index of sustainable development.

14.5.2 Fuzzy Modelling Phase

In this study indices of the antropogeneous loading and ecosystem resistance are represented by a triangular-shaped membership function, known as triangular fuzzy numbers (TFN) (Kaufmann and Gupta 1988). TFN have been widely used in practical applications because of their simplicity and ability to describe experts' subjective information. TFN are the simplest type of membership functions that are linear on either side of the peak that is equal to 1. TFN can be described by the three values on the abscissa: *a, b* and *c* (Figure 14.1).

The fuzzy resistance of ecosystem *R* and fuzzy antropogeneous load *L* are both described by TFNs. *R* and *L* are both defined by experts. The difference *Z* between *R* and *L* can be obtained by subtraction:

$$Z = R - L \qquad (8)$$

It should be noted that the resulting failure surface, *Z*, will also be a TFN. Sustainable development will be impossible when *L* equals or exceeds *R*, that is, when $Z < 0$. The event $Z > 0$ implies safety or reliability. To define a measure to quantify the reliability under such circumstances, Bogardi *et al.* (1996) proposed a fuzzy reliability index, *Co*, as follows:

$$Co = \frac{\int_{x>0} \mu_z(x)dx}{\int_x \mu_z(x)dx} \tag{9}$$

where $_z(x)$ is the membership function of fuzzy number Z.

The definition of a fuzzy reliability index is shown on Figure 14.6. *Co* is the ratio of area between the abscissa and curve of $_z(x)$ where $x > 0$ (shaded area on Figure 14.6) to the area between the abscissa and curve of $_z(x)$. The fuzzy reliability index, *Co,* in this context is interpreted as the index of sustainable development. *Co* varies from 0 to 1.

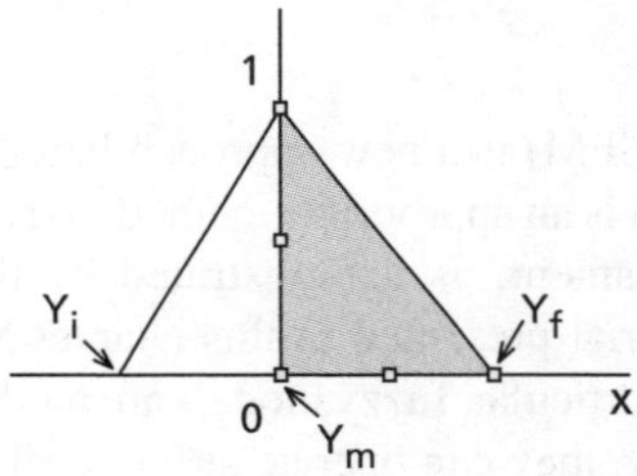

Fig 14.6: Definition of fuzzy reliability index, *Co*

14.5.3 Evaluation Phase

The area under consideration is a large swampy territory where there are polluted plots of land (Figure 14.7). In this study two particular land plots were selected using the opinions obtained through consultation with land users and ecologists.

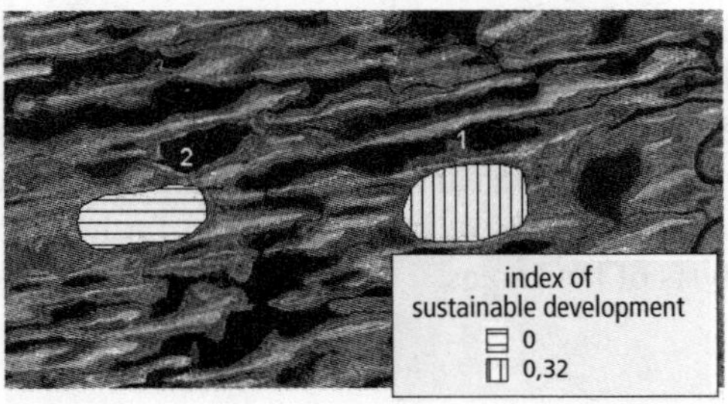

Fig 14.7: Index of sustainable development for selected land plots

The indices of antropogeneous loading, L, and ecosystem resistance, R, have been assessed by an expert panel as TFNs. It is well known that uncertainty is inherent in processes of assessment of both R and L, which involve data and model uncertainty. In connection with this, the application of TFNs for definition of R and L is a suitable way to take into account this uncertainty. The index of sustainable development, Co (Figure 14.7), shows current state of environmental situation on areas under consideration. It is easy to see that, on the first land plot, sustainable development of ecosystem is possible but on the second land plot, where $Co = 0$, it is impossible.

14.6 Conclusions

GIS Fuzzy Modelling (GISFM) is a new approach based on interfacing fuzzy and crisp modelling with GIS. It is an appropriate methodology to support location choice and land suitability assessment, as demonstrated by the examples of practical implementation. The material presented in this chapter has also demonstrated the advantages of GISFM. In particular, fuzzy models are modelling constructs featuring two main properties: firstly, they can operate at the level of linguistic terms (fuzzy sets); and secondly, they represent and process uncertainty. Thus, by the use of fuzzy models, it is possible to describe both subjective (knowledge-driven) and deterministic (data-driven) information.

As with most parametric methods, the greatest difficulties come with the identification of fuzzy sets and the choice of membership functions. As a rule there is distance between points in original space and the space induced by the fuzzy set and the membership function. The methodology for the definition of fuzzy sets and membership functions is not sufficiently advanced. The success of GISFM applications is related to the competence and experience of planners because a statement of the meaning of a criterion is based on an expert's subjective perception. The added difficulty is that the methodology of defuzzification is still imperfect.

Several trends in GISFM can be anticipated, one of which involves the development of a scheme of fuzzy modelling that emphasises information processing at different levels of information granularity. The dominant tendency will be given to develop methods of defuzzification. Moreover, the GISFM approach can be applied in other ways. For example, it is an effective tool for evaluation of the quality of attribute databases. It allows the calculation and mapping of the measure of data uncertainty. GISFM will find undoubtedly have practical use in further social, economic and environmental analysis of land areas.

References

Bogardi, I., Bardossy, A., Mays, M.D. and Duckstein, L. (1996) Risk assessment and fuzzy logic as related to environmental science, *SSSA Special publ.* No 47.

Borisov, A.N., Krumberg, O.A. and Federov, I.P. (1990*) Prinytie resheniy na base nechetkikh modeley,* (In Russian) Zinatre Publ, Riga.

Banai, R. (1983) Fuzziness in geographical information systems: contributions from the analytical hierarchy process, *International Journal of Geographical Information Systems*, 7: 315-329.

Burrough, P.A. (1986) *Principles of Geographical Information Systems for Land Resource Assessment*, Oxford University Press, New York.

Burrough, P.A. (1989) Fuzzy mathematical methods for soil survey and land evaluation, *Journal of Soil Science*, 40: 477-492.

Burrough, P.A. and McDonnell, R.A. (1998) *Principles of Geographical Information Systems*, Oxford University Press, New York.

Burrough, P.A., MacMillan, R.A. and van Deursen, W. (1992) Fuzzy classification methods for determining land suitability from soil profile observations and topography, *Journal of Soil Science,* 43: 193-210.

Carver, S. (1991) Integrating multi-criteria evaluation with geographical information systems, *International Journal of Geographical Information Systems*, 5: 321-339.

Corne, S., Murray, T., Openshaw, S., See, L. and Turton, I. (1999) Using computational intelligence techniques to model sub-glacial water systems, *Journal of Geographic Systems*, 1: 37-60.

Dubois, D., Prade, H. and Yager, R.R. (eds.) (1993) *Readings in Fuzzy Sets for Intelligent Systems*, Morgan Kaufmann Publishers.

Dueker, K.J.D. and Barton, P. (1990) GIS in land development planning process: balancing the needs of land planners and real estate developers, *Journal of the American Planing Association*, 56: 483-491.

Eastman, J.R., Kyem, P.A.K., Toledano, J. and Jim, W. (1993) *GIS and Decision Making. Explorations in Geographic Information System Technology*, 4 UNITAR, Geneva.

Eastman, J.R. and Jiang, H. (1995) Fuzzy measures in multi-criteria evaluation. In: *Proceedings, Second International Symposium on Spatial Accuracy Assessment in Natural Resources and Environmental Studies*, May 21-23. Fort Collins, Colorado, pp 527-534.

Geertman, S. and Toppen, F. (1990) Regional planning for new housing in Randstad Holland. In: Scholten, H.J. and Stillwell, J.C.H (eds) *Geographical Information Systems for Urban and Regional Planning*, Kluwer Academic Publishers, Dordrecht.

Gopal, S. and Woodcock, C. (1994) Theory and methods for accuracy assessment of thematic maps using fuzzy sets, *Photogrammetric Engineering and Remote Sensing*, 60(2): 181-188.

Hartkamp, A., White, J. and Hoogenboom, G. (1999) Simulation and modeling: interfacing Geographic Information Systems with agronomic modeling, *Agronomy Journal*, 91:761-772.

Hall, G.B., Wang F. and Subakyono (1992) Comparison of Boolean and fuzzy classification methods in land suitability analysis by using geographical information systems, *Environment and Planning A*, 24: 497-516.

Jager, R. (1995) *Fuzzy Logic in Control*, Delft TU Publisher, Delft, The Netherlands.

Jansen, R. and Rietveld, P. (1990) Multi-criteria analysis and geographical information systems: an application to agricultural land use in the Netherlands, in Scholten H. and Stillwell, J. (eds) *Geographical Information Systems for Urban and Regional Planning*, Kluwer Academic Publisher, Dordrecht.

Jankowski, P. (1995) Integrating geographical information systems and multiple criteria decision making methods, *International Journal of Geographical Information Systems*, 9: 251-273.

Jankowski, P., Nyerger, T.L., Smith, A., Moore, T.J. and Horvath, E. (1997) Spatial group choice: a CDSS tool for collaborative spatial decision making, *International Journal of Geographical Information Systems,* 11: 566-602.

Jankowski, P., Andrienko, N. and Andrienko, G. (2001) Map-centred exploratory approach to multiple criteria spatial decision making, *International Journal of Geographical Information Science*, 15(2): 101-127.

Jankowski, P. and Nyerges, T. (2001) *Geographic Information Systems for Group Decision Making*, Taylor & Francis, New York.

Jiang, H. and Eastman, J.R. (2000) Application of fuzzy measures in multi-criteria evaluation in GIS, *International Journal of Geographical Information Science*, 14(2): 173-184.

Joerin, F., Thériault, M. and Musy, A. (2001) Using GIS and outranking multicriteria analysis for land-use suitability assessment, *International Journal of Geographical Information Science*, 15(2): 153-174.

Kaufmann, A. and Gupta, M.M. (1988) *Fuzzy Mathematical Models in Engineering and Management Science,* Elsevier Science Publ., Amsterdam.

Kurtener, D. and Badenko, V. (1999) Questions of integration of some ecological models into geoinformation system, in *Proceedings of UDMS'99 on Information Technology in the Service of Local Government Planning and Management*, UDMS Press, Venice, Italy.

Kurtener, D., Yakushev, V., Badenko, V. and Pourabbas, E. (1999a) Development of methodology of multiple assessment of landscape parcels on the base fuzzy models integrated into GIS environment, *SPBISTRO Special publ. No 1,* St. Petersburg, Russia.

Kurtener, D., Badenko, V. and Cudlip, W. (2000) Development of the methodology of multiple assessment of burned areas in forest regions for the planning of restoration actions, in Yakushev, V. and Kurtener, D.(eds) *Agrophysical and Ecological Problems of Agriculture in the 21st Century*, SPBISTRO Press, St. Petersburg, Russia, 2: 54-62.

Kurtener, D. and Badenko, V. (2000a) Precision agriculture experimentation on the base of fuzzy models synthesised with GIS, *Aspects of Applied Biology*, 61: 139-143.

Kurtener, D. and Badenko, V. (2000b) Development of the methodology of assessment of site-specific residue management actions on the basis of fuzzy models integrated into a GIS environment, in Morrison, J. (eds) *Proceedings of 15th ISTRO conference on agroecological and ecological aspects of soil tillage*, ISTRO Press, Fort Worth, Texas.

Kurtener, D. and Badenko, V. (2000c) Applications of GIS knowledge management for decision making in the field of land market and land consolidation, in *Proceedings of UDMS'2000*, UDMS Press, Delft, The Netherlands.

Kurtener, D. and Badenko, V. (2000d) Methodological framework based on fuzzy set theory for land use management, *Journal of the Brazilian Computer Society*, 6(3): 26-32.

Kurtener, D., Rossi, L. and Badenko, V. (2000e) Development of fuzzy direction of GIS knowledge management with the use of Eurimage products, in Yakushev, V. and Kurtener, D. (eds) *Agrophysical and Ecological Problems of Agriculture in the 21st Century,* SPBISTRO Press, St. Petersburg, Russia, 2: 14-26.

Kurtener, D. and Badenko, V. (2001) Applications of GIS knowledge management for spatial planning of water resources for sustainable development of European agriculture, in *Proceedings, 19th European Regional Conference on Sustainable Use of Land and Water*, 4-8 June, Brno, Czech Republic.

Mays, M.D., Bogardi, I. and Bardossy, A. (1997) Fuzzy logic and risk-based soil interpretations, *Geoderma*, 77: 299-315.

Malczewiski, J. (1999) *GIS and Multicriteria Decision Analysis*, Wiley & Sons, New York.

Pereka, J.M.C. and Duckstein, L. (1993) A multiple criteria decision making approach to GIS-based land suitability evaluation, *International Journal of Geographical Information Systems,* 7: 407-424.

Pedrycz, W. and Gomide, F. (1998) *An Introduction to Fuzzy Sets: Analysis and Design,* MIT Press, Cambridge.

Ross, T.J. (1995) *Fuzzy Logic with Engineering Applications,* McGraw-Hill, New York.

Senes, G. and Toccolini, A. (1998) Sustainable land-use planning in protected rural areas in Italy, *Landscape and Urban Planning*, 42: 107-117.

Smith, P.N. (1992) Fuzzy evaluation of land-use and transportation options, *Environment and Planning B*, 19: 525-544.

Scholten, H.J. and Stillwell, J.C.H. (eds) (1990) *Geographical Information Systems for Urban and Regional Planning*, Kluwer Academic Publisher, Dordrecht.

Voogd, H. (1983) *Multicriteria Evaluation for Urban and Regional Planning*, Pion, London.

Wang, F., Hall, G.B, and Subakyono (1990) Fuzzy information representation and processing in conventional GIS software: database design and application, *International Journal of Geographical Information Systems,* 4: 261-283.

Xiang, W.N., Gross, M., Fabos, J.G. and Macdougall, E.B. (1992) A fuzzy group multi-criteria decision making model and its application to land-use planning, *Environment and Planning B*, 19: 61-84.

Yakushev, V.P., Kurtener, D.A., Badenko, V.L., Kudashev, E.V. and Cudlip, W, (2000) Methodology of multiple assessment of landscape parcels on the base of fuzzy set theory models integrated into geographic information systems (GIS), *Russian Agricultural Science,* 4: 42-43.

Yen, A.G. and Hong, C.M. (1996) An integrated GIS and location-allocation approach to public facilities planning – an example of open space planning, *Computers, Environment and Urban Systems*, 20: 339-350.

Zadeh, L.A. (1965) Fuzzy sets, *Information and Control*, 8: 338-353.

Zadeh, L.A. (1994) Wh*y the success of fuzzy logic is not paradoxical.* IEEE Expert – A Fuzzy Logic Symposium, 9(4): 43-46.

Zimmermann, H.J. (1985) *Fuzzy Set Theory and its Applications*, Kluwer Academic Publishers, Dordrecht.

PART 3

SUPPORT SYSTEMS FOR STRATEGIC PLANNING

15 A Migration Modelling System to Support Government Decision-making

Tony Champion[1], Glen Bramley[2], Stewart Fotheringham[1], James Macgill[3] and Philip Rees[3]

[1] Department of Geography, University of Newcastle upon Tyne, Newcastle upon Tyne, NE1 7RU, United Kingdom
[2] School of Planning and Housing, Edinburgh College of Art/Heriot-Watt University, 79 Grassmarket, Edinburgh EH1 2HJ, United Kingdom
[3] School of Geography, University of Leeds, Leeds LS2 9JT, United Kingdom

Abstract: The planning support system described in this chapter is a prototype internal migration modelling system developed for the UK's Department of Environment, Transport and the Regions during 2000. It forms part of the Government's drive towards evidence-based policy-making. It is designed to be run by civil servants on stand-alone desk-top computers primarily in order to gauge the likely impact on between-area migration flows of alternative economic and policy scenarios. The system is based on a two-stage representation of the migration process, the first stage predicting out-migration from each of 100 areas of the UK, and the second predicting the distribution of these migrants between destinations. A user-friendly front-end allows the alteration of the levels of the determinant variables for running scenarios and the easy visualisation of the migration impacts through tables and maps. The system is now being refined in a second phase of work for the client, which includes the preparation of a scoping report on future enhancements. The chapter outlines the policy context of this work and the client's requirements, describes the modelling approach and the basic structure of the initial model, and presents an example of running a trial scenario.

15.1 Introduction

The planning support system (PSS) described in this chapter is the prototype of a migration modelling system developed for the UK's Department of the Environment, Transport and the Regions (DETR, restructured and renamed Department of Transport, Local Government and the Regions in June 2001). It represents an attempt to model flows of people between 100 areas of the UK, with the primary aim of investigating the quantitative impacts of alternative economic and policy scenarios. This work was commissioned because central government has identified migration as critical to a number of the issues that it is currently addressing in the areas of regional development, urban regeneration, housing, countryside planning and land use policy. The PSS is designed for internal departmental use by civil servants advising Ministers about the factors driving migration and about the migration impact of proposed policies. The study was carried out over the 12 months ending December 2000 by a team of researchers drawn from a total of four institutions (see Acknowledgements).

This chapter concentrates on the rationale behind this PSS and on the output end of the model as experienced by users (for a more detailed account of the methods of model development and of the characteristics of the model, see Rees *et al.* forthcoming). The first section below describes the context of the policy issues and initiatives that have led to central government's current interest in population movements. The second section gives DETR's requirements of the migration model as indicated to potential contractors in the tender documentation. The third outlines the model's content and function, emphasising its two-stage representation of the migration process (decision to leave and choice of destination) and summarising its main features (in terms of what model calibration indicates to be the principal determinants of internal migration). Thereafter, the system's user-friendly front-end is presented, followed by a demonstration of its operation using one scenario as an example. The concluding section looks at the ways in which it is proposed to move forward from the prototype model.

15.2 The Context of Policy Interest in Migration

The policy salience of migration in the UK can be linked to the growth of central government interest in spatial issues, boosted by the election of a Labour administration in 1997. In many ways, the current level of interest in broader regional and urban planning strategy resembles that of the late 1960s and early 1970s before attention shifted both upwards in scale to the challenge of an uncompetitive national economy and downwards to the 'inner city' problem. In particular, the 1980s had seen the virtual dismantling of strategic planning and a move towards highly spatially focused and largely *ad hoc* decision-making, as exemplified by the uncoordinated consideration of private proposals for building new 'country towns' and by the targeting of inner-city regeneration incentives on very tightly defined areas. Even before 1997, however, the Conservative Government had accepted the need for wider strategic thinking, this being mainly under pressure from two sets of traditional Tory supporters – the residents of 'middle England', where the bulk of new urban development was projected to occur, and big business, notably the large supermarket chains and volume house builders. They were keen to protect their investments in property and land banks and began to realise that they preferred the relative certainty provided by strong policy guidance. Moreover, since 1997, this reviving interest in spatial strategy, which includes revised Regional Planning Guidance, new regional institutions and the European Spatial Development Perspective, has been combined with a strong commitment to evidence-based policy.

These developments have been particularly evident in terms of urban regeneration. Here the key step was the setting up of an Urban Task Force in 1998, with its Mission Statement being to *"identify causes of urban decline in England and recommend practical solutions to bring people back into our cities, towns and urban neighbourhoods"* (Urban Task Force 1999, title page). Chaired by the architect Lord Rogers of Riverside, the Task Force published its findings in *Towards An Urban Renaissance* (Urban Task Force 1999). This was followed by the Government's proposals for action in *Our Towns and Cities: The Future – Delivering an Urban Renaissance* (DETR 2000a). Emphasising the commitment to evidence-based policy, this so-called 'Urban White Paper' was accompanied by two further research-based documents, *The State of English Cities* (Robson *et al.* 2000) and *Living in Urban England: Attitudes and Aspirations* (Todorovic and Wellington 2000).

This renewed interest in spatial policy and evidence-based decision-making can also be seen in relation to initiatives on the countryside, housing, sustainable development and regional balance. The Urban White Paper was followed closely by its rural equivalent *Our Countryside: The Future – A Fair Deal for Rural England* (DETR 2000b), which developed out of the earlier consultation document *Rural England: A Nation Committed to a Living Countryside* (DETR 1998). The Government's thinking on housing issues reached the consultation, or 'green paper', stage in *Quality and Choice: A Decent Home for All* (DETR/DSS 2000). Its broad framework for promoting sustainable development was set out in *A Better Quality of Life* (DETR 1999a), while the Prime Minister's concerns over regional trends led to the preparation of *Sharing the Nation's Prosperity: Variation in Economic and Social Conditions Across the UK* (Cabinet Office 1999).

Internal migration has emerged as a major concern across this spectrum of policy arenas. The most frequently cited aspects include:

- the resumption of the north-to-south drift of population since the early 1990s, building up to its long-term average of a net migratory flow of some 30,000 residents a year;
- the net exodus out of the main English conurbations to the 'shire counties' of some 90,000 a year on average, half of this from Greater London alone and with higher rates of loss for better-paid and more skilled people;
- the continuing overall suburbanization of population in both conurbations and other large cities and towns, despite some well-publicised examples of successful city-centre housing schemes;
- the growing pressure for the expansion of towns and cities into greenfield development in spite of, or perhaps because of, the existence of underused land and buildings within these urban areas;
- the acute problems of housing affordability in more accessible rural communities as local people are being outbid by the better-off incomers; and

- the wasteful use of natural resources and increase in pollution as those who move outwards travel greater distances to get to work, shops and the places where they spend their leisure time.

The extent to which these issues were central to DETR's commissioning of work on the development of a migration model can be seen from the document accompanying the invitation to tender (DETR 1999b). In particular, this stated:

"There are concerns that current trends in migration may run counter to sustainable development objectives. Population and jobs have tended to shift from urban areas to suburbs and from suburbs to rural areas. In addition, there has been a continuing drift of population from the north to the south, which can result in imbalances of economic activity. Particularly in the south, there are pressures for development on greenfield sites and concerns about the affordability of housing. At the same time, especially in the north, existing capacity can be underused, exemplified by the problem of unpopular housing. Even prosperous cities and regions, including London, contain areas of deprivation." (DETR 1999b, paragraph 3).

Therefore, "*Ministers are keen to gain a better understanding of regional migration*" (DETR 1999b, paragraph 2). DETR "*wishes to develop a (migration) model ... in order to inform the development of policy*" (DETR 1999b, paragraph 1), within the Government's over-arching policy objectives which are "*to enhance economic growth and social cohesion throughout England and to promote high quality and sustainable development*" (DETR 1999b, paragraph 1).

15.3 User Requirements for the PSS

DETR's expectations of the modelling system were very clearly set out in the contract tender documentation (DETR 1999b). The primary aim was to provide DETR with an *initial* migration model aimed at informing policy. This should enable the investigation of the first-round quantitative impacts of alternative economic and policy scenarios on gross flows of population between regions. For the longer term, the intention was that the model should be capable of further enhancement, notably in order to gain a better understanding of the role of migration in influencing household change and also to be developed to handle finer spatial scales. The work on this model subsequently proceeded to a second phase of development, but as this work had not been completed at the time of writing, the account below is based entirely on the initial phase.

In terms of specific objectives, particular emphasis was given to the quantification of the role of migration determinants. DETR drew attention to the fact that it had previously commissioned some related studies that should be built on. The most relevant of these was the report *The Determinants of Migration Flows in England: A Review of Existing Data and Evidence* (Champion *et al.* 1998). This had identified the principal factors influencing the scale and nature of all migration flows affecting the distribution of England's population. As such, it was a qualitative review and assessment of previous work. Model development should also bear in mind the lessons learned from two other studies, one on the economic determinants of household formation (Bramley *et al.* 1997) and the other on the economic modelling of the demand and need for social housing (Peterson *et al.* 1997).

Secondly, the research specification made it clear that the running of alternative scenarios was central to the work. Once a quantitative model was developed and tested, it should be capable of being used to assess the impact on migration and population distribution of different policy options. Three examples of the types of questions it would be expected to address were included in the tender documentation, as follows:

- *"What would be the implications for north-south drift if a regeneration programme in the North East were to succeed in generating 10,000 additional jobs?*
- *If house prices in London and the South East were to rise by 2% per annum faster than those in the South West over the next ten years, what would be the implications for inter-regional migration?*
- *If regional policies were successful in boosting GDP growth in the North West by 0.5% per annum relative to the rest of the UK, how would this affect migration?"* (DETR 1999b, paragraph 7).

This requirement had important implications for the nature of the migration model. Most notably, the model should be transparent, without 'black box' elements. As such, it should be trying to replicate the real world as far as possible, both in terms of the way in which people make decisions about moving house and in relation to the factors that affect migration. Given that the existing literature had indicated the importance of such variables as house price, land supply, employment and GDP in influencing migration behaviour and that most of these were potentially amenable to policy intervention, these types of variables should be given priority in model development and calibration. Moreover, bearing in mind that the outputs from running such scenarios would be scrutinised by politicians and officers in central and local government, it was essential that all elements in the final model were readily defensible in both intuitive and statistical terms.

The research specification also made it clear that the model should be user-friendly and well documented. The contractor's role would end with the running of some trial scenarios that would demonstrate the viability and reliability of the model. Thereafter, it would be the task of DETR's own staff to run the model as and when required, using whatever combination of policy-related assumptions were of relevance at that juncture. This placed two requirements on the contractor. First, the model should be capable of being run on the type of desktop computers possessed by DETR, namely Pentium II machines running Microsoft Windows NT 4.0 (Service Pack 3) with 64 MB of memory and operating at 266 MHz. In second place, the model should have a front-end which allowed easy setting of alternative assumptions and provided output in readily accessible form including statistical tables, charts and maps and the automatic aggregation of output to higher-level geographies.

In terms of the geography of the migration model, DETR required that, *"The main outputs should be in terms of gross flows of population between regions"* (DETR 1999b, paragraph 12). Although the term 'region' was not actually defined, it could be inferred (for instance, from the examples of policy scenarios mentioned above) that this meant the division of England into a few large regions such as the Standard Statistical Regions (SSRs) and the Government Office Regions (GORs), the latter having being introduced for administrative purposes in 1996 and subsequently adopted for statistical reporting by the Office for National Statistics. While these two sets of regions are essentially similar, they differ in at least one important respect, namely that whereas under the SSR system the South East formed a single region covering most of the area within commuting range of central London, the GOR system treats Greater London as a separate region and splits up the rest of the former South East into two segments – a new smaller 'South East' and an element which is combined with former East Anglia to form the 'East of England' GOR. In addition, DETR's specification also stipulated that, *"The model should ideally be able to provide some advice ... on urban-rural shift"* (DETR 1999b, paragraph 12). Clearly, given this latter requirement combined with the subdivision of the South East macro-economic region, the system had to be capable of handling not only conventional inter-regional migration (normally seen as primarily determined by labour-market factors) but also intra-regional flows (for which housing and environmental considerations also loom large in migration decisions).

Further guidance was also given on the geographical scope of the migration to be modelled. In particular, most attention was to be given to migration flows within England. This reflects DETR's main area of responsibility. Even before the setting up of national assemblies for Wales, Scotland and Northern Ireland in the late 1990s, planning matters in the latter two countries had been the responsibility of separate executives based in Edinburgh and Belfast and answerable to their own Secretaries of State rather than to the Secretary of State for the Environment, Transport and the Regions based in Whitehall. Wales had been in a somewhat anomalous position, with planning responsibilities shared between Cardiff and London, but is now also completely separate. Nevertheless, the contractor was encouraged to give *"some consideration to flows to and from Scotland, Wales and Northern Ireland"* (DETR 1999b, paragraph 11). Finally, while the model was not required to make predictions of migration between the UK and other parts of the world, it should be capable of incorporating alternative scenarios for the overall level and distribution within the UK of international migration.

Lastly, various parts of the tender document stressed the exploratory nature of the contract and emphasised the need for pragmatism. Thus, this 'initial' model should *"be regarded as potentially a first step towards creating a more advanced framework of analysis"* (DETR 1999b, paragraph 6). Data limitations should not delay progress; instead, the contractor "*should propose pragmatic but defensible ways for filling data gaps and advise on the scope for subsequent research in order to refine the model's performance*" (DETR 1999b, paragraph 9). Also, it was recognised that there would not be enough time to conduct a full range of diagnostic tests nor investigate counterintuitive results in any detail, with such work being left to a potential further phase of the project. Moreover, DETR "*did not require the model ... to internalise the feedback that, for instance, could be expected to occur between migration and other variables such as house prices*" (DETR 1999b, paragraph 6).

Basically, DETR was keen to get some early results in order to help inform its decision-making in relation to the policy issues outlined above, but clearly it also wanted to assess progress before committing itself to longer-term work on a more sophisticated version of the model. The corollary was that the model produced by this study needed to be capable of further enhancement. In particular, in a fuller form, it should be able to go beyond the first-round impacts of the scenarios by internalising the feedback between migration and the determinant variables. Secondly, it was important that it could be developed to produce outputs to a finer spatial scale. Thirdly, "*the model must be of a form which is compatible with DETR's ongoing development of the next set of household projections, in order to keep open the option to use an enhanced version of the migration model to assist in developing these projections*" (DETR 1999b, paragraph 5).

15.4 Form and Content of the Modelling System

The main feature of the modelling approach is the use of a two-stage model that recognises that migration results from two factors: the desire to leave an origin and the choice of destination. The system therefore consists of a model to predict outflows from each zone ('Stage 1 model') and a model to allocate the flows of migrants from each origin zone to each of the alternative destination zones ('Stage 2 model'). The results of the two stages are brought together to implement the prediction of migration flows in a user-friendly front end (see Figure 15.1 for the basic modelling plan). This section describes the principal dimensions of the modelling framework and provides a brief summary of the two stages, leaving the front end to the next section.

This prototype model is based on a division of the UK into 100 zones (Figure 15.2). These correspond to the 98 pre-1996 health service areas for England and Wales together with Scotland and Northern Ireland. The former, officially known as Family Health Service Authority areas, comprise the 47 'shire county' areas of England and Wales (as set up in 1974 and existing through to the subsequent local government reorganisation of 1996-98), 35 metropolitan district areas (all individual local government districts apart from Knowsley and St Helens which were combined) and 16 groups of the 32 London boroughs.

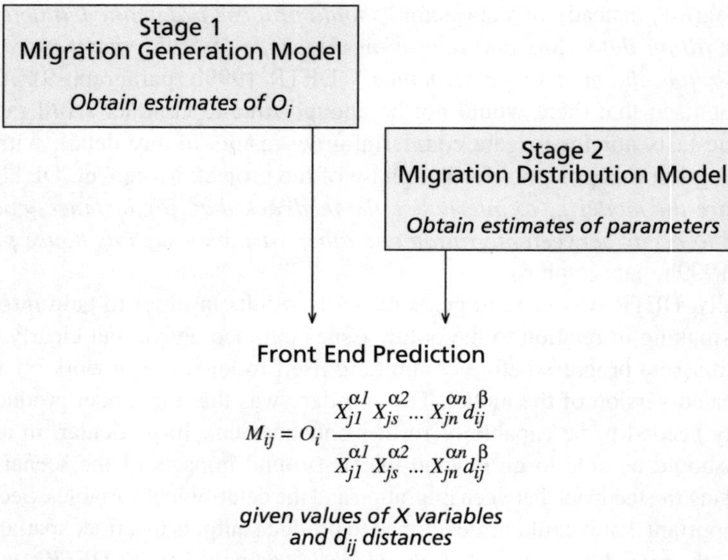

Fig 15.1: Basic modelling plan for the migration model

Fig 15.2: The 100 origin-destination zones of the migration model

1 Northern Ireland
2 Scotland
3 Gateshead
4 Newcastle
5 North Tyneside
6 South Tyneside
7 Sunderland
8 Cleveland
9 Cumbria
10 Durham
11 Northumberland
12 Barnsley
13 Doncaster
14 Rotherham
15 Sheffield
16 Bradford
17 Calderdale
18 Kirklees
19 Leeds
20 Wakefield
21 Humberside
22 North Yorkshire
23 Derbyshire
24 Leicestershire
25 Lincolnshire
26 Northamptonshire
27 Nottinghamshire
28 Cambridgeshire
29 Norfolk
30 Suffolk
31 London: Barking & Havering
32 London: Barnet
33 London: Bexley & Greenwich
34 London: Brent & Harrow
35 London: Bromley
36 London: Camden & Islington
37 London: City, Hackney, Newham & Tower Hamlets
38 London: Croydon
39 London: Ealing, Hammersmith & Hounslow
40 London: Enfield & Haringey
41 London: Hillingdon
42 London: Kensington, Chelsea & Westminster
43 London: Lambeth, Southwark & Lewisham
44 London: Merton, Sutton & Wandsworth
45 London: Redbridge & Waltham Forest
46 London: Richmond & Kingston
47 Bedfordshire
48 Buckinghamshire
49 Essex
50 Hertfordshire
51 Berkshire
52 East Sussex
53 Hampshire
54 Isle of Wight
55 Kent
56 Oxfordshire
57 Surrey
58 West Sussex
59 Avon
60 Cornwall
61 Devon
62 Dorset
63 Gloucestershire
64 Somerset
65 Wiltshire
66 Birmingham
67 Coventry
68 Dudley
69 Sandwell
70 Solihull
71 Walsall
72 Wolverhampton
73 Hereford & Worcester

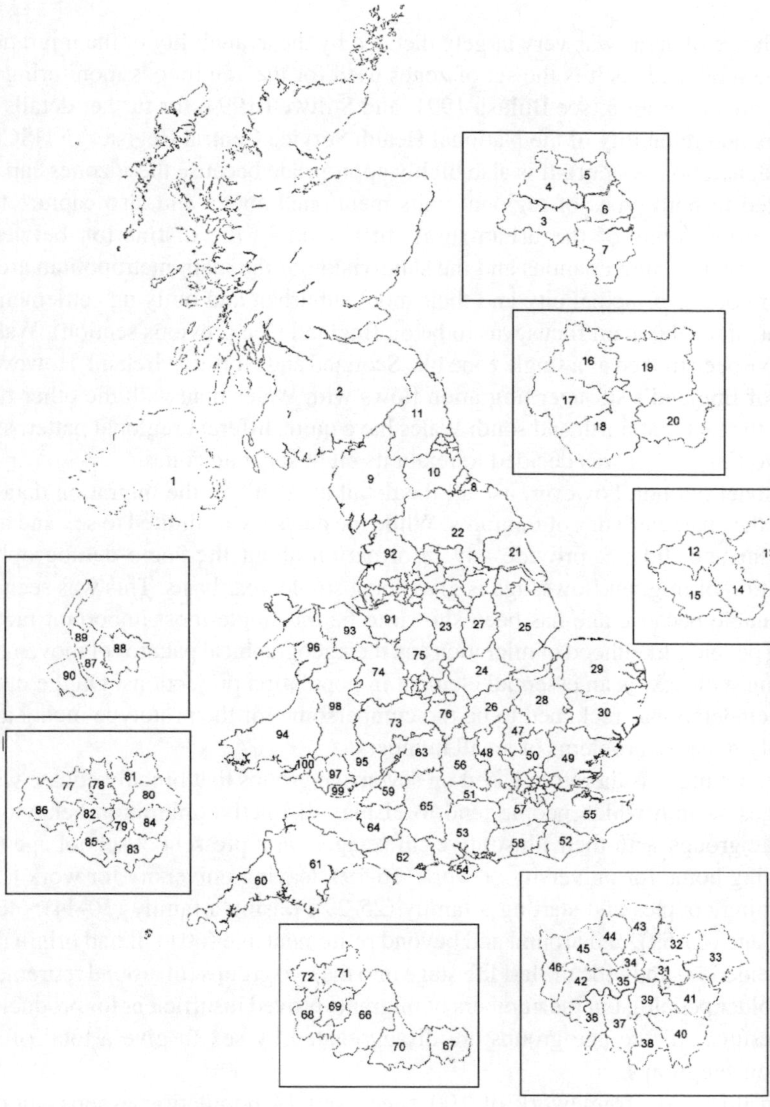

74 Shropshire
75 Staffordshire
76 Warwickshire
77 Bolton
78 Bury
79 Manchester
80 Oldham
81 Rochdale
82 Salford

83 Stockport
84 Tameside
85 Trafford
86 Wigan
87 Liverpool
88 St Helens & Knowsley
89 Sefton
90 Wirral
91 Cheshire

92 Lancashire
93 Clywd
94 Dyfed
95 Gwent
96 Gwynedd
97 Mid Glamorgan
98 Powys
99 South Glamorgan
100 West Glamorgan

Fig 15.2: Continued

This choice of areas was very largely dictated by the availability of the migration data to be modelled, as it is the set of zones used for the continuous monitoring of between-area migration (see Bulusu 1991, and Stillwell 1994, for further details of the nature and reliability of the National Health Service Central Register (NHSCR) patient registration system). It is also highly appropriate because these zones can be aggregated to both types of regional units mentioned above and also capture the primary dimensions of the urban-rural shift, namely the distinction between metropolitan and shire counties and the subdivision of the main metropolitan areas into their core or principal city and their more suburban and outlying settlements. Note that, since the main focus was to be on England (see previous section), Wales could have been treated as a single zone like Scotland and Northern Ireland. However, in view of England's stronger migration flows with Wales than with the other two and also the fact that north and south Wales have quite different regional patterns of linkage to England, it was decided to retain its eight separate zones.

The model did not, however, use all the detail available in the migration data in terms of the characteristics of migrants. While the dataset was limited to sex and age of migrant and did not provide any information about the socio-demographic composition of migrant flows, age is given on a single-year basis. This was seen as very valuable because age has been shown to be the single most important factor affecting people's likelihood of migration and their geographical patterns of movement and, along with sex, is an essential element in population projections. On the other hand, such detail was reckoned to be too cumbersome for the prototype model and also likely to cause problems of small numbers.

The modelling was therefore based on seven age groups that broadly equate with 'life stages' within which people tend to display distinctive migration behaviour. These age groups with their illustrative 'life stages' are: pre-school/school age (0-15), leaving home for university or work (16-19), leaving university for work (20-24), forming couples and starting a family (25-29), raising a family (30-44), older working age (45-59), and around and beyond retirement age (60+). It had originally been intended to subdivide the last life stage into the two groups (of around retirement age and older people), but the numbers of migrants proved insufficient for producing robust results. All the age groups are disaggregated by sex to give a total of 14 population subgroups.

Within this basic framework of 100 zones and 14 population groups, model calibration was carried out in the two stages mentioned above. The Stage 1 model (for predicting population outflows from each zone) was calibrated using a panel dataset covering the 14-year period 1984-85 to 1997-98 and the 98 zones of England and Wales. The Stage 2 model (a spatial interaction model for allocating out-migrants between the alternative destinations) was calibrated for the single year 1996-97 – taken as representative of the conditions prevailing in recent years – and for all 100 zones apart from Northern Ireland. The exclusion of Northern Ireland from both stages and Scotland from Stage 1 was because the data that could be assembled on migration determinants in the time available was not sufficiently comprehensive and/or reliable in these cases.

The data assembly requirements for this work were indeed formidable. As outlined above, it was intended that the system should model all types of migration besides international moves to and from the UK. Previous research, as reviewed by Champion *et al.* (1998), had identified seven categories of factors influencing migration: demographic, cultural and social, labour market, housing, environmental, public policy and impedance. In all, data was needed on four types of variables: the general state of the nation (especially phase in the business cycle), the factors that would affect the attractiveness of each zone as a place to stay/leave or move to, the conditions in the surrounding region (that would influence its 'pull' on the residents of the origin zone), and the compositional effect at each origin (i.e. the innate mobility of the residents as indicated by their occupation, ethnicity, etc.). In the end, despite DETR's call for pragmatism, it proved possible to build a dataset of nearly 150 variables, 139 of which were devised for use in Stage 1 modelling and 69 for Stage 2.

What did the final versions of the two stages look like in this prototype model? For Stage 1, the model form determined by reference to the results of separate regression analyses for the 14 population groups contained 58 of the 139 explanatory variables for which data was available. When calibrated for each group independently to produce 14 sets of 58 parameters each, the general picture that emerged was as follows. In terms of national-level time-series variables, out-migration was found to be significantly higher when mortgage interest rates were lower and the real price of new housing was higher, allowing for inflation. The characteristics relating to each origin zone that were found to be most strongly associated with a higher rate of out-migration are shown in Table 15.1.

The results for the Stage 2 element of the model are more difficult to summarise because a separate model form was determined for each of the 14 population groups and then calibrated separately for all the 99 included origins (i.e. excluding Northern Ireland). Given that each of the 1,386 (14 x 99) models contained between 20 and 30 explanatory variables (selected from the 69 offered in the regressions), the result was a total of around 35,000 parameter estimates based on the migration patterns observed for the year 1996-97. Reference to the means of the parameter estimates suggested that, other things being equal, destinations which attract larger numbers of migrants are those which:

- are in closer proximity to the area from which the migrants originate;
- have a larger population;
- have less competition from nearby destinations;
- have a higher rate of employment growth;
- have a lower rate of new house building in the private sector;
- have higher GDP per capita;
- have lower numbers of offences per capita;
- are in relatively non-urbanised areas;
- have larger numbers of listed buildings; and
- have lower proportion of their employment in manufacturing.

Table 15.1: Origin-zone characteristics relating most strongly associated with a higher rate of out-migration

demography
larger proportion of non-white persons
larger proportion of non-white persons in surrounding areas relative to origin
larger proportion of two-adult pensionable households
smaller proportion of one-adult pensionable households
lower proportion of divorced people
economic
higher level of car ownership in the surrounding areas relative to that at origin
lower level of deprived households in surrounding areas relative to origin
employment
lower proportion of working-age people in work
housing
lower average house price
lower average house price in surrounding areas relative to that at origin
larger proportions of social rented housing
larger proportions of social rented housing in surrounding area relative to origin
social
higher values of the occupational migration index (1)
environmental
more rainy days
lower rate of visitors per head
other
greater accessibility to airports
lower levels of the 'bright lights indicator' (2)
higher expenditure on local government services relative to Standard Spending Assessment

Notes: (1) an index of the expected level of out-migration given the occupational mix of the origin's population; (2) an index based on people's access to theatres and concert halls.

Most of these results were as expected, but one was not, namely that the migration attractiveness of a destination zone was higher where the rate of new private-sector house building was lower. Also, it should be noted that the included variables differed substantially between age groups, again with most being as expected but with some cases of counterintuitive results.

In sum, the final version of this prototype model embraces several dimensions of space, time and population cohort. The basic framework consists of 100 origin-destination areas, 14 years for modelling out-migration flows from origin areas, seven years for modelling the distribution of migrants between destination areas (though only one was used in the final calibration), and disaggregation of migrant flows by sex and seven 'life-stage' age groups. Further details of the model determination and calibration procedures can be found in Rees *et al.* (forthcoming), along with a fuller interpretation and discussion of what the results say about the factors influencing the observed migration patterns.

15.5 Running the Migration Model

Given that one of the most important requirements of the model was that it should be user-friendly, a great deal of effort went into the construction of a 'front end' that would contain all the programming needed to run the model. All the user should have to do is, firstly, to set the levels of the determinant variables to produce results for any situation different from that of the default year 1996-97 and, secondly, to specify the ways in which the output should be arranged and presented. In this prototype version, all that was demanded of the model was that it should be capable of being run for a single year. The incorporation of feedback effects on the levels of the determinant variables was to be left to a subsequent phase of work on the migration model.

The front end consists of three inter-linked units: a set of model inputs, a model processor that runs a prediction and a set of model outputs. The *model inputs* consist of the following elements:

- Stage 1 determinant variables for 1996-97 for the 98 zones in England and Wales. Out-migration rates from Scotland and Northern Ireland were not modelled but are replaced by their observed 1996-97 values;
- Stage 1 parameter estimates (partial regression coefficients in a multiplicative model equivalent to elasticities) for 14 age-sex groups;
- Stage 2 determinant variables for 1996-97 for 99 zones (98 in England and Wales plus Scotland). The flows to and from Northern Ireland were not modelled but are replaced for reporting purposes by their observed 1996-97 values; and
- Stage 2 parameter estimates (partial regression coefficients in a multiplicative model equivalent to elasticities) for 14 age-sex groups.

Baseline predictions of migration are produced when the migration model is run with these four inputs, together with the average populations of the reference year 1996-97. For each of the 14 age-sex groups, the model predicts out-migration totals from the 98 origin zones and then allocates the flows from each origin across all the destination zones. Scenario predictions of migration are produced when the migration model is run with inputs (1) and (3) modified to reflect a proposed policy or new trend, together with the fixed Stage 1 and Stage 2 parameter estimates (2) and (4).

Additional information contained in the system, which may be viewed, is as follows:

- observed migration outflows for the 100 origins for the 14 age-sex groups, for periods 1983-84 to 1997-98 are stored. These data were used in the calibration of the Stage 1 model in migration rate form;
- observed origin-destination migration flows 1990-91 to 1996-97 for the 14 age-sex groups are stored. These data were used in calibrating the Stage 2 model parameters; and
- Stage 1 and Stage 2 parameter files are connected to the viewing system so that the full model results can be inspected on-line.

The *model processor* consists of a Stage 1 model for each age-sex group that applies to all origins and a Stage 2 model for each age-sex group and each origin modelled. The Stage 1 and Stage 2 model parameters, together with the values of the determinant variables and zonal populations, are used by a prediction program, written in Java.

The *model outputs* consist of the following elements:

- predicted migration outflows from 98 origins for each age-sex group;
- predicted migration outflows from 99 origins to 99 destinations for each age-sex group; and
- routines for summing origin-destination flows by destination to give predicted migration inflows for each age-sex group, for computing: a) net migration, b) differences between predicted and observed migration for outflows, inflows, net flows and origin-destination flows, and c) differences between baseline predictions and scenario predictions.

The system provides two means for viewing these outputs: a tabulating function that presents user selected information in tables, and a mapping function which enables the user to display selected variables shown in the tables in map form. For each of these, the user can choose to look at the output for each of the 14 age-sex groups separately or can select the level of aggregation required, for instance all those aged 16-59 or all males.

Figure 15.3 shows an example of an output table from the baseline prediction. This table shows summaries of the model predictions for out-migration, in-migration and net migration for males aged 16-19 (one of the 14 age-sex groups). The table can be scrolled vertically to show other zones than the sample selected. The zone names are listed in the first column for easy reference and the numbers are presented as rounded whole numbers for ease of interpretation. Note that rounding is only used at the display stage.

The table in Figure 15.4 presents the origin-destination flows predicted by the model. Note that the diagonals of the flow matrix are zero: the model does not predict intra-zonal migration because the NHS Central Register provides no observed migration data on this.

The system also enables the user to map the variables presented in the tables. It uses a mapping program called *GeoTools*, developed by Ian Turton and James Macgill of the Centre for Computational Geography at the University of Leeds. Maps are normally in colour, but can be produced in black and white for printing. There are two alternative classification methods: classes with equal intervals between the maximum and minimum value, and classes with equal numbers of observations in each class (quantile maps). The number of classes for mapping can be chosen by the user.

Figure 15.5 provides an example of the mapping of the outflows generated by Stage 1 of the model. It shows the total outflows from each of the zones as predicted by the baseline model for males aged 16-19 for a single year. These outflows are a product of the population size of each zone and the predicted propensity for out-migration.

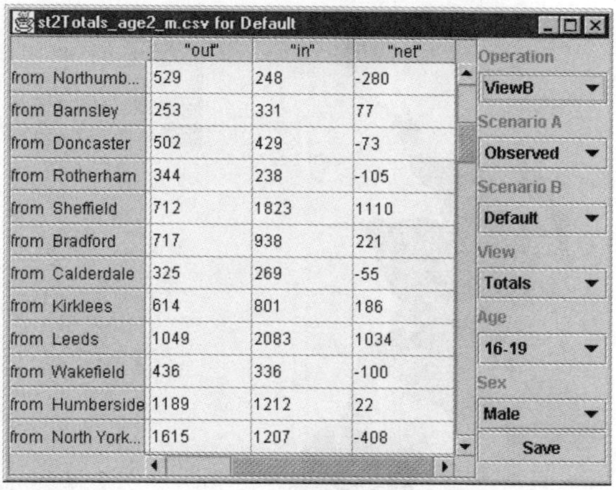

Fig 15.3: A sample screen showing baseline model predictions for 1996-97 for males aged 16-19: total inflows, outflows and net flows

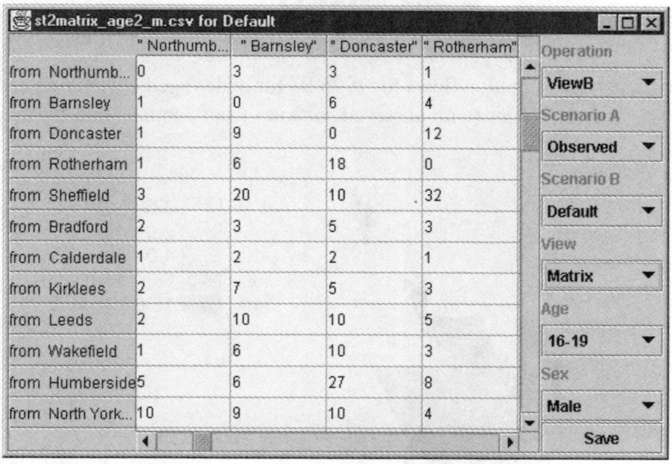

Fig 15.4: A sample screen showing baseline model predictions for 1996-97 for males aged 16-19: origin to destination flows

Figure 15.6 is an example of a map of the origin to destination flows, based on the allocation of the outflow across all the other zones predicted by Stage 2 of the model. This is again for males aged 16-19 and is the one centred on Leeds (shown in white in central northern England). In other words, it shows where out-migrating young men of Leeds go to: it is an *out-migration field* map. The system contains another 99 similar out-migration field maps, which can be accessed 'on the fly' by locating the mouse pointer in the relevant zone on the map. In-migration fields (e.g. all the young men aged 16-19 who are predicted to come to Leeds in the year) can also be viewed.

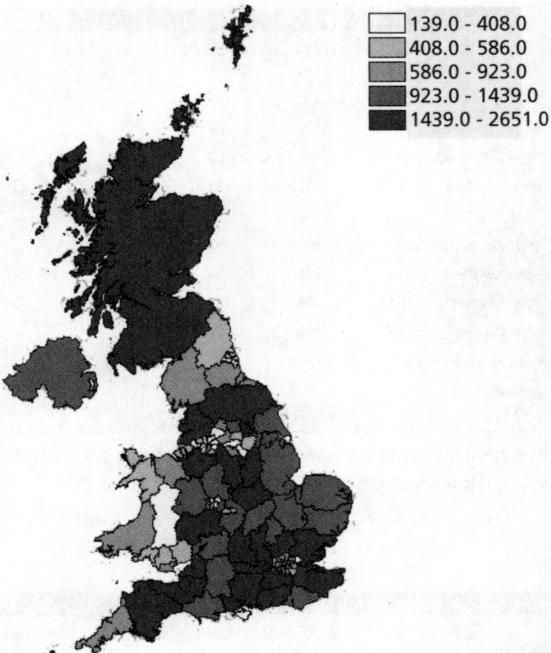

Fig 15.5: Predicted out-migration flows for 1996-97 for males aged 16-19 from each of the 98 zones in England and Wales (with observed outflows from Scotland and Northern Ireland)

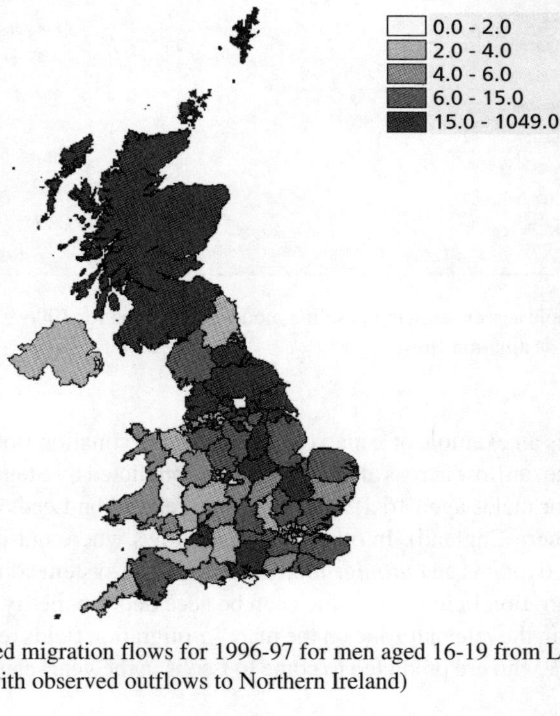

Fig 15.6: Predicted migration flows for 1996-97 for men aged 16-19 from Leeds to the other 98 zones (with observed outflows to Northern Ireland)

At the same time, it is recognised that there are a number of features of the prototype version of the front end that could be improved. One is that the setting of the levels of the determinant variables in the running of alternative scenarios could be made less onerous than in the present system, which involves manual alteration or the use of a subroutine to alter values across a set of zones by a fixed percentage amount. In particular, there needs to be an automated method for recalculating the values of the 'regional' variables that in the Stage 1 model measure the overall state of a determinant variable in the zones around any particular origin zone. In terms of output features, there is the need to incorporate a menu-driven system for automatically aggregating zones to a selection of higher-level geographies such as metropolitan and non-metropolitan, north and south, and other regional groupings such as the GORs and SSRs. In addition, a fuller range of options for graphs and charts could be provided.

15.6 Scenario Running

As stressed at the outset, the primary purpose of the migration model is to test out the impact on migration of alternative economic and policy scenarios. A scenario is implemented by changing the levels of the determinant variables that are called upon in the running of Stages 1 and 2 of the model. The predicted migration flows are then compared with the outputs from the baseline results of the model that uses the levels for the reference year 1996-97. A scenario could involve changing the level of just one of the determinant variables, but this would normally be done only if the purpose was to get a firmer feel of the properties of the model. In a 'real life' scenario, it would usually be appropriate to alter the values of a larger set of related variables. Here there is only enough space to describe the running of one such scenario.

The following example is based on the first of the types of the questions that DETR required the model to be able to handle, namely the effect of generating extra jobs in northern England. The scenario is based on the assumption that a 5 per cent increase in jobs takes place across the main urban areas in this part of the country, namely the five former metropolitan counties (Greater Manchester, Merseyside, South Yorkshire, Tyne and Wear and West Yorkshire) plus former Cleveland. This group of areas constitutes the largest block of 'problem' areas in the British urban system, in the sense that they are generally losing population to surrounding less urban areas and to more southerly regions. The scenario is set up so as to include consequential changes in the levels of variables relating to economic activity, incomes and unemployment at both sub-regional and national level.

The full list of specific assumptions is as follows:

- the employment/working age ratio rises by 5 per cent points in the affected areas;
- the 3-year rate of employment growth rises by 5 per cent points in the affected areas;
- household income rises by 3 per cent in the affected areas;

- GDP per capita rises by 3 per cent in the affected areas;
- economic activity rises by 2.5 per cent in affected areas;
- national real GDP change rises by 1 per cent point in all areas;
- national unemployment rate falls by 1 per cent point in all areas; and
- national real personal disposable income index rises by 1per cent in all areas.

In terms of the way the model is designed, the three national variables and the economic activity variable affect only Stage 1 and GDP per capita and household income affect only Stage 2. The first two variables in the list affect both stages. The main driver of this scenario – the 5 per cent extra jobs in the affected areas – is equivalent to around 169,500 additional jobs compared with the baseline level in the model.

Table 15.2 presents a summary of the results. These are expressed as differences in the number of migrant persons per year, as compared with the baseline model prediction. The numbers shown are the all-persons total (i.e. for all age groups and both sexes combined). The top three rows show the numbers for (a) the five core cities of the metropolitan counties (Manchester, Liverpool, Sheffield, Newcastle upon Tyne and Leeds), (b) the target area as a whole (five metropolitan counties plus Cleveland), and (c) the rest of the three northern regions of North East, North West and Yorkshire and the Humber. Underneath are shown the figures for all of the Government Office Regions (including the three northern ones), Wales and Scotland, plus a check total for Great Britain. Northern Ireland is excluded from this set of results, because its observed flows were used in both baseline and scenario output.

Table 15.2: Results of a scenario based on 5 per cent more jobs in the main urban areas of northern England

areas	change (persons p.a.)			% of baseline			% of total population		
	out	in	net	out	in	net	out	in	net
Northern England									
5 Core Cities	2868	4990	2121	2.8	5.4	-18.4	0.118	0.205	0.087
Target area	9730	12919	3182	3.2	4.5	-28.2	0.107	0.142	0.035
Rest of 3 N Regions	-192	-751	-555	-0.1	-0.5	-18.2	-0.004	-0.014	-0.010
Government Office Regions of England									
North East	1482	1908	426	2.2	2.8	50.7	0.057	0.073	0.016
Yorkshire & Humber	3134	4587	1451	2.0	3.0	-86.2	0.062	0.091	0.029
North West	4922	5673	750	2.2	2.6	-10.1	0.071	0.082	0.011
East Midlands	-1011	-1446	-435	-0.8	-1.2	-22.9	-0.024	-0.035	-0.010
West Midlands	-1682	-1993	-305	-0.9	-1.1	4.1	-0.032	-0.037	-0.006
South West	-1319	-2655	-1333	-0.9	-1.4	-3.6	-0.027	-0.055	-0.027
East of England	-1814	-2262	-446	-1.0	-1.2	-121.9	-0.034	-0.043	-0.008
South East	-2916	-4079	-1159	-1.0	-1.2	-4.1	-0.037	-0.051	-0.015
Gtr London	-5285	-3210	2068	-1.1	-0.8	-3.2	-0.075	-0.045	0.029
Rest of Great Britain									
Wales	-638	-1061	-422	-0.8	-1.1	-3.3	-0.022	-0.036	-0.014
Scotland	0	-588	-588	0.0	-1.1	-26.8	0.000	-0.011	-0.011
Check total	-5127	-5126	7	-0.3	-0.3	0.0	-0.009	-0.009	0.000

It can be seen that the numerical size of migration changes is quite modest compared with the scale of the job change itself: an extra 13,000 gross in-migrants, and around 3,200 net migrants per year. It can, of course, be argued that these migrations could continue over several years, so that the ultimate impact would be multiplied several times, but this cannot be quantified at present because of the prototype model not being fully geared to estimating a trajectory over several years. Some factors would persist, whilst others (change indicators) would drop out after between one and three years.

One of the most striking findings, which is to be expected from migration theory, is that the improved economic/job performance of the northern urban areas would result in a substantial rise of out-migration, so that the net effect of attracting more in-migrants is substantially muted. Most of these extra out-migrants would have destinations in the same regions, judging by the in- and net migration falls shown in the remaining regions. The target urban areas would only attract a small number of extra net migrants from the remainder of the northern regions (550 per annum).

An unexpected finding is that out-migration falls in all the remaining regions. This must be driven by the changes applied to the national variables, and suggests that these have counterintuitive signs in the Stage 1 models. Another curiosity is that London actually gains in net migration terms.

Migration changes may also be expressed as percentages of baseline migrant numbers (see Table 15.2, data columns 4-6). The target areas see a 4.5 per cent rise in in-migration accompanied by a 3.2 per cent rise in out-migration. However, the change in net migration is more significant, representing a 28 per cent fall in the level of net out-migration. From this viewpoint, the impact of the scenario is quite substantial, and encouraging for policy makers. The net impact on the core cities is rather less in absolute and percentage terms, because these five zones account for all of the net out-migration of around 11,500 per annum in the baseline model. Inspection of the individual zone results shows that the core cities generally see larger gains, whilst some zones within the target area actually see slight negative movements in net migration.

Expressed as a percentage of total base population (final three columns of Table 15.2), the migration impacts are pretty small. In net terms, the effect of the scenario is a positive impact of 0.035 per cent per annum. If the impact were to persist over five years, this could mean an increase in population of 0.175 per cent. It is proportionately greater for the core cities, with an increase of 0.435 per cent if sustained over five years.

These results could be interpreted as suggesting that, while employment growth would slow the rate of population loss in the northern urban areas, the scale of extra job growth required to eliminate net out-migration altogether would need to be substantially larger than in this scenario (about three-and-a-half times bigger in fact). This would equate to close to 0.6 million extra jobs, whilst assuming that there was no change in the number of jobs elsewhere in the country or in any other circumstances besides those altered in the scenario. Some caution is, however, needed when extrapolating a change of this magnitude.

15.7 Concluding Comments

This chapter has described the prototype of a migration model developed for a central government department primarily for use in quantifying the impact on within-UK migration of alternative economic and policy scenarios. The above account of running one such scenario clearly demonstrates that the migration prediction and user interface works satisfactorily both in technical terms and in giving results that, for the most part, conform to expectations. This outcome is therefore seen as providing a sound basis for moving forward towards a more robust and sophisticated version.

The first priority is to undertake a fuller scientific verification than was allowed for in the 'pragmatic' approach required of the development work on this initial model. This needs to include exploration of the degree of inter-correlation between the explanatory variables used in the model, tests on the economic time-series data and checking for non-stationarity, more detailed analysis of model goodness-of-fit, fuller examination of the accuracy and reliability of parameter estimates, and sensitivity analysis depicting the impact of changes in reach variable on the outcomes. In addition, as mentioned above, further improvements are planned in the user interface, notably in the handling of the 'regional' variables and in the presentation of results.

Beyond this, there is the task of model enhancement in the ways suggested at the outset by DETR. As noted above, this includes applying the same modelling approach to smaller geographical units such as the 354 post-1998 local government districts of England. Secondly is the task of capturing the impacts of migration on the characteristics of the individual localities in such a way that they can in turn be fed back into the migration model. Thirdly, given its increasing importance in both national and sub-regional population growth, there is the question of whether international migration can be included as part of the migration being modelled rather than being used merely as a determinant of within-UK migration. Finally is the possibility that the migration model could contribute directly to population and household projections.

In summary, this embryonic planning support system represents the most ambitious attempt yet undertaken to model the UK's internal migration. As such, the work has already generated a great deal of interest, as witnessed by the number and range of individuals and agencies attending presentations of the prototype model. The latter is also a reflection of the salience of migration as a policy issue in Britain today, as mentioned at the outset. Potential users of the outputs of this system, or subsequent developments of it, include other branches of government such as other central governments besides the client, the devolved national governments for Wales, Scotland and Northern Ireland, the regional agencies in England, and local government. A substantial number of other public and private sector organizations also attach great importance to trends in the distribution of population and households, not least the house-building industry and infrastructure providers. Given central government's abandonment of the traditional 'predict and provide' philosophy and its emphasis on evidence-based policy, it is crucial that the system is intuitively plausible and can withstand rigorous academic scrutiny. At the same time, the development work on this planning support system is throwing up new questions for migration researchers. The end result should be a much better understanding of migration, as well as a wider appreciation of its complexity and a clearer feel for the degree to which it is amenable to policy interventions.

Acknowledgements

This chapter is based on work carried out for the Department of the Environment, Transport and the Regions (DETR, restructured and renamed Department of Transport, Local Government and the Regions in June 2001) under Contract RADS 5/9/22 by a research team based at four institutions – University of Newcastle upon Tyne, University of Leeds, Heriot-Watt University/Edinburgh College of Art, and Greater London Authority (formerly London Research Centre). The authors gratefully acknowledge the support of other members of the team in data assembly, modelling and scenario development: Seraphim Alvanides, Hilary Anderson, Adrian Bailey, Martin Charlton, Mike Coombes, Oliver Duke-Williams, Heather Eyre, Tania Ford, John Hollis, Simon Raybould, John Stillwell and Tom Wilson. They would also like to thank members of DETR's steering group for their thoughtful and constructive inputs, and especially Andrew Parfitt for his clear and efficient work as DETR's manager for this contract. In addition, many people were extremely helpful in identifying and providing data. Their names are listed in *Development of a Migration Model: Final Report*, submitted to DETR in March 2001.

References

Bramley, G., Munro, M. and Lancaster, S. (1997) *The Economic Determinants of Household Formation: A Literature Review*, Department of the Environment, Transport and the Regions, London.

Bulusu, L. (1991) A review of migration data sources, *OPCS Occasional Paper 39,* Office of Population Censuses and Surveys, London.

Cabinet Office (1999) *Sharing the Nation's Prosperity: Variation in Economic and Social Conditions across the UK,* A report to the Prime Minister, Cabinet Office, London.

Champion, T., Fotheringham, S., Rees, P., Boyle, P. and Stillwell, J. (1998) *The Determinants of Migration Flows in England: A Review of Existing Data and Evidence,* Department of Geography, University of Newcastle upon Tyne.

DETR (1998) *Rural England: A Nation Committed to a Living Countryside,* Department of the Environment, Transport and the Regions, London.

DETR (1999a) *A Better Quality of Life: A Strategy for Sustainable Development for the United Kingdom*, The Stationery Office, London.

DETR (1999b) *Development of a Migration Model*, Invitation to tender for a research project: specification of requirements. Department of the Environment, Transport and the Regions, London, mimeo.

DETR (2000a) *Our Towns and Cities: The Future – Delivering an Urban Renaissance,* Department of the Environment, Transport and the Regions, London.

DETR (2000b) *Our Countryside: The Future – A Fair Deal for Rural England,* Department of the Environment, Transport and the Regions, London.

DETR/DSS (2000) *Quality and Choice: A Decent Home for All*, Department of the Environment, Transport and the Regions and Department of Social Security, London.

Peterson, W., Pratten, C. and Tatch, J. (1997) *An Economic Model of the Demand and Need for Social Housing,* Department of the Environment, Transport and the Regions, London.

Rees, P.H. *et al.* (forthcoming) *Migration modelling for policy analysis*, in Clarke, G. and Stillwell, J. (eds.) *Applied GIS and Spatial Analysis,* Wiley, London.

Robson, B., Parkinson, M., Boddy, M. and Maclennan, D. (2000) *The State of English Cities*, Department of the Environment, Transport and the Regions, London.

Stillwell, J. (1994) Monitoring intercensal migration in the United Kingdom, *Environment and Planning A,* 26, 1711-1730.

Todorovic, J. and Wellington, S. (2000) *Living in Urban England: Attitudes and Aspirations*, Department of the Environment, Transport and the Regions, London.

Urban Task Force (1999) *Towards an Urban Renaissance*, Final report of the Urban Task Force, E & F Spon, London.

16 Spatial Decision Support Systems for Petrol Forecourts

Mark Birkin[1], Peter Boden[2] and Julia Williams[2]

[1] School of Geography, University of Leeds, Leeds LS2 9JT, United Kingdom
[2] GMAP Consulting, 15 Blenheim Terrace, Leeds LS2 9HN, United Kingdom

Abstract: The chapter considers the experience of GMAP Limited, and other market planning businesses, in providing decision support to major oil companies. Such systems can typically offer many of the basic functions of a geographic information system. However, in order to provide effective solutions to real planning problems, some form of spatial modelling capability is also required. Site ratings, spatial interaction and network optimisation models are amongst the techniques that can help to meet these requirements. The use of such models within a decision support framework is illustrated with reference to the value that can be added by these applications within an applied planning context.

16.1 Introduction and Background

The retail or 'upstream' petroleum industry (as distinct from the production or 'downstream' end of the petroleum supply chain) has a long history of the use of location planning tools and spatial decision support systems (SDSS). MPSI is the most well established independent player, having begun trading in the mid-70s and reaching a turnover of over $30million at its peak in the late 1980s and early 1990s. Newer entrants to this marketplace include NeuroCorp, GMAP-Consulting, and Knowledge Support Systems (KSS). In this chapter, we will draw on our experience with GMAP to provide an overview of the capabilities of SDSS when applied within this industry sector. We will also draw on known experience of the other competing organisations, where appropriate.

At current turnover levels, the combined revenue of these location-planning companies within the petrol industry is probably in the order of $20 million per annum. Nevertheless, these organisations share relationships with all of the major multinational oil companies. Furthermore, it should be borne in mind that each of the oil companies employs dedicated teams in location planning and site assessment across many countries. Major management consultancies, such as Deloitte and Touche, Mercers, and Booz Allen, are also known to be highly active in this field. The real value of SDSS to the global retail petroleum industry is probably measured in several hundreds of millions of dollars per annum.

The importance of the global market for SDSS is illustrated by excitement in the London Stock Exchange over the flotation of KSS (Independent 2000). The company raised £35 million for expansion, capitalising the business at over £250 million, at a time when the annual revenue stream was yet to exceed £2 million. It is also interesting to note that in addition to GMAP, both KSS and MPSI are businesses with strong academic roots. KSS was a spin-off from the Mathematics department at UMIST, while the principal of MPSI (Ron Harper) has been an associate of David Huff (author of one of the seminal papers of twentieth century location theory – Huff 1963) since the formation of MPSI thirty years ago. Huff remains a non-executive director of MPSI to this day.

Several particular reasons for the importance of SDSS in this sector may be highlighted. Firstly, it is a highly globalised sector. This tends to emphasise the need for high quality, consistent and objective information to support centralised planning and resource allocation between geographical markets. Secondly, the relationship between customer and supplier is fundamentally geographical. Unlike many other retail and service businesses, it is impossible to conceive of a situation in which electronic transmission replaces the physical interaction of drivers and service stations. Thirdly, relative location is highly significant. Having the right kinds of facilities, being in the right part of town, being on the right road, and even being on the right side of the road, can all have a crucial impact on the performance and profitability of sites. Fourthly, the oil companies have huge investments within their retail infrastructure. For example, and in contrast to other retail activities, a petrol station may not be readily sub-let for other purposes should a site no longer be deemed to be sustainable. This dictates that considerable care be exercised in the selection and management of sites. Finally, network planning is becoming increasingly important as margins from the core petrol business are squeezed, and the oil companies are seeking to maximise the value of their sites as real estate. This has led to much interest over the last five years in the use of forecourt space for ancillary activities, such as the forecourt shop, car wash, quick service restaurants and ATMs.

In the second section of this chapter, we will describe the features of the SDSS developed by GMAP for its retail petroleum industry clients. Section 16.3 considers the applications and benefits that can accrue from the deployment of SDSS in this sector. These applications stretch from the global and strategic right through to support for micro-locational decisions. The fourth section of the chapter will consider some less well developed applications which are nevertheless of considerable theoretical interest and have considerable potential for future development. Sections 16.3 and 16.4 are illustrated with examples from the work of GMAP and, occasionally, others. In order to protect client confidentiality, we have avoided references to specific projects in these sections, but feel confident that we have provided the reader with a clear indication of the way in which the technology is being used and the value which can be extracted from it. Conclusions from our experience in the retail petroleum sector will be offered in Section 16.5.

16.2 *ForecourtPlanner*: A Global Site Assessment Capability

16.2.1 Overview

GMAP first began work in the petrol sector in the early 1990s with a project for BP Australia to forecast fuel volumes at petrol stations across the city of Brisbane. Since this time, GMAP has undertaken further projects for BP, Esso, Conoco, Total-Fina-Elf, Sainsburys and AGIP in more than 20 countries on all continents, with work stretching from volume estimation to shop potential modelling, price management, investment prioritisation, cash flow projection and format optimisation. These applications are discussed in further detail in Sections 16.3 and 16.4 below.

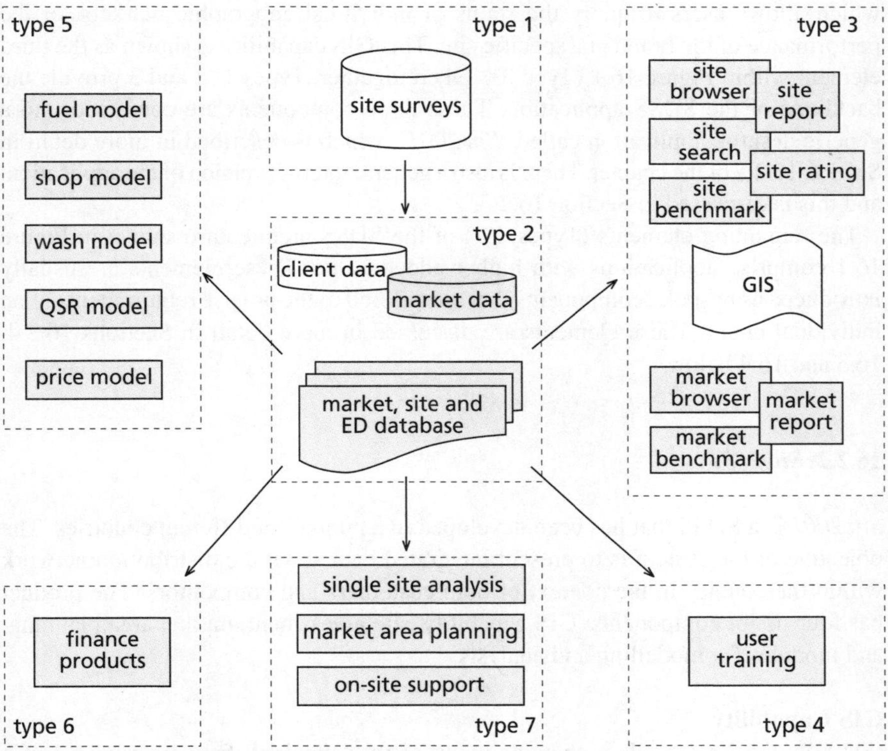

Fig 16.1: SDSS architecture of *ForecourtPlanner*

The components of GMAP's SDSS capability are illustrated in Figure 16.1, where we have identified seven modular components to an SDSS application for the retail petroleum industry. The data that sits at the core of the application is usually of two types. Data relating to specific outlets is often collected by survey teams operating within individual oil companies. These data usually comprise detailed facility audits, from a count of the number of fuel pumps on a site and the volume of traffic that passes, to an evaluation of the layout and product-ranging within the forecourt shop. Site surveys are shown in the 'Type 1' box in Figure 16.1. The collection of these data is hugely expensive and is increasingly sub-contracted by the oil companies to specialist agencies, such as Catalist Ltd in the UK and New Image Marketing in North America. The second type of data relates to customers – their behaviour, the demographics of the market, brand preferences and spending characteristics. These data are largely drawn from third party sources such as market research or census data, although specific client data from loyalty cards or customer surveys may also be included within this category.

Site surveys may be combined with customer and market data (Type 2) into a GIS which allows users to query the status of individual geographic markets, or the performance of the brand at a specific site. This GIS capability is shown as the third element within Figure 16.1 (Type 3). Taken together, Types 1, 2 and 3 provide the backbone of the SDSS application. These three components are combined into a generic desktop application called '*Site2001*', which is described in more detail in Section 16.2.2 of the chapter. There is also a generic Internet version of the application, and this is discussed in Section 16.2.3.

The remaining elements (Types 4-7) of the SDSS architecture shown in Figure 16.1 comprise applications with higher added value. These elements are usually introduced as bespoke components that are tailored to the precise requirements of an individual client. These elements are discussed in more detail in Sections 16.2.4, 16.3 and 16.4 below.

16.2.2 *Site2001*

Site2001 is a SDSS that has been developed in a number of different countries. The objective of the system is to present integrated data about the distribution network within the country in the context of both customers and competitors. The product has four major components: GIS capability, site assessment, market area planning, and modules for modelling and analysis.

GIS Capability

GMAP's *Site2001* product contains many of the 'standard' features required of a desktop GIS. For example, Landis (1993) describes nine capabilities:

- Presentation and thematic mapping. Users are able to create regional maps, for example, to show the distribution of market share; or site maps to show the location of sites relative to one another and to the geography of the area.

- Data query. It is possible to 'show me' the distribution of all Texaco sites within a selected area.
- Spatial query. It is possible to 'tell me' where I can find all of the sites that are within five miles of a hypermarket.
- Database integration and updating. The system combines spatial data (towns, roads, rivers) with site attribute data (sales, facility characteristics) and customer data (customer type, spending power). All of the data are updated on a regular, usually quarterly, cycle.
- Buffering. It is possible to compute site profiles within circular catchments around a site or by selecting specific zones for aggregation and profiling.
- Overlay. Site performance data can be overlaid against market data or customer data, for example, to show the turnover of each site overlaid against the customer expenditure per site in the local area.
- Distance, adjacency and proximity analysis. It is possible to calculate the distance (straight-line or drive time) between any pair of outlets. It is also possible to find the nearest petrol station to any other petrol station or some other feature, such as a residence, workplace, shop, hospital or school.
- Point-in-polygon methods. Often necessary as a base for the other applications; for example, if it is required to produce an overlay of provision, then it would be necessary to count the number of petrol stations within each small geographical area.
- Routing and minimum path analysis. This is the one capability that is not really required in this application, although this would be relevant to a customer who wished to study the routing of tankers to replenish petrol stations, for example. Applications of this type have been implemented by GMAP for the retail and automotive sectors.

Site Assessment

At the heart of the application is a series of ratings for each site, which combines the site attributes into a rating score which is highly correlated against the business potential of the site. An example is shown in Figure 16.2. The ratings comprise two components – a location score, and a facility score. The location score is made up from attributes characteristic of the site regardless of the way in which it is developed, and includes factors such as local demographics, traffic flows and local competition. The facility score is made up of attributes that are specific to the layout and composition, such the plot size, number of pumps and site opening hours.

The ratings are produced using simple statistical analysis, primarily linear multiple regression, at three levels:

- Using all the available data, identify which attributes are the most significant drivers of the location and facility scores. While the application of Figure 16.2 shows 16 attributes in use within the scoring model, these are selected from a database holding well over 100 candidate attributes.

- For each attribute, fix the scores associated with the attribute categories. Some of the attributes may be 'logical' (i.e. present or absent), for example car wash. This is a straightforward calculation, since the site will attain the weighting if the condition is true, and zero otherwise. Other attributes may be nominal (such as the number of pumps) or categorical (e.g. access, which might be poor, medium, good or very good). In this case, each attribute category or group of values needs to be assigned an intermediate value between zero and the maximum weighting for the attribute.
- Calibrate the relative importance of each attribute, to which a weighting is attached. In Figure 16.2, the ratings are weighted equally between location attributes and facility attributes, with each contributing 20 points to the total score for each site. In effect, therefore, each site has three key rating scores: a 'location score', between 0 and 20; a 'facility score', between 0 and 20; and a compound score, the 'site rating', which is the sum of the location and facility scores, and therefore runs from 0 to 40. From Figure 16.2 we can see that the most important location attribute is the traffic passing each site. Estimated traffic volumes – the number of vehicles per hour passing the site – are known for each of the 14,000 plus outlets in the UK network. These volumes are assigned a score such that the lowest volume sites get no points, and the highest volume sites get 7 points. Intermediate volume sites are assigned an integer value between 0 and 7, although half point scores (i.e. 0.5, 1.5 etc) are also allowed. Fractional values, other than multiples of 0.5, are not permitted.

site details

site:	Skyrack Service Station
address:	256 Otley Road LEEDS West Yorkshire
postcode:	LS16 5DJ
brand:	Oilco
facility type:	petrol station
ownership:	company
dev't space?:	no
location type:	high density

catchment size = 2km

performance assessment

location	16 / 20
facility	14 / 20
fuel rating	30 / 40
performance	0.79
est. fuel volume	3,800 kl

fuel rating composition

location attributes	value	score
traffic	24	6 / 7
cars in catchment	12,441	2.5 / 3
households in catchment	18,685	2 / 2
visibility	good	1.5 / 2
accessibility & no. kerb	medium	1.5 / 3
competition	0.75	3 / 3
location score		16 / 20

facility attributes	value	score
plot area (m^2)	2800	2 / 2
fuelling positions	8	3 / 4
quality	good	1.5 / 2
ultra-modern facility	no	0 / 1
development year	1975	1 / 3
no. shop services (0-8)	6	2.75 / 3
24 hours?	yes	2 / 2
car wash	auto	2 / 2
ATM?	no	0 / 1
facility score		14 / 20

Fig 16.2: Site ratings report

Figure 16.2 shows the example of a site in North Leeds with an excellent location and high quality, although somewhat dated, facilities. The site is believed to dispense 3.8 million litres of fuel per annum. This is somewhat less than would be expected from a site with a rating of 30 points, as shown by the 'performance' measure which is constructed as the ratio of the expected fuel volume at this site, to the estimated volume across all sites with a similar score. This performance statistic must be interpreted in the light of factors not included within the general rating model. In this case, it may be a reflection of the fact that the catchment of the site is dominated by students with relatively low rates of fuel consumption.

Market Area Planning

In addition to site level data and ratings, *Site2001* contains aggregate intelligence and reports for market areas for use in strategic planning applications. A sample report for a market area, combining demographics, market share and market performance data is shown as Figure 16.3. In this example, we see an area with a relatively affluent population (car ownership is 5 per cent higher than the national average) with slow growth (0.6 per cent in the ten years to 2006). The fuel market is highly competitive, with high provision of petrol stations and discount pricing. Oilco sites are generally effective, as shown by above average ratings and efficiency greater than unity – except for two dealer-owned sites, which perform extremely poorly.

OILCO

Market Area: **Poshtown**
Region: **MIDLANDS**

	Market	Region	National
Population	560,519		
Change 1991-1996 (%)	0.6		
Change 1996-2006 (%)	0.6		
Population/Hypermarket	93,420	154	151
Hypermarket Risk	5		
Cars	229,454		
Cars/Household	1	112	105
Cars/Outlet	2,250	132	134

	Market
Competitor Ratio	123
Planning	15
Conversions	11
Investment	8
Market Pricing	-0.6
Fuel Cards	0

	Volume (kl)	Share of National
Demand	330,236	0.91%
Supply	430,769	1.13%
Difference	-100,533	
Difference %	-30.4	

	number	volume (kl) total	volume (kl) average	shares (%) site	shares (%) Volume	effect-iveness	average rating	efficiency ratio
all sites	102	430,769	4,223	100.0	100.0	1.00	22.7	0.9
company-owned	64	260,440	4,069	62.7	60.5	0.96	23.9	0.9
dealer-owned	32	105,829	3,307	31.4	24.6	0.78	18.9	0.8
hypermarkets	6	64,500	10,750	5.9	15.0	2.55	29.8	1.9
Oilco	15	68,608	4,574	14.7	15.9	1.15	23.9	1.1
company-owned	13	67,950	5,227	12.7	15.8	1.24	24.8	1.1
dealer-owned	2	658	329	2.0	0.2	0.08	13.0	0.1

Fig 16.3: Market area summary report

Analysis Modules

Two straightforward analytical modules are supported within the application. The *Analogue* procedure takes a site or group of sites, and identifies a target group of sites that are the most similar with respect to a range of attributes which may be selected by the user. The target sites could be existing petrol stations, or greenfield sites awaiting development. This technology might be used to help understand the characteristics of a poorly performing site in relation to similar petrol stations at other locations; or to roll out a new format from a successful pilot to other appropriate locations.

The *Optimise* procedure finds a group of sites that score most heavily for a range of characteristics. In effect, this allows the user to define a customised rating system and find an ideal network of locations using these criteria. This technology might be used as a filter to find the best locations for new development before producing detailed proposals for new building or site acquisition.

The data for *Site2001* can be provided from a variety of sources. Historically, the major oil companies have maintained an inventory of both their own sites, and those of the competition using information provided by surveyors in the field. However as these databases are increasingly difficult and expensive to maintain, the industry has begun to look to third party suppliers to provide site data. In the UK and, increasingly, throughout Europe, Catalist Ltd provides comprehensive site audits regarding the location of petrol stations and associated facilities (Catalist 2001). In developing markets, there are opportunities for automatic data capture using a combination of palm-top computers and, for instant processing, mobile telecommunications technology.

16.2.3 *ForecourtPlanner*

ForecourtPlanner is a web-enabled version of *Site2001*. The technology is equally capable of supporting applications via:

- internet – general access to users from any organisation anywhere in the world;
- intranet – support for applications within an oil company on that company's internal corporate IT network; and
- extranet – secure access for users within a single oil company through a server maintained by GMAP or in association with a third party.

The most compelling rationale for the development of a web-enabled version of the technology is logistical. For one of its clients, GMAP maintains versions of *Site2001* within at least 20 countries, with each system being used both centrally (i.e. within either a global or regional HQ) and locally (i.e. in each country). The data within each country are typically upgraded on a quarterly basis. The system itself may also be subject to periodic upgrades and enhancements. New releases are customarily provided on CD, but ensuring that the correct version of each file is held in the right place on each of the client machines is a time-consuming task that is inevitably subject to multiple errors and confusion. How much simpler if GMAP were able to control upgrades and new releases through a central server, which can also be accessed through 'thin client' browsers, eliminating the need for expensive local processing and unnecessary duplication of files.

For individual client applications, either an intranet or extranet platform would be used. As for all users of SDSS, maintaining the confidentiality of data within the system is a key concern. *ForecourtPlanner* has been developed as a secure, encrypted application through a partnership between GMAP and energis-squared, formerly Planet Online, a blue chip provider of hosting for web applications.

A possible spin-off from the technology is an ability to provide universal access to versions of the system that have been stripped of confidential customer data. Thus it is possible that the owner of a portfolio of three or four petrol stations in Portugal, for example, might be interested in acquiring the sites of a neighbouring family business on retirement of the owner. *ForecourtPlanner* might provide a means to acquire individual reports about each of the available sites at a modest cost of a few hundred pounds, at most. In this way, it could be possible to reach out from the lucrative but bounded market of major oil industry players to a diffuse international customer base. The purchase of individual reports within *ForecourtPlanner* is supported by an electronic merchanting capability, although this system has not yet been released for general access.

16.2.4 Other Components

Referring back to Figure 16.1, we have now made extensive reference to the data, GIS and ratings components of the SDSS architecture. The remaining components sit outside the main *Site2001/ForecourtPlanner* environment. For completeness, we will comment on each of these in turn.

Type 4: User training. This is usually provided face-to-face with a group of users from a client organisation. Although reasonable prices can be charged for user training, the courses are usually time-intensive to prepare and deliver, and not necessarily hugely profitable. However, they are an excellent opportunity to meet and develop relationships with the users of the system who will provide the key to long-term development of the business relationship.

Type 5: Model applications. Reference has been made to some of the basic capabilities in Section 16.2.2 above. More advanced modelling applications are described in Section 16.4.

Type 6: Investment model. Many of the modelling applications are focused on the relationships between customer behaviour and site performance. However, to understand the benefits of a particular decision (e.g. to build a new petrol station, or re-brand an existing site), then the costs involved also need to be evaluated. GMAP has developed a suite of investment models that allow site turnover assessments to be balanced against development costs, so that profits and the return on investment may also be quantified. Applications of the investment model are discussed in Section 16.3.

Type 7: Single Site Analysis and Market Area Planning. These are discussed in Section 16.3 below. On-site support may be provided on a short- or long-term basis where clients lack either the people or the capability to use the SDSS for problem-solving applications once it has been developed. For example, following oil company mergers, as between BP and Mobil (1997) and Total-Fina-Elf (2000) it is typically necessary to produce network plans to determine potential rebranding of sites, possible adjustment or harmonisation of formats across sites and decommissioning or onward sale of sites.

Plans need to be produced quickly and accurately at a time when resources are already stretched to their limit by the logistics of the merger process (integration of the supply chain, support services, management, and so on). The process could be supported by the placement of consultants on-site with the client to assist or manage the production of appropriate network plans.

16.3 Applications and Benefits

In this section, five classes of application will be considered: capital allocation, sales forecasting, investment modelling, format optimisation and representation planning.

16.3.1 Capital Allocation

At the strategic level, major oil companies are operated and managed as multi-national businesses. This means that they must continually address questions such as whether it is more important to create a network in Poland or to upgrade sagging infrastructure in Austria. GMAP's 'Site Ratings' provide a powerful means for the crystallisation of capital priorities between countries. For example, suppose that $200 million have been set aside for investment in improvements to existing facilities within western Europe. Priority sites are those with considerable location potential and poor existing facilities.

Table 16.1 shows a count of the number of sites in which the location ratings exceed 12 but the location ratings are no better than eight. These counts are shown in the third column of the table (labelled 'Target sites') alongside a count of the total number of petrol stations in each country (labelled 'All sites'). In the absence of high quality network information, it might be recommended to allocate expenditure in strict proportion to the total number of sites in each country. These apportionments are shown in the column labelled 'Base Spend'. However, if the available spend is distributed between countries in proportion to the number of target sites then the allocations would be as shown in Column 5 (labelled 'Adjusted spend'). This produces considerable variations in the volume of capital allocated per site, as illustrated in Figure 16.4.

Similar considerations also operate at the level of market areas within countries. For example, Table 16.2 shows network performance data for selected customer marketing areas in Germany. The number of sites for a selected brand ('Brand sites'), and the total number of petrol stations in each area ('Total sites') are shown in columns 1 and 2. The next column shows the share of sites for the brand in each area ('Brand share'). For each area, the average site rating has been calculated to indicate the quality of sites ('Brand rating'). This has been normalised against the average rating of 21 points in the next column ('Brand quality').

Table 16.1: European capital allocation model

country	all sites	target sites	base spend	adjusted spend
Austria	604	79	5,573	7,190
Belgium	962	87	8,875	7,879
Denmark	481	53	4,437	4,823
Finland	346	12	3,195	1,056
France	3,293	238	30,396	21,519
Germany	3,197	585	29,508	52,961
Ireland	510	9	4,704	814
Italy	4,808	375	44,373	33,962
Luxembourg	46	3	426	238
Netherlands	769	38	7,100	3,478
Norway	481	58	4,437	5,230
Portugal	481	44	4,437	4,008
Spain	1,442	236	13,312	21,396
Sweden	692	53	6,390	4,812
Switzerland	769	82	7,100	7,390
United Kingdom	2,788	257	25,737	23,244
total	21,669	2,208	200,000	200,000

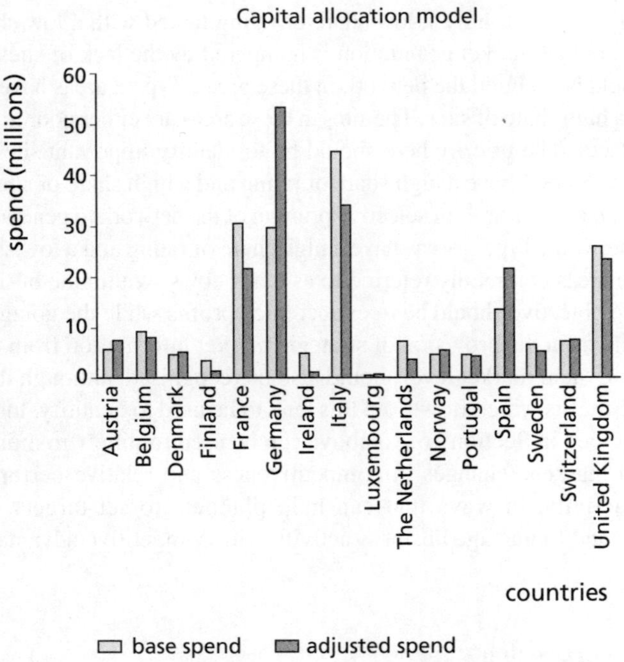

Fig 16.4: European capital spend

Table 16.2: Key indicators for German market areas

	brand sites	total sites	brand share %	brand rating	brand quality	area type
Berlin E	15	275	-55	16.3	-4.7	1
Berlin W	37	309	0	19.3	-1.7	2
Bremen	9	75	0	20.7	-0.3	2
Dortmund	21	131	34	23.3	2.3	3
Dresden	13	124	-13	22.5	1.5	4
Dusseldorf	31	159	62	18.3	-2.7	2
Erfurt	7	93	-37	14.1	-6.9	1
Frankfurt	11	80	15	19.3	-1.7	2
Freiburg	5	104	-60	20.2	-0.8	1
Hamburg	5	73	-43	25	4.0	4
Hannover	25	170	23	22.3	1.3	3
Koln	2	58	-71	22.5	1.5	4
Leipzig	28	142	64	16.7	-4.3	2
Magdeburg	9	74	1	22.3	1.3	3
Mannheim	14	102	14	23.9	2.9	3
Munchen N	16	119	12	23.4	2.4	3
Munchen S	17	159	-11	25.1	4.1	4
Nurnberg	18	123	22	18.3	-2.7	2
Rostock	12	94	6	28	7.0	3
Stuttgart	10	111	-25	22.4	1.4	4
Trier	12	91	10	23.3	2.3	3

Figure 16.5 shows the distribution of share of outlets against site quality. This form of representation is popularly referred to as a 'Boston Chart' in management literature. The chart can be used to illustrate four different area types, each with a different strategic imperative. Type 1 areas have a low share of rating linked with a low share of sites. The chances are that market penetration is hampered by the lack of sites. A strategic objective would be to build the network in these areas. Type 2 areas have a low share of rating but a high share of sites. The sites in these areas are either poor in quality or in the wrong places. The priority here should be to identify important sites for facility upgrades. Type 3 areas have a high share of rating and a high share of sites. In mature markets, there may be a case for selective pruning of the network to generate efficiency gains in the network. Type 4 areas have a high share of rating and a low share of sites. These are the areas commonly referred to as 'cash cows' within the business studies literature. The objective should be to extract juicy profits while the going is good.

The benefits from the provision of strategic market information from the SDSS at both the country and market level should also be recognised, although the monetary value of the benefits which flow from this tend to be hard to quantify. Intelligence of the type described in Section 16.2.3 above can be used routinely to monitor growth or decline in markets, changes in competitiveness and relative performance, and competitor activity, in ways that can help planners to set targets, to monitor performance, and to manage business activity with competitive advantage.

16.3.2 Sales Forecasting

Site assessment techniques within the retail petroleum industry have historically been focussed around the concept of 'dirt strength'. Using this approach, company surveyors would assess the strength of proposed petrol station locations on the basis of the latent characteristics of each site – plot size, accessibility, traffic counts, market demographics and so on. The greater the dirt strength, the higher the volumes that can potentially be extracted.

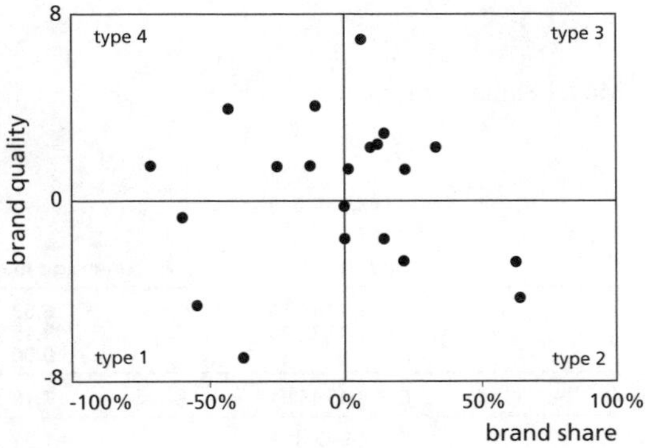

Fig 16.5: Boston Chart of German market area performance

In effect, dirt strength is analogous to the location score within the GMAP Ratings model. The facility score provides further insight by focusing on the effectiveness with which the potential of the location can be realised. When location and facility scores are combined into a Site Rating using parameters that are calibrated against real data, then the ratings will generally be highly predictive of trading at the site. Although the ratings model is less sophisticated than some of the alternatives that are considered in Section 16.4.3 below, it is common to explain about 75 per cent of the variance in trading patterns from the ratings. The ratings may therefore be used to predict the potential sales from a new site, to predict the impact of a new site on existing sites or to compare alternative rejuvenation strategies for existing sites.

16.3.3 Investment Modelling

The principle behind the investment model is straightforward in that one seeks to combine outputs from the sales forecasting process, as discussed in Section 16.3.2, with data and assumptions about the associated costs. Figure 16.6 shows an example in which an investment has been proposed to improve the forecourt at a petrol station. The impact of this investment is shown in two summary tables. The 'NPV Scenario Summary Report' (Table 16.3) shows the costs and revenues associated with the investment.

NPV is an abbreviation of 'Net Present Value' – the value of a stream of revenues or earnings over a period of time, which are discounted to allow for the fact that an income of 5,000 Euros in 5 years time is worth less than 5,000 Euros now. This is partly because future earnings are less certain than current earnings; and partly because current earnings can themselves be reinvested to yield a further return, while future earnings cannot. The 'discount rate' reflects the relative valuation of current versus future earnings. For example, if the discount rate is 10 per cent, then expected profits of 1,000 Euros in one year's time would have a net present value which is reduced by 10 per cent, i.e. 900 Euros.

Table 16.3: Investment Model Summary Report

annual turnover		**£48,736,643**	
		NPV at 15%	**average margin %**
food	£	2,426,746	6.03
non-food	£	1,072,545	9.15
other	£	-	0.00
total earnings	£	3,499,291	7.18
retail overheads	£	-547,499	-1.01
supply and distribution	£	-683,783	-1.44
property costs	£	-288,434	-0.65
third party fees	£	-373, 624	-0.75
credit card costs	£	-282,552	-0.77
total costs	£	-2,175,892	-4.70
net income	£	1,323,399	2.48
net income	£	1,323,399	
investment	£	-313,331	
tax	£	-372,952	
working capital	£	-75,598	
	£	461,518	
net cash flow	£	461,518	
DCFR			15.2%

The net present value of a specific capital project, such as the investment of funds into a petrol station upgrade, is sometimes referred to as the Return on Capital Employed (ROCE). NPV is usually stated as a gross amount, whereas ROCE usually has a percentage value. Returns on capital are often appraised using the discounted cash flow rate (DCFR), which is rather like the capital equivalent of the interest earned on a bank or building society account. If the DCFR is 20 per cent, then for every 1,000 Euros invested, we expect to get 1,200 Euros back at present values. A full discussion of the concepts of return on capital employed, discounted cash flow and net present value can be found in Pike and Neale (1993).

The NPV Scenario Summary Report shows that the baseline annual turnover of the site is £48.74 million. The proposed development is a significant extension and refit of the existing forecourt shop to provide a better range of retail products. The sales forecasting process shows a long-term sales boost, with a net present value just short of £3.5 million, or 7.18 per cent of current turnover. More than two-thirds of the extra sales (£2.4 million) come directly from new food sales within the shop. The remainder (just over £1 million) comes from non-food sales, which may be from the shop, or from some extra fuel sales due to an increase in the overall attractiveness of the enhanced site. These net present values are calculated at a discount rate of 15 per cent, which reflects the internal cost of capital within the business – in other words, approved projects are expected to yield a return on capital employed of at least 15 per cent.

An investment of just over £2 million is proposed, as shown in the 'total costs' line of the report. The investment is split between 'retail overheads' (the additional cost, e.g. extra staff costs, of maintaining an improved facility); 'supply and distribution' (the additional cost of delivering extra products to the facility e.g. supplying more bread to keep the shop well-stocked); property costs (the cost of new building work e.g., cost of new land to be acquired, and the costs associated with extending and refitting the premises); third party fees (e.g., surveyors, architects, lawyers, accountants involved e.g., with the design of the project, planning permission and so on); and credit card costs (costs associated directly with the extra business generated by the project).

The 'Investment Appraisal Model' (Figure 16.6) allows the user to evaluate the project outcome against appropriate investment targets.

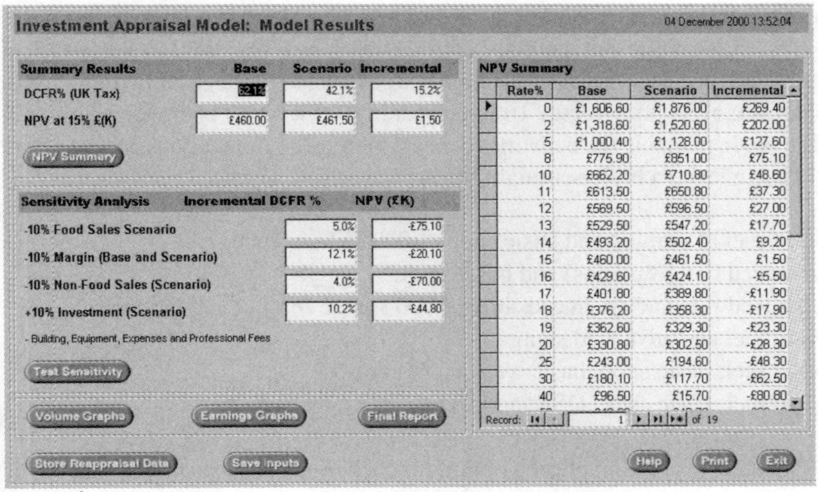

Fig 16.6: Investment Appraisal Model results

The net income generated from the project comprises additional earnings less costs. However, further allowances need to be made against income in respect of investment, tax and working capital. These considerations yield net cash flow of £461,518 with a discounted cash flow rate (DCFR) of 15.2 per cent. These figures are shown at the bottom of Table 16.3, and are then transferred across to Figure 16.6, where net cash flow appears in the 'scenario' box, and the DCFR appears in the 'incremental' box. The 'base' value against NPV shows that a return of £460,000 is required to achieve a target return of 15 per cent. Thus, the project meets its target with £1,500 to spare

– as shown in the 'Incremental' and 'NPV' box. The dialog box to the right, labelled 'NPV Summary', shows the incremental cash flow to this project under a variety of discounting assumptions (shown in the column labelled 'Rate%'). We can see that if a less demanding rate of return is adopted, then the project appears to be even more worthwhile. However if the target rate of return is lifted above 15 per cent, then the project quickly becomes non-viable.

The boxes labelled 'sensitivity analysis' show what happens to the project returns under a variety of adverse model assumptions – a 10 per cent shortfall in food sales against expectations; a similar shortfall in non-food sales; loss of margin on sales; or an overspend in the project against planned investment. Any of these scenarios will depress the DCFR below 15 per cent, and send the net present value of the project below zero.

16.3.4 Format Optimisation

We have already discussed the use of rating systems for both fuel volume and retail turnover assessment. In addition, GMAP has also devised rating systems for car wash facilities, QSR and ancillary facilities such as ATM. We saw earlier (Section 16.3.2) that for each class of application it is possible to evaluate numerous alternative scenarios e.g. 24 hour versus 16 hour opening; own brand or franchised branding; and so on. In planning the network, there are therefore a tremendous variety of format options e.g. even a basic set might begin:

- basic fuel (4 pumps) and basic shop (1,000 square feet);
- advanced fuel (8 pumps) and basic shop;
- advanced fuel and advanced shop (2,000 square feet);
- basic fuel and advanced shop;
- basic fuel; basic shop; basic wash (spray only);
- basic fuel; basic shop; QSR; and so on.

Systematic evaluation of each option for each of many thousands of sites within a network to identify the ideal formats has been undertaken by GMAP in a number of European countries.

16.3.5 Representation Planning

There are a number of ways in which location modelling can support effective representation planning. For example, many UK oil companies have recently been establishing partnerships with grocery retailers in order to develop the potential of their forecourt shops e.g. Esso and Tesco; BP and Safeway; Shell and Sainsburys. These relationships might take the form that the partners will agree a list of sites for development and the oil company will pay the grocer a fixed sum for the right to position a branded outlet on each site and an appropriate share of the profits. If the potential of each location is evaluated with accuracy, then it is possible for the oil company not only to identify the prime sites for extraction of profits, but also to negotiate the most favourable contractual terms with its partner.

A second example of a representation planning application would be to optimise facias across a network. Many companies at present adopt snack food formats at busy locations such as motorway and trunk road sites; groceries and fresh food in urban locations with a defined local catchment; CTN formats in busy rush hour locations; and often a mixture of other formats in between. One of the problems is to distinguish between locations that may appear to have the characteristics of more than one of these location types, or none at all. Many organisations have undertaken local market analysis, including geodemographic or psychographic classifications, in order to characterise sites for this purpose. In a sense this is a network planning version of the format optimisation model described in Section 16.3.4.

An advanced representation planning example is discussed in Section 16.4.4 below. Further insights into the use of location models within representation planning are provided by Birkin and Culf (2001), and Birkin *et al.* (2002).

16.4 Advanced Modelling Techniques

16.4.1 Introduction

In Sections 16.2 and 16.3 above we have shown that a wide range of beneficial applications can be supported by a SDSS that integrates a range of data sources using a straightforward method for site assessment based on the concept of ratings. The beauty of the Site Rating as a concept lies to a great extent in its simplicity, which makes it relatively easy to implement in standard fashion across varied geographical markets which of course have heterogeneous data sources as well as varied market behaviour. Nevertheless it is also appropriate to recognise that there are situations in which more sophisticated modelling applications may also be embedded within the SDSS framework. Some advanced applications are reviewed in this section.

16.4.2 Price Modelling

Knowledge Support Systems (KSS) Limited is a company based in Manchester. The company was formed in 1993 by two researchers at UMIST *"to exploit their original scientific research in intelligent tactical decision making"* (KSS 2001). The systems are described as 'intelligent' *"because they generate 'optimal' pricing structures or resource allocation in accordance with the client's goals or strategy"*. The models are founded in mathematical optimisation and are said to have a 'learning capacity' that allows the models to be adaptive to changing market conditions. The KSS technology has a predictive capability, which allows users to evaluate the impact of price changes at a petrol station, or across a network; an optimisation capability, which allows profits to be maximised; and a forecasting capability that allows future sales to be estimated in the light of current sales and price performance. It is affirmed that the PriceNet product has yielded annual profit improvements of up to £10,000 per petrol station (Business Standard 2000). The product is licensed to companies such as Repsol, Chevron and Shell. MPSI also offers a variety of price modelling solutions.

The question of price modelling for petrol stations has attracted a certain amount of interest within the geographical literature. Haining *et al.* (1996) have developed a theoretical model of price variations in space and time, while Plummer *et al.* (1998) report on its empirical application within St Cloud, Minnesota. The geographical dimension to variation in prices is a strong one. For example, it is common practice for individual petrol station managers to maintain price in relation to defined 'marker' sites within the local catchment. This suggests that any change in the price of petrol at a single location can have a 'ripple' effect on neighbouring outlets. In a similar way, the effects could continue to ripple throughout a network. The effect is perhaps not unlike the diffusion of innovations first hypothesised by Hagerstrand (1953). It is fair to assume that geographical effects are not fully embraced by the practical models of KSS and MPSI; but equally that geographical models are not yet available to real users of price decision support systems. It is possible, therefore, that further work in this area could lead to improved models.

16.4.3 Ratings Optimisation

We have seen in Section 16.2 above that it is possible to develop Site Ratings for petrol forecourts using methods that are analytically grounded, but also conditioned by practical and intuitive criteria. The ratings have to use whole numbers; they have to be balanced between location and facility criteria; and they need to biased towards the sorts of criteria which have traditionally been viewed as the most important within the industry.

It is a question of both theoretical and applied relevance as to whether the ratings can be appreciably improved through the application of advanced statistical methods. GMAP has derived an approach to this problem using a type of genetic algorithm (GA). An outline of the strategy now follows (see Birkin and Culf 2001; George *et al.* 1997; or Hurley and Moutinho 1996, for a more detailed discussion of genetic algorithms within location planning). Suppose there are 20 explanatory variables – a combination of facility and location variables. A solution to the problem is a series of 20 weightings (say between zero and ten) which define the relative importance of each of the variables. The goodness of fit or 'fitness' of each solution can be evaluated by applying the rating system, and correlating the ratings that are derived with known performance (turnover) of the petrol station. This process is repeated for a number of candidate solutions, say 100. Now the candidate solutions are updated through a combination process that exchanges weights between pairs of solutions. The better the fitness of a candidate solution, then the more likely it is to be selected as a 'parent' of a solution in the next generation.

Some results from the GA are illustrated in Table 16.4. The improvement on the baseline solution might be described as 'significant' rather than 'dramatic'. There could be merit in using the enhanced ratings within some of the applications described in Section 16.3, particularly those such as sales forecasting which require optimum predictive accuracy. On the other hand, the ratings are messy and more difficult to comprehend than the base ratings. In practice, the main outcome of this analysis might be that the base ratings should be re-examined with a view to adjusting the balance between location and facility scores, since in the base ratings the location score accounts for 50 per cent of the rating score, but this jumps to around 80 per cent for the optimised ratings.

Table 16.4: Optimised Site Ratings from a genetic algorithm

location	score		facility	score	
	version 1	optimised		version 1	optimised
traffic	18	7	quality + ultra modern	12	19
plot area	10	3	pumps	12	18
access	8	2	attended hours	8	8
kerb cuts	2	1	shop area	4	3
visibility	5	1	shop type	3	5
traffic builder	4	0	wash type	3	5
avarage speed	3	0	MPDs	3	6
			HGV pumps	3	6
			operations method	2	0
			development year	0	3
total	50	14		50	73

An alternative approach to statistical optimisation would be to employ artificial neural networks. This methodology has been employed by NeuroCorp (2001) in the development of its NeuroLocator product. The NeuroLocator uses a 'Profile Screen' which allows users to profile petrol stations according to a variety of characteristics, which includes 'Operations', 'Additional Profit Centres' and 'General' in addition to facility and location characteristics. Neural network technology is used to create a series of scores (Location Score, Operations Score, Facilities Score) which are predictive of site performance. The profile has a C-Store and QSR component in addition to 'Gasoline'. The neural network approach shares the strengths and weaknesses implied in the discussion of the genetic algorithm above. It can be expected that the neural network will produce a statistical model that explains a significant amount of the variance in site turnover. However, there is also a 'black box' dimension to the technology which users are likely to find less appealing.

16.4.4 Advanced Location Modelling

Whilst the ratings approach described above is ideally suited to problems of site location and assessment, further difficulties may be encountered when questions of network planning or market area planning start to come into focus. Consider for example a situation in which a new petrol station is to be opened in Chandigarh. Data has been collected about other petrol stations within the market area and an appropriate ratings model has been devised. This allows the potential of the new site to be assessed under a variety of format options for the new station. What it does not tell us is anything about the *impact* on surrounding sites which is of fundamental interest to the computation of returns on investment (see Section 16.3.3).

In order to address this problem, MPSI have developed a tool branded as Huff's Market Area Planner which is described on the corporate web-site as *"... a powerful tool used for formulating and evaluating strategic decisions ... (which) integrates the new Huff model, a statistical package and a mapping engine that makes it possible for you to analyze and predict consumer patronage patterns geographically."* (MPSI 2001). It is unclear what makes the 'new Huff model' different from its predecessor, although it seems a fair bet that the model centres around some variation of the gravity model originally propounded in Huff's famous paper (Huff 1963).

Similar techniques have been applied by GMAP in the development of a spatial interaction modelling approach to fuel turnover modelling. These applications present unusual problems for at least three reasons. The first difficulty is the element of dislocation between supply and demand, particularly for important trunk road sites. Thus, with a supermarket or retail outlet, it is possible to develop a modelling approach which relates outlet performance to customer demand within a local catchment area. However, for petrol stations on major roads, a high proportion of customers may be drawn from highly dispersed 'transient' customers whose residences are far away from the natural local catchment of the site.

A second difficulty is that petrol is a well known 'distress purchase'. People do not like buying it – still less will they make a dedicated trip for this purpose, again in contrast to retailing. Thus it is impossible to understand the way the market operates in isolation from other processes such as journey-to-work behaviour, leisure activities, and indeed shopping trips. Thus, petrol buying is usually associated with another trip of this type.

Conventional models can be constructed from two elements – small geographical areas for customers and retail outlets or centres for suppliers. Fuel modelling requires a third component of the road network – it is not possible to understand the fuel generation capabilities of a site without understanding the passage of traffic on the particular network link(s) on which the site is located, and the demand associated with that traffic.

In spite of these difficulties, GMAP has successfully developed spatial interaction modelling applications in the UK, Europe, Australia and elsewhere. The most recent and most successful of these applications was developed for study areas in Sicily in the late 1990s. At this time, the Italian retail petroleum network was heavily over-provided, with around 25,000 outlets. A government-approved plan sought to reduce this number through closure of between five and six thousand small outlets, mostly within urban areas, together with around 1,000 new outlets on high volume traffic routes.

GMAP's task was to develop a model to underpin representation planning within this dynamic market environment. The application was tested in a study area of approximately 700,000 people and 270 petrol stations. GMAP was provided with known fuel volumes for 70 client sites, and volume estimates for a further 190 competitor sites. The fuel volumes for a further eight client sites were withheld as a test on the robustness of the modelling approach.

In order to address the difficulties described above, a model with three component sub-models was constructed. These components comprise a residence-based sub-model; a work-based sub-model; and a transient sub-model. Each of the sub-models is driven by a separate demand equation:

$$O_i^1 = (\lambda_i^1 / (\lambda_i^1 + \lambda_i^2))\,(1 - \mu_i)\,O_i^* \qquad (1)$$

is the demand within the residence-based sub-model, where:
O_i^* is the demand for fuel within small area *i*, which is modelled as a function of the demographic and economic characteristics of that area (i.e., car ownership, family structure, affluence, activity rates and so on);
$_i$ is the proportion of fuel spending in zone i through transient sites;
^1_i is the parameter associated with residence-based expenditure in zone *i* and ^2_i is the parameter associated with work-based expenditure;

$$O_i^2 = \sum_j (\lambda_j^2 / (\lambda_j^1 + \lambda_j^2))\,(1 - \mu_j)\,O_j^*\,p_{ji} \qquad (2)$$

is the demand in the work-based sub-model, where:
p_{ji} is the probability that individuals living in zone *i* work in zone *j*

$$O_i^3 = \sum_j O_j^*\,p'_{ji} \qquad (3)$$

is the demand in the transient sub-model, where:
p'_{ji} is the probability that transient demand from zone *j* is allocated to zone *i*.

The residence-based and work-based sub-models now take the form of a conventional production-constrained spatial interaction model:

$$T_{ij}^{km} = A_i^k O_i^k W_j^m \exp(-\beta^k c_{ij}), \qquad k = 1,2, \qquad (4)$$

where c_{ij} is the drive time from *i* to *j*, and ^{k} is a distance deterrence parameter;
A_i^k is a balancing factor:

$$A_i^k = 1 / \sum W_j^m \exp(-\beta^k c_{ij}) , \quad k = 1,2, \qquad (5)$$

and W_j^m is the attractiveness of petrol station *j* of brand *m*:

$$W_j^m = f\,(pumps, quality, ownership, hours, brand, price) \qquad (6)$$

The transient sub-model takes the form of an allocation model between petrol stations that fall on major routes:

$$T_{ij}^{3m} = O_i^3 \gamma_{ij} W_j^m \phi_j^m / \sum_k \gamma_{ik} W_k^m \phi_k^m \qquad (7)$$

where $_{ij}$ is a probability that station *j* is fed by transient demand from zone *i*, and ^m_k is a measure of competitive density around station *k*. The sales at each outlet can be accrued as the sum of the flows into each location:

$$D_j^m = \sum_{ik} T_{ij}^{km} \qquad (8)$$

The model was calibrated and implemented using a combination of small area demographics and spending data, site data, known sales and volume estimates, traffic flow data and travel-to-work information. The results of the volume estimation tests are reproduced in Table 16.5. The model volume estimates explain just over 81 per cent of the variance in actual sales volumes. Of more interest to commercial users are the relative estimation errors for each individual site, as shown in the 'performance' column of Table 16.5. The restriction of modelling errors to less than 10 per cent in six out of nine cases is a creditable achievement. Thus the model may be used with reasonable confidence for planning applications.

Table 16.5: Model performance against known volumes: Sicily

site location	site volume actual	site volume predicted	performance
Corso Tukory	1510	1663	-9 %
Piazza Giacchery	1860	1600	16 %
Via Uditore	1002	1050	-5 %
Viale Regina Margherita	1424	1000	42 %
Corso Dei Mille	956	900	6 %
Viale Michelangelo	1089	1150	-5 %
Via Aurispa	763	600	27 %
Via Sciuti	1980	2141	-8 %
Niscemi	1250	1350	-7 %

A number of representation plans and scenarios were evaluated using the model. One of the preferred scenarios is illustrated in Table 16.6. A systematic reconfiguration has been applied to an initial network of 71 outlets to provide a better balance of representation between nine functional areas of the city of Palermo. Five site closures are counterbalanced by six new developments, in addition to five site upgrades. The scenario yields a significant improvement in profitability and market effectiveness, as shown in Table 16.7 against key indicators of market share, sales volume and sales per outlet.

Table 16.6: Representation planning scenario for Palermo

macro area	outlet share %	market share %	scenario changes open	close	upgrade
Centro Storico	32.0	29.5			
Litoranea	24.2	26.8			
Viale regione Siciliana	22.5	31.9	2		
Zona Ind. Bracaccio	22.7	16.1	1		
Policlinico - Iniversita	29.2	20.5		2	1
Zona Malfitana	33.3	18.1		3	2
Zona Sperlinga	26.4	19.7	1		
Monte Pellegrino	33.3	31.2			2
Pa Nord	21.4	19.6	2		

Table 16.7: Results of the Palermo representation plan

	baseline	scenario
outlets	71	72
sales volume	56638	63810
market share	25.2%	28.4%
sales/outlet	798	886

16.5 Conclusions

In this chapter, we have considered the development of SDSS for oil companies with particular reference to petrol forecourts. The move from GIS to SDSS is characterised by the incorporation of a spatial modelling capability that adds considerable value to the base geographic data. This capability may take several forms, from relatively straightforward site scoring models to complex spatial interaction models. It has been argued that the introduction of a spatial modelling capability can support many important business planning decisions, relating to areas such as capital allocation, sales forecasting, investment modelling, configuration of outlets and representation planning. For this reason, we can expect with confidence that applications of this type will become increasingly prevalent within the business world.

References

Birkin, M., Clarke, G., Clarke, M. and Wilson, A.G. (2002) *Retail Intelligence and Network Planning*, Wiley, London (in press).

Birkin, M. and Culf, R, (2001) Optimal distribution strategies, in Clarke, G. and Madden, M. (eds.) *Regional Science in Business*, Springer, Berlin, pp. 223-241.

Business Standard (2000) India: A Calculating Professor, Business Standard, 28/10/2000: p. 9.

Catalist (2001) Corporate Web Site, *http://www.catalist-uk.com/*.

Independent (2000) The Investment Column – KSS, Independent Newspaper, 12/9/2000: p 19.

George, F., Radcliffe, N., Smith, M., Birkin, M. and Clarke, M. (1997) Algorithms for solving a spatial optimisation problem on a parallel computer, *Concurrency: Practice and Experience*, 9: 753-780.

Hagerstrand, T. (1953) *Innovation Diffusion as a Spatial Process.* Trans: Pred, A. (1967), University of Chicago Press.

Haining, R., Plummer, P. and Sheppard, E. (1996) Spatial price equilibrium in interdependent markets: prices and sales configurations, *Papers of the Regional Science Association,* 75: 41-64.

Huff, D. (1963) A probabilistic analysis of consumer behaviour, *Papers and Proceedings of the Regional Science Association*, 7: 81-90.

Hurley, S. and Moutinho, L. (1996) Approximate algorithms for marketing management problems, *Journal of Retailing and Consumer Services,* 3: 145-154.

Knowledge Support Systems (2001) Corporate Web-Site, *http://www.kssg.com/*.

Landis, J. (1993) GIS capabilities, uses, and organisational issues, in Castle, G. (ed.) *Profiting from a Geographic Information System*, GIS World, Fort Collins, pp. 23-53.

MPSI (2001) Corporate Web Site, *http://www.mpsisys.com/*.

Neurocorp (2001) Corporate Web Site, *http://www.n-corp.com/*.

Pike, R. and Neale, B. (1993) *Corporate Finance and Investment: Decisions and Strategies*, Prentice Hall, Englewood Cliffs, NJ.

Plummer, P., Haining, R. and Sheppard, E. (1998) Spatial pricing in interdependent markets: testing assumptions and modelling price variation. A case study of gasoline retailing in St Cloud, Minnesota, *Environment and Planning A* , 30: 67-84.

17 An Urbanisation Monitoring System for Strategic Planning

JanJaap Harts[1], Kees Maat[2] and Henk Ottens[1]

[1] Urban Research Centre Utrecht, Utrecht University, P.O. Box 80.115, 3508 TC Utrecht, The Netherlands

[2] OTB Research Institute for Housing, Urban and Mobility Studies, Delft University of Technology, P.O. Box 5030, 2600 GA Delft, The Netherlands

Abstract: Efficient use of land is a major goal in urban planning strategies. In densely populated regions, where land available for urban expansion has become scarce, urbanisation policies promote compact urban development along with good spatial quality. National policy in the Netherlands seeks to reduce the consumption of urban land. This chapter describes and discusses the design, implementation and application of an instrument to monitor these strategic planning goals. It is based on current, generally available administrative and statistical data sets and on geo-information technology. The Urban Pattern, Density and Diversity (UPDD) Monitor makes it possible to translate the policy goals into indicators and to measure and visualise urban development for ex-post policy evaluation. However, some problems with the interpretation of the indicators' values could not be completely solved. Moreover, future policy changes will make periodic adjustment of the instrument inevitable.

17.1 Introduction

Right from the birth of the technology, back in the 1960s and 70s, geographical information systems (GIS) have been ascribed a great potential for practical urban and regional planning (Steinitz 1993). Innes and Simpson (1993, p. 230) stated that *"this developing technology offers an opportunity to transform and empower planning practice"*. But reality has once more proven to be more unruly than expected. In spatial management and geospatial research, the technology is much more accepted than in day-to-day spatial planning and decision-making circles. Within the planning sector itself, administrative tasks like planning control are more easily supported by computers than strategic decision-making (Masser and Ottens 1999). Because of the nature of urban and regional planning, computer support should be introduced carefully, and dedicated systems for specific tasks will be inevitable (Ottens 1990, Geertman 1999).

Studies of how GIS can meet the information requirements for planning show that this technology can only handle a part of all the information that is necessary for complex planning and decision-making processes. GIS is particularly useful for generating descriptive and prescriptive information (Webster 1993). The data pre-processing, integration, management and visualisation (mapping) functions of a GIS are particularly useful for monitoring spatial development. The geographical modelling functions (location- and topology-based object and layer manipulations) are effective in applications like *ex-ante* and *ex-post* policy evaluation and suitability analysis. Therefore, spatial policy monitoring systems – particularly tools that make use of a combination of descriptive and prescriptive methods – should benefit from modern geo-information technology.

This chapter presents a prototype for a GIS-based system that monitors aspects of urbanisation that are subject to spatial policy (Harts *et al.* 2000). First, we briefly describe the context in which the system is currently operational, namely the national policy for spatial planning in the Netherlands. Then we examine the data and methods that have been incorporated into the system. The following section highlights the major results of an application: evaluating urban policy versus actual urban development. Finally, we draw some conclusions and discuss their implications.

17.2 Urban Policy and Policy Evaluation

A formal system of physical planning in the Netherlands (Cammen *et al.* 1988, Wusten and Faludi 1992) has been in place since the 1960s, but its roots go farther back. Dutch spatial policy stems from the organised drainage and reclamation of land and the system of water management – practices that started in late medieval times – and from the social housing legislation introduced 100 years ago. After the Second World War, spatial planning was focused on controlling spatial urban development and reducing car mobility – an obvious strategy in one of the world's most densely populated countries. Urban planning over the past 50 years has revolved around the Randstad/Green Heart doctrine, which has three basic tenets: preservation of landscape diversity; perpetuation of a ring of distinct cities in the western, most urbanised part of the country; and containment of urban sprawl. During the 1970s and 80s, urban growth was accommodated in new towns and growth centres. Since 1990, it has been channelled into large-scale, high-density expansion schemes at the edges of the large cities (the so-called 'compact city' model). Despite all these efforts, suburbanisation has proceeded in a selective and dispersed form, especially during periods of strong economic growth. The policy for urbanised areas has sought to improve both the quality and the functionality of urban space. To that end, extensive, and heavily subsidised, urban renewal and restructuring schemes have been operational since the 1970s.

The *Urban Pattern, Density and Diversity (UPDD) Monitor* was developed to facilitate the periodic evaluation of policy addressing the spatial aspects of urbanisation. It was first used in an *ex-post* evaluation of some key elements of the urbanisation policy that was presented in the Fourth National Physical Planning

Report, which came out in 1991 (Kreukels 1992). This document set forth three policy goals regarding urban form and functions:

- reducing the growth of urban land consumption by promoting higher urban density;
- reducing car mobility by coordinating urban and transportation planning to reinforce bundled urban growth at nodes of public transportation networks; and
- improving spatial quality by increasing the diversity of milieus within urban areas.

The need to monitor actual urban developments in relation to these goals was clear from the outset. Indeed, evaluation studies have been conducted periodically, the findings being published in comprehensive evaluation reports. However, the envisioned monitoring process, which was to be both continuous and robust, was still lacking. The *UPDD Monitor* was developed to fill that gap.

17.3 The *Urban Pattern, Density and Diversity Monitor*

17.3.1 Project Aims and Assumptions

The Ministry of Housing, Spatial Planning and the Environment contracted the Universities of Delft and Utrecht to develop an urban monitoring system. The aim was to produce a system that could track changes in land use patterns, functional mix, and building density within urban areas and then relate these changes to the policy goals. All data processing was based on ESRI's ArcInfo and Microsoft's Access software packages, both of which are available to planners at the Ministry, as are the data sets. The *Monitor* is not fully automated; human intervention is required between the successive calculations and visualisations. The *Monitor* can be operated by casual users who know how to use a PC and a desktop GIS.

When designing this system, the first step was to identify indicators that could be extracted from current data sources; this was done to make sure the system could also be used in future evaluations. Two quantitative indicators were selected: the urban density index (UDI) and the function mix index (FMI). These two indicators are directly related to the first policy goal, namely to reduce urban land consumption by promoting higher densities, whereby the FMI measures the spatial bundling of different functions. The other two policy goals – to reduce the growth of car mobility and improve the quality of local environments – are indirectly related to the UDI and the FMI. Any changes that occur are revealed by comparing the values for subsequent cross-sections over time. The third indicator is the functional-morphological character of a small urban area. In light of its qualitative nature, a multivariate classification of urban environments had to be generated, the types being derived from empirical observations. This classification is used to determine (changes in) physical-environmental diversity within urban areas.

Because the classification covers the whole country, it is possible to identify particular cells as urban. This entails selecting spatial units with a relatively high concentration of urban land uses and functions. Variables like housing, non-agricultural jobs, services and non-rural land use (which includes city parks, public sport fields, cemeteries and military facilities) are used for this purpose.

17.3.2 Data Preparation and Spatial Units

The monitoring system was built with geographical information technology (GIS software) using digital land use maps and automated registers of housing, jobs and shops. The data from the registers was geocoded based on x-y coordinates of data points representing postcode areas. The number of jobs is derived from a source (LISA employment) that combines several regional registers of employment. The numbers of dwelling units and retail outlets, including an indication of the type of building, are derived from a register of the national postal service (LBV real estate). A time series of digital land use data is published by Statistics Netherlands.

As density and mixture measurements for spatial units are sensitive to differences in shape and size, administrative and statistical divisions (e.g. districts or postcode areas) are less suitable for this application. Therefore, the data were converted to grid cells measuring 250 by 250 metres. The database holds information for approximately 800,000 cells. Table 13.1 lists the data used to construct the classification and to calculate the indices.

Table 17.1: Data sources and variables used in the system

variable	source	year	classification	density/mixture indices
number of dwellings	LBV real estate	1989, 1996	x	x
percentage multi-story units	LBV real estate	1989, 1996	x	
number of shops	LBV real estate	1989, 1996	x	x
number of persons employed	LISA employment	1990, 1997	x	x
percentage business area	land use statistics	1989, 1996	x	
percentage services sites	land use statistics	1989, 1996	x	
percentage green & sports facilities	land use statistics	1989, 1996	x	
percentage infrastructure lands	land use statistics	1989, 1996	x	

Any spatial subdivision creates a sharp and arbitrary delineation of regions. Shifting the boundaries will often influence the values. This drawback is overcome by using a grid-based GIS, whereby a 'spatially moving average' can be calculated for each cell (i.e. the average of the value of the cell itself plus the values of the eight adjoining cells). This procedure smoothes out the abrupt transitions across cell borders and compensates for the arbitrary spatial configuration of postcode areas – which were, of course, created for mail delivery purposes. Consequently, for each of the eight variables, the initial values of the cells are replaced by the average value of the cell and its eight neighbours, an area of 750 by 750 metres. This smoothing operation also gives a better representation of the way land use and density of small urban areas will be perceived by the inhabitants and users of the city. The awareness of the direct local environment is generally limited to the distances that people commonly walk around an address location, for instance 300-400 metres. Figure 17.1 shows an example of how of this filtering operation affects the variable 'number of dwelling units' for the Amsterdam region.

a) Initial density pattern

c) Density pattern after spatial smoothing

Fig 17.1: Housing density in Amsterdam (for urban cells 250 by 250 metres)

The variables are measured in either percentages or absolute numbers. Those values must be normalised in order to give them equal weight in the calculation of the three indicators. All cell values are expressed as a percentage of the maximum cell value for the Netherlands. As a result, all cell values lie in the range of 0 to 100. The data pre-processing steps mentioned here are schematised in Figure 17.2.

The same procedures have to be followed for the data from an earlier point in time (t-1), which in the prototype case is the year 1990. But there is one exception: normalising is done by expressing the cell values as a percentage of the maximum value at time (t) in order to determine the exact changes between the two years. The procedure takes on this general form:

$$y_{ij} = \frac{x_{ij}}{\max_i x_{ij}} * 100 \quad (1)$$

where x_{ij} denotes the number of elements in cell i of attribute type j (e.g. number of dwellings) and y_{ij} denotes the proportion of the maximum value.

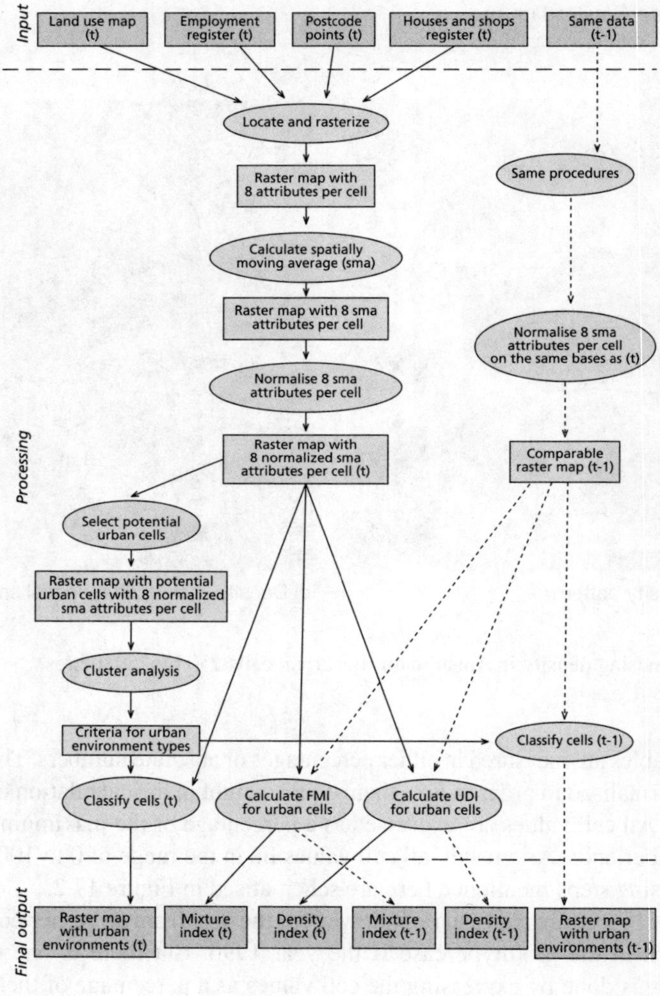

Fig 17.2: Procedure for the classification of urban cells

17.3.3 Urban Pattern Changes

As noted above, the *Monitor* is also supposed to identify land use changes within urban areas. This is accomplished by formulating classification rules on the basis of the eight variables in the database. Applying these rules, the cells are classified according to a functional-morphological typology of small urban areas. Comparison of the results for two points in time (t and t-1) yields a policy-relevant picture of changes in urban land use and density patterns.

The classification rules are derived from a cluster analysis. A k-means analysis is performed (also called segmentation, partition or taxonomy analysis). This technique allocates cells to clusters, the number of which is determined *a priori*. In an iterative process, groups of cells are formed. Internally, these groups should be as homogeneous as possible; at the same time, the differences between groups should be maximised. Group membership is optimised by calculating the multivariate Euclidian distance to and between the group means. Figure 17.2 shows the steps in this procedure.

The cluster analysis described here started with the situation at time t=1. First, urban and rural cells were distinguished. If none of the eight variables for a cell had a value greater than 10, the cell was considered to be potentially rural. Thus, the spatial differentiation analysis could be restricted to cells that are potentially urban; these were subsequently grouped using the k-means cluster procedure. Two solutions were then evaluated: one with 25 clusters and one with 16. The result of the 16-cluster solution was meaningful and useable. Of these 16 clusters, 15 were definitely urban in character and one predominantly rural. The urban clusters could be reasonably well interpreted as classes in a typology of urban environments. Finally, the cluster allocation criteria generated by this analysis (expressed as ranges for the normalised values of the variables in the data set) were used for the definitive classification of all cells in the Netherlands at both points in time (t=1 and t-1). Section 17.4 gives some examples of the measured changes in urban patterns.

17.3.4 Urban Density and Function Mix

The density and mixture indices express key aspects of urban development in easily comprehensible indicators. The urban density index (UDI), which can be calculated for each cell, expresses the composite density of housing, jobs and shops. In this way, it differs from sectoral measures like population or housing density. The calculation is based on the normalised attribute values (equation 1). Housing, jobs and shops contribute equally to the index value. Although the UDI can be calculated for every cell, it is only applied to the urban cells. The normalised values y are summed and divided by the number of categories that occur (three in this case) to arrive at the density value, as denoted by d_i. The theoretical maximum is 100, whereby one cell would have the greatest number of dwellings as well as the most jobs and shops. The higher the UDI value, the higher the 'combined density' will be.

$$d_i = \left(\sum_j y_{ij} \right) / j \qquad (2)$$

The second indicator is the function mix index (FMI), which measures the diversity of functions in a cell. A crucial issue is how to determine the set of reference values against which diversity is measured. This problem can be solved by using the specialisation index (Hoorn 1981) or the Gini coefficient (Robinson 1998) to compare the proportions of the functions in a cell with those for the area under study. In this monitor, the proportions of housing, employment and shopping in the country as a whole are used as the reference mix. If the variety of functions in a cell deviates from that in the study area, it is assumed that this cell is specialised – that is, it is less mixed than the average urban cell. This is expressed as follows:

$$s_i = \frac{1}{2} \sum_j \left| \frac{x_{ij}}{\sum_j x_{ij}} - \frac{\sum_i x_j}{\sum_i \sum_j x_{ij}} \right| * 100 \qquad (3)$$

where s_i is the FMI value. A value of zero indicates that the mixture of functions in the cell equals the distribution for the entire study area. Higher values indicate the level of deviation from the urban average in the Netherlands. However, the function is asymptotic, which means that the value of one is not reached. The FMI in a cell is calculated for all urban cells with at least one dwelling unit, job or retail outlet. Therefore, some of the urban cells – green areas, for instance – have no FMI values.

Both the UDI and the FMI index combine unweighted, normalised values for cell characteristics based on both functional and morphological units. As a consequence, one might argue that housing, which accounts for half of all urban land use, is somewhat under-represented in the determination of the index values. On the other hand, the retail sector may be over-represented. However, the heterogeneity of the functions makes it difficult to weight them in an objective manner.

It was not surprising, given the sets of variables used in the analysis, to find strong relationships between the types of urban milieus and the values for the density and function mix indices (Table 17.2). In the city centres, functions are found at higher densities than on the periphery or in suburban settings. UDI values are low in those parts of the cities or towns that have many green spaces or much land devoted to infrastructure. On the other hand, urban areas dominated by a single type of land use – the so-called mono-environments – are very specialised. In general, the residential areas in the city, especially those built at higher densities, have the highest degree of function mix.

Table 17.2: Urban environments: extent, density, specialisation and development, 1990-1996

urban environment type	urban area		density index		specialisation index	
	1996	1990-96	1996	1990-96	1996	1990-96
	[km2]	[%]		[%]		[%]
metropolitan centre	45	-1.5	27.8	-1.3	35.3	-3.7
urban centre	102	10.5	16.7	3.0	24.5	-4.0
weakly urbanised centre	305	1.5	7.3	3.7	16.2	-3.1
concentration of services	149	10.9	6.0	6.4	28.5	6.1
metropolitan residential area	71	3.6	23.7	2.3	13.6	-0.7
urban residential area	123	6.0	14.1	1.7	14.4	-0.6
moderately urbanised residential area	281	9.9	10.8	3.4	15.3	-3.9
weakly urbanised residential area	863	6.8	5.2	5.7	18.6	-1.3
business areas	972	16.3	1.1	8.1	50.5	3.3
green space and sports areas	1456	5.3	0.5	1.4	26.5	0.4
infrastructure lands	359	3.4	1.7	10.3	34.3	5.7
residential area and green space	321	5.6	6.2	2.4	18.8	-1.5
business and dwellings	183	6.2	7.8	-0.4	26.9	1.5
business and green space	296	14.2	3.3	2.9	25.7	-1.0
services in rural areas	444	-1.6	0.4	7.0	40.1	1.5
total urban area	5970	7.1	4.0	2.9	27.9	2.1
total rural area	34869	-1.1	–	–	–	–

17.4 The *UPDD Monitor* in Action

This section treats the dynamics of urban pattern, density and diversity in the Netherlands, as described by the *Monitor*. The changes were analysed by comparing the situations in 1990 and 1996. Rather than elaborating on urban development, the purpose of this discussion is to demonstrate the functionality of the *Monitor*.

First, urban and rural areas were distinguished. The Netherlands proved to be a strongly urbanised country, with 13 per cent of the land classified as urban. This outcome is in line with the findings of Needham (1996) and the National Spatial Planning Agency (NSPA 1996). As a result of the cluster analysis, urban space was allocated to 15 types of urban environment, specifying centre, residential, mono- and combination milieus. The centre environments are characterised by relatively large proportions of jobs and shops. Differences in density, the number of multi-storey buildings, the proportion of retail outlets for shopping goods and the amount of land used for services determine whether the centre environment is said to be strongly, moderately or weakly urbanised. Obviously, residential environments are dominated by housing functions. They differ from each other mainly with respect to density, the proportion of multi-storey housing and the mix of workplaces and shops. The mono-environments, accounting for a large total area, are dominated by one specific activity; this may be a business estate, green space and sports facility, infrastructure or some such use. The remaining types, which are combination environments, are zones of transition or mixtures of different types of environments.

Comparison of the two cross-sections (1990 and 1996) shows that urban land use continues to expand at the expense of rural landscapes. Although a national population increase of only four per cent was recorded in that period, the urban areas grew by 7.1 per cent (400 sq. kilometres), thereby encroaching on 1.1 per cent of the countryside. Nevertheless, urban density – as measured by the number of dwellings, shops and jobs – increased by 2.9 per cent (Table 17.2). Thanks to this densification process, an even higher rate of growth for urban land consumption was prevented.

The analyses suggest that the hierarchical structure of centre environments had decreased. Furthermore, there seems to be less differentiation in residential areas. Conversely, a tendency towards more spatial separation is discernible, due to the large increase of the specialised mono-environments. These developments are illustrated by the following examples.

The hierarchy of centres is blurring due to differences in growth recorded by metropolitan centres, second-tier urban centres and peripheral service concentrations. Figure 17.3, depicting Amsterdam's downtown area, shows that 18 cells have changed from the metropolitan centre type to metropolitan residential areas, while 10 cells have changed the other way around.

Another striking outcome is that employment has declined in the metropolitan centres. This change is in stark contrast to the spectacular growth recorded in specialised service centres on the periphery of the urban regions. Figure 17.4 shows that this growth was concentrated along motorways and main railway lines. This type of spatial specialisation increased along with the growth in business areas, again on the periphery of the cities. It appears that the pattern of urban employment in the Randstad is gradually being turned inside out. The phenomenon of suburban edge cities (Garreau 1991) is still much less pronounced in the Netherlands than in the United States, though Dutch urban development seems to be moving in that direction.

Business areas grew by 16.3 per cent, accounting for one third of the overall urban growth. In fact, urban land use in peripheral provinces outside the Randstad is characterised by a large share of mono-environments such as business estates. Clearly, more space is available there than in the crowded Randstad, the urban heartland of the Netherlands. Figure 17.5 is an example of urban corridor development along the Almelo-Enschede axis (in the eastern part of the country near the German border). Here, new business areas fill the open space between the cities.

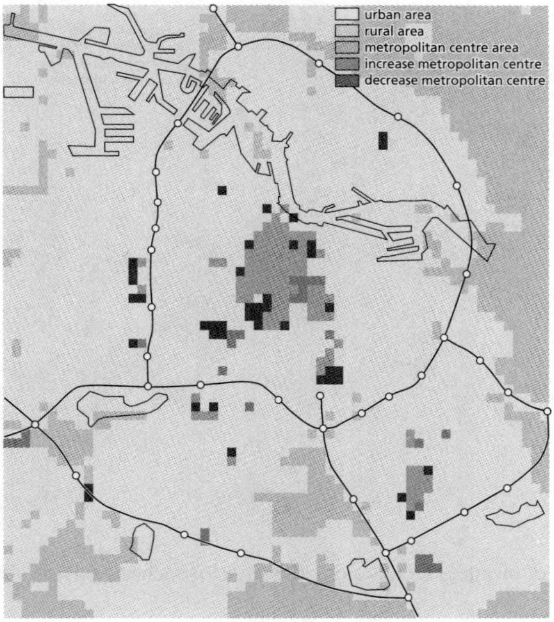

Fig 17.3: Shift of metropolitan centre and residential area of Amsterdam, 1990-96

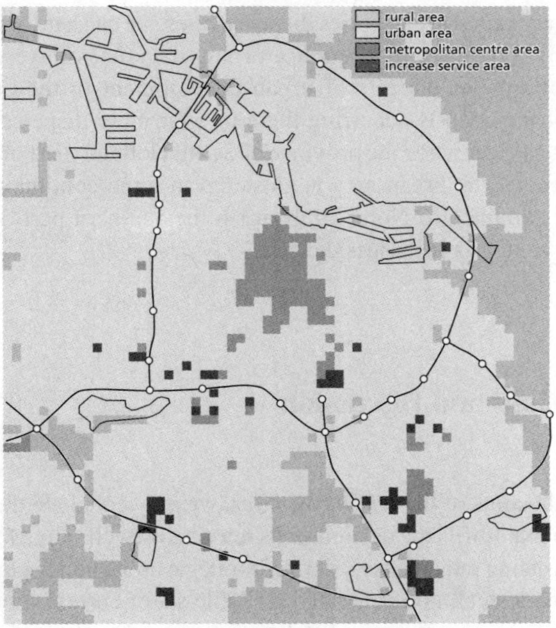

Fig 17.4: Increase of service areas in Amsterdam, 1990-96

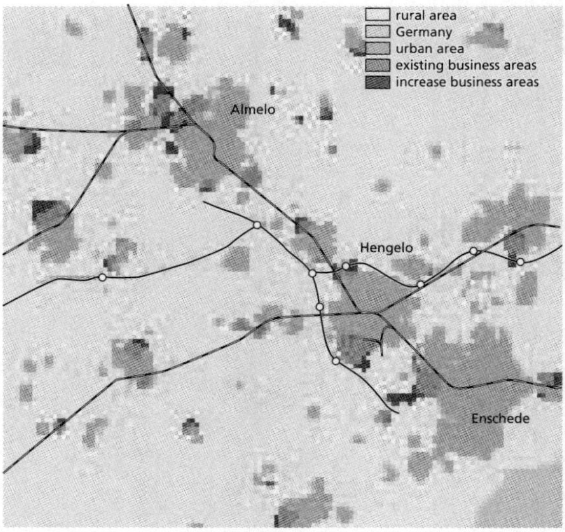

Fig 17.5: Growth of business estates along the Almelo-Enchede axis, 1990-96

Finally, separate urban settlements in the Randstad appear to be much larger than elsewhere in the Netherlands. To complicate matters, they are also amalgamating, spatially as well as functionally. Figure 17.6 depicts the extent of the distinct urban settlements in three main size categories. In many cases, we can hardly discern separate cities and towns anymore, as complex urban structures have emerged. The predominance of smaller pockets of urban development in the other provinces indicates that urbanisation is occurring there in a far more dispersed fashion. The difference becomes clear when the province of South Holland (part of the Randstad), where 74 per cent of the urban area is clustered in large continuous stretches, is compared to the province of North Brabant (in the southern part of the country), where, at 32 per cent, this is far less the case.

17.5 Conclusions and Discussion

With respect to the aims of the *UPDD Monitor*, we may conclude that it is possible to build a system that fulfils the desired tasks reasonably well. The *Monitor* has been successful in translating rather loosely defined strategic urban policy goals into indices. They are based on a small and generally available set of current official databases. Moreover, the indices are simple and transparent enough to be comprehended and trusted by policy-makers. In light of these characteristics, the *Monitor* is suitable for replication elsewhere. As usual, the possibility to rapidly, easily and convincingly visualise results cartographically is a major asset. It has been demonstrated that the *Monitor* offers detailed insight into both quantitative and qualitative aspects of urban sprawl. In fact, the *Monitor* has been used as a source of information for the recently published assessment of spatial quality in the Netherlands (Ministerie van VROM 2000). In the future, the *Monitor* might be extended to allow an assessment of spatial developments in rural areas; this possibility is now under consideration.

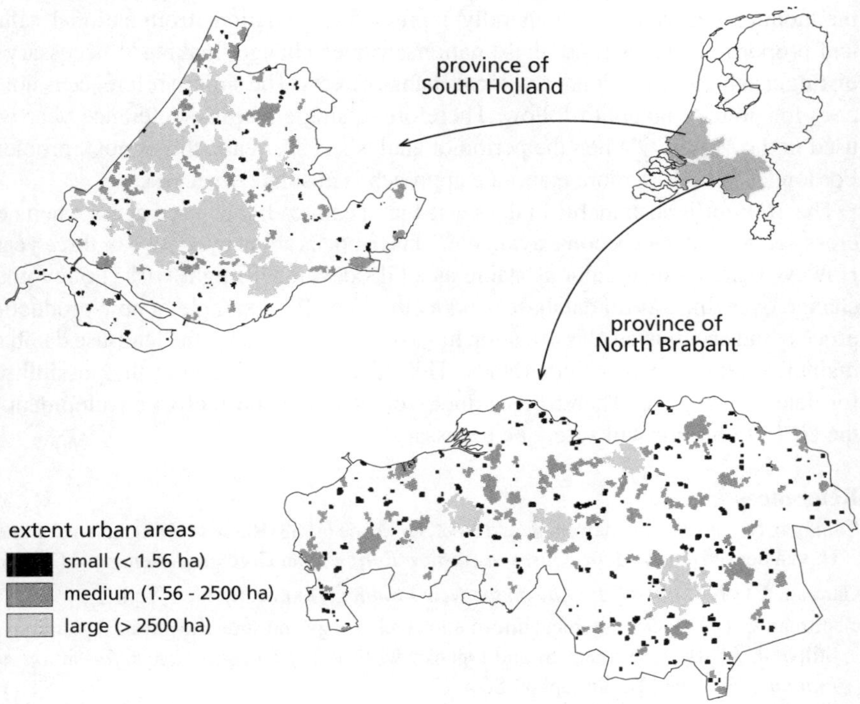

Fig 17.6: Urban areas in three size categories

The most problematic – and most debated – elements of the data analysis scheme used for the *Monitor* are the following: the moving average, the urban milieu classification and the change detection method. Use of the spatially moving average relates the resulting cell values more realistically to perceived urban density and diversity. However, the original intrinsic values no longer play a role in the analysis. Nonetheless, as these values remain present in the database, it is always possible to refer back to these initial values. This may be done, for example, to calculate national or regional statistics. Furthermore, in thematic maps with smoothed-out data, topographic referencing is not always easy, especially when the user has limited cartographic experience. This drawback can be largely overcome by adding topographic framework features like roads and boundaries.

The urban milieu classification is empirically derived and is not always easy to interpret in terms of urbanisation theory. Yet that did not create any problems for this practical, policy-oriented application. The resulting types of urban milieu corresponded only partially to the typologies that have been used thus far in urban structure plans. But the empirical clustering provided some useful pointers on ways to adjust and refine these normative typologies.

It is not always obvious how changes between cross-sections can be measured. In the monitor, cell values are generally expressed as deviations from national values and proportions. Over time, these national values change, making it necessary to distinguish between national and local change effects. The shift in reference is not so easy for policy-makers to follow. Therefore, a single national reference year was used in the *Monitor*. When the period of analysis is six years, this is not a problem. For longer periods, a more elaborate approach is absolutely necessary.

The *Monitor* is sustainable in the sense that it can readily be used again when new cross-sections of data become available. This happens about every two or three years. However, situations are not as stable as a GIS developer might wish. Policy goals change over time, as do databases and technology. For example, a new production process and an adjusted classification have been developed for the land-use database maintained by Statistics Netherlands. This adaptation will restrict the possibilities for detecting change. Therefore, further technical and substantive development of the *UPDD Monitor* will surely be necessary.

References

Cammen, H. van der, Groeneweg, R. and Hoef, G. van de (1988) Randstad Holland, in Cammen, H. van der (ed) *Four Metropolises in Western Europe*, Van Gorcum, Assen, pp. 117-176.

Garreau, J. (1991) *Edge City: Life on the New Frontier*, Doubleday, New York.

Geertman, S. (1999) Geographical information technology and strategic physical planning, in Stillwell, J.C.H., Geertman, S. and Openshaw, S. (eds.) *Geographical Information and Planning*, Springer, Berlin, pp. 69-86.

Harts, J.J., Maat, C. and Zeijlmans van Emmichoven, M. (2000) *Monitoring Stedelijke Milieus: Menging en Dichtheid*, Delft University Press, Delft.

Hoorn, F. van (1981) *Basisstudie Spreiding en Menging: de Begrippen*, Geografisch Instituut, Utrecht.

Innes, J.E. and Simpson, D.M. (1993) Implementing GIS for planning: lessons from the history of technological innovation, *APA Journal,* 59: 230-236.

Kreukels, A. (1992) The restructuring and growth of the Randstad cities: current policy issues, in Dieleman, F.M. and Musterd, S. (eds.) *The Randstad: A Research and Policy Laboratory*, Kluwer, Dordrecht, pp 237-262.

Masser, I. and Ottens, H.F.L. (1999) Urban planning and geographic information systems, in Stillwell, J.C.H., Geertman, S. and Openshaw, S. (eds.) *Geographical Information and Planning*, Springer, Berlin, pp 25-42.

Ministerie van Volkshuisvesting, Ruimtelijke Ordening en Milieubeheer (1990) Vierde Nota over de Ruimtelijke Ordening Extra: Op weg naar 2015, Deel 1, SDU Uitgeverij: The Hague,

Ministerie van Volkshuisvesting, Ruimtelijke Ordening en Milieubeheer (2000) *Balans Ruimtelijke Kwaliteit*, VROM, The Hague.

Needham, D.B. (1996) Ruimte en stedeling, in Rijksplanologische Dienst (ed.) *Ruimte als Voorraad,* VROM, The Hague.

NSPA, National Spatial Planning Agency (1996) *Ruimtelijke Verkenningen*, VROM, The Hague.

Ottens, H.F.L. (1990) The application of geographical information systems in urban and regional planning, in Scholten, H.J. and Stillwell, J.C.H. (eds.) *Geographical Information Systems for Urban and Regional Planning*, Kluwer, Dordrecht, pp. 15-22.

Robinson, G.M. (1998) *Methods and Techniques in Human Geography*, Wiley, Chichester.

Steinitz, C. (1993) Geographical Information Systems: a personal historical perspective, *GIS Europe,* 2: 42-45.

Webster, C.J. (1993) GIS and the scientific inputs to planning, *Environment & Planning B*, 20: 709-728.

Wusten, H. van der and Faludi, A. (1992) The Randstad: playground of physical planners, in Dieleman, F.M. and Musterd, S. (eds.) *The Randstad: A Research and Policy Laboratory*, Kluwer, Dordrecht, pp. 17-38.

18 An On-line Planning Support System to Evaluate Urban and Regional Planning Scenarios

Christopher Pettit, Tung-Kai Shyy and Robert Stimson

School of Geography, Planning and Architecture, University of Queensland, St Lucia 4072, Queensland, Australia

Abstract: This chapter describes the development of a planning internet site incorporating a suite of spatial decision support system tools to undertake multi-scaled planning analysis. Current functionality allows users to perform dynamic analyses of regional demographic and employment trends to formulate thematic map and associated tabular data results. At the local urban level, users can review the results of a number of 'what-if' planning scenarios based on a multiple criteria evaluation approach. The site has been built at the regional level for the Wide Bay-Burnett region in Queensland, Australia, and within it, at the local urban level for the rapidly growing and expanding Hervey Bay local government area.

18.1 Introduction

More effective use of spatial information can assist the process of urban and regional development, planning and management. Since the 1960s, researchers in geography, economics, planning and regional science have developed a suite of analytical techniques to build spatial decisions support systems to improve our understanding of the complex relationships existing between the economic, social, environmental and land use phenomena associated with growth and decline in cities and regions. Such techniques include demographic analysis and population projections, location quotient analysis, shift-share analysis, input output analysis, multiple criteria evaluation (MCE) and visual assessment.

Developments in geographical information systems (GIS) and internet (WWW) technology allow us to view the results of urban and regional analysis in a spatially integrated environment (Stillwell *et al.* 1999, Shyy *et al.* 2000). Planners, decision-makers and members of the public can interrogate databases to perform both spatial and non-spatial queries through accessing a dedicated on-line planning support system (PSS). The Internet now permits public access to a wide variety of static and dynamic information and integration of different types of spatial data at different scales in a GIS environment using spatial modelling analytic tools and the visualisation of output of such analysis and modelling. For example, the results of a shift-share analysis of regional industry sector employment dynamics can be viewed for a whole region and its sub-regional components. Connected to the same internet site might be the result of spatially disaggregated population projections and the testing of potential urban growth scenarios to accommodate the associated urban development within a region, the latter having being formulated using techniques such as MCE or spatial optimisation modelling.

This chapter discusses the development of an internet site, incorporating a number of analytic planning tools, for Queensland's Wide Bay-Burnett Region and, within it, the rapidly growing Hervey Bay local government area (LGA), the largest urban component of the region. For illustrative purposes, the focus is on the application of a shift-share methodology for regional scale analysis of employment trends in the Wide Bay-Burnett Region and on a MCE for urban scale analysis of development in Hervey Bay. By providing access to multi-scaled analytic tools, the user can better understand the performance of a region and the implications of regional planning policy for settlement and land use patterns under an urban growth scenario. The chapter provides examples of these kinds of analytic techniques showing how they have been applied using a GIS-based PSS accessible via an internet site. It also discusses the future development of the prototype on-line PSS. The prototype system referred to is accessible at the following URL: *http://www.ahuri.uq.edu.au/wide_bay/default.htm.*

18.2 Background and Study Area

18.2.1 Purpose

A customised GIS provides a powerful information technology environment for the development of a PSS and its associated analytic tools. GIS offers an environment where a wide range of spatial data may be collected, integrated, manipulated and analysed by the user in order to obtain a better understanding of spatial patterns, trends and relationships between spatially referenced phenomena. A customised planning based GIS – generically referred to as a spatial decision support system (SDSS) (Armstrong and Densham 1990; Jankowski and Richard 1994; and Pettit 1999), or more specifically referred to as a PSS (Klosterman 1997, 1999a, 1999b; Harris and Batty 1993) – can be designed so that planners and decision-makers may (a) assign priority weightings to decision criteria; (b) evaluate the suitability of alternative scenarios and (c) visualise the results of those choices (Jankowski 1995).

Many planning decisions are made in committee boardrooms and offices to which the public has no direct access (Carver and Peckham 1999). By integrating the development of a PSS with internet technology, the opportunity exists to further open up the planning process through increased public awareness and involvement.

Placing a suite of planning tools on a public domain internet site can facilitate wider input into the planning process, not only by all decision advocates including planners and decision-makers, but also by local businesses and members of the community who ultimately will be affected by planning decisions.

18.2.2 Study Area

The Wide Bay-Burnett region is situated along the east coast of Australia in the State of Queensland (Figure 18.1). It was chosen as the case study area to develop this type of internet accessible, GIS-based PSS with its associated analytical tools.

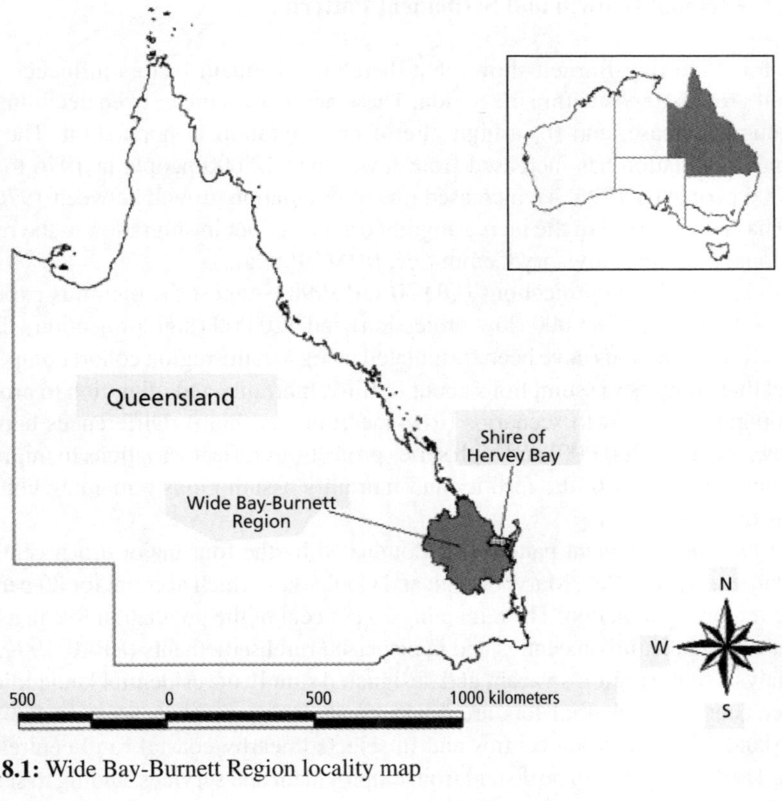

Fig 18.1: Wide Bay-Burnett Region locality map

This region was selected as the case study area for a number of reasons, primarily because it is experiencing high levels of population growth, economic growth at a lesser level, but a declining quality of life on some measures. The region is characterised by:

- considerable biodiversity (examples include: rainforest on Fraser Island; canefields in Maryborough and Bundaberg; whale watching on Hervey Bay; and beef cattle grazing in the Miriam Vale area);
- most growth being located within 50kms of the coastline, which is indicative of the population distribution trends of Queensland where 85.2 per cent of people live within 50km of the coast (QDLGP 1997); and
- availability of a large range of spatial and non-spatial data sets, including physical data layers made available by the Queensland Department of Local Government and Planning (QDLGP) and the on-line Wide Bay 2020 project resource which can be assessed on the WWW at *http://www.dcilgp.qld.gov.au/*

18.2.3 Regional Growth and Settlement Pattern

Data for Wide Bay-Burnett shows that there are two main factors influencing the overall growth trends within the region. These are: (a) a static or even declining rate of natural increase; and (b) a high rate of net migration in population. The total regional population has increased from fewer than 10,000 people in 1976 to over 200,000 people by 1996. An increased rate of population growth between 1976 and 1996 has been related to the increasing importance of net in-migration to the region (Regional Planning Advisory Committee, RPAC 1999a).

Official population projections (QDCILGP 1998) suggest the region is expected to grow to between 285,000 (low projection) and 320,000 (high projection). These population projections have been formulated using a multi-region cohort component model that combines assumptions about fertility, mortality and migration to produce three population growth scenarios (low, medium, and high). Differences between the low, medium (300,000) and high series projections reflect variations in migration assumptions only, with the fertility and mortality assumptions remaining constant across the three series.

In 1996, the settlement pattern was dominated by the four major urban centres – Bundaberg, Hervey Bay, Maryborough and Cooloola – which account for 80 per cent of the regional population. The remaining 20 per cent of the population live in a large number of smaller urban centres and in dispersed rural settlements (RPAC 1999a).

Analysis of the region's vacant and designated supply of residential land indicates 40 per cent of residential lots are being created in non-urban areas in the rural hinterland of larger urban centres and in selected nearby coastal rural centres and areas. These areas tend to be distant from employment and services, adding to service delivery difficulties (RPAC 1999). Thus, the preferred settlement pattern for future development outlined in the Regional Growth Management Framework (RGMF), the official regional planning document, proposes a greater emphasis on achieving a compact form of urban growth to minimise encroachment on the region's economic and environmental resource areas and to bring about a more effectively delivery of infrastructure services (RPAC 1999b).

18.2.4 Pressure and Challenges Facing the Region

The RGMF (RPAC 1999b) identifies a number of social, economic, and environmental pressures are impacting on the region. These include the following:

- Protecting the region's natural and cultural heritage resources from unnecessary encroachments by urban development which needs to be managed in an ecologically sustainable manner to ensure their special value and economic potential are retained for future generations.
- Efficient planning for future water needs and the sustainable management and allocation of water resources in the region on a catchment basis. This is of the highest importance due to the region's limited water supply, a major constraint to economic development and urban growth.
- Creating of new wealth generating initiatives building on the region's economic strengths and complementing population-led employment growth. This is urgently required to combat the growing gap between population growth and the growth in jobs (RPAC 1999b).

Population growth rates and unemployment rates are often seen as important indicators of socio-economic performance as indicated in the Australia: State of Environment Indicator Report (Environment Australia 1998). Analysis of population data for the region shows that over the decade 1986-1996, the Wide Bay-Burnett statistical division (SD) increased its share of national population by 0.18 per cent, making it the fifth ranked region for the greatest increase in national population share from among Australia's 58 SD regions (Stimson *et al.* 1998). But the Wide Bay-Burnett statistical division was also prominent among those SDs with the lowest ratio of total persons employed per 1,000 resident population in 1996 (309.38), ranking the forth lowest among the nation's 58 SDs. Hence, despite strong population growth, the region has been experiencing deteriorating rates of employment.

Within the Wide Bay-Burnett region, LGAs situated along the Queensland coastline are experiencing the highest rates of population growth, while unemployment rates are lowest within the inland areas but increase within LGAs situated along the coast. Thus, an inverse relationship exists between population growth and unemployment rates across the region. This represents a challenge for regional development planning.

18.3 A Prototype PSS for a Multi-scaled Planning Analysis Approach

We now turn to demonstrate how a prototype PSS has been developed for multi-scale planning analysis. The use of two spatial analytic techniques that have been incorporated into the prototype on-line PSS site for the Wide Bay-Burnett region are illustrated. The first is an employment shift-share analysis of the entire region and its LGAs over the decade 1986 to 1996. The second technique, MCE, is applied to analysis of the growth centre of Hervey Bay, one of the 21 LGAs within the Wide Bay-Burnett region.

By providing access to both spatially aggregated and disaggregated socio-economic data, land and other essential statistical information, linkages between regional and urban scales of analysis can be made. For example, an employment shift-share analysis may be undertaken for all LGAs comprising the region, or the pattern for population growth in the future can be analysed at an LGA level using the population projection data referred to earlier. LGAs shown to be experiencing considerable growth, as in the case of Hervey Bay, may then be targeted for further analysis at a land parcel level, using techniques such as MCE, to construct land suitability maps for future urban development.

18.3.1 Shift-share Analysis of the Wide Bay-Burnett Region

Shift-share analysis is a widely used analytic tool for examining regional and local differences in employment growth and decline, either in aggregate or by industry sectors. Shift-share analysis has been carried out for the Wide Bay-Burnett region to evaluate the employment performance of various industry sectors over the decade 1986 to 1996 by Stimson and Davis (1999) using the Haynes and Dinc (1997) methodology. The shift-share technique identifies three components of employment shift for a locality in a region.

- *National share*: indicates what the employment growth in a LGA for each industry would have been if it had grown as the same total employment growth rate for all industries for all of the State.
- *Industry mix*: indicates what the employment growth in a LGA for an industry would have been if it had grown at the employment growth rate for each individual industry for all of the State.
- *Regional or differential shift*: indicates the employment growth (or decline) for each industry attributable to the LGA's own natural advantages or disadvantages.

Each of these components has been evaluated based on the Queensland State average as compared to each LGA within the Wide Bay-Burnett region for the 1986-1996 timeframe and are accessible via the on-line Wide Bay-Burnett Regional Information System (WBBRIS).

For total employment with respect to the total national share, within the region it was found that areas such as Monto, Perry and Mundaberra are experiencing employment growth less than the State average. The summed industry mix for the Wide Bay-Burnett LGAs highlights the numbers of jobs that can be attributed to the industry structure of the region. A positive industry mix is found in areas such as Hervey Bay and Bundaberg. The differential (or regional) shift component of the shift-share analysis represents the number of jobs explained by an area's own advantages or disadvantages; in other words, it identifies endogenous growth. Access to the on-line PSS suite of tools allows the user to generate the results of shift-share analysis dynamically, as shown in the example in Figure 18.2, where it can be seen that only a small component of the region exerts a strong positive regional shift, (the Burnett and Hervey Bay LGAs) in total employment. Of course, disaggregated industry sector analysis shows that there are considerable variations in the performance of LGAs between industry sectors with respect to the components of the shift-share analysis.

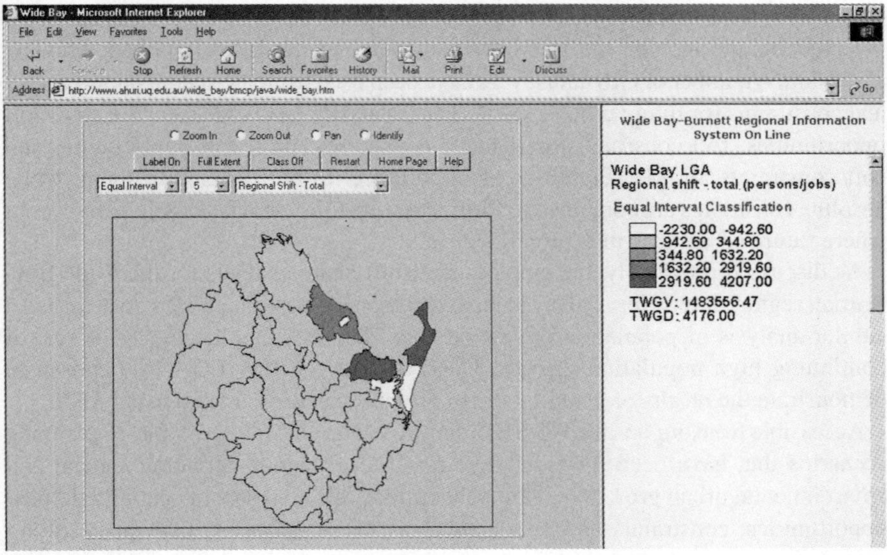

Fig 18.2: Regional shift-share effect, 1986-1996

18.3.2 MCE Analysis for Hervey Bay

In the 1970s, a methodology emerged that seemed to be well suited to support the public decision-making process. This was MCE (Nijkamp and Van Delft 1977, Nijkamp and Spronk 1981, Voogd 1983). MCE involves *"the weighting of independent criteria in terms of judged relative importance or judged relative value"*(Smith 1980, p.1). Basically MCE provides a framework for derived weighted values (Pettit and Pullar 1999). In the planning and GIS context, this means that weighted values can be assigned to spatial data layers. The specific MCE technique used here is known as weighted linear combination (WLC), which is similar in nature to regression. Using the following equation (Eastman *et al.* 1995), land use suitability may be determined:

$$S = \quad (w_i x_i) \quad c_j \qquad (1)$$

where: S refers to suitability, i is a decision factor, j is a constraint, c_j is the criterion score of constraint; w_i is the weight such that a value of 1 is important down to a value of 0; x_i is the criterion score of factor i;   is the sum; and   is the product.

WLC has been incorporated into a number of planning decision support tools developed including: the spatial understanding and decision support system (SUDSS) (Jankowski and Richard 1994); the 'What-if?' collaborative planning support system (Klosterman 1999b); the IDIRISI decision support module (Eastman *et al.* 1995) the MCE Planning Arcview extension (Pettit 1999, Pettit and Pullar 1999), and more recently, the Model Builder Arcview GIS extension released by ESRI.

GIS tools used in the formulation of land suitability maps derived using the MCE WLC technique include the ESRI Arcview 3.2 and the Model Builder Arcview extension. A number of GIS data layers have been used to create spatial opportunities and constraints for future urban growth. The input data layers are categorised as soft opportunities, soft constraints or absolute constraints. Both soft opportunities and soft constraints allow 'weightings of importance' to be assigned to them, while absolute constraints are combined to form a 'restricted' spatial layer delineating areas where future urban growth is prohibited.

As discussed previously, the employment shift-share analysis for the Wide Bay-Burnett region shows Hervey Bay to have experienced significant jobs growth, and a similar analysis of population projection data identifies this locality to be one of continuing high population growth. Thus, the Hervey Bay LGA was chosen to demonstrate the results of potential urban growth scenarios created using MCE.

Accessible from our on-line WBBRIS site are the results of three 'what-if' planning scenarios that have been formulated so far. These include economic, social and environmental urban growth scenarios that utilise different sets of spatially defined opportunities, constraints and restricted areas. Note, however, that the resulting scenarios are currently accessible via static html pages, which link future urban growth maps and supporting decision tables for each of the three scenarios. Here, only one of these possible 'what-if' planning scenarios is illustrated, namely the economic-based urban growth scenario.

18.3.3 An Economic-based Urban Growth Scenario for Hervey Bay

This scenario is based on the supposition that the most important factors pertaining to the location of future urban growth in Hervey Bay will be associated with the proximity to existing urban infrastructure, which includes water mains, sewers, roads, railways and existing urban development. These data layers were obtained through the Hervey Bay City Council (HBCC), the QDLGP and the Queensland Department of Natural Resources (QDNR). The scenario is based on an objective of achieving urban consolidation and promoting a more compact urban form that contests urban sprawl and is therefore in accordance with the preferred settlement pattern outlined in the RGMF for the region's future development and planning (RPAC 1999b).

A number of factors have been entered into the GIS and comprise the decision table shown in Table 18.1, which represents the exhaustive set of mutually exclusive conditional statements within a pre-specified problem area (Vanthienen and Dries 1994, and Witlox *et al.* 1997). The proximity factors in the decision table – for example, distance to water mains – have been generated within Arcview using the buffer command. All of the factors contained within the decision table have been converted to discrete grids themes and assigned weightings of importance. Positive values have been assigned to opportunity factors and negative values to constraint factors. The constructed decision table is accessible from the on-line WBBRIS site. This enables users to view the variables and their assigned importance with respect to the generated 'what-if' planning scenario and has the advantage of making the planning process more open for public examination.

Table 18.1: Decision table for the economic-based urban growth scenario, Hervey Bay

factors	type*	variable (* 100)	w	variable (* 100)	w	variable (* 100)	w	variable (* 100)	w	variable (* 100)	w
distance to water mains	opp	0 <x> 10	5	10 <x> 20	4	20 <x> 30	3	30 <x> 40	2	40 <x> 50	1
sewer areas	opp	within area	5								
distance to roads	opp	0 <x> 1	5	1 <x > 2	4	2 <x> 3	3	3 <x> 4	2	4 <x> 5	1
distance to railways	opp	0 <x> 10	5	10 <x > 20	4	20 <x> 30	3	30 <x> 40	2	40 <x> 50	1
distance from exciting development	opp	0 <x> 5	5	5 <x> 10	4	10 <x> 15	3	15 <x> 20	2	20 <x> 25	1
significant wetland areas	con	within area	-5								
national parks	con	within area	-10								
state forest	con	within area	-5								
distance to waste disposal site	con	0 <x> 5	-5								
distance from mine	con	0 <x> 5	-5	5 <x> 10	-4	10 <x> 15	-3	15 <x> 20	-2	20 < x > 25	-1
roads	rest	within area									
existing development	rest	within area									
coal mines	rest	within area									
waste disposal site	rest	within area									
waterways & dams	rest	within area									

* type: opp=opportunity; con=constraint; rest=restricted

Note: all distance are in metres and w = weighting

The resulting economic base urban growth map for Hervey Bay is also accessible via the on-line WBBRIS site. Figure 18.3 illustrates the map produced in the Arcview GIS, based upon the factors outlined in the decision table. The final land suitability map shows restricted areas, land least suitable for future urban development with values between -1 and -3 and land most suitable for future urban development with values between 1 and 4. It is noted that these classification ranges were derived using WLC functionality available through Model Builder where originally classes between -9 and 9 were specified by the authors. Model Builder proceeded to perform various standardisations of map overlay values in order to derive the final range of suitability classes. Figure 18.3 shows that the most suitable land for urban growth occur mainly in those areas adjacent to the existing township of Howard and Hervey

Bay City. Areas least suitable for development occur in the southwest section of the shire and the land comprising Fraser Island, a World Heritage listed area.

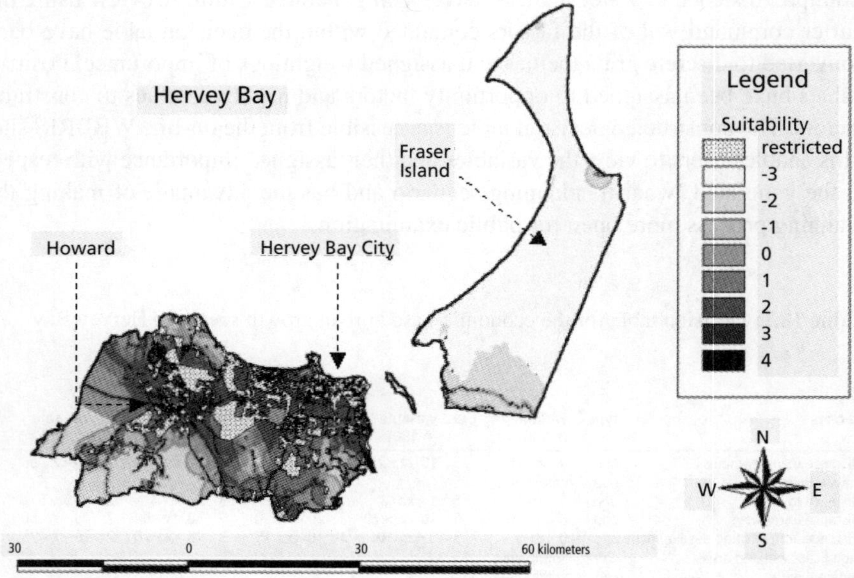

Fig 18.3: Economic-based urban growth map, Hervey Bay

18.3.4 Implications of Findings from Wide Bay-Burnett Regional and Urban Analysis

Using the prototype PSS described above, our analyses of demographic and shift-share data for Wide Bay-Burnett shows that two major industry clusters exist, tourism and agriculture. Tourism is predominant within LGAs located along the coastline including Miriam Vale, Burnett, Isis, Hervey Bay and Cooloola. Agriculture is predominant in the Bundaberg, Burnett and Isis LGAs. LGAs that are experiencing the greatest increase in population and employment are located along the coastline. The analyses conducted to locate future urban growth is in accordance with the compact development policy set down in the Wide Bay-Burnett RGMF (RPAC 1999b). This needs to be undertaken at the individual LGA level.

So far urban growth scenarios have been produced only for the Hervey Bay LGA, which is high on the priority of the Queensland Government for developing strategies to better manage growth. The use of MCE in the formulation of 'what-if' planning scenarios has yielded results based on parameters set by the authors, which show that, for all three scenarios created so far, future development should occur within close proximity to the township of Howard and City of Hervey Bay and along corridors in close proximity to existing water mains. Future urban development should be prohibited from occurring on Fraser Island, because of its World Heritage national park status. Thus the protection of the region's natural resources and efficient planning for future water needs are adhered to in these 'what-if' planning scenarios; and they are in accordance with policies outlined in the Wide Bay-Burnett RGMF (RPAC 1999b). Further urban analysis needs to be undertaken for other coastal areas experiencing significant socio-economic growth, especially for the shires of Miriam Vale, Burnett and Cooloola.

18.4 How to Use the On-line Wide Bay-Burnett Regional Information System

18.4.1 The System

The internet GIS application for the prototype on-line PSS described in this chapter is served by a DELL PowerEdge Server running ESRI's MapObjects Internet Map Server (IMS) and the Microsoft Internet Information Server (IIS). MapObjects, which contains an ActiveX map control and a set of ActiveX objects, such as data access objects and map display objects, are used in a Visual Basic programming environment that runs on a Microsoft Windows NT platform. These objects can be combined inside a Visual Basic form to develop customised end-user GIS applications. The mapping capabilities developed within the application include zoom, pan, select polygons, identify features and label features. MapObjects can present a map highlighting the areas of interest when the user clicks on those areas on the map. MapObjects IMS, which provides software components that link MapObjects mapping applications to the web server (ESRI 1998), is used to deploy applications on the Internet. The shapefile format that defines the geometry and attributes of geographically referenced features (ESRI 1999) is used to store map data.

Planners, decision-makers and members of the public can access the on-line WBBRIS using a web browser (such as Internet Explorer or Netscape Navigator). Users may then send their requests for information through the Internet to the web server. The web server interprets and distributes dynamically driven shift-share or demographic requests to the specified IMS application. The IMS application performs its tasks and sends results back to the Web server. Requests of static driven analysis are also handled through the web server, though there is no link to the IMS application. Finally the web server returns the requested results to the user's browser as shown in Figure 18.4. This is an example of a server-side application, which depends on a powerful GIS server to perform the dynamic GIS data management and processing.

A client-side application component has been built the using a platform-independent Java applet tool. The Java applet is embedded in the html and is downloaded to the client's computer in order to allow faster interactivity through the user interface. A review of the server-side and client-side approaches to internet GIS development can be found in Peng (1999).

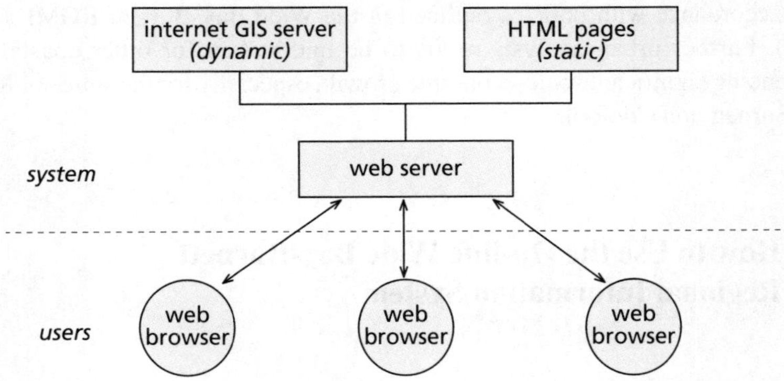

Fig 18.4: WBBRIS system architecture

18.4.2 Using the System

Once a user makes a Uniform Resource Locator (URL) request to the Internet GIS server, the WBBRIS homepage is displayed as shown in Figure 18.5. The homepage provides a locality diagram, accompanied by a series of links to a number of dynamic and static PSS tools, including the shift-share analysis (dynamic), demographic analysis (dynamic), MCE (static), visual assessment (static), location-quotient analysis (static) and input-output analysis (static). Additional links from the homepage are available to the three 'what-if' planning scenarios supporting decision tables and resulting land suitability maps. The underlying MCE methodology as applied in the Model Builder extension is accessible via the WBBRIS homepage via the *'What-if' Overlay Approach* hyperlink as shown in Figure 18.6.

The WBBRIS comprises a number of static pages, as illustrated in Figures 18.5 and 18.6, while also maintaining an interactive component. The shift-share and demographic analysis tools link to a suite of dynamic mapping functions. Options of functionality remain constant throughout the user's visit to the map server (*http://www.ahuri.uq.edu.au/wide_bay/bmcp/java/wide_bay.htm).* The first four options allow navigation around the Wide Bay-Burnett regional map, namely: Zoom In, Zoom Out and Pan (which centres the map at the point where a user clicks with the mouse and changes the map extent), and Identify (which provides summary statistics for a selected LGA). Other functionality includes LabelOn which displays the name of each LGA on the map. LabelOff turns the label off. The FullExtent option displays a full extent map. The Restart option takes the user back to the opening page and the Help option provides a description of the interactive functionality available.

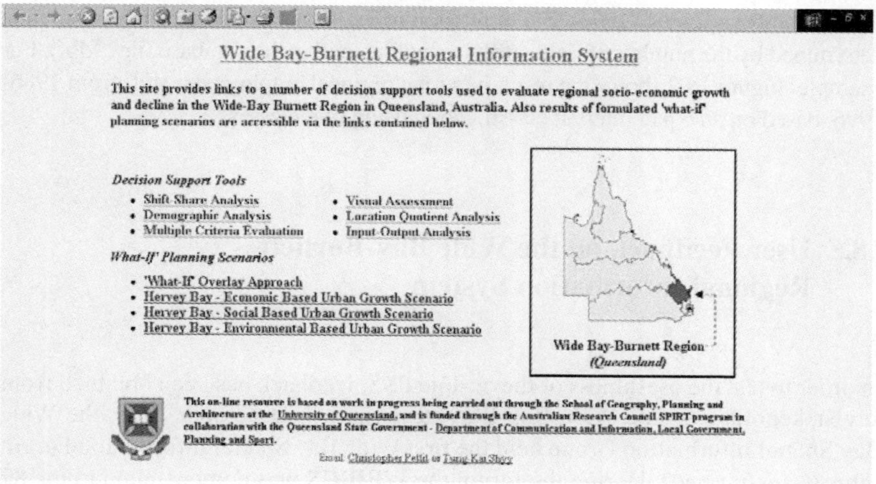

Fig 18.5: Wide Bay-Burnett regional information system homepage

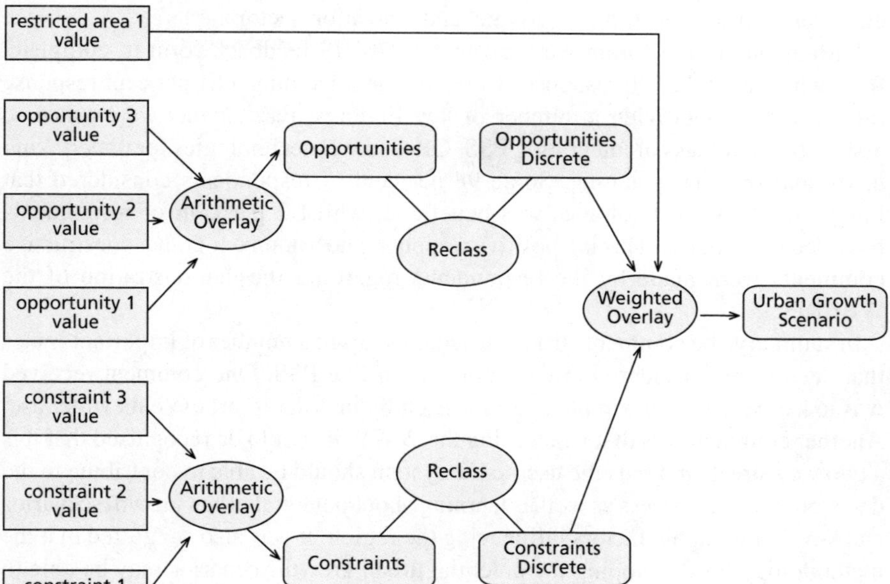

Fig 18.6: 'What-if' overlay approach web page

The remaining options available to the user are the classification tools. Once the user selects the Classification option, a number of drop-down menu items are made available. These include Method (equal interval or quantile), Number of Intervals (3-16), and the field category to be classified (for example, population 1996, unemployment 1996, or regional shift total 1986-1996). The limitation of variables available to be classified is determined by the number of field attributes maintained in the database file (*dbf). For example, Figure 18.2 shows the result of a total regional employment shift from 1986-1996, based on an equal interval classification using five classes.

18.5 User Feedback on the Wide Bay-Burnett Regional Information System

In order to test the usefulness of the on-line PSS, feedback has been obtained from key stakeholders in the Wide Bay-Burnett region. In December 2000, the Wide Bay Spatial Information Group held the first Wide Bay Spatial Information Forum (*http://widebay.net/*). During the forum the WBBRIS was demonstrated to the 82 delegates who attended. The delegates comprised urban and region planners, decision-makers and various other stakeholder from the region – including surveyors, demographers, cartographers, information and asset managers and engineers – from government, private, and education sectors.

Participants to the forum were given a WBBRIS feedback form to complete, from which a total of 50 responses were obtained, yielding a 61 per cent response rate. The results provide a number of key findings. Participants were asked to assess the usefulness of integrating PSS, GIS and web technologies for undertaking urban and regional planning. Some 98 per cent of respondents considered that integrating these technologies was beneficial, whilst 2 per cent of respondents provided no response. This is a positive response, and a number of other constructive comments were made by the respondents regarding the demonstration of the WBBRIS.

In summary, the comments from participants raised a number of important issues that need to be considered in designing an on-line PSS. One comment received was to keep the system simple and user-friendly in order to promote its wider use. Another comment was that when using the WWW it should be recognised that it is a two way street, and thus the user to the system should be able to contribute to the decision-making process as well as learning about potential urban growth scenarios and key demographic factors influencing the region. It was also suggested that the methodology for developing the potential urban growth scenarios may be able to be applied to other study areas. Overall it can be concluded from the positive feedback obtained from stakeholders within the study area that the online WBBRIS site is considered a useful tool for evaluating urban and regional planning scenarios in the Wide Bay-Burnett Region.

Consequently, the WBBRIS has been incorporated by the QDLGP as an integral component to the Wide Bay-Burnett Regional Managers Forum web site (*http://widebay.net/*). The purpose of the Forum is to provide an opportunity for information exchange and the development of joint delivery projects. The WBBRIS serves as the primary co-operative information sharing initiative development so far. Feedback from regional managers using this site has yet to be collected and analysed.

18.6 Conclusions and Directions for Future Research

This chapter highlights some of the possible uses and advantages of the on-line PSS and its analytical tools. Advantages of using an on-line PSS, such as the WBBRIS include increased transparency in the planning process and improved access to spatial analysis techniques for planners, decision-makers and members of the community. By allowing access to interactive web tools and supporting planning documentation and maps, public participation within the planning process can be improved. Through the development of a suite of on-line PSS analytic tools available from a dedicated internet site, an environment is created where multi-scale analysis may be conducted to provide input into complementary planning processes. For example, the results of modelling population projections for a region or of a regional shift-share analysis can be used as inputs to deciding where future urban growth may occur. A MCE technique may then be applied (at a micro/urban scale of analysis) to these predicted growth areas in formulating possible 'what-if' based land suitability scenarios.

The further development of the WBBRIS will include the addition of other data sets, including a regional input-output table and location quotient results. A visual assessment tool linking photographs and other images to spatial locations is currently being developed and will offer the advantage of providing spot location descriptions for landmark features existing within the region. The development of the MCE tool as an interactive component that allows users to select their own variables and weight them accordingly, in order to formulate their own 'what-if' planning scenarios is also being examined. Finally, the dynamic linking of multi-scale planning analysis tools, such as shift-share analysis and MCE is being further investigated.

Acknowledgement

This chapter is based on research undertaken by Project #9600222 funded through the Australian Research Council SPIRT Scheme, with the Queensland State Government Department of Local Government and Planning as the supporting industry partner.

References

Armstrong, M.P. and Densham, P.J. (1990) Databases organisational strategies for spatial decision support systems, *International Journal of Geographical Information Systems*, 4(1): 3-20.

Carver, S. and, Peckham, R. (1999). Using GIS on the Internet for planning, in Stillwell, J., Geertman, S. and Openshaw, S. (eds.) *Geographical Information and Planning*, Springer, Berlin, pp. 372-389.

Eastman, R.J., Jin, W., Kyem, A.K. and Toledano, J. (1995) Raster procedures for multi-criteria/multi-objective decisions, *Photogrammetric Engineering and Remote Sensing*, 61(5): 539-547.

Environment Australia (1998) *Environmental Indicators for National State of the Environment Reporting: Human Settlements*, Department of Environment, Canberra.

ESRI (1998) *Using MapObjects on the Internet,* Environmental Systems Research Institute, Redlands.

ESRI (1999) *Building Applications with MapObjects*, Environmental Systems Research Institute, Redlands.

Harris, B. and Batty, M. (1993) Locational models, geographical information, and planning support systems, *Journal of Planning Education and Research*, 12: 184-198.

Haynes, K.E. and Dinc, M. (1997) Productivity change in manufacturing regions: a multifactor/shift-share approach, *Growth and Change,* 28(Spring): 201-221.

Jankowski, P. (1995) Integrating geographical information systems and multiple criteria decision-making methods, *International Journal of Geographical Information Systems*, 9(3): 251-273.

Jankowski, P. and Richard, L. (1994) Integration of GIS-based suitability analysis and multicriteria evaluation in a spatial decision support system for route selection, *Environment and Planning B*, 21: 323-340.

Klosterman, R.E. (1997) Planning support systems: a new perspective on computer-aided planning, *Journal of Planning Education and Research*, 17: 45-54.

Klosterman, R.E. (1999a) Theme Issue: New perspectives on planning support systems, *Environment and Planning* B, 26: 317-392.

Klosterman, R.E. (1999b) The What if? collaborative planning support system, *Environment and Planning B*, 26: 393-408.

Nijkamp, P. and Spronk, J. (1981) *Multiple Criteria Analysis: Operational Methods,* Aldershot, Gower.

Nijkamp, P. and Van Delft, A. (1977) *Multi-criteria Analysis and Regional Decision-making*, Martin Nijhoff Social Sciences Division, Leiden.

Peng, Z.R. (1999) An assessment framework for the development of Internet GIS, *Environment and Planning B*, 26: 117-132.

Pettit, C. (1999) Multiple Criteria Evaluation in Planning, *Master of Regional and Town Planning Thesis*, University of Queensland.

Pettit, C. and Pullar, D. (1999) An integrated planning tool based upon multiple criteria evaluation of spatial information, *Computers, Environment and Urban Systems,* 23: 339-357.

QDCILGP (1998) *Population Projections for Queensland – 1998 Edition*, Queensland Department of Communication and Information, Local Government and Planning, Brisbane.

QDLGP (1997) *Recent Population and Housing Trends in Queensland*, Queensland Department of Local Government and Planning, Brisbane.

RPAC (1999a) *Wide Bay Regional Profile*, Wide Bay 2020 Regional Planning Advisory Committee. Bundaberg.

RPAC (1999b) *Wide Bay 2020 Regional Growth Management Framework*, Wide Bay 2020 Regional Planning Advisory Committee, Bundaberg.

Shyy, T., Davis, R., Stimson, R. and Murray, A. (2000) An Internet geographical information system for revolutionizing access to regional development research, Western Regional Science Association, Thirty-Ninth Annual Meeting, Hawaii.

Smith, P. (1980) A review of some methods for weighting criteria in the evaluation of multi-dimensional alternatives, *Planning Research Papers, No. 4*, University of Queensland, Brisbane.

Stillwell, J., Geertman, S. and Openshaw, S. (1999) *Geographical Information and Planning*, Springer-Verlag, Berlin.

Stimson, R. and Davis, R. (1999) Decade 1986-1996, Western Regional Science Association, 38th Annual Meeting, California.

Stimson, R., Shuaib, F. and O'Connor, K. (1998) *Population and Employment in Australia: Regional Hot Spots and Cold Spots, 1986-1996*, University of Queensland Press, Brisbane.

Vanthienen, J. and Dries, E. (1994) Illustration of a decision table tool for specifying and implementing knowledge based systems, *International Journal on Artificial Intelligence*, 3: 267-288.

Voogd, H. (1983) *Multicriteria Evaluation for Urban and Regional Planning*, Pion Ltd, London.

Witlox, F., Arentze, T. and Timmermans, H. (1997) Constructing and consulting fuzzy decision tables, in Timmermans, H. (ed.) *Decision Support Systems in Urban Planning*, E & Fn Spon, Melbourne.

Web Sites

WBBRIS: *http://www.ahuri.uq.edu.au/wide_bay/default.htm*

WBBRIS Java Applet: *http://www.ahuri.uq.edu.au/wide_bay/bmcp/java/wide_bay.htm*

Wide Bay 2020 Project: *http://www.dcilgp.qld.gov.au/*

Wide Bay Spatial Information Forum: *http://widebay.net/*

Wide Bay-Burnett Regional Managers Forum: *http://widebay.net/*

PART 4

SUPPORT SYSTEMS FOR LAND USE AND INFRASTRUCTURE PLANNING

19 Models for Assessing the Effects of Community Change on Land Use Patterns

Cary Gaunt[1] and Laura Jackson[2]

[1] Science Applications International Corporation (SAIC), 11251 Roger Bacon Drive, Reston, VA 20190, USA
[2] U.S. Environmental Protection Agency, Office of Research and Development B205-01 Research Triangle Park, NC 27711, USA

Abstract: The United States Environmental Protection Agency (EPA) recently developed a selective inventory and evaluation of 22 leading land use change models currently in use or under development. This chapter outlines the necessity for planning models and provides a synopsis of the computer-based land use change models that are contained in the EPA's report. General and technical details of each model are presented and the capabilities of each model in relation to potential community questions and issues are compared.

19.1 Introduction

Many communities throughout the United States and the rest of the world are struggling to balance the competing demands of economic growth, environmental sustainability and quality of life. Many places are facing tremendous growth pressures. Accompanying this boom in population and economic growth are unprecedented changes in land use. Indeed, changes in land use and land cover are considered to be some of the most significant global changes along with increasing levels of atmospheric carbon dioxide and modifications to the global nitrogen cycle (Vitousek *et al.* 1994, Agarwal *et al.* 2000). In the United States, the boom is occurring in the suburbs surrounding many metropolitan areas. For example, more than 80 per cent of new home construction in the 1990s occurred in suburban areas. Accompanying this suburban growth are significant losses in open space, in part because the amount of land consumed in new development is outpacing population growth. From 1960 to 1990, the amount of developed land in the nation's metropolitan areas (including suburbs) more than doubled, while population grew by less than 50 per cent. In some areas, the discrepancy is much higher. For example, the multistate Chesapeake Bay watershed saw population grow by 50 per cent between 1950 and 1980, while land consumption grew by 180 per cent. This kind of growth is referred to as sprawl development, a kind of unplanned growth that often strains community resources and may detract from the quality of life. The National Research Council (1998) defines sprawl as unplanned, uncontrolled, and uncoordinated single-use development that does not provide for an attractive and functional mix of uses and/or is not functionally related to surrounding land uses. Communities associate sprawl with traffic congestion, increased noise, environmental degradation, fiscal strains, a loss of special places and community character, and a host of other problems. In the last few years, sprawl has become a hot political issue.

The changing land uses associated with sprawl have the potential to contribute to significant environmental degradation. In general, there are two *direct* ways a project may affect land use: (i) by consuming open space, or (ii) by converting existing land uses to alternative ones. A project also may seed *indirect* actions that could affect how land uses change. For example, a new road may need to be built or an existing one widened; additional housing may be needed for new employees; or other businesses may need to be brought in to support the new development. The construction process for new development can disturb soils and natural hydrologic regimes, increasing flow and sediment concentrations to receiving waters. Increased loads of storm water runoff and sewer discharges can strain outdated collection and treatment systems. Natural habitats and human landscapes can be altered and fragmented, stressing local inhabitants.

Clearly, as pressures from the 'three-legged stool' of economy, community and environment become more complex to balance, communities need improved tools for decision-making. Fortunately, with the steady advances in computer technology, these tools are beginning to exist. One such tool is computer models that evaluate the consequences of population and land use changes on environmental quality, infrastructure (e.g. transportation, energy, water and sewage), and revenues and expenditures. Unfortunately, many different potential models exist and it is often difficult to identify the most useful ones, let alone understand their applicability, strengths and limitations in an easy fashion. At present, very few sources exist that summarize and evaluate potential land use change models. The United States Environmental Protection Agency (EPA) recently undertook to fill this gap by developing a selective inventory and evaluation of the leading land use change models currently in use or under development. EPA's Office of Research and Development (ORD) spear-headed this effort with support from Science Applications International Corporation (SAIC).

This chapter provides a synopsis of the EPA's efforts. The full report entitled *Projecting Land-Use Change – A Summary of Models for Assessing the Effects of Community Growth and Change on Land-Use Patterns* (EPA 2000) can be downloaded at *www.epa.gov/ecocommunity/tools/reportfinal3.pdf.* The full guide provides a summary of 22 computer modelling tools that can be used to assess the impacts of community actions and policies on land use, and the reciprocal effect of land use changes on certain community characteristics. By using the guide, the reader will have a starting point to assess a model's applicability, data and resource requirements, strengths and limitations, and costs. The guide also provides information on why there is a need for planning models, a brief overview of the types of planning models, steps on how to choose the right model for users' needs, and the essential general and technical information on 22 leading land use change models. Neither the guide nor this chapter covers all models pertaining to growth-related issues, such as transportation or environmental quality models. The discipline of modelling transportation/land use interactions is well established and several compendia already exist that document and compare leading models of this type (e.g. Southworth 1995, Miller *et al.* 1999, Parsons *et al.* 1999). As for environmental quality models, EPA is sponsoring development of a companion document to the current guide to characterize models projecting the environmental impacts that may result from land use change.

19.2 Deciding to Use Models

Trying to determine how community growth and change affect land use patterns can be very difficult. Over time, communities have employed all kinds of techniques ranging from manual cartographic techniques and mylar overlays to computer-based geographic information systems (GIS) and models. Communities must decide what approach is best for their available resources. When faced with such a situation – either as the decision-maker or one who seeks to inform or influence the decision-making – it is worth considering a five-step approach to selecting the best land use model:

- understanding the project;
- asking the right questions;
- identifying information needs;
- assessing internal capabilities; and
- choosing the right model.

This approach begins by clearly defining the whole scenario involved in the decision-making (e.g. a proposal for a new development project) and concludes with applying criteria for model selection. The appropriate model will vary from scenario to scenario and from community to community. The model needed today may be completely different from the model needed for tomorrow's project. This variation between projects requires an understanding of the diversity of models available in order to determine which is the best model to use in a given situation.

The most critical component of an effective community decision-making modelling tool is its ability to address specific analytical needs and to model alternative planning approaches. The model's competence in integrating and analyzing diverse data to quantify the benefits of a particular approach, followed by its ability to project the outcomes of that approach with the desired level of precision, are invaluable when trying to build community consensus, understand the costs and benefits of a proposed action and move new plans toward implementation.

It is important to note that while current land use change models serve as powerful tools to assist communities in making sound planning decisions, many are limited in their capabilities. Few, if any, attempt to model fully all of the various factors affecting land development patterns, and none can account for all of the factors that affect land markets and travel behaviour. Thus, no model can perfectly replicate reality. But, as technology continues to advance and land use changes continue to be of concern, more and better models will be developed to answer the specific policy questions of citizens, land use planners, and decision-makers. EPA recognizes this is a dynamic field. The guide presents a starting point, but users should be aware of continual changes and advances in the field.

19.2.1 Step 1: Understanding the Proposal

The first step in selecting a model is comprehending the proposed project. The community should be able to answer questions about the project's type, scale and fiscal implications. For example, questions might include: What is the exact nature of the project? Is the consideration a new discount chain store that wants to locate on the edge of town, a new residential development? Where exactly do they want to locate the new project? How large is it? What are its spatial boundaries? When reviewing the proposal, it is important to determine the project's entire geographic scope. The proposal must be stated as succinctly, objectively and as accurately as possible in order to begin the decision-making process. In addition, it is imperative that all involved decision-making parties have a common understanding of the proposal. It is key that all decision-makers and others involved in reviewing the project clearly understand and are in agreement about its vision and components.

19.2.2 Step 2: Asking the Right Questions

Once the proposal has been clearly defined, the project scope must be broken down into detailed pieces. At this point, essential questions must be asked in order to have a complete understanding of the full costs and benefits of the proposal. Consideration of the direct and indirect, synergistic and cumulative impacts that the project will bring to the community is needed. Each community's situation is unique, so the questions from project to project will be unique as well. Questions will cover topics associated with impacts on cost of services, potential for new revenues, other impacts on local and regional economies, effects on existing transportation networks, environmental consequences, and social impacts. Examples include: What are the potential impacts of expanding infrastructure to a new area to meet project needs? How much new traffic will be added to existing streets? What are the consequences of a revised business tax strategy? Does the community desire to preserve environmental resources? If so, where and what are the potential impacts of the project to these resources? How will the project change the character of the community?

19.2.3 Step 3: Identifying Information Needs

Once the questions are formulated in step 2, a determination should be made as to what types of data are needed to answer the questions. If one question is where to locate a proposed project to minimize impacts on sensitive environmental areas, information is needed on where those areas are. If another question is the potential consequences of a new mall on surrounding land uses, information may be needed on trends from similar projects, projections on the number of shoppers and employees, and estimates of potential daily traffic. At this stage, it may not be possible to identify all of the informational needs (some answers will come after a decision has been made as to which model to use because the model will require certain inputs), but the process of identifying how well internal capabilities can find answers to the questions should begin.

19.2.4 Step 4: Assessing Internal Capabilities

The next step in selecting a model requires a clear understanding of what internal capabilities can be accessed to acquire and use the model. This includes assessing the following:

- *Financial resources:* How much can be afforded?
- *Staff resources:* What are the extent and talent of staff available to use the tool? Does additional help or consulting expertise need to be hired?
- *Computer resources:* Do the proper hardware, software, and computing power resources exist on-site to run models?

It is important to be realistic in this assessment since a shortage of resources could result in ineffective installation and maintenance of the modeling tool, essentially rendering it useless. The community also may evaluate whether they can hire outside consulting services to fill key gaps.

19.2.5 Step 5: Choosing the Right Model (Using Selection Criteria)

Once the first four background steps have been completed, the final step is assessing and selecting the best model to meet identified needs. Before choosing, however, each option should be thoroughly analyzed against selection criteria. Thirteen suggested selection criteria are provided and explained below as guidance. They are listed in an order that follows the likely thought process of a community that is considering a range of models. This is not an all-inclusive list of selection criteria. Other criteria may be important to consider based on the particular results of steps 1 through 4. The criteria may be weighted, based on level of importance, to guide the decision-making process. Certainly, these suggested criteria can and should be modified or enhanced to meet the unique needs of the community.

- *Relevance:* Does the model provide pertinent information that meets the analytical needs of the community?
- *Resources:* Are the model and the computer requirements (hardware, software) and staff (number of people and their time) needed to support the system within the community's budget and infrastructure?
- *Model access:* Is the model easy to research, locate, and obtain?
- *Model support:* Do the model developers, or does the model itself, provide sufficient support needed to understand and implement the model (e.g. model documentation, user discussion groups, and training)?
- *Technical expertise:* Does the community have the technical expertise required to use, calibrate and interpret the results of the model?
- *Data requirements:* Does the community have, or can they obtain, the data necessary to run the model?
- *Accuracy:* Are the projections generated by the model reliable to a degree that is useful to the community?
- *Resolution:* What amount of land and what level of detail can be modelled in a single scenario?
- *Temporal capabilities:* Can the model project outcomes for multiple time periods?
- *Versatility:* Can the model project outcomes for multiple variables (i.e. land use, transportation, employment, housing, and environmental)?
- *Linkage potential:* Can the model be linked to other models currently in use by, or of interest to, the community?
- *Public accessibility:* Can the model be run in an interactive public environment and display the results in a manner that is comprehensible to the general public?
- *Transferability:* Can the model be applied to locations other than the one(s) for which it was developed?
- *Third party use:* How extensively has this model been used in 'real-world' situations?

19.3 Models in the Inventory

In this section of the chapter we present a snapshot of some of the systems contained within the inventory. Table 19.1 lists the models considered, their developers, and their internet web site addresses. Thereafter, short summaries for each model are presented. The complete guide has much more information, presented as detailed factsheets.

Table 19.1: Summary of land use change models, software, developers and web sites

California Urban Futures (CUF) Model: CUF-1 John Landis, Institute of Urban and Regional Planning, University of California at Berkeley *http://www-dcrp.ced.berkeley.edu*	MEPLAN Marcial Echenique & Partners Limited. Contact: Ian Williams *http://www.meap.co.uk*
California Urban Futures (CUF) Model: CUF-2 John Landis, Institute of Urban and Regional Planning, University of California at Berkeley *http://www-dcrp.ced.berkeley.edu*	METROSIM Alex Anas & Associates *http://www.bts.gov/tmip/papers/landuse/compendium/dvrpc_ch1.htm#1.4.3*
California Urban and Biodiversity Analysis Model (CURBA) John Landis, University of California at Berkeley *http://www-dcrp.ced.berkeley.edu*	Sub-Area Allocation Model-Improved Method (SAM-IM) (formerly LAM) Planning Technologies, LLC *http://www.all4gis.com*
DELTA (formerly DSCMODE) David Simmonds Consultancy *http://www.davidsimmonds.com/*	SLEUTH (formerly Clarke Cellular Automata) Keith C. Clarke, Department of Geography, University of California at Santa Barbara *http://www.ncgia.ucsb.edu/projects/gig*
Disaggregated Residential Allocation Model of Household Location and the Employment Allocation Model (DRAM/EMPAL) S.H. Putman and Associates, Inc. *http://dolphin.upenn.edu/~yongmin/intro.html*	Smart Growth INDEX® Criterion Planners/Engineers, Inc. (with Fehr & Peers Associates, Inc.) *http://www.crit.com*
Growth Simulation Model (GSM) Maryland Department of Planning, Baltimore, Maryland. Contact: Joe Tassone *http://www.mdp.state.md.us*	Smart Places Electric Power Research Institute (EPRI). Contact: Paul Radcliffe *http://www.smartplaces.com*
INDEX® Criterion Planners/Engineers, Inc. *http://www.crit.com*	TRANUS Modelistica *http://www.modelistica.com/tranus/*
IRPUD Model (formally Dortmund) Michael Wegener, Institute of Spatial Planning, University of Dortmund, Germany *http://irpud.raumplanung.uni-dortmund.de/irpud/index_e.htm*	UGrow Wilson W. Orr, Prescott College *http://www.prescott.edu*
Land Transformation Model (LTM) Dr. Bryan C. Pijanowski, Michigan State University *http://www.ltm.msu.edu*	UPLAN Robert Johnston, Dept. of Environmental Science and Policy, University of California at Davis (e-mail: rajohnston@ucdavis.edu) Planned website: *http://ice.ucdavis.edu*
Land-Use Change Analysis System (LUCAS) Michael W. Berry, et al., Department of Computer Sciences, University of Tennessee *http://www.cs.utk.edu/~lucas*	UrbanSim Paul Waddell, Daniel J. Evans School of Public Affairs, University of Washington *http://www.urbansim.org/*
Markov Model of Residential Vacancy Transfer Philip Emmi and Lena Magnusson *http://www.geog.utah.edu/faculty/*	What if? Dr. Richard E. Klosterman (as Community Analysis and Planning Systems, Inc) *http://www.what-if.com*

19.3.1 California Urban Futures Model (CUF-1)

This model simulates alternative development scenarios for cities and counties, based on specified policy changes at various levels of government. Units of analysis are political jurisdictions and developable land parcels. The California Urban Futures (CUF-1) model projects population growth from past trends in the study area, then allocates future development based on these projections plus the profitability of each parcel if developed, and user-specified development regulations and incentives.
Inputs: Required data include existing land use, environmental features, and jurisdictional boundaries (ArcInfo coverages) and numerical data on historic population growth by jurisdiction and on development costs per parcel.
Outputs: Vector-based maps (ArcInfo coverages) and numerical data on developed acreage.
Note: CUF-1 is limited to projecting residential development, and does not generally designate new growth for infill or redevelopment. It requires a UNIX platform, programming language compilers and SPSS statistical analysis software.

19.3.2 California Urban Futures Model, Second Generation (CUF-2)

CUF-2 was developed to address some of the theoretical holes in CUF-1. It performs many of the same functions, with the following augmentations: employment projections; more growth variables such as proximity to transportation networks and job centres; spatial bidding for sites among four types of new development and three types of redevelopment; and a raster format with one hectare resolution. CUF-2 uses econometric procedures and may be linked to a travel demand model.
Inputs: See CUF-1 (graphical data in ArcInfo grids).
Outputs: Development and redevelopment acreage by land use type (ArcInfo grids and numerical data).
Note: It requires SAS statistical analysis software.

19.3.3 California Urban and Biodiversity Analysis Model (CURBA)

This model helps the user to calibrate equations in a binomial logit model to explain historic growth in a study area, and then to apply these equations to project future development. A policy module, enabled through a user checklist, simulates the effects of various policies and constraints on urban growth and ecological habitat quality. A study area typically comprises one or more counties, with a minimum mapping unit of one hectare.
Inputs: Spatial data may include current land uses, slope and elevation, transportation networks, jurisdictional boundaries, water bodies, wetlands and wildlife habitats. Exogenous population or household growth projections are required.

Outputs: ArcView shapefiles (raster and vector), graphs and tables.
Note: CURBA does not differentiate among urban land use categories. The model runs in ArcView using Avenue scripts and requires SAS or SPSS statistical analysis software.

19.3.4 DELTA

This model projects change in multiple urban land uses, including specific household and employment types, at city and regional scales. Projections may be driven by economic, demographic and policy scenarios, and can also incorporate transportation and environmental variables when linked with external models. The user must specify the scenarios to be modelled, and provide coefficients to describe agent behavior such as sensitivity to rent.
Inputs: These are user defined. Examples include household location by type, car ownership by household, job location by sector, amount of floorspace by category and rent by location.
Outputs: Projections may be imported into maps, databases and spreadsheets.
Note: It requires developer consulting services.

19.3.5 DRAM/EMPAL

This model projects the interactions and distribution of employment, housing and transportation networks based on past patterns, projected growth or decline, and various growth constraining and attracting scenarios.
Inputs: Users must provide information (Excel or .dbf files) at two scales: citywide and zonal. Examples include total population, person-trips by purpose, commuting rate and projected total employment by sector (city wide); and households by type, employment by sector, total land area, and zone-to-zone travel times and costs (zonal).
Outputs: Examples include employment projections by sector by zone, household projections by income by zone, and projected land consumption by zone (Excel or .dbf files and ArcView shapefiles), as well as statistical results.
Note: DRAM/EMPAL is currently the most widely used housing and employment forecasting application worldwide. Purchase costs range from $30,000 to $60,000, which include consulting services, maintenance and training.

19.3.6 Growth Simulation Model (GSM)

This model uses exogenous population, household and employment projections to estimate demand for residential and commercial development. Demand is then allocated based on zoning and other development regulations, conservation ordinances, and recent, locally-specific development patterns and trends that reflect the market for land. GSM can operate at various levels of resolution and extent.

Input: In addition to land use and management areas, spatial data may include soils, sewer service and watershed boundaries, stream networks and other environmentally sensitive areas (ArcInfo coverages and relational databases).
Outputs: Projections are in the form of ArcInfo maps, relational database files, spreadsheets, graphs and reports.
Note: It requires Windows NT or Unix platform and relational database software and skills.

19.3.7 INDEX

Developed to visualize exogenous growth alternatives, this tool can map development from the parcel to the regional scale based on population projections and policy constraints and incentives. INDEX can also simulate interactions between land use and transportation if run in tandem with a four-step travel demand model. Growth scenarios may be scored as to expected environmental performance.
Inputs: These are user defined. Examples include current land uses, zoning, housing and employment units by type, street centrelines, transit routes and stops and environmental features (ArcView shapefiles). Users can also specify coefficients of environmental impacts such as rates of water consumption and waste generation by housing type, and greenhouse gas emissions by building type.
Outputs: Example maps (ArcView shapefiles) and database results are residential and employment densities, street connectivity, availability of open space, imperviousness, air pollutant emissions and residential water consumption.
Note: Purchase costs range from $15,000 to $75,000.

19.3.8 IRPUD

IRPUD is an integrated land use/transportation modelling system that evaluates supply and demand of land and labor, industry profitability, worker preferences, and policies at citywide and localized scales to project future land use. Consisting of six sub-models, IRPUD is a complex model that requires significant modelling and computing expertise.
Inputs: Users must provide information (text files) in four categories: model parameters, and regional, zonal and network data. Examples include employment and immigration forecasts (regional data) and current population, employment, land uses and rents (zonal data).
Outputs: Example results include the distribution of households and employment by type, mean travel times and costs, daily trips by mode, transport expenses per household and CO_2 emissions per capita by mode (text and WordPerfect graphics files).
Note: This software is available only in conjunction with developer consulting services. It requires Windows NT, a programming language compiler, GIS and statistical analysis software.

19.3.9 Land Transformation Model (LTM)

In addition to estimating land transition probabilities from alternative socioeconomic and policy scenarios, LTM describes the resulting effects on ecological condition and economic sustainability. The model contains six interacting modules and is also linked to an artificial neural network package that learns how past changes in land use are driven by various social, political and environmental factors. LTM is designed for watershed-scale analysis and requires significant modelling and computing expertise.
Inputs: Minimum requirements include current and past land uses, elevation and population (ArcInfo grids), and road and hydrologic networks (ArcInfo coverages).
Outputs: Transition probability maps (ArcInfo grids) and data summaries (Excel).
Note: Residential land use is the only urban category addressed. It requires Windows NT or Sun Solaris, and additional software including S-Plus and SAS, Stuttgart Neural Network Simulator and C compiler.

19.3.10 Land Use Change Analysis System (LUCAS)

This model projects the conversion of land in natural resource usage to residential and commercial uses, based on parcel ownership, slope and elevation; and proximity to transportation networks and population centres. LUCAS is intended as a landscape and watershed analysis tool for natural resource managers and researchers.
Inputs: Geographic features must be provided as raster files (GRASS grids).
Outputs: Examples include size and shape of habitat patches, amount of edge, edge-to-area ratio and water-quality impacts from development. These data are provided in map (GRASS grids) and statistical (SAS) formats.
Note: It requires a UNIX platform.

19.3.11 Markov Model of Residential Vacancy Transfer

This model permits the exploration of changing supply and demand for various types of residential housing as demographic subpopulations transition through a community over time. It can be used to help rezone residential areas and mitigate areas of potential future decline based on expected population changes. The model is a mathematical formula based on linear algebra and ordinary differential equations, and requires the use of statistical analysis software.
Inputs: Residential addresses and occupant profiles by housing sector.
Outputs: Household moves between sectors across a projection period and probabilities of residential mobility by household type.
Note: The model addresses only residential land uses.

19.3.12 MEPLAN

MEPLAN is an integrated land use/transportation modelling package that considers supply and demand in both land and transit to project the effects of planning decisions on where people will choose to live and work and how their transit needs will develop. MEPLAN is highly flexible, with many user-specified parameters and requires significant modelling expertise.
Inputs: Examples of useful data include current land use, floor space, real-estate prices, transportation network, household car ownership and trip types.
Outputs: MEPLAN can produce projections of employment by sector, population by income, households by car ownership, land area and floorspace by activity, and price by floorspace and land type, among other parameters.
Note: It requires Windows NT, MapInfo GIS software, and training by a MEPLAN consultant; training costs are in addition to the $25,000 purchase price.

19.3.13 METROSIM

This model projects changes in land use and travel flow across a metropolitan area, based on alternative policy, employment and income scenarios. Cost-benefit analyses can also be produced. An economic model, METROSIM will also benefit from importing a regional input/output model and land and property values by zone and land use type, if available.
Inputs: The required data include the U.S. Census Bureau's transportation planning package (urban elements 1, 2, and 3), the transportation network's link-node description and land uses of interest by zone.
Outputs: Examples include industry, residential, vacant and other real-estate distribution by zone and type; rents and market prices by zone and type; vacancy rates; and travel assignment on the network.
Note: Purchase price ranges from $20,000 to $30,000, plus maintenance and training costs. The model does not currently have a mapping component, although the developer can create one for a fee. A UNIX platform is recommended.

19.3.14 SAM-IM

The Sub-Area Allocation Model-Improved Method (SAM-IM) is a powerful and highly flexible planning tool that spatially allocates growth according to user-specified scenarios of population growth, land value, probability of conversion, and land use plans. In addition to projecting future land use patterns, SAM-IM allows users to modify ArcView land use layers using on-screen graphical editing and drawing tools.

The software also produces input datasets for travel demand models, and can compare scenarios against user-specified performance indicators.
Inputs: The required data are existing and proposed land use layers (ArcView shapefiles), employment sectors, and housing and employment densities by land use category (text files). Optional data include environmentally sensitive and other growth-constrained areas.
Outputs: Distribution of future land uses (in a variety of formats).
Note: This software is available only in conjunction with developer consulting services. Total cost may range from $30,000 to $100,000.

19.3.15 SLEUTH

This model simulates the transition from non-urban to urban land uses by representing the project area as a grid of cells, each of whose land use state is dependent upon local factors (e.g. slope and proximity to roads and existing urban cells), temporal factors (past trends influence future growth), and random factors. It operates by Monte Carlo simulation and is typically used for metropolitan area analyses.
Inputs: Current and historic land use and transportation network, exclusion areas and slope (.gif files).
Outputs: Cell transition probabilities for a given year, cumulative probabilities of urban conversion over time (.gif files), animated imagery of growth and statistical tables.
Note: It requires UNIX operating system and gnu C compiler. SLEUTH is currently being adapted to address national and global land use change.

19.3.16 Smart Growth INDEX

Described by its developers as a sketch-level planning tool, this software helps communities evaluate alternative land use and transportation scenarios by scoring them with environmental indicators. It is capable of operating in two modes – projections over time at neighbourhood resolution (driven by exogenous population forecasts and user-selected policies), and snapshots at a single point in time at parcel resolution (driven by a pre-defined land use plan). Smart Growth INDEX contains a travel demand sub-model, and can also link with external travel models.
Inputs: Required data are planned land uses by category, existing housing by type, existing employment by number of jobs and a transportation network (ArcView shapefiles). Optional data include features that constrain and attract development, traffic volume by segment, existing and planned infrastructure and traffic analysis zones.
Outputs: Housing and employment density, proximity to transit, vehicle miles travelled, energy and water consumption, and greenhouse gas emissions are among the range of possible output variables.

19.3.17 Smart Places

This planning tool allows users to design development alternatives and evaluate their environmental and economic impacts. It runs as an ArcView extension and accommodates skill levels from beginner to advanced. Highly flexible, it receives user-specified and user-attributed data layers for map overlays and analyses across all or parts of an urban study area. Users must customize the land-use change and impact algorithms, which may be written in Avenue, C++, and Visual Basic, among other languages.
Inputs: Examples include current land use, zoning, transportation networks, infrastructure plans, construction plans, environmental features and aerial photographs. Depending on desired analyses, feature parameters may include dwellings per acre, persons per dwelling, and solid waste generated or electricity consumed per building per month.
Outputs: Maps (ArcView shapefiles) and projections of impacts such as units of incinerated and recycled waste, energy consumption, and emissions.

19.3.18 TRANUS

TRANUS is an integrated land use/transportation model designed for professional planners. It simulates human behaviours in response to a broad range of policies, and projects the consequent changes in locations and interactions of transportation and land use, and in the real-estate market, at urban, regional and national scales. A powerful and sophisticated model, TRANUS also provides indicators that quantify the benefits of alternative scenarios.
Inputs: Basic input data include network nodes, links and routes; and activities and land uses by zone. Users may also specify attractor functions such as environmental quality, neighborhood quality, crime rates, slopes and floodplains.
Outputs: In addition to activity location and land consumption by category, TRANUS projects numerous transportation variables by mode including travel paths, transit demand and capacity, total and average travel times, distances, and costs, and transit revenues and maintenance costs. Numeric results are provided as tab-delimited files, and graphical results are displayed in vector format by TRANUS software.

19.3.19 UGrow

This system is a suite of models that spatially allocates housing for the projected urban population, based on various zoning and land preservation policies. Working with the developer and stakeholder groups, users define and then adjust the system's settings for ten policy categories that encourage or discourage efficiencies in, for example, housing density, energy consumption, transportation and business activity.

Inputs: Current land use, housing and economic data, environmentally sensitive areas and vehicle use.
Outputs: GIS maps, 3-D 'fly-through' visualizations, and tables and graphs of projected variables including quality of life, energy use and climate change impacts.
Note: While the UGrow software is free, the developer must adapt the model to a particular community at a cost ranging from $30,000 to $200,000. Also it requires ArcInfo, ArcView and Powersim software.

19.3.20 UPLAN

This is another highly flexible visualization tool that runs in the ArcView environment. Users evaluate alternative development scenarios based on adaptable, raster-based data layers. UPLAN permits users to generate and assign weights to existing land uses, current land use and infrastructure plans, development exclusion areas, and attraction areas such as freeway ramps. Weights influence future growth, and can represent expected population changes, employment projections and market forces such as subsidies and taxes.
Inputs: Additional data may include persons per household, housing and employment densities per land use category, average parcel size per density class, transportation networks and environmental features.
Outputs: Raster-based maps and reports, as well as spreadsheets for data exchange with other models.
Note: Users must program the model to their needs in Avenue and can create optional additional grids in ArcInfo.

19.3.21 UrbanSim

This model simulates interactions among land use, transportation and public policy to project the effects of alternative scenarios on land use change. It is designed to interact with four-step and activity-based travel models. Projections arise in part from analysis of real-estate supply and demand; model calibration requires the use of econometric software.
Inputs: The required spatial data include parcels, travel zones, policy boundaries and environmentally sensitive areas (ArcView shapefiles); tabular data include business information, Census household data and travel impedance from travel models (ASCII files).
Outputs: Results may be reported by parcel, grid cell, or zone and include land use acreage and price by category; businesses and employment by sector, housing units and square footage by type and household characteristics (ASCII files).
Note: Future versions of the model will address environmental impacts such as water demand and nutrient loads.

19.3.22 What if?

A self-contained visualization tool, What if? accepts user-defined spatial data, growth rules and parameters to map land-use allocation alternatives. Users operate the system primarily through check boxes, text boxes, and standard Windows buttons.
Inputs: Options include existing land uses, current land use and infrastructure plans, environmental features, population and employment projections, and user-defined alternative development scenarios (ArcView shapefiles).
Outputs: Maps of development suitability and allocation, and analysis reports.

19.4 Model Comparisons

Tables 19.2 and 19.3 provide a quick way to determine if a model will answer essential questions. Table 19.2 states whether a model assesses the impacts of community decisions on land use patterns, such as impacts of tax code changes on development. Table 19.3 addresses whether a model explores how land use changes might affect community characteristics. An example is whether land use change will affect demand on infrastructure or open space.

19.5 Methods of Report Compilation

The EPA's first step in developing the full report on land use change models was to convene a workgroup comprising academic modellers and federal staff who serve as community liaisons. This workgroup framed the report's content and format according to perceived user needs. EPA then contracted with Science Applications International Corporation (SAIC) to collect appropriate information and compile the findings. Using the workgroup's list of 14 model features for characterization, SAIC developed a standard worksheet to examine identified models. Model features researched by SAIC included the following:

- model name;
- contact information for the model developer;
- description;
- resource requirements;
- theoretical framework;
- data inputs;

- model outputs;
- strengths;
- limitations;
- required expertise;
- geographic and temporal scale;
- transferability to multiple locations;
- capability of linking with other models; and
- gaps that need to be addressed.

Table 19.2: Impacts of community decisions on land use patterns: comparative matrix

model name	transportation infrastructure	local zoning	city and county master plans	other local fiscal policies	developer impact fees	property taxes	municipal sewer and water fees	subsidies	road tolls	parking fees	fuel and sales taxes	VMT[1]	registration fees
CUF-1		✓	✓		✓		✓						
CUF-2	✓	✓	✓		✓		✓						
CURBA	✓	✓	✓										
DELTA	✓	✓	✓						✓[2]	✓[2]	✓[2]	✓[2]	✓[2]
DRAM/EMPAL[3]	✓	✓	✓	✓	✓	✓	✓	✓	✓	✓	✓	✓	✓
GSM	✓[4]	✓	✓	✓[5]									
INDEX		✓	✓										
IRPUD										✓	✓	✓	✓
LTM	✓		✓										
LUCAS	✓	✓	✓										
Markov		✓	✓										
MEPLAN	✓	✓	✓	✓	✓	✓	✓	✓	✓	✓	✓	✓	✓
METRO-SIM	✓	✓	✓	✓	✓	✓	✓	✓	✓	✓	✓	✓	✓
SAM-IM	✓	✓	✓										
SLEUTH	✓	✓	✓	✓									
Smart Growth INDEX	✓	✓	✓										
Smart Places[6]	✓	✓	✓	✓	✓	✓	✓	✓	✓	✓	✓	✓	✓
TRANUS	✓	✓	✓	✓	✓	✓	✓	✓	✓	✓	✓	✓	✓
UGrow													
UPLAN	✓	✓	✓		✓	✓	✓	✓		✓			
Urban-Sim	✓	✓	✓	✓					✓[7]	✓[7]	✓[7]	✓[7]	✓[7]
What if?	✓	✓	✓										

1 Vehicle miles travelled.
2 Yes, but only if these can be modelled in an associated transport model integrated with DELTA.
3 Any may be addressed by DRAM/EMPAL when linked to the right model. Without linking, most can not be.
4 Under development.
5 No fiscal policies are pre-set in the model. However, if the user can provide specifications on the impact of the revenue source, then the policy can be incorporated.
6 Smart Places can be customized to evaluate the impact of changes in land-use patterns based on user-supplied criteria.
7 Included only through interaction with travel models.

Table 19.3: Impacts of land use patterns on community characteristics: comparative matrix

model name	travel demand	changes in infrastructure costs	changes in local tax revenue	other fiscal impacts	open space	nutrient loading	increases in storm water runoff	other nonpoint source water pollution	other water quality impacts	changes in criteria pollutants	changes in green-house gasses	other air quality impacts
CUF-1		✓			✓	✓[1]	✓[1]	✓[1]	✓[1]	✓[1]	✓[1]	✓[1]
CUF-2					✓	✓[1]	✓[1]	✓[1]	✓[1]	✓[1]	✓[1]	✓[1]
CURBA						✓[1]	✓[1]	✓[1]	✓[1]	✓[1]	✓[1]	✓[1]
DELTA	✓[2]									✓[2]	✓[2]	✓[2]
DRAM/ EMPAL	✓	✓	✓	✓	✓	✓	✓			✓	✓	
GSM	✓[3]				✓	✓	✓		✓			
INDEX						✓	✓			✓	✓	
IRPUD	✓T				✓							✓[4]
LTM									✓			
LUCAS					✓							
Markov		✓	✓	✓								
MEPLAN	✓	✓	✓	✓	✓[5]					✓	✓	
METROSIM[6]	✓	✓	✓	✓	✓					✓	✓	
SAM-IM	✓				✓							
SLEUTH			✓[7]	✓[7]	✓[7]	✓[7]	✓[7]	✓[7]	✓[7]	✓[7]	✓[7]	✓[7]
Smart Growth INDEX	✓				✓	✓	✓		✓	✓		
Smart Places[8]	✓	✓	✓	✓	✓	✓	✓	✓	✓	✓	✓	✓
TRANUS	✓	✓	✓	✓	✓					✓	✓	✓
UGrow	✓				✓						✓	
UPLAN	✓[9]	✓	✓	✓	✓	✓[10]		T[10]		✓	✓	
UrbanSim	✓	✓	✓	✓	✓	✓						
What if?					✓	✓[1]	✓[1]	✓[1]	✓[1]	✓[1]	✓[1]	✓[1]

1 Model does not directly address these issues. However, model results may be applicable as inputs into appropriate impact models to determine effects of urbanization and land use change on other systems.
2 When DELTA is integrated with transport and environmental models.
3 Currently under development.
4 The IRPUD can forecast CO_2 emissions as a function of forecasting transportation-related indicators. Environmental submodels that calculate traffic noise and air pollution indicators are under development.
5 Open space is addressed in MEPLAN only when linked to the right model.
6 Effects on air and water pollution can be treated if METROSIM is interfaced with any add-on environmental package.
7 Any fiscal or environmental impact which can be estimated as a function of urbanized area could be developed for the output of SLEUTH, but the model does not do so directly.
8 Smart Places can be customized to evaluate these impacts based on user-specified criteria.
9 UPLAN does not address travel demand directly but can when linked to any travel model.
10 UPLAN is under development to address nutrient loading or sedimentation in surface waters and other nonpoint pollution.

SAIC conducted independent research on these factors using journal publications, internet web sites, user's guides, and demonstration models. Researchers used information collected from these sources to complete a summary worksheet for each model. These worksheets formed the basis for two narrative factsheets (general and technical) per model and eleven comparative matrices across models. One limitation to this approach was that SAIC was not able to obtain and use copies of all the models. Some provided demonstration copies, while others did not (e.g. the model was available only through the consulting services of the model developers). While this prevented actual hands-on use, SAIC approached the project from the aspect of a community – what could a community find out about a model? What kind of information was available to help communities make decisions about models? The research team found it was often difficult to obtain key information or contact the model developer. This could be a deterrent to many communities seeking models for themselves.

Upon completion of this research and factsheet preparation, SAIC provided the worksheets and factsheets to the model developers, or appropriate designated colleagues, for verification of technical accuracy and supplemental information. All information on models included in the report was subject to this review process and was approved by the model developers. Once again, SAIC and EPA recognize this potential limitation to the study in that independent, objective reviewers were not consulted. Unfortunately, project constraints prohibited that approach. The SAIC/EPA research team compensated by providing information on how to learn more about each model. This included additional references, information on how to obtain preview copies of the model if possible, information on case studies (including contacts), and the number and location of sites where the model had been applied. With this information, users considering a model would get a sense of whether it had been used a great deal or not very much at all, and could learn more from another source if desired.

19.6 General Conclusion and Next Steps

One challenge in completing this study was obtaining timely information on the models in a quick and efficient fashion. The frustrations facing the research team would likely effect communities as well. Many of these models were developed for specific regions or clients (even though they are applicable to broader areas), so mass marketing was not a consideration. Therefore, the model developers did not make efforts to advertise or provide background materials on the models. For most of the models, it was next to impossible to obtain a demonstration copy or some hands-on tool to enable potential users to evaluate it for themselves. Most of the models were not developed or envisioned as commercial 'off-the-shelf' products. Thus, communities may have a hard time finding out about them. Furthermore, many of theses models are not intended as software packages to be easily 'plugged in' to the computer. Most do not exist in that form and thus require the consulting services of the model developer. Some may be available as a software product, but a fairly sophisticated end user is expected. In other words, at this stage, land use change models are not a particularly user-friendly product either to locate or to employ. It is not like purchasing the latest copy of a standard PC operating system.

These models are just now beginning to integrate with geographic information systems (GIS) and other models. Thus, it is very difficult to create a truly integrated system of linked models and GIS to enable comprehensive decision-making. The day of a complete PSS is still fairly far off. Indeed, the day of a truly user-friendly model that is robust enough to sufficiently address varied community needs is still in the distance. A community can gain some information from modelling, but it will need to be prepared to supplement modelling results with additional investigations. Many improvements and advances are being made in this dynamic field and model developers should be applauded for their efforts. There needs to be continued support and focus on developing PSS that can integrate population, transportation, land use, economic and environmental considerations. Indeed, communities are longing for a comprehensive community decision-making tool.

Acknowledgements

Many collaborators participated in the development of the EPA report summarized in this chapter. The authors particularly appreciate the research and writing contribution of Mary O'Kicki and Christine Garrow of Science Applications International Corporation. We also wish to highlight the efforts of the following people in framing the contents of the report: Bob Johnston of the University of California at Davis, Dick Klosterman of the University of Akron and Jim Kreissl of US EPA.

References

Agarwal, C.,Green, G.L., Grove, M., Evans, T. and Schweik, C. (2000) *A Review and Assessment of Land-Use Change Models – Dynamics of Space, Time, and Human Choice*, Center for the Study of Institutions, Bloomington, Indiana.

Environmental Protection Agency (2000) *Projecting Land-use Change – A Summary of Models for Assessing the Effects of Community Growth and Change on Land-use Patterns*, EPA.

Miller, E., Kriger, D. and Hunt, J. (1999) Integrated urban models for simulation of transit and land-use policies, *Transit Cooperative Research Program Report 48*, National Academy Press, Washington, DC.

National Research Council (1998) The Costs of Sprawl Revisited, Transportation Research Board, *Transit Cooperative Research Program Report 39*, National Academy Press, Washington, DC.

Parsons Brinckerhoff Quade and Douglas, Inc. (1999) Land Use Impacts of Transportation: A Guidebook, *National Cooperative Highway Research Program Report 423A*, National Academy Press,Washington DC.

Southworth, F. (1995) A Technical Review of Urban Land Use-Transportation Models as Tools for Evaluating Vehicle Travel Reduction Strategies, *Oak Ridge National Laboratory Report 6881*, U.S. Department of Energy.

Vitousek, P.M., H.A. Mooney, J., Lubchenco, and Melillo, J.M. (1997) Human domination of earth's ecosystems, *Science*, 277:494-499.

Web Sites

Environmental Protection Agency Report on Project Land-use Change:
http: www.epa.gov/ecocommunity/tools/reportfinal3.pdf
CUF 1: *http://www-dcrp.ced.berkeley.edu*
CUF2: *http://www-dcrp.ced.berkeley.edu*
CURBA: *http://www-dcrp.ced.berkeley.edu*
DELTA: *http://www.davidsimmonds.com*
DRAM/EMPAL: *http://dolphin.upenn.edu/~yongmin/intro.html*
GSM: *http://mdp.state.md.us*
INDEX: *http://www.crit.com*
IRPUD: *http://irpud.raumplanung.uni-dortmund.de/irpud/index_e.htm*
LTM: *http://ltm.msu.edu*
LUCAS: *http://www.cs.utk.edu/~lucas*
Markov Model: *http://www.geog.utah.edu/faculty/*
MEPLAN: *http://www.meap.co.uk*
METROSIM: *http://www.meap.co.uk*
SAM-IM: *http://www.all4gis.com*
SLEUTH: *http://www.ncgia.ucsb.edu/projects/gig*
Smart Growth: *http://www.crit.com*
Smart Places: *http://www.smartplaces.com*
TRANUS: *http://modelistica.com/tranus/*
UGrow: *http://www.prescott.edu*
UPLABN Planned Website: *http://ice.ucdavis.edu*
UrbanSim: *http://www.urbansim.org/*
What if?: *http://www.what-if.com*

20 The New Jersey (USA) Growth Allocation Model: Development, Evaluation and Extension

James Reilly

New Jersey Office of State Planning, 33 West State Street, PO 204, Trenton, NJ 08625, USA

Abstract: A land use planning support system (PSS) developed by the New Jersey Office of State Planning is described along with on-going and proposed revisions to the application. Current uses of the PSS include both policy testing for state level capital facility and regulatory agencies and the development of regional (countywide) land use master plans. Acceptance of the model has been greatly facilitated by its independent verification and validation by the Rutgers University Center for Urban Policy Research. Knowledge-based and other complexity-related issues related to the model are discussed since these problems are likely to be common to all similar computer applications. To facilitate the use and political acceptance of decision support applications, the creation of an international organization is proposed and the roles such a group might provide are recommended.

20.1 Introduction

The New Jersey Office of State Planning (NJOSP) has developed a computer program to facilitate regional and local land use planning. Given any exogenous, statewide forecast of population and employment, the Growth Allocation Model (*GAMe*) first produces a historically-driven forecast of growth for each of the State's 566 municipalities. These municipal forecasts are then reviewed and calibrated by the model user to reflect new growth patterns, new development approvals, recently enacted zoning changes or changes to the municipal political environment to encourage or discourage development.

The resulting calibrated 'trend forecast' is not considered to be deterministic (absolutely correct), but is thought to represent a realistic forecast of relative municipal growth within the county or region. It demonstrates what and where growth is likely to occur if existing and known development trends continue (and the exogenous regional growth forecast is reasonably accurate). The purpose of the 'trend forecast' is to specify a municipal forecast which the user feels is reasonable, and against which alternative 'plan' results can be compared.

The computer program then allows users to quickly prepare an alternative 'plan' development scenario by allowing the user to select and refine policies found in the New Jersey State Development and Redevelopment Plan (NJSDRP) (New Jersey State Planning Commission 2001), the official land use policy document intended to guide growth in the State. *GAMe* contains enough NJSDRP variables to allow the user to construct a very large number of distinct plan scenarios. *GAMe* takes the user-selected policies and uses them as assignment rules to produce a new set of 566 municipal (plan) growth estimates (for the same exogenous statewide forecast). Additionally, the application forecasts the following municipal-specific variables for each 'trend' and 'plan' scenario: mean household income, the number of households in each of eight income categories, average household size and the distribution of households by six household size categories.

GAMe is referred to as a planning support system (PSS), which is a land use-specific type of a more general model form termed either decision support systems (DSS) or integrated assessment models (IAM), since *GAMe* also forecasts various impacts associated with each 'trend' or 'plan' scenario. The program projects the following impacts: capital costs associated with public school buildings (kindergarten through secondary); costs for public sewer systems and local roads; the amount of land developed and left for further development; and the operational costs to local and state government in the forecast year. Using *GAMe*, the model user can learn if various land use policies achieve their growth management objectives and learn some of the associated cost impacts of the resulting land use pattern.

Unlike many types of computer programs that are designed to produce a correct or optimized answer, *GAMe* does not produce a recommended or 'best' land use plan. It is a tool that allows users to explore various land use policies in light of both the existing conditions in their town or region according to the values and goals of the model user. The tool simply takes user-selected policy choices and forecasts how these policies would affect the location of growth (or decline) and the costs associated with the resulting land use pattern. The program does not think, nor does it make judgments or recommendations. It does not know, for example, if an increase in local government costs might be an acceptable outcome of a policy to enhance natural resources or if a policy to preserve a particular resource, which results in a shift of growth to other areas, produces an overall desirable land use pattern.

The model user is the thinking part of *GAMe.* It is up to the user to decide how successfully any development scenario might be. Like all PSS, the application is designed for iterative use, which means that once a 'trend' and a 'plan' are run, the user evaluates the impacts of each. If the user is unsatisfied with the performance of a 'plan' scenario, the user can develop a new scenario by selecting other policies from those included within the application or by modifying some of the development criterion (such as residential density) from the recommended range found in the NJSDRP. Much like learning the rules of a game by playing it over and over, users can learn if, and how, various land use policies impact their town and region. Through this iterative process of trial, evaluation and policy modification, the user discovers a preferred set of land use development policies which produce a land development pattern (with its associated costs) which are acceptable and compatible to the goals and values of the person (or group) running the DSS program. This iterative process is illustrated in Figure 20.1.

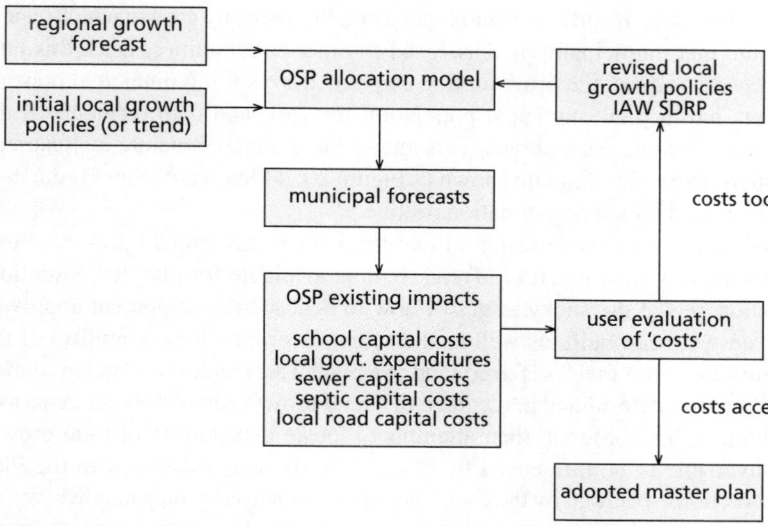

Fig 20.1: Schematic of the NJOSP DSS iterative process

GAMe is designed to be used by a citizen group, such as a local planning board, who has the fiduciary responsibility to make political/land use judgments. *GAMe* is a tool to inform the thoughtful deliberations of such groups and to assist the land use professionals who support the deliberations of such groups. The program can also be used by various interest groups, such as conservation organizations, and by government regulatory agencies to simulate the effects of proposed policies that regulate land uses.

20.2 Model Structure

Although the program is easy to use and quick to run, it has a complicated structure. Initially, the program converts exogenous population forecasts into forecasts of housing need. It then makes a preliminary assignment of the change in housing (growth or decline compared to the base year housing inventory) to the State's 566 municipalities using a mathematical model (Reilly 1997a). The exogenous statewide forecasts of population and job (by type) and the resulting preliminary municipal housing assignments are used to produce a preliminary forecast of future incomes, including mean municipal income. Next, the program makes a preliminary assignment of job growth or decline based on another statistical model (Reilly 1997b). Finally, in Trend, these preliminary assignments of housing and employment are tested to determine if sufficient land is available in each municipality to accommodate the preliminary forecasts. If sufficient land is not available, then any growth which cannot be fitted into that municipality is reassigned to other communities. Once this land 'fitting' process is done, the program first converts the resulting municipal housing assignments into a final municipal population forecast, and then recalculates the final forecast of municipal incomes. This inter-related model structure is illustrated in the systems dynamics diagram shown as Figure 20.2. Please note that Figure 20.2 does not include the income estimation routine.

Plan assignments are made using a more mechanical assignment process. First, 'Plan' asks the user how much of 'Trend' will be available for plan redistribution. This question gets at the sticky issues of how to deal with development approvals that have not yet been built, as well as the future determination (or ability) of the municipality to secure their preferred land use plan. Therefore, the Plan simulation begins with a user determined percentage of Trend growth (up to 100 per cent) as a given condition. The program then attempts to locate that portion of total growth which is available to be influenced by Plan, to locations consistent with the Plan policy instructions specified by the user. Plan offers numerous policy handles, which are sufficiently varied and robust so that the user can simulate sprawl and/or various forms of concentrated growth. Policies can emphasize: redevelopment of areas with existing development; infill growth into suburban areas; or development into more rural areas. Policies choices also allow the user to experiment with a very large number of programmatic characteristics, such as residential and employment density, job-to-household ratios, and growth capacities for new developments. Programmatic development characteristics can be municipal-specific. The user can specify up to six major planning areas (areas with similar development characteristics ranging from urban to rural in each municipality) and specify several types of development within each municipal-specific planning area.

Similar to the Trend fitting process, the Plan policy-dictated preliminary growth allocation is then tested to insure that sufficient land is available, with any excess growth re-allocated to other municipalities. As in Trend, the allocation process is continued until all plan-affected growth is assigned, at which point (as in Trend), the program converts assigned housing to population, estimates future municipal income and other housing and demographic characteristics of the municipal population.

Despite its current sophistication, *GAMe* was not conceived as a single unified concept; rather it evolved, driven both by circumstances as well as by evolving concepts regarding the application's potential utility. Since its inception, *GAMe* has undergone substantial changes that reflect response to users' comments, the research needs of OSP and other state agencies, as well as the amazing hardware and software changes that have occurred in the past decade. The original MultiPlan program was written as an attempt to develop a demographically sophisticated method to forecast housing need. It then occurred to NJOSP researchers that the housing needs model could be linked to newly developed (hand-planimetered) land availability information to test the capacity of a region's supply of available land to accommodate growth forecasts. The first prototype growth assignment model was completed by May 1989. We soon realized that the growth assignment/land accommodation process could be expanded by adding a set of growth assignment rules that simulated plan policies. By early 1990, a Trend allocation, a single hard-coded plan scenario and several impact models were programmed using Excel ML4.

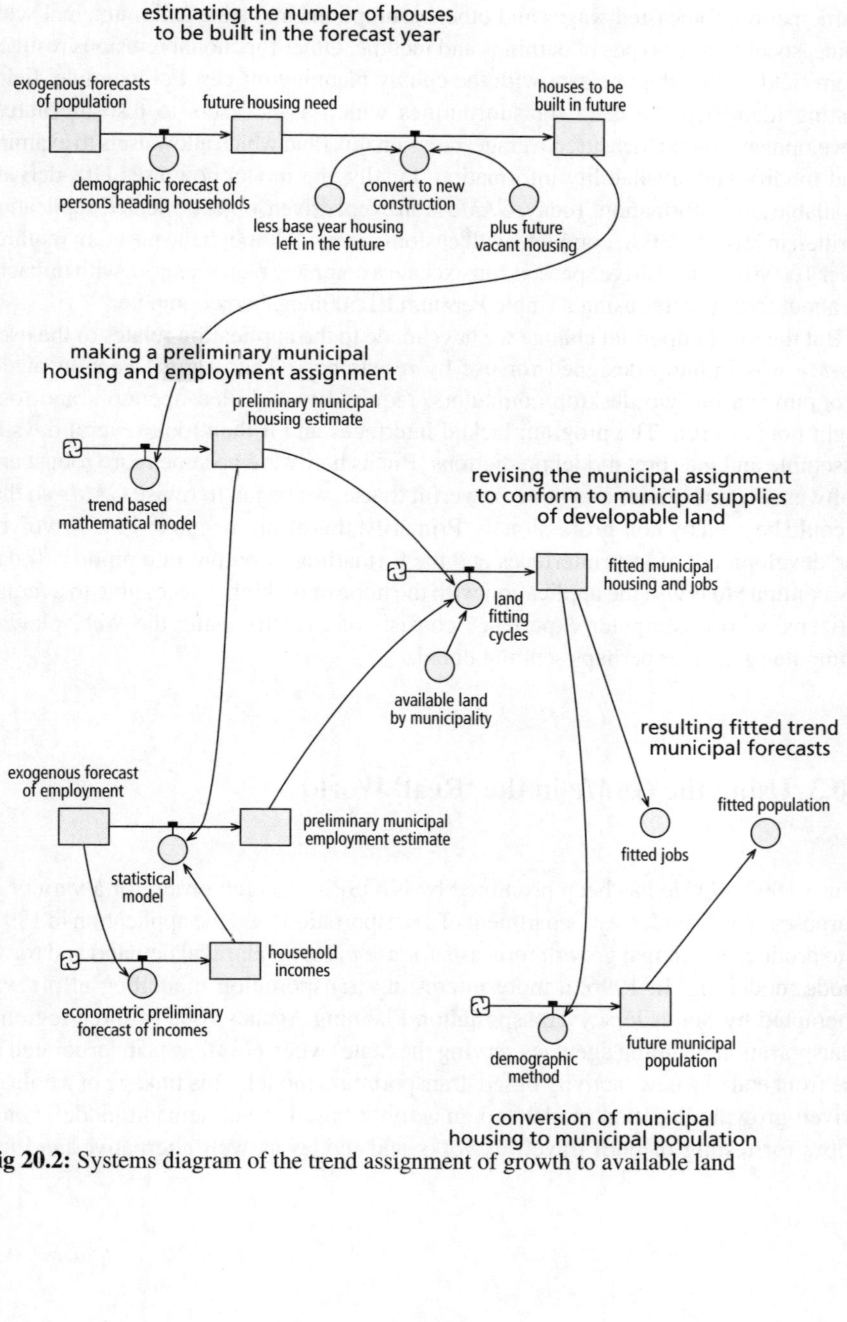

Fig 20.2: Systems diagram of the trend assignment of growth to available land

Other substantive changes have been made over the years. First order 'intuitive' or 'rule-of-thumb' algorithms were replaced with documented statistical (or other forms of mathematical modelling) methodologies. Also, the base program was expanded to include new functionality. For example, in 1993, the New Jersey Department of Transportation (NJDOT) funded several new features, such as the ability to forecast municipal householders by household size category. We also incorporated a variant of the OBERS (Office of Business and Economic Research and Statistics) economic model, allowing the program to convert forecasted employment by type, labour force participation, forecasted wages and other assumptions into state and municipal scale forecasts of various types of earnings and income. Other functional revisions resulted from field-testing the program with the county planning offices. For example, field-testing identified the need for subroutines which assist users to test alternative development criteria (density, coverage, program mix) and which allow users to examine and modify land availability information. Finally, the model now uses GIS-derived available land information. Today, *GAMe* is an event driven, object-oriented application written in VB and VBA, complete with custom menus. Although the program requires over 100MB of hard drive space, it can execute a complete plan scenario, with impacts, in about four minutes using a single Pentium III 500megahertz computer.

But the most important change we have made to the application relates to the user. *GAMe* was initially designed for use by researchers. The original version of the program ran on two desktop computers, required two skilled operators and took eight hours to run. The program lacked interfaces and it then took several days to assemble and interpret model predictions. But as hardware became more robust and software became easier and more powerful to use, we began to revise *GAMe* so that it could be used by non-professionals. Primarily, this effort, begun in 1995, involved the development of user interfaces and the formatting of output into reports. Today we continue to revise the application with the hope of making it accessible to average citizens, whose computer experience consists of casually using the Web, playing computer games or perhaps sending email.

20.3 Using the *GAMe* in the 'Real' World

Since 1993, *GAMe* has been promoted by NJOSP's research group for a variety of purposes. The New Jersey Department of Transportation used the application in 1993-4 to produce municipal growth forecasts for use in regional travel demand and travel mode modelling. In 1996, a more interesting transportation modelling effort was attempted by South Jersey Transportation Planning Agency (one of three regional transportation planning agencies serving the State) when *GAMe* was incorporated as the front end of a new 'activity based' transportation model. This linkage of a policy-driven growth allocation model and an activity-based travel demand model would allow for testing of both travel networks and modes as well alternative land use configurations. Unfortunately, the project proved to be too ambitious and the consultant developing the activity-based travel demand model defaulted. In 1998, NJOSP encouraged county planning agencies to use *GAMe* as part of their formal review of proposed revisions to the State Plan, a process in New Jersey called 'cross acceptance'. In 1999, *GAMe* was used to simulate environmental policies intended for application along the coastal region of New Jersey.

These field trials have met with a mixed degree of success. Most users found the application difficult to navigate, its results difficult to readily interpret and some of the variables highly technical in nature. Another problem related to bug reporting and getting the latest 'bug fix'. Most profoundly, the least satisfying outcomes have resulted when *GAMe* was used to develop regional land use patterns guided by regional policies. Despite numerous iterations, most users have been disturbed by the persistent (apparent) inability of *GAMe* to produce anticipated land use patterns, given what appear to be rational policies to guide development. These enigmatic results have caused many users to lose faith in *GAMe* simulations. Therefore, *GAMe* has not been widely used, or if used, unexpected predictions have been frequently discounted as 'counter intuitive' and therefore erroneous. In general, these counter intuitive predictions appeared to be so erroneous that users were unwilling to spend the time needed to closely examine the validity of the application's prediction. We began to worry that a great deal of public money had been spent on a flawed program.

Fortunately, in 1998-99, the State Planning Commission completed a draft revision of the State Plan. By act of Legislature, such revisions are required to be independently evaluated to insure that the revised State Plan continued to produce public benefits. Therefore, a year long analysis, completed in September 2000, was performed by the Rutgers University's Center for Urban Policy Research under the direction of Robert Burchell. The Rutgers team chose to use *GAMe* to assist in their analysis based on their preliminary review of the program structure and a review of the application's statistical models. However, once again, *GAMe* predictions looked to be counter-intuitive and just plain wrong. However, the Rutgers team took the time to closely examine the program's logical structure and to carefully re-examine many model equations by examining the statistical research performed by NJOSP and, in some cases, by solving the equations manually. This meticulous re-evaluation of the application demonstrated that the application's structure was logical and reasonable, that its models were statistically valid and that its predictions reasonable. This close examination validated *GAMe* and also demonstrated that, like many sophisticated IAMs, *GAMe* produces forecasts that are the result of interactions which are so numerous, interrelated, complex and (frequently) nonlinear that reasonable forecasts produced by the computer program appear to be irrational to the model user.

As a direct result of the Rutger's validation, NJOSP is encouraging counties to use *GAMe* to prepare regional land use plans. This effort is being fueled by a grant program administered by NJOSP, which supplies funding to the counties to subsidize strategic (Smart Growth) planning costs. At present, several counties, or their consultants, have expressed interest in using *GAMe*. Additionally, NJOSP has stated its intention to use *GAMe* internally to evaluate regional plans submitted to the State Planning Commission for review and approval (called endorsement in NJSDRP speak).

20.4 Believability: A Critical Problem for all PSS Applications

Our experience demonstrates that although program bugs and poorly designed interfaces can produce short term erosions in user confidence, knowledge-related problems are at the core of the mismatch between user's expectations and the forecasts produced by applications of PSS/DSS/IAM. At the heart of this verification, darkness is the conflict between human capabilities and those of sophisticated computer simulations.

One cause of this problem results from the fact that most people use 'shallow knowledge' to construct models of the future while scientific programs use causal models. Despite the seemingly derogatory name, the use of shallow knowledge models is a very desirable skill for one to acquire and it is taught in most educational systems – the current buzz word for it is 'critical thinking'. Shallow knowledge is based on experience (Giarratano and Riley 1994). For example, one might observe from experience that people tend to live in four person households. This information is then generalized into a model to forecast the number of future households (given a future population forecast). Such an experiential shallow model might lead one to the expectation that a declining population might need fewer housing units in the future. Causal models are used in science to simulate a more detailed understanding of how a particular system or event actually works (the model simulates the circumstances which cause the event to occur). For example, *GAMe*'s housing model uses a relatively complex methodology for converting population forecasts into household populations. Therefore, *GAMe*'s causal model accounts for the ageing of the population at the forecast year, likely changes in life span, the tendency for single parent households, the tendency for persons to delay marriage and the preference to have children. Because of these factors, *GAMe* forecasts the need for more houses in the future, despite an assumed decline in population – a seemingly counter intuitive forecast compared to the shallow knowledge model.

A second problem relates to the limited capability for humans to remember large amounts of detailed information that affect growth allocations and impacts. For example, both Trend and Plan preliminary assignments are delimited by the local supply of available developable land and, in the case of 'Plan' scenarios, the concurrence of available land in areas where policies encourage growth. Additional available land can be created using the redevelopment subroutine in the DSS. There are 566 municipalities, each of which may have up to six planning areas, each with specific amounts of land available for development. Under Plan, each planning area may be designated to accept growth and may have up to five different types of existing or proposed development 'centre', as well as policies pertaining to infill growth and land which is neither centre nor infill (areas termed 'environs' in the NJSDRP). These local conditions greatly influence growth forecasts. The computer easily handles the problem of correctly remembering all of this data.

A third problem is that most New Jersey municipalities are planned as if they were separate and distinct islands. Increasingly, municipalities act to exclude perceived problem land uses (increasingly seen as certain types of residential growth which is assumed to demand a high level of public services). To avoid this problem a rural municipality, with lots of available land, might select Plan policies to encourage job growth and to suppress residential growth (such as requiring very large lots). Such a strategy may or may not play out work in a *GAMe* simulation. For example, if other towns in the region also choose to restrict growth, perhaps by vastly increasing public ownership of otherwise developable land, then their preliminary growth assignment might be greater than their land capacity and the unfitted portion of the regional growth allocated to another municipality in the region. Despite the policy efforts of our rural municipality to avoid growth, a seemingly counter intuitive serge of sprawl residential growth might be the *GAMe* forecast.

A fourth problem relates to *GAMe*'s use of layers of models and the fact that predictions are frequently based on econometric, multivariate equations using values input by the user or forecasts previously generated by the application. For example, the user-selected (or entered) forecast of statewide population that directly affect housing needs, is a variable used in estimating future income forecasts, is a variable used in the distribution of employment-based growth and in the estimation of local road costs in each municipality. Additionally, some of *GAMe*'s statistically derived equations use stochastic modelling routines, where appropriate.

All of these knowledge and complexity problems contribute to user dissatisfaction with *GAMe* forecasts. This widespread dissatisfaction reminds one of Lee's *Requiem for Large Scale Models* (Lee 1973) where he argued that until certain hardware or software improvements were made, the large scale mainframe applications of that period would not be accepted. There remains a problem of acceptance, but the cause today is fundamentally different. Lee's concerns have been largely resolved in the past 20 years.

Today, our highly complex models produce forecasts which are valid, but which can be startling. If these models are to be accepted and used, non-scientific decision-makers need to be comfortable with model predictions, especially when they appear to be counter intuitive. There is a great deal of political theatre related to this problem. Opponents of a particular position will produce their expert who will announce that 'only liars use statistics' and that his analysis has shown that the model predictions are wrong. The policy-maker will initially feel comfortable sensing the antagonistic nature of the argument and will then loudly ask his scientist if the model's prediction might be wrong. The scientist, not being a politician, will proudly announce that there is only a 1 in 1,000 (or so) chance that this prediction is wrong – at which point the policy-maker will get visions of lawyers and lawsuits and the difficulties of re-election should the model projection be the 1 in 1,000 chance that is wrong. Given the troubling uncertainty of the prediction, the policy-maker will begin to compromise, and quietly swear never to trust computer simulations again. So, should we hold a coup and require that only scientists make land use decisions? Perhaps we should send all the policy-makers back to university for a course in statistics or modelling? Do these folks need to be reminded that only death and taxes are certain? Clearly none of these options are likely to occur, no matter how desirable or entertaining they may be. Ultimately, if applications like *GAMe* are to be widely used, they must be validated by independent academic organizations, a process much like the Rutgers review of *GAMe*, to provide the texture of credability for model results so craved by political decision-makers.

20.5 Current and Future Revisions to the Application

In 1998, we commenced a formal evaluation of *GAMe* with the invaluable assistance of large number of academic researcher from both the United States and Europe. As part of this evaluation, frank discussions were held to discover what features were most needed to assist citizen planners to improve master plans. A critical look at the capabilities and the weaknesses of *GAMe* was undertaken. Workshops were conducted where leading researchers in visualization techniques and quality of life modelling were invited to show their latest ideas and to recommend changes to *GAMe*. This effort produced several major revision recommendations. First, *GAMe* should be expanded from its regional planning scale to allow municipal scale master planning. Second, a more extensive and balanced impact assessment component should be developed to include a comprehensive environmental impacts suite, quality of life impact measures and a more comprehensive transportation impact element. Finally, it was recommended that the application use visualization techniques for both input and output to facilitate user friendliness and user comprehension of *GAMe* findings. At present, and for the next several years, we are implementing these recommendations.

20.5.1 Extending the Scale of *GAMe*'s Forecast to the Municipal Planning Level

Both Trend and Plan assign regional growth to municipalities, but the forecasts do not consider where within the municipality the growth actually occurs in Trend and it only knows the location of growth in a general way in the Plan simulation. In Plan, assignments are done by planning areas and to centres, the locations of which are mapped.

The major revision we are making to Trend is the addition of a site scale development assignment routine, patterned after CURBA developed by John Landis and his team at the University of Southern California at Berkeley (Landis 1998). Landis freely offered NJOSP copies of this model and papers, as well as providing time to answer our questions and provide guidance. Landis' assistance to our project was crucial and we are in his debt. In CURBA, Landis used a time series of land use/ land cover to identify locations that developed. Statistical probability methods then were used to relate the likelihood of any parcel becoming developed as a function of various locational or other descriptive attributes associated with these sites (each site in the Landis analysis was a 100 metre square grid). The resulting statistical findings were then used to construct a 'development probability overlay' that associated site attributes with development probabilities ranging from 0 to 100 per cent. Landis then developed a locational assignment model to use this development probability overlay.

NJOSP contracted Richard Brail, Rutgers University, to apply the CURBA methodology to New Jersey, using a statewide land use/land cover time series published by New Jersey Department of Environmental Protection (NJDEP). Brail found that many of the forcing variables reported by Landis (proximity to other development, proximity to certain hierarchical types of roads) also were explanatory in New Jersey, but that the coefficients were slightly different than those found in California. Brail's team also developed an improved stochastic growth assignment routine. At present, this application, termed the Regional Allocation Model (*RAM*) by Rutgers, is an ArcView application. Figure 20.3 displays the main *RAM* interface. Burlington County, New Jersey is the place displayed.

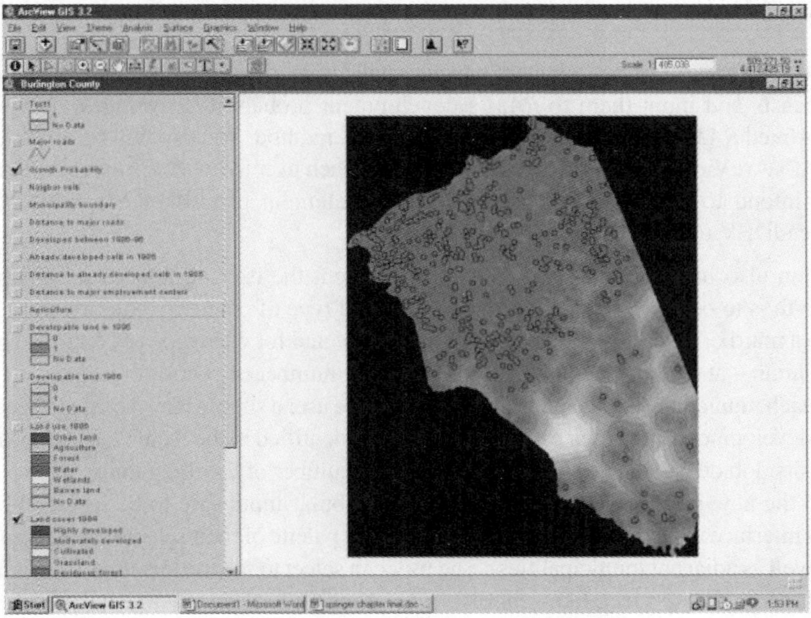

Fig 20.3: *RAM* interface displaying land uses in Burlington County, NJ USA

We feel that *RAM* is useful for making inter-municipal assignments, but that the *GAMe* methodology is better suited for making inter-regional (municipal) allocations of an exogenous regional forecast. We intend to link these program in a manner illustrated in the systems diagram shown in Figure 20.4.

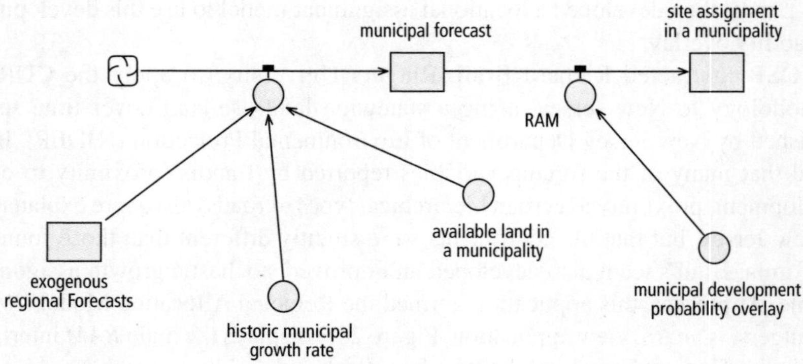

Fig 20.4: Systems diagram of future relationship between *GAMe* and *RAM*

To incorporate this new element into our application, we will use *GAMe*'s municipal forecasts and input them to *RAM*'s development probability coverages, using the improved *RAM* allocation stochastic assignment method. We also will be discarding *RAM*'s ArcView environment, since it requires each user to own expensive software. We intend to implement our new *RAM*-like element use either MapObjects or DBcadDEV to support the GIS environment.

Plan also has to be re-programmed. At present the user designates areas where growth is to occur by specifying the number and type of centre by entering numbers into a matrix (for each municipality there is a row and for each type of centre, there is a column – at the intersection, the user enters the number of centres of a specific type for each municipality). The program also offers the user a simple input screen to specify the development criteria for each centre. Criteria identified in the State Plan are housing density, job density, job to house ratio and total number of dwelling units.

In the new interfaces, these rather abstract tabular inputs are to be replaced by a GIS interface in which the user can view an aerial palette of their selected municipality (as well as adjacent municipalities). The user can select to see the locations of features, such as public services like schools, or to view a set of environmental suitability maps and to take this information into account in locating growth. The program will allow the user to draw polygons to define each centre (and, by exclusion, to define the environs in each municipality). When a centre is drawn, the program would display various data, such as total area, total developable acres and prompt the user to define the development criteria (residential density, job density, *et cetera*) for the centre. The program would display the development capacity of both the individual centre and for the municipality as a whole, and prompt the user to define the development criteria for the environs portion of the municipality. Figure 20.5 displays a GIS interface that has some of the features we will include into *GAMe*. This interface, developed by ABACO, for the European Union, allows the user to define crop areas (red hatched area). Once a crop area is defined, the program produces a text box which prompts the user to record crop characteristics. By using our new GIS input screens and the addition of *RAM*, both Trend and Plan routines will make predictions at a sub-municipal scale suitable for the development of municipal plans.

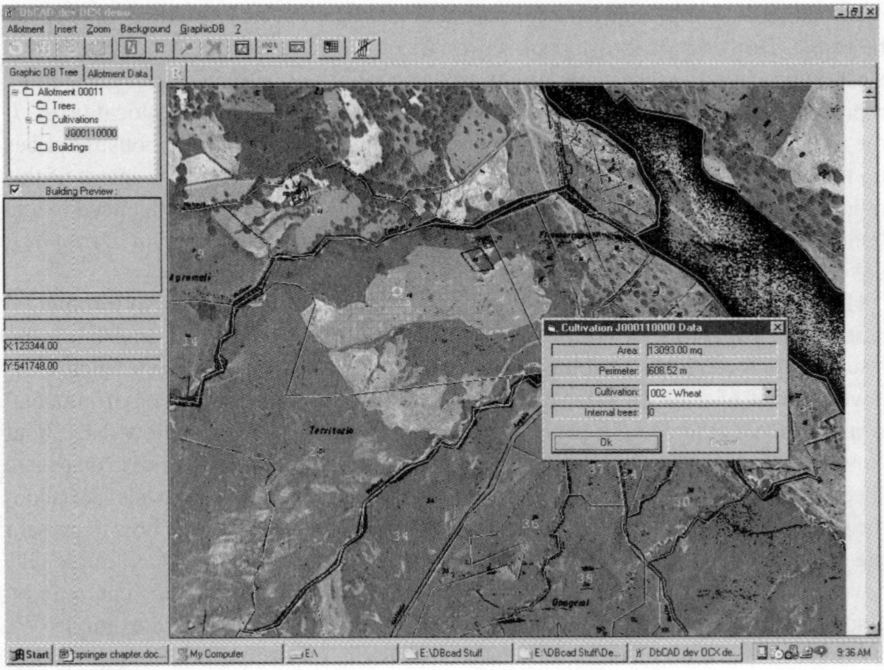

Fig 20.5: Display of sample DBcadDEV GIS interface

20.5.2 Developing More Extensive Impact Assessment Capabilities

The current version of *GAMe* projects impacts that are primarily fiscal. Even the sewer and school impact models forecast capital costs. We anticipate providing the user with a more balanced set of assessment tools, including environmental, infrastructural and quality of life information. The development of this suite of assessment routines is expected to take several years.

We have begun discussions with the NJ Department of Transportation (NJDOT) to seek improved linkages between *GAMe* and the agency's transportation models. Most recently, NJDOT experimented with altering land use and development mixes as a methodology to relieve congestion (as opposed to adding to highway capacity). NJDOT found encouraging results and we expect to collaborate with them to improve this approach. The quality of life component is expected to be a gaming environment in which municipalities, within the same region, can propose policies to change the levels of municipal services and to encourage or discourage other private sector or quasi-public amenities. For example, if a municipality chooses to increase spending on municipal schools, it might expect student test scores to increase (may or may not be true!). However, if the student scores do increase, and/or the school system is flush with amenities, the town might prove to be overly attractive and grow more rapidly that anticipated – causing fiscal disorder and possibly higher local taxes. If the town is largely built, the high quality school system might drive housing prices up to the point that existing residents find the town unaffordable. The intent of this model is to simulate these policy outcomes. This model is largely influenced by the controversial, but very interesting systems dynamic model, *Urban Dynamics*, developed by Jay Forrester (Forrester 1969).

However, our initial work emphasis has been to develop the environmental assessment capability of *GAMe*. We began by identifying a list of environmental issues, which might be developed into assessment tools, and by commencing the development of a fresh water quality impact model. To guide our environmental work we have assembled a panel of environmental scientists to review data about New Jersey and to review available impact models for appropriateness (there is a special concern regarding scale, both of the data used to prepare models and about the scale of the prediction). Another topic for this group is the issue of how to present the information. Some environmental data can usefully be displayed as simple GIS coverages (such as a map of wetlands). Other environmental features can only be assessed with the development of models, water quality being a good example. We also have assembled a planning advisory committee to insure that the scientific models address local planning issues in ways that are easy to understand.

NJOSP also has entered into a cooperative agreement with U.S. Geological Survey to develop a model relating water quality to land use change. The initial work effort, undertaken by USGS, was to develop a hydrological model (using TopModel) for each of the approximately 800 sub-watershed in the State (or which serve the State, but begin or end out of the State). Other models newly developed for this work include a climate/precipitation model, by USGS, and an impervious surface model, by NJOSP.

Once the hydrology model is calibrated, changes to watershed attributes (land use/land cover) can be associated with changes in stream ecology, through regression-based equations. With this model, which is sensitive to locational and programmatic attributes of development, *GAMe* users will be able to see streams within their municipal watershed change colour (to indicate a change in water quality) as they draw polygons designating new areas for growth and define development criteria.

Other products from this work effort that may find utility are related to hydrologic suitability. As an example, Figure 20.6 maps wetness index values, derived from TopModel, found in the Mulhockaway Creek watershed. Areas with values higher above 10 are more likely to generate high levels of surface runoff. This information can be used to establish a local program to mitigate runoff.

While the provision of such GIS information to guide virtual development decisions may seem obvious, it represents a major paradigm shift for our application. The initial PSS structure was one of posing a development option that was evaluated by various impact programs. Adding the GIS interface allows the user to view environmental data to facilitate development site selection. We feel that this paradigm change will greatly simplify using the application and greatly enhance the user's understanding of impacts produced by the application. The provision of this information during the plan development process also creates the opportunity to educate the user about how various natural and man-made systems work and to introduce the concept of impact mitigation via zoning code provisions.

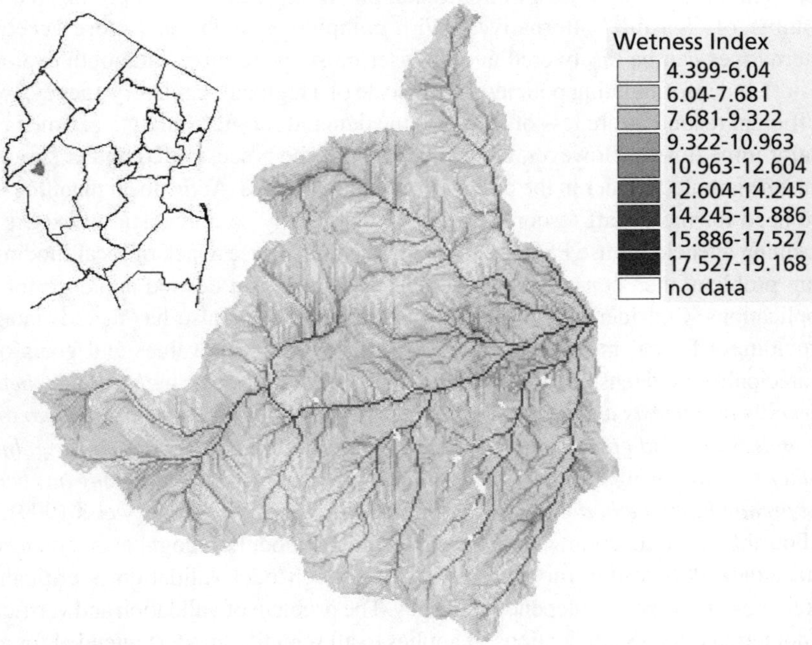

(Source: USGS)

Fig 20.6: Suitability for development in a watershed from a non point source perspective

20.6 Conclusions

NJOSP has been developing and using a PSS or over a decade. During this time we have learned many lessons about designing and appropriately using a PSS. We offer the following observations to other agencies or researchers interested in affecting a PSS.

Firstly, start with a simple model. It is very easy to accomplish very little at great expense. Design your model structure to be flexible and easy to code. Remember, most planning agencies are not software companies for a reason. Secondly, recognize that the development of a PSS is a long term very costly business. Today, there are programs that can be freely studied and perhaps amended to serve your locale. All of our work at NJOSP is freely available to the public. Many of our statistical equations only require modified coefficients to be very well suited for use in other locales. However, even with these cost saving opportunities, PSS are long term, high cost ventures.

Thirdly, computer simulations do not reduce the need for local resources. The paradox of many computer applications is that they ultimately require more of the user's time, rather than less. In traditional planning, reductionism produced a small number of 'feasible' alternatives. With computer simulations, more acceptable alternatives can be discovered and the user must spend more time both evaluating results and reformulating policies. At the scale of a regional regulatory agency, (where sufficient resources are less of an issue) the demand for more analytical time is not a major impediment. However, it does suggest that the process of computer simulation must be initiated sooner in the policy formulation process. At the local planning scale, the need for more staff resources suggests that it may be unrealistic to expect local planning agencies to use PSS and DSS applications. Since a lack of local funding is a core problem, then consultants and citizens might be the desired audience for such applications. Consideration of citizens as the target users also has the advantage of providing a formal method to incorporate the diversity of values and goals of the municipality's citizens into the land use plan. *"Modern society, with its overwhelming diversity of interests and development and its ever growing complexity, can no longer be understood and governed by the paradigm of centralized decision-making. In fact, rather than following a monolithic approach, distributed decision-making has become the predominant methodology to handle complex systems"* (Schneeweiss 1999).

Fourthly, in connection with the need to validate models, recognize that plan critics will attack PSS results for a variety of reasons. Model validation is critical, and likely best done by an independent agency. The problem of validation and verification is not unique to PSS applications. It applies to all scientific models intended for use to guide public policy. It is likely that it would be too expensive for smaller governments to pay for such analysis. It also is very expensive to develop PSS applications. Therefore, if PSS/DSS/IAM models are to be increasingly developed and widely used, their acceptance and usefulness would be enhanced by the creation of a national or international review and validation body, whose general roles would be:

- to supply validated algorithms to modellers (with proper metadata describing the appropriate use of the equation or systems model);
- to publish model guidelines incorporating best practices;
- to conduct outreach and training programs with the various planning organizations in the world; and
- to review and validate individual applications.

Such an organization should also encourage the development of both regional science models and a variety of impact models. At present, we feel that most academic PSS research publications are highly focused on locational theories. This research must be continued and reported by these excellent journals. However, the very nature of PSS is that they provide the user with indications of the impact of various development patterns. Unfortunately, many of the existing natural resources and other models, which are available to modellers (such as the USEPA Basins Model) have been developed at a scale which is not suited to regional, let alone municipal land use planning purposes. A rigorous research agenda to produced balanced, validated allocation models and of impact models, at a scale appropriate for land use planning, needs to be made an international priority. Additionally, there might be the need for more interdisciplinary research journals to properly inform the PSS development community of validated models and other research of interest. There also is the need for an internet-based service to collate abstracts from numerous existing journals into a single PSS publication.

References

Forrester, J.W. (1969) *Urban Dynamics*, Productivity Press, Cambridge MA.

Giarratano, J, and Riley, G. (1994) *Expert Systems – Principles and Programming*, PWS Publishing Company, Boston.

Landis, J. (1998) Development and pilot application of the California Urban and Biodiversity Analysis (CURBA) model, *IURD monograph 98-01*.

Lee, D.B. Jr. (1973) Requiem for large-scale models, *Journal of the American Institute of Planners*, 39: 163-178.

New Jersey State Planning Commission (2001, forthcoming) *The New Jersey State Development and Redevelopment Plan*, New Jersey Office of State Planning, Trenton, NJ.

Reilly, J. (1997a) A method to assign population and a progress report on the use of a spatial simulation model, *Environment and Planning B*, 24: 725-739.

Reilly, J. (1997b) A method to assign regional employment to municipalities, *Computers, Environment and Urban Systems*, 21(6): 407-424.

Schneeweiss, C. (1999) *Hierarchies in Distributed Decision Making*, Springer Verlag, Berlin.

21 Using an Operational Planning Support System to Evaluate Farmland Preservation Policies

Richard E. Klosterman, Loren Siebert, Mohammed Ahmadul Hoque, Jung-Wook Kim and Aziza Parveen

Department of Geography and Planning, University of Akron, Akron, Ohio 44325-5005, USA

Abstract: This chapter describes the use of a fully operational, commercially available, planning support system, What if?™, to evaluate alternative policies for preserving farmland in a rapidly urbanizing area in the midwestern United States. The chapter describes the study area, Medina County, Ohio, and the What if? model. It then describes the procedures that were used to obtain the required data and to apply the model to Medina County. The chapter concludes by considering the implications which the study has for efforts to preserve farmland and accommodate growth in the County.

21.1 Introduction

Academic and professional planners have attempted for more than thirty years to develop computer models to support the planning process (Klosterman 2000). While a large number of appealing prototype models have been developed, efforts to develop tools to support planning practice have been largely unsuccessful (Klosterman 1998). Fortunately, a combination of factors has come together recently which suggest that the decades-old dreams of using computers to support professional planning may soon be realized.

Computer hardware and software have become fast, convenient, powerful, reliable and relatively inexpensive. A wealth of spatially referenced digital data is becoming available at the national, state and local levels. Legislative requirements and popular concern with issues of urban sprawl, the loss of open space and environmental degradation are creating a demand for models that can consider the potential impacts of alternate growth management policies. And popular demand for direct involvement in the policy-making process is creating a need for easy-to-use and easy-to-understand models that will allow non-professionals to address issues of local concern. These developments have led to the emergence of planning support systems (PSS), which combine GIS data, urban models and computer-based visualization techniques to support community-based planning (Brail and Klosterman 2001).

This chapter will describe the use of a fully operational, commercially available PSS, *What if?™*, to consider alternative policies for preserving open space in a rapidly urbanizing county in the midwestern part of the United States. It begins by describing the study area, Medina County, Ohio, and the issues facing this rapidly developing area. The chapter then briefly describes *What if?* and the data and procedures required to apply the model to Medina County. The third section of the chapter describes the processes by which the required data were obtained for Medina County. The fourth section describes the application of the model to Medina County and the study results. The final section considers the implications that the study has for the residents of Medina County.

21.2 The Study Area

Medina County, Ohio, is a largely rural county with 151,000 residents living in three towns and several smaller communities. The County is approximately 110,000 hectares in size, of which 65,000 hectares (54 per cent) are currently devoted to agricultural uses, 39,000 hectares (33 per cent) are in forest, and only 7,700 hectares (6 per cent) are developed (Figure 21.1).

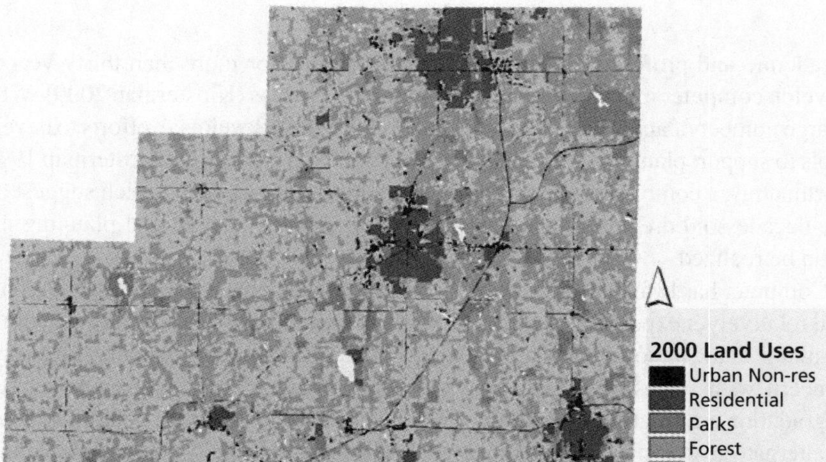

Fig 21.1: Current land uses in Medina County, Ohio

Medina County is located in close proximity to two larger metropolitan counties, Cuyahoga County (containing 1.4 million people and the City of Cleveland) to the north and Summit County (with 540,000 residents and the City of Akron) to the east. As shown in Figure 21.2, two interstate highways provide ready access to the County from its neighbours, making it easy for people to live in Medina County and work in Cleveland or Akron.

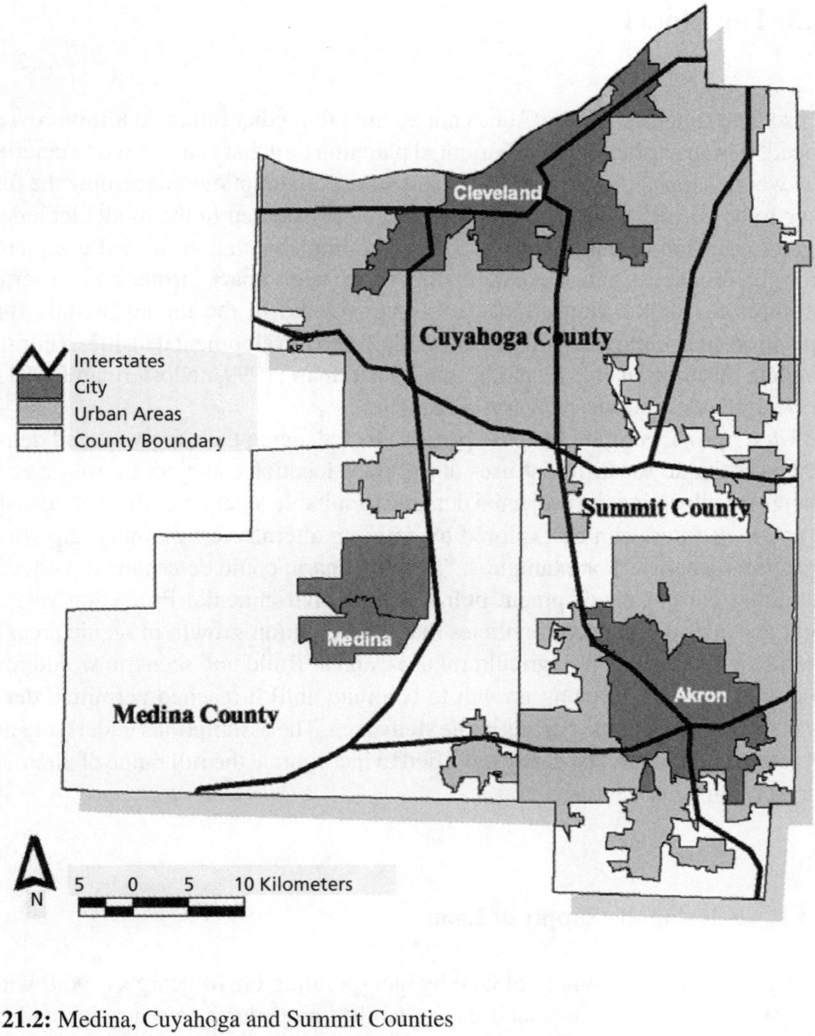

Fig 21.2: Medina, Cuyahoga and Summit Counties

The County's convenient location and rural amenities have generated substantial population growth. The County's population grew by 23 per cent in the last decade,

33 per cent in the last twenty years, and over 82 per cent over the last thirty years. A recent analysis (Bier *et al.* 1998) suggests that if these trends continue, virtually all of the County's vacant land will be developed by 2045. *What if?* will be used to determine if these projections are reasonable and to evaluate alternative policies for accommodating continued growth while protecting the County's rural character.

21.3 The Model

As its name suggests, *What if?* does not attempt to predict future conditions exactly. Instead, it is an explicitly policy-oriented planning tool that can be used to determine *what* would happen *if* policy choices are made and assumptions concerning the future prove to be correct. Policy choices that can be considered in the model include the staged expansion of public infrastructure, the implementation of land use plans or zoning ordinances, and the establishment of open space protection programs. Assumptions for the future that can be considered in the model include future population and employment trends and anticipated development densities. (For more complete descriptions of the model see Klosterman (1999), Klosterman (2001), or the *What if?* web site (*www.What-if-PSS.com*).

What if? projects future land use patterns by balancing the supply of, and demand for, land suitable for different uses at different locations, subject to user-specified policies for allocating the projected demand to suitable locations. Alternative visions for an area's future can be explored by defining alternative suitability, growth and allocation scenarios. For example, a 'Trend' scenario could determine the effects of continuing current development policies; an 'Environmental Protection' scenario might consider the impact of policies that severely limit growth in scenic areas and on land highly suitable for agricultural uses; and a 'Build out' scenario would reveal the implications of allowing growth to continue until it reached permitted density levels for all developable parcels in the study area. The assumptions underlying these and other scenarios can be easily modified to incorporate the full range of alternative visions for an area's future.

21.3.1 Analyzing the Supply of Land

What if? considers the supply of land by incorporating Ian McHarg's (1969) widely used 'weighting and rating' land use suitability procedures into a quick and easy computer-based process. The model provides a familiar Windows point-and-click interface that allows the user to define or modify scenarios determining the relative suitability of different locations for alternative land uses.

The user must first select the *factors* that they feel should be considered in determining the suitability of different locations for a particular land use. Thus, for example, the user may specify that slopes, septic soils and locations with respect to flood plains should be used to determine the most suitable locations for residential development. The factors which can be considered in a particular area are determined by the data that are available for that area and can include the full range of natural features, e.g. slopes, soils, flood plains, scenic vistas and hazardous areas.

The user must then specify suitability *factor weights*, i.e. numerical scores indicating the relative importance of different factors for determining the suitability of different locations for a particular land use. Thus, for example, the 'slope factor' could be given a factor weight of 2 and the 'septic soils factor' could be given a factor weight of 1 to indicate that slope is considered to be twice as important as septic soils in determining a site's suitability for residential uses. Factor weights are assigned on a three-point scale from 1 (low) to 3 (high).

Next, the user must specify *ratings* for each factor type, i.e. numerical values that indicate the relative suitability of locations with a particular factor type for locating a specified land use. The factor types are rated on a six-point scale from 5 (high) to 1 (low) and 0 (excluded). The 'excluded' category identifies areas from which development is to be excluded, regardless of its suitability with respect to other factors. For example, the slope factor may have three slope types: low slopes, medium slopes, and high slopes. In this case, the factor ratings are used to indicate the relative suitability of locations with each type of slope for locating, say, residential development. For instance, areas with low slopes could be given a rating of 5 and areas with moderate slopes could be given a rating of 3 to indicate that areas with low slopes are more suitable than areas with moderate slopes. Zero ratings could be given to areas with high slopes to indicate that residential development should be excluded from these areas, no matter how desirable they may be otherwise.

Finally, the user can specify land uses that may be converted from their current use (e.g. agriculture) to another use (e.g. residential) during the land use allocation process. If no land uses are identified as conversion candidates, only currently undeveloped land will be available for satisfying the projected land use demands.

After the user specifies the suitability factors, factor weights, factor scores and permitted land use conversions for each land use, the model multiplies the user-specified factor weights and ratings for each suitability factor and sums these products. The resulting *suitability scores* indicate the relative suitability of all locations for each land use when all of the suitability factors have been considered. Areas with high suitability scores are assumed to be highly suitable for a particular land use; areas with lower scores are assumed to be less suitable. *What if?* then generates: (i) a series of maps showing the relative suitability of different locations for each land use; (ii) tables reporting the quantity of land within each suitability class for all land uses; and (iii) reports identifying the assumptions which underlie a suitability scenario.

21.3.2 Projecting the Demand for Land

What if? considers the demand for land by converting the five main categories of land use demand – residential, industrial, commercial, preservation and locally-oriented uses -- into the equivalent future land use demand. More than one land use is permitted within each general land use type. For example, the model can accommodate up to ten different types of residential land uses, up to five different types of industrial uses, and so on. The projected demand for each land use is computed for three future years and for buildout, providing a staged development process in which future development is based on previous development patterns.

What if? computes the demand for residential land by allowing the user to specify: (i) the projected number of residential households by year; (ii) the future breakdown by housing type; and (iii) future housing densities by housing type. The demand for non-residential (e.g. industrial, retail and office uses) is derived from user-specified values for the projected employment in each year and the assumed employment densities for each non-residential use. The model also allows the user to specify the amount of land that should be set aside for preservation land (e.g. for agriculture or open space conservation) and per-capita standards for local land uses (e.g. local retail and neighbourhood parks) in each projection year.

The model uses the user-specified growth assumptions to project the demand in each projection year for all land uses. It then generates reports on the projected demand for all land uses in each projection year and on the assumptions that underlie a particular growth scenario.

21.3.3 Allocating Future Land Use Patterns

What if? projects future land use patterns by allocating the projected land use demands, derived from a user-selected growth scenario, to different locations on the basis of their relative suitability, as defined by the assumptions in a user-selected suitability scenario. If desired, the growth allocation can be controlled by user-selected land use controls (e.g. land use plans or zoning restrictions) or plans for providing public infrastructure such as sewer and water service. *What if?* then generates a series of maps showing the projected land use patterns in each projection year. It also generates reports showing the projected land use acreage in each projection year and the assumptions that underlie a scenario.

21.4 Obtaining Required Data

Four main types of data are required to apply *What if?* to a given study area. The first – and often most difficult to obtain – piece of information is a digital map showing the area's current land uses. The second piece of required information is a series of digital maps showing the distribution of different suitability factors – slopes, soils, floodplains, and so on – in the study area. Third, projections for the area's future residential population and employment are required to project future land use demands. Finally, digital map layers are needed that describe alternative land use control policies such as land use plans, zoning ordinances, or infrastructure expansion plans. The procedures used to obtain each of these kinds of data for Medina County are described briefly below.

21.4.1 Existing Land Use Information

Any projection for the future must be based on the present and the past. As a result, the first piece of information required to project an area's future land use patterns is a GIS layer showing the area's current land use patterns. In some areas, this information is readily available from parcel-level GIS data maintained by local governments. However, in many areas, including Medina County, reliable and up-to-date information on current land uses is not available, requiring this information to be obtained indirectly. In this study, a combination of remote sensing and population census data were used to determine the County's current land uses.

Remotely Sensed Data

A Landsat Enhanced Thematic Mapper (ETM) image with 30-metre resolution was used to produce a land cover map for the County. An unsupervised classification was conducted in ERDAS Imagine 8.4 using the ISODATA method to classify the image into fifty spectral clusters. Recent aerial photography, GPS-linked air video (Siebert *et al.* 2000) and other reference maps were then used to interpret the spectral clusters and assign them to the following generalized land cover categories:

- 'Water', including all water bodies such as lakes, ponds, and rivers;
- 'Residential', including residential lots, neighbourhood roads, low-density buildings and similar land areas;
- 'Urban', including high-density residential, commercial and industrial areas, roads and quarries;
- 'Forest', including deciduous and conifer forests; and
- 'Farmland', including open space, grassland, and bare and vegetated farmland.

The raster land cover map was smoothed twice by using a majority filter to facilitate converting it into ESRI's coverage format for use with other data layers. The resulting land cover map for the northeastern Ohio area, including Medina County, is shown in Figure 21.3.

Population Census Data

It has long been recognized that remote sensing imagery is appropriate for collecting *land cover* information but less appropriate for collecting *land use* information. For example, some areas that contain a mixture of rooftops, roads, lawns and trees may be correctly identified as residential neighbourhoods, but other residential areas that have more trees or larger lawns might be misidentified as farmland or forest rather than residential. Areas that are classified as 'farmland' because they consist largely of vegetated areas with few roads or buildings may also be unfarmed open space or parks. And areas that are classified as 'urban non-residential' because they are barren of vegetation may also be highways, quarries, or recently tilled bare fields.

The classified remote sensing data were found to adequately identify water bodies, urban non-residential areas, and areas that are forested or devoted to agricultural uses in the County. However, it was determined that the large-lot residential areas with a mix of grass and trees found throughout the County were frequently classified as 'Forest' or 'Farmland'. To address this problem, the 2000 Population Census data were used to compute the residential densities for all of the Census blocks in the County. Locations that had residential densities between one to four dwelling units per acre were classified as low-density residential; areas with four to eight dwelling units per acre were classified as medium-density residential; and areas with more than eight dwelling units per acre were classified as high-density residential. These data were combined with the remote sensing data on water, urban, forest, and farmland areas to provide the required land use information.

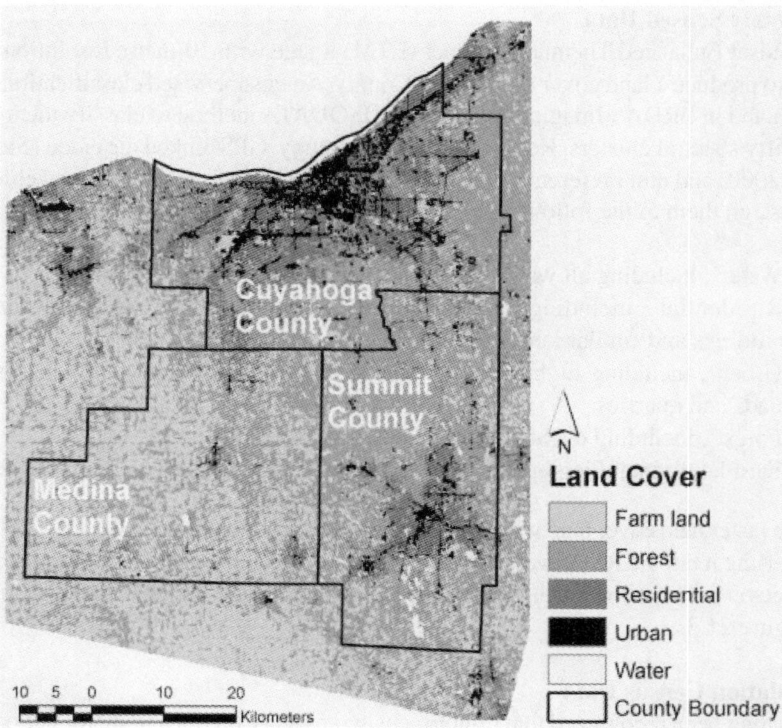

Fig 21.3: Generalized land cover map for northeast Ohio

In addition, data available from the regional park district were used to identify park areas, and 15 and 30-metre buffers were used to estimate rights of way along interstate and other major highways. Together, these procedures yielded the following land use categories: (1) non-residential urban; (2) high-density residential; (3) medium-density residential; (4) low-density residential; (5) parks; (6) forest; (7) farmland; (8) right of way; and (9) water.

21.4.2 Land Suitability Data

Like most rural areas in the United States, Medina County does not have a detailed parcel-level GIS system, greatly limiting the availability of spatially related information. Fortunately, substantial quantities of digital data are available on-line for the United States via the World Wide Web. This allowed the following suitability factors to be considered for Medina County: (i) slopes, (ii) agricultural soils, (iii) flood plains, (iv) wetlands, and (v) river buffers.

The slope data were derived from Digital Elevation Models (DEMs) available on the Web from the U.S. Geological Survey. ArcInfo was used to convert the DEMs into a TIN model and to derive five slope categories (<5%, 5-<10%, and so on). Soil data available on the Web from the State of Ohio Department of Natural Resources were used to identify areas that had slight, moderate or severe limitation for agricultural use. Data available on-line from the Department of Natural Resources were also used to identify wetlands and areas that were inside the 100-year flood plain. ArcInfo was used to convert DEMs into five slope categories.

ArcInfo's UNION command was used to combine the existing land use layer and the five suitability map layers into a single coverage. The coverage contained a large number of irregularly shape polygons called 'uniform analysis zones' (UAZs). The UAZs were uniform internally with respect to all of the factors considered in the model, i.e. all points inside a UAZ have the same current land use, the same slope and the same soil type. These small polygons, which ranged in size from one-half hectare to several hectares in size, comprised the spatial units for which the suitability analysis was conducted and future land use patterns were projected.

21.4.3 Projected Population and Employment

Two sets of projections were prepared for the future residential population and employment of Medina County. The 'High Growth' projections assumed that the 23 per cent growth rate observed between 1990 and 2000 would continue for the next 45 years. Under this assumption, the County's residential population would grow from 151,000 in 2000 to over 390,000 in 2045 (i.e. by more than 2.5 times) and its employment would grow from 50,000 to 129,000 (i.e. also by more than 2.5 times).

The 'Low Growth' projections assumed that the 33 per cent growth rate for the 20-year period between 1980 and 2000 would continue until 2045. Under this assumption, the County's residential population in 2045 would be 290,000 (nearly double the current population) and its employment would be 96,000 (again nearly twice what it is currently).

21.4.4 Public Policy Choices

Two public policies for preserving farmland and open space in Medina County were considered. The first assumed that policies would be enacted which prohibited development in areas with good agricultural soils, wetlands, in the 100-year flood plain, or near streams. This 'Preservation' policy was incorporated into the suitability portion of the model by assuming that residential and non-residential development is prohibited in all locations that either: (i) had soils with slight limitations for agricultural uses; (ii) were located in the 100-year flood plain; (iii) were wetlands; or (iv) were located within 30 metres of a stream. Development was also prohibited in areas with high slopes. The alternative 'Development' policy did not consider the environmental impacts of development and prohibited development only in areas with high slopes.

The second 'Growth Controls' policy assumed the County enacted a growth policy which limited development to areas that have public water and sewer service. It also assumed that the County enacted a staged infrastructure expansion policy in which sewer and water service would be made available to rural areas in a series of three stages. The first stage would provide sewer and water service to all areas within existing municipalities by the year 2015. The second stage would provide services to areas that were within 2.4 km of the current municipal boundaries by the year 2030. The third stage would make sewer and water service available to all areas within 4.8 km of the current municipal boundaries by the year 2045. The alternate 'No Growth Controls' policy assumed that no limitations were placed on the location and timing of future development.

21.5 Applying the Model

21.5.1 Suitability Analysis

Two suitability scenarios were created. The 'Preservation' scenario assumed that the preservation policies were enacted and development was prohibited in areas that had good agricultural soils, were wetlands, in the 100-year floodplain or near streams. The 'Development' scenario assumed that development was permitted in all of these areas. As one would expect, the two suitability scenarios assume that dramatically different quantities of land will be available to accommodate future development. As shown in Table 21.1, under the Development scenario roughly 85 per cent of the County is suitable for non-residential or residential development. Under the Preservation scenario, roughly 50 per cent of the County is suitable for development.

Table 21.1: Suitability analysis results

land use	suitable area (1,000 hectares) development scenario		preservation scenario	
urban non-residential	89.8	(82.1%)	56.10	(51.3%)
high density residential	95.5	(87.3%)	55.80	(51.0%)
med. density residential	95.5	(87.3%)	54.80	(50.2%)
low density residential	94.9	(86.3%)	52.80	(48.2%)

This is revealed clearly in Figures 21.4 and 21.5, which map the suitability of different locations in the County for low-density residential development. Under the Development scenario (Figure 21.4), almost all areas of the County that are not currently developed are highly suitable for development. Under the Preservation scenario (Figure 21.5), large portions of the southern part of the County, which have prime agricultural soils (shown in white), are assumed to be not suitable for development. The implications that these two suitability assumptions have on the County's future development patterns will be considered below.

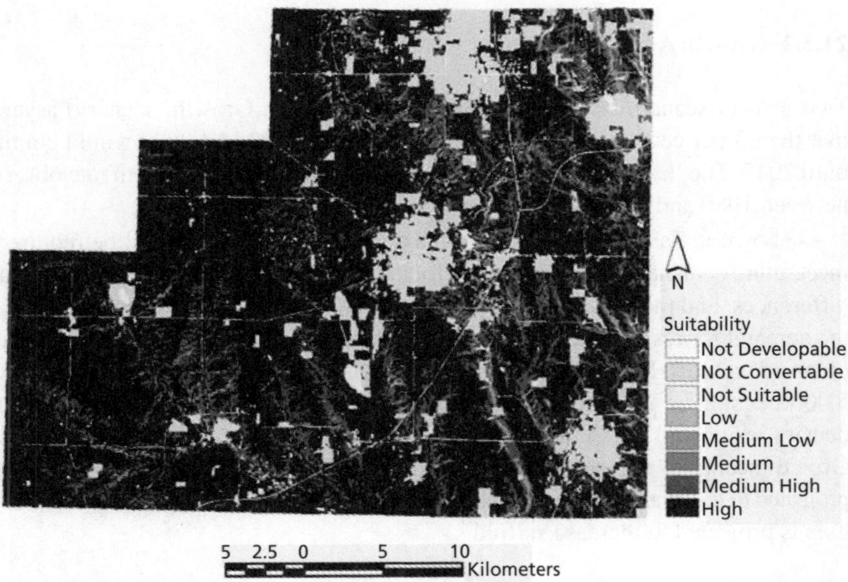

Fig 21.4: Low-density residential suitability with the Development scenario

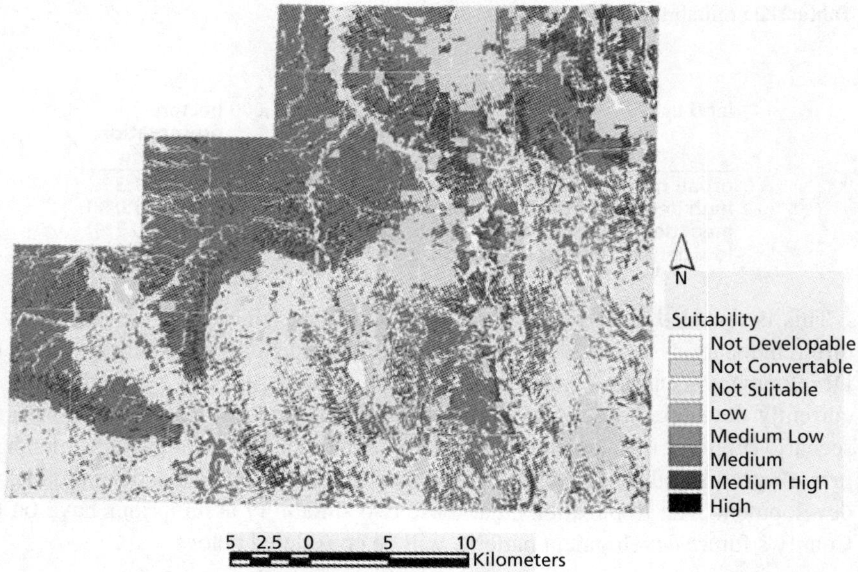

Fig 21.5: Low-density residential suitability with the Preservation scenario

21.5.2 Growth Analysis

Two growth scenarios were then prepared. The 'High Growth' scenario assumed that the 23 per cent growth rate observed between 1990 and 2000 would continue until 2045. The 'Low Growth' scenario assumed that the lower growth rate observed between 1980 and 2000 would continue until 2045.

As shown in Tables 21.2 and 21.3, the quantities of land that will be required to meet future demand differ substantially for the two growth scenarios. The most notable differences, and the ones that will have the largest impact on land consumption, are the demand for residential land uses. Under the High-Growth scenario, the quantity of land devoted to low-density residential uses is projected to nearly quadruple, from 5,000 hectares in 2000 to 19,700 hectares in 2045; the land devoted to medium-density residential uses is projected to increase by nearly ten fold. Under the Low Growth scenario, the quantity of land devoted to low-density residential uses is projected to less than triple by 2045 and the land devoted to medium-density residential uses is projected to increase sixfold.

21.5.3 Allocation Analysis

As was pointed out previously, *What if?* allocation scenarios are created by combining: (i) the land use supply assumptions defined in a suitability scenario, (ii) the land use demand assumptions defined in a growth scenario, and (iii) rules for allocating land use demand as defined by one or more public policies. A number of allocation scenarios were prepared which combined the High Growth and Low Growth demand scenarios with a wide range of suitability and allocation assumptions. All of the scenarios indicated that the High Growth assumptions could be accommodated in 2045 and beyond. As a result, the more conservative Low Growth scenario assumptions were not considered further.

Table 21.2: Current and projected land uses with the High Growth scenario

land use	area (1,000 hectares)			
	2000	2015	2030	2045
urban non-residential	1.6	1.8	2.1	2.5
high density residential	0.0	0.2	0.5	0.8
med. density residential	0.4	1.2	2.4	3.9
low density residential	5.0	8.4	13.2	19.7

Table 21.3: Current and projected land uses with the Low Growth scenario

land use	area (1,000 hectares)			
	2000	2015	2030	2045
urban non-residential	1.6	1.7	1.9	2.1
high density residential	0.0	0.1	0.3	0.5
med. density residential	0.4	0.9	1.6	2.4
low density residential	5.0	7.2	10.0	13.5

Four allocation scenarios were then prepared to determine the likely effects of implementing the two growth policies described above. The two Preservation scenarios prohibited growth in areas that have prime agricultural soils, are wetlands, in the 100-year floodplain, or near streams. The two Development scenarios imposed none of these limitations on future development. The two 'Growth Control' scenarios assume that a staged infrastructure expansion program will be enacted and development will be limited to areas that have public water and sewer service. The 'No Growth Control' scenarios assume no limitations will be placed on the location and timing of future development. The implications of implementing these two farmland preservation policies, both individually and together, are illustrated in Table 21.4 and Figures 21.6-9.

As shown in Table 21.4, the two Preservation scenarios maintain all of the County's prime farmland and lose less farmland than the two Development scenarios. This is not surprising because, as the suitability maps in Figures 21.4 and 21.5 indicate, the Preservation scenarios exclude large quantities of land, and all of the land with prime agricultural soils, that could be developed under the Development scenarios.

Table 21.4: Farmland losses, 2000-2045

scenario	area lost, 2000-2045 (1,000 hectares) farmland	prime farmland
preservation – no growth controls	10.9 (19%)	0.0 (0%)
preservation – growth controls	11.7 (20%)	0.0 (0%)
development – no growth controls	12.3 (21%)	5.9 (24%)
development – growth controls	13.5 (23%)	6.5 (26%)

What is somewhat surprising is the indication in Table 21.4 that the two Growth Control scenarios (which limit growth to areas near the existing cities) lose substantially more farmland than the No Growth Control scenarios, which impose no limits on future development. Casual inspection of Figures 21.6 and 21.7 (which both incorporate the Preservation suitability assumptions) suggests that far more farmland is being consumed without Growth Controls (Figure 21.6) than with Growth Controls (Figure 21.7). However, as Table 21.4 indicates, this impression is incorrect and, in fact, more farmland is being consumed with the growth controls; these losses are just concentrated near the cities. An identical situation occurs for the two scenarios that incorporate the Development suitability assumptions (Figures 21.8 and 21.9).

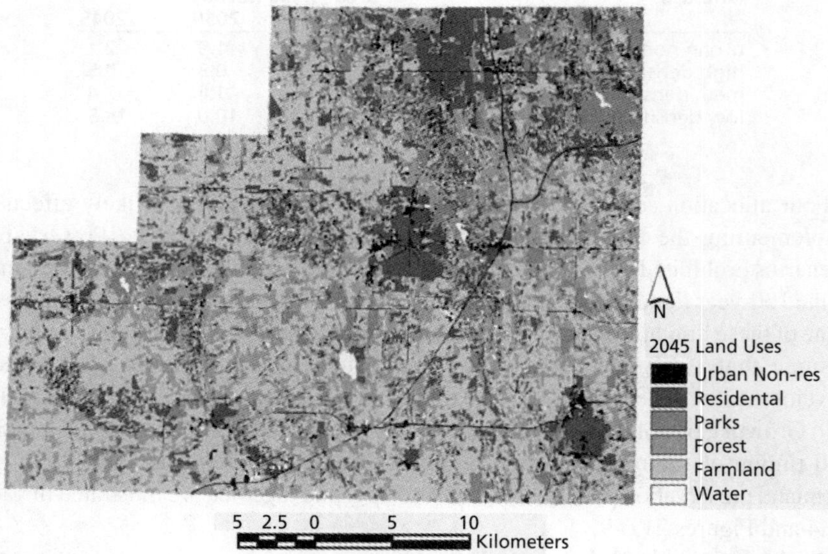

Fig 21.6: 2045 land uses under the Preservation scenario without growth controls

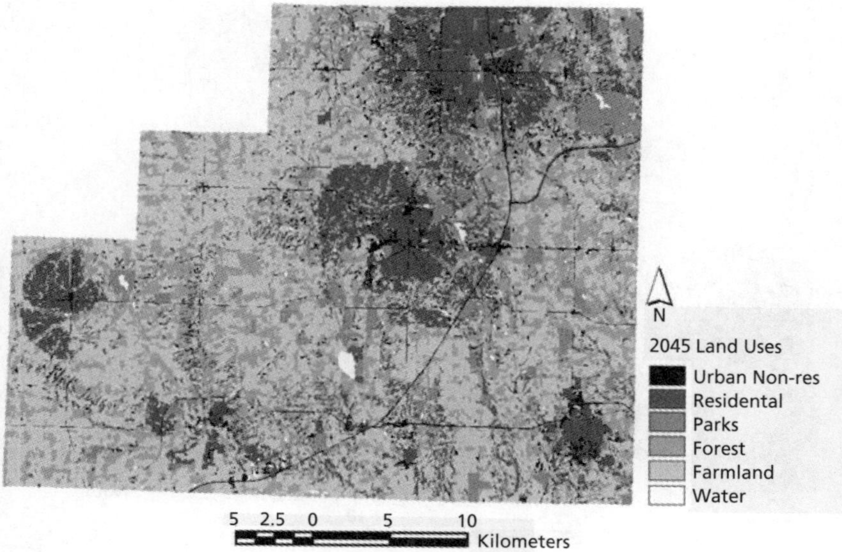

Fig 21.7: 2045 land uses under the Preservation scenario with growth controls

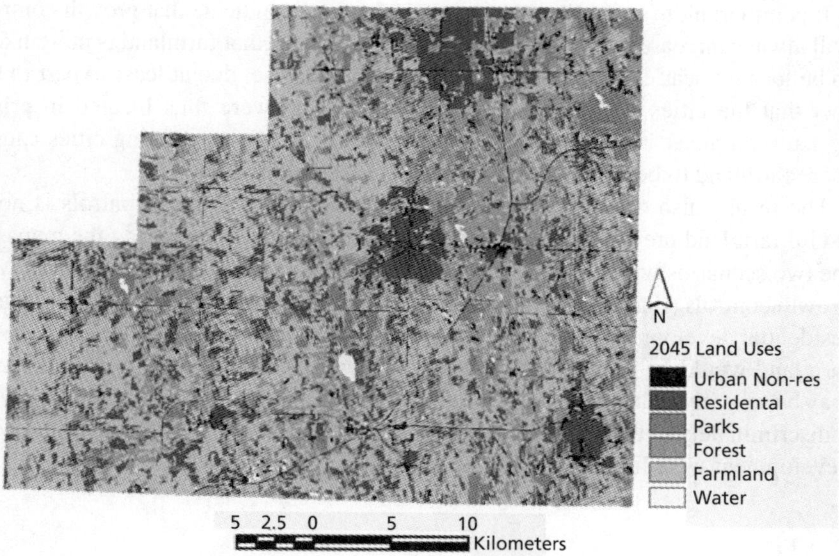

Fig 21.8: 2045 land uses under the Development scenario without growth controls

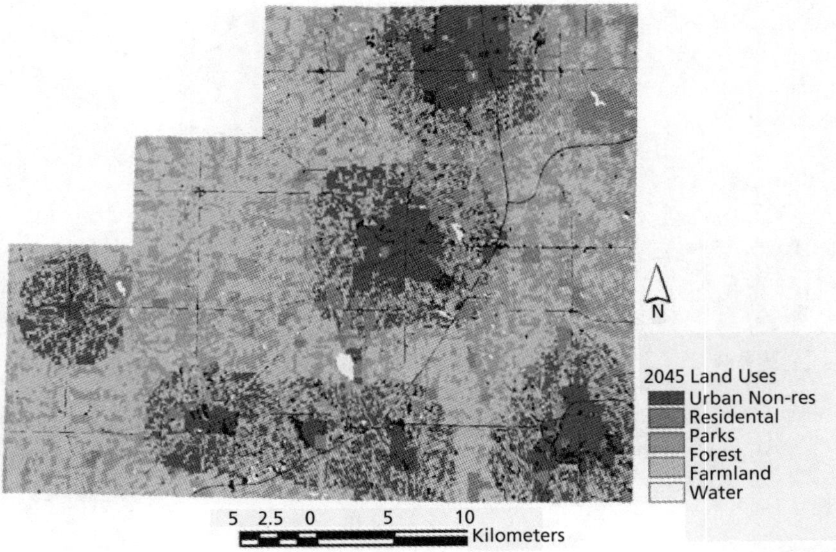

Fig 21.9: 2045 land uses under the Development scenario with growth controls

It is important to recognize that these results do not indicate that growth controls will always increase the farmland loss. It only indicates that farmland is more likely to be located near existing cities in Medina County case, due at least in part to the fact that the cities began as farm communities and were thus located in prime agricultural areas. As a result, limiting growth to areas near existing cities causes more farmland to be lost than might be otherwise in this situation.

The results also do not indicate that the imposition of growth controls is not a useful farmland preservation strategy for Medina County. Comparing the maps for the two scenarios without growth controls (Figures 21.6 and 21.8) to the ones with growth controls (Figures 21.7 and 21.9) indicates that without these controls, future residential development is scattered throughout the County. As a result, while more farmland would be preserved without these controls, the County's rural character may be severely reduced. In addition, the costs of providing sewer and water service indiscriminately to the entire county will be substantially higher than it would be if development were limited to areas near existing cities.

21.6 Implications of the Study

These results are extremely useful for the residents of Medina County who want to maintain the County's rural character, while accommodating future growth. It first indicates that an earlier analysis that suggested that the County would be totally developed by the year 2045 was overly pessimistic. Instead, it indicates that over three-quarters of the County's farmland will remain, even if the substantial growth rates observed over the last decade continue for the next 45 years.

Nevertheless, the results indicate that the people of Medina County face a major choice about the kind of community they, their children, and grandchildren will occupy in the year 2045. If they do nothing, 12,300 hectares of farmland and 6,500 hectares of prime farmland will be lost and residential development will be scattered throughout the County (as shown in Figure 21.8). If they establish policies for preserving prime farmland and do not control growth (as shown in Figure 21.6), they can save 1,400 hectares of farmland and 5,900 hectares of prime farmland but the county will lose much of its rural character. And if they enact policies for preserving prime farmland and for controlling growth they would lose more farmland than they would without controls but would maintain the County's current agricultural character (as shown in Figure 21.7). The choice between these alternative futures is one that only the people of Medina County and their elected officials can make.

Acknowledgements

This research was conducted with the financial support of the U.S. Geological Foundation and the Fannie Mae Foundation.

References

Bier, T., Post, C., Weizer, W. and Burgess, P. (1998) *Development trends in Medina County, Ohio*, The Urban Center, Cleveland State University, Cleveland, Ohio.

Brail, R.K. and Klosterman, R.E. (eds.) (2001) *Planning Support Systems: Integrating Geographic Information Systems, Models, and Visualization Tools,* ESRI Press, Redlands, CA.

Klosterman, R.E. (1998) Computer applications in planning, *Environment and Planning, B: Planning and Design Anniversary Issue*: 393-404.

Klosterman, R.E. (1999) The What if? collaborative planning support system, *Environment and Planning, B,* 26: 393-404.

Klosterman, R.E. (2000) Planning in the information age, in Hoch, C.J., Dalton, L.C. and So, F.S. (eds.) *The Practice of Local Government Planning*, 3rd ed. International City/County Management Association, Washington, DC.

Klosterman, R.E. (2001) The What if? planning support system, in Brail, R.K. and Klosterman, R.E. (eds.) *Planning Support Systems: Integrating Geographic Information Systems, Models, and Visualization Tools*, ESRI Press, Redlands, CA.

McHarg, I. (1969) *Design with Nature. Garden City*, Natural History Press, Garden City, NJ.

Siebert, L., Stabler, B. and Parveen, A. (2002) Using GPS-linked air video for satellite image interpolation, *Geocarto International*, 17(1): 5-11.

Web Site

What if?: *http://www.what-if-pss.com*

22 Development of the Klang Valley Regional Planning Support System

Ruslan Rainis[1], Kamalruddin Shamsudin[2], Mohd Nazari Jaafar[3] and Hadi Harman Shah[4]

[1] Geoinformatic Unit, School of Humanities, Universiti Sains Malaysia, 11800 Penang, Malaysia
[2] Development Program Division, Federal Department of Town and Country Planning, 20th. Floor, Wisma Sambanthan, Jalan Sulaiman 50000, Kuala Lumpur, Malaysia
[3] Department of Geography, Universiti Malaya, Kuala Lumpur, Malaysia
[4] Faculty of Development Science, Universiti Kebangsaan Malaysia, Bangi, Selangor

Abstract: The regions in Malaysia are facing rapid changes due to globalisation. These changes call for a new approach in local and regional planning in the country, an approach that requires the ability to continuously monitor and even predict the dynamics of each region. Previous regional plans that focus on end-states as opposed to processes have been deemed not to address these changes. A test case for such a need is the Klang Valley Region. The development of the Klang Valley Regional Planning Support System (KV-RPSS) is a move towards addressing this issue. Using technology adapted to local needs, the focus of this planning support system is to assist federal level regional planners to work with local authorities in identifying and predicting urban changes.

22.1 Introduction

The Klang Valley Region is Malaysia's largest and most dynamic region in the country. In the past, numerous regional planning studies have been undertaken for the region, but the application of spatial analysis tools, such as plan generation techniques, have only recently been developed and applied to the region. The coordinating agency, the Federal Territory Development and Klang Valley Planning Division (FTDKVPD) of the Prime Minister's Department has taken the initiative to incorporate this spatial analysis approach. Unlike previous studies, this study puts strong emphasis on the applications of information technology especially geographical information system (GIS) in an effort to make the regional plan a 'living document'. Therefore, a major component of the study is the development of a planning support system (PSS) called the *Klang Valley Regional Planning Support System (KV-RPSS)* to assist the study team in various activities within the planning process, such as generating scenarios, analysing impacts and monitoring plan implementation. This chapter describes the structure and components of the *KV-RPSS* and records the experiences of using such a system in generating three development scenarios for the region.

22.2 The Study Area

The Klang Valley Region is the most central region in the country. It is where Kuala Lumpur, the capital city, is located (Figure 22.1). The region comprises the Kuala Lumpur Federal Territory and four other districts in the State of Selangor. The size of the region is about 2,843 square kilometres (284,342 hectares) and there are eight local authorities. In 1988, about 13.5 per cent of the region consisted of built-up areas, whereas in 1998, the built-up areas were 36 per cent of the territory, an increase of about 190 per cent in 10 years. Most of the built-up areas are residential. Due to this unprecedented growth, environmentally sensitive areas are presently under great pressure for development. Large tracts of land have already been developed within these areas, and the picture looks bleak unless immediate corrective measures are undertaken.

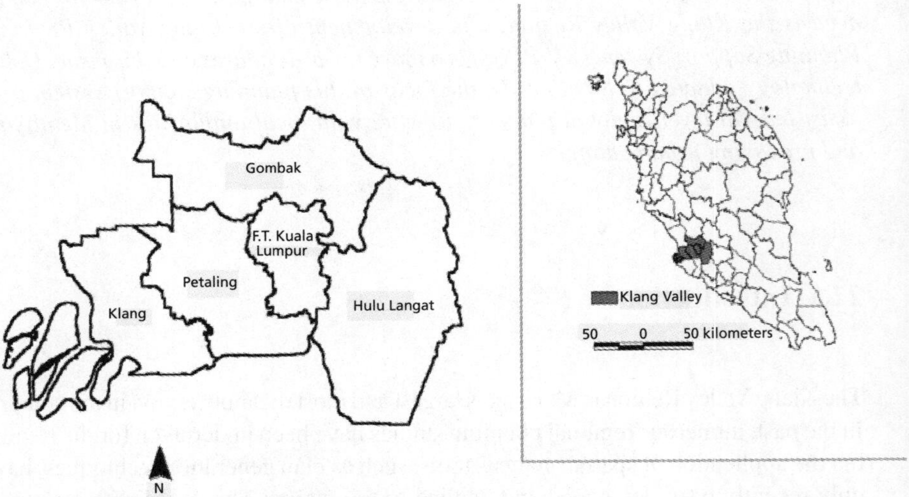

Fig 22.1: The location of the study area: the Klang Valley Region

22.3 History of GIS Implementation

A GIS-based system, beginning with simple data query and property level information, was established in 1995 through a pilot project for a small area in Kuala Lumpur. This was later extended in 1996 to cover the whole of Kuala Lumpur, focusing on government land. The whole exercise basically showed the strength of simple GIS functions like data query, tabulation of data and display of required maps.

22.3.1 From Property-based Data System to Regional Planning

Subsequently in 1997, the system was re-examined in scope and content to focus on the Division's regional functions, which included planning, coordination and monitoring development in the region. Region-based applications were soon developed through a session of brainstorming (involving selected planning authorities and university professors with GIS expertise) and numerous follow-up discussions with the appointed consultants. The brainstorming session was held in Port Dickson where a paper entitled 'GIS for Regional Planning' was presented and parallel sessions held to discuss relevant regional questions. This subsequently led to the setting up of a major GIS applications project in the Division, commonly known as AGISwlk (Applications of GIS for the Klang Valley Region).

This shift in approach was more in keeping with the role of the secretariat. This allows data to be collected for required analysis rather than large quantities of only marginally useful data. Considerations as to the number of purposes for which the data can be used, the availability of proxies and statistical techniques for reducing data requirement were considered. Numerous workshops were organised with various stakeholders, enabling the various sectoral analyses to be examined in regard to methodology, usage of relevant variables and the actual data to be used in the analysis. Such stakeholders' workshops continue to be the main activity of the current GIS system.

22.3.2 Regional Planning Support System for the Klang Valley

Scenario generation or plan generation is a vital component of the planning process. It allows the reassessment of growth strategies. Conventional non-IT approaches, like sieve mapping (Forbes, 1969), are usually too laborious to cater for numerous alterations from various stakeholders inputs. Besides lacking replicability, it is almost impossible to generate scenarios for various planning and implementation intervals (e.g. every five years). The advent of geographical information science has progressively been developed and utilised for spatial analysis, including for plan generation. Two plan generation approaches has been widely discussed elsewhere, i.e. the 'What-if' approach by Klosterman (1997) and the 'California Urban Futures Model' approach by Landis (1994, 1995). The latter approach in particular goes beyond conventional approaches by allowing planners to test the effects of alternative growth-management policies by considering precise GIS databases and private market logic (Godschalk *et al.* 1995).

So when the Klang Valley Regional Study (2000-2020) was initiated in 1997, it was intended that GIS technology would play a major part. Great concerns were placed on its analytical capability for scenario generation and impact analysis as well as its tracking module; these three modules were the major components of the regional planning support system (RPSS), a departure from conventional approaches for regional planning in Malaysia. Answering issues of replicability was an implicit concern, given the fact that reworking from a manual approach was virtually impossible due to lack of transparency in the documentation of techniques.

A conference with local experts in GIS was held in 1997 to examine the role and functionality of GIS analytical capability for regional planning. Generally, it was felt that in theory such implementation was possible but it required much collaborative input from sectoral consultants appointed to the study and the translation of forecasting models *vis-à-vis* a GIS environment. In the past, no such integration of analytical tools had ever been attempted successfully and was worthy of emulation. At the same time, it was noted that no available expertise in GIS was listed with the professional institute or with the treasury. Only a few disparate firms doing mapping with some CAD or mapping software were available in the country. So invitations were made to experts in the local universities for their involvement in the proposed study. We realised too that coordination may be an issue and lecturers cannot provide full-time attention to the study. Nonetheless, lecturers from Universiti Sains Malaysia, Universiti Malaya and Universiti Kebangsaan Malaysia took up the challenge.

22.4 The Structure and Component of the *KV-RPSS*

The *KV-RPSS* comprises of three main components or modules for scenario generation, impact analysis and planning monitoring (Rainis *et al.* 2001). Figure 22.2 shows the interrelationship between the three components and other GIS applications in the Division. Each of the components is described in detail in the following sections, with greater emphasis on the scenario generation module.

22.4.1 The Scenario Generation Module

The purpose of the scenario generation module is to support the generation of various alternative development land use plan scenarios. It comprises three major sub-modules for growth (demand) projection, land use suitability analysis and land use allocation. The modelling process of the scenario generation module is quite similar to the 'What-if' PSS developed by Klosterman (1997, 1999). We did not use the 'What-if' PSS for this project for several reasons. At the time that this study was initiated in 1997, the 'What-if' methodology was not readily available commercially. Furthermore, the modelling approaches (especially the land projection procedures) adopted by the 'What-if' approach were different from those used by our consultants. Projections were based on a five-year interval and the study area is comprised of five districts, which place greater complexity on the land allocation process.

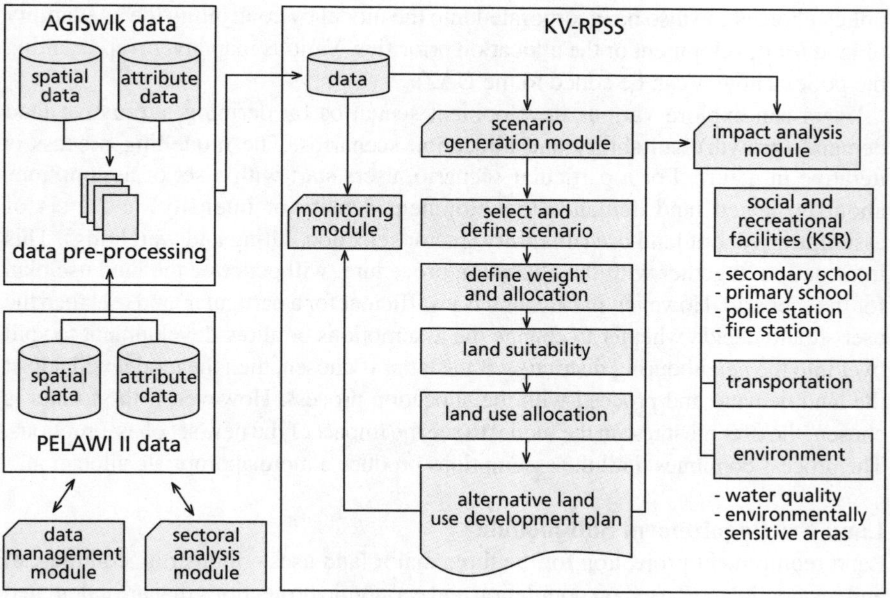

Fig 22.2: The relationship between the three components of *KV-RPSS* (scenario generation, impact analysis, monitoring) and other applications at the FTDKVPD

The plan scenario generation process involves about 20 factors. Overlaying these maps takes a lot of time. Therefore, in a manner similar to the 'What-if' approach, the model for scenario generation begins with pre-processed homogeneous uniform analysis zones (UAZs). These UAZs are the resultant polygons formed by a series of GIS overlays of all the factors taken into consideration. All areas within a UAZ have the same slope and land value, are within the same distance from rivers, roads and built-up areas *et cetera*. However, due to the size of the region and the many factors considered in the study, the number of UAZs was very large (more than 200,000 small polygons). Most of the UAZs were not critical to the scenario generation process.

Therefore, to reduce processing and modelling speed, only UAZs potentially available for development were extracted for further processing. These UAZs are subsequently known as developable land units (DLUs). In other word, DLUs are a subset of UAZs that are available for development. The DLUs will then be merged back with the rest of the UAZs for processing in the impact analysis module.

At the present time, *KV-RPSS* can only handle three broad land use categories: industrial, commercial and residential. The model projects future land use patterns by balancing the supply of, and demand for, land suitable for different uses at different locations. It considers alternative policy choices in a number of ways. Firstly, by applying them directly to the land demand projection; for example, by increasing or decreasing development density or intensity located in a particular district. Secondly, policy choices can also be incorporated into the model by controlling the availability of land for development or the allocation priorities. Various map layers representing the policy choices can be added to the UAZs.

Users can explore various development scenarios by defining alternative land demand (growth), suitability and allocation scenarios. The modelling process is iterative in nature. For a particular scenario, users start with a set of assumptions about projected land demands (development density or intensity) and a set of assumptions about land use suitability (factor selection, rating and weighting). This information, together with the allocation procedure, will generate the land use plan for the scenario. However, in case land is insufficient for a particular land use, then the user should decide whether to change the assumptions or allow development to spill over into the neighbouring district(s). If the latter is chosen, then the model will adjust the land demand and proceed with the allocation process. However, if the former is chosen, the user must re-run the model to see the impact of the new set of assumptions. The process continues until the assumptions produce a more appropriate allocation.

Land Use Requirement Sub-module

Land requirement projection for the three major land uses – industrial, commercial and residential – is based on population and economic projections that are calculated outside of the model. This means the user is allowed to choose any existing population and economic projection scenarios (as well as modify them appropriately). The model calculates land use requirements by district for every five years until 2020. The procedure for computing residential land involves three components: housing parameters, the percentage of other development and the rate of redevelopment.

There are five residential parameters considered in the model: backlog, immediate replacement, additional household, normal replacement and vacancy requirement. Backlog is the amount of housing units required to achieve a desired average household density per housing unit at the base year. The goal is to have one housing unit for each household. The backlog is the difference between the target and the existing stock. The need for immediate replacement is calculated based on the percentage of houses with the ‘temporary’ status. The need of additional households is calculated based on the average household size. Normal replacement refers to the amount of unit required to replace existing housing stock. Vacancy requirement is the amount of unit required to maintain a reasonable vacancy rate. Apart from residential houses, residential land use also includes other amenities such roads, social facilities, *et cetera*. These other uses usually constitute about 60 to 70 per cent of the total acreage. Similarly, the user is allowed to assume a certain percentage of redevelopment may be taken out of the total requirement. The procedure for computing commercial and industrial land demand is based on the employment projection. The user then specifies the assumed employment density (employees per hectare) for each land use category.

Land Use Suitability Sub-module

The land use suitability sub-module provides the capabilities to identify land supply or areas suitable for development for the three land uses. Like other studies, land use suitability analysis in this *KV-RPSS* adopts a multi-criteria evaluation (MCE) approach. Basically this involves five major steps: identifying the criteria, standardising factors, determining factor weights, identifying land use type to be considered for conversion and conducting the multi-criteria evaluation.

Establishing the Criteria: Factors and Constraints: A criterion is some basis for a decision that can be measured and evaluated (Eastman 1999). It is the evidence upon which an individual can be assigned to a decision set. Criteria can be of two kinds: factors and constraints. A factor is a criterion that enhances or detracts from the suitability of a specific alternative for the activity under consideration, and most commonly measured on a continuous scale. A constraint serves to limit the alternatives under consideration. In many cases, constraints will be expressed in the form of a Boolean (logical) map, areas excluded from consideration being coded as 0 and those open for consideration being coded as 1. In some instances, the constraint will be expressed in terms of some characteristic that the decision set must possess (Eastman 1999). For example, the total area of lands selected for residential development must be less than 100 hectares.

Based on discussions with various lead agencies and site requirements as provided for in planning guidelines, a total of 14 factors were identified as being relevant to the siting of the three land uses (Table 22.1). These factors are basically divided into two categories: site factors and location factors. The site factors taken into consideration include slope, and environmentally sensitive areas (areas prone to flood, soil erosion and river reserve), social-economic criteria (land ownership status, land value) and distance from the existing built-up area. The location factors take into consideration accessibility to a main road, access to a major interchange, proximity to a transit station, existing housing, industries, commercial areas and town centres. These criteria reflect the attractiveness of any particular zone for a certain land use. As shown in Table 22.1, there are seven factors for each set. The user must first identify which factors are to be considered for a particular land use. Different factors may be selected for different land uses. Suitability factors must be selected for each land use considered in the study, otherwise the default choice will be used.

Table 22.1: Factor types rating for each development scenario

factors considered	factor types	disperse			compact			transit		
site factors		**I**	**C**	**R**	**I**	**C**	**R**	**I**	**C**	**R**
1. slope	1. <= 0.5°	3	3	3	3	3	3	3	3	3
	2. 0.5 - 5.7°	5	5	5	5	5	5	5	5	5
	3. 5.7 - 12°	0	1	4	0	1	4	0	1	4
	4. 12 - 25°	0	0	2	0	0	2	0	0	1
	5. 25 - 35°	0	0	1	0	0	1	0	0	1
2. susceptibility to flooding	area susceptible to flooding	0	0	0	0	0	0	0	0	0
	area not susceptible to flooding	5	5	5	5	5	5	5	5	5
3. land ownership	government	5	1	3	5	1	3	5	1	3
	private	4	5	5	4	5	5	4	5	5
4. erosion risks	very low	5	5	5	5	5	5	5	5	5
	low	4	4	4	4	4	4	4	4	4
	medium	3	3	3	3	3	3	3	3	3
	high	0	0	2	0	0	2	0	0	2
	very high	0	0	1	0	0	1	0	0	1
5. distance to river	0 - 50 m	0	0	0	0	0	0	0	0	0
	50 - 100 m	0	0	1	0	0	1	0	0	1
	100 - 300 m	1	1	2	1	1	2	1	1	2
	300 - 500 m	3	3	3	3	3	3	3	3	3
	>500 m (luar buffer)	5	5	5	5	5	5	5	5	5
6. distance to existing built-up areas	<500m	1	5	5	1	5	5	1	5	5
	500 - 1500 m	2	4	4	2	4	4	2	4	4
	1500 - 3000 m	5	3	3	5	3	3	5	3	3
	3000 - 5000 m	4	2	2	4	2	2	4	2	2
	>5000 m	3	1	1	3	1	1	3	1	1
7. land values	RM0 - RM50	5	5	5	5	5	5	5	5	5
	RM50 - RM150	3	4	4	3	4	4	3	4	4
	RM150 - RM300	1	3	3	1	3	3	1	3	3
	RM300 - RM500	0	2	2	0	2	2	0	2	2
	>RM500	0	1	1	0	1	1	0	1	1
locational factors		**I**	**C**	**R**	**I**	**C**	**R**	**I**	**C**	**R**
8. proximity to major roads	1. 0 – 500 m	5	5	5	5	5	5	5	5	5
	2. 500 – 1500 m	4	3	4	4	3	4	4	3	4
	3. 1500 – 3000 m	3	1	3	3	1	3	3	1	3
	4. 3000 – 5000 m	1	1	2	1	1	2	1	1	2
	5. > 5000m	1	1	1	1	1	1	1	1	1
9. proximity to major road intersections	1. 0 – 500 m	5	5	3	5	5	3	5	5	3
	2. 500 – 1500 m	4	4	5	4	4	5	4	4	5
	3. 1500 – 3000 m	3	3	4	3	3	4	3	3	4
	4. 3000 – 5000 m	1	1	2	1	1	2	1	1	2
	5. > 5000m	1	1	1	1	1	1	1	1	1
10. proximity to transit/ LRT stations	1. 0 – 500 m	5	5	5	5	5	5	5	5	5
	2. 500 – 1500 m	3	4	5	3	4	5	3	4	5
	3. 1500 – 3000 m	1	3	3	1	3	3	1	3	3
	4. 3000 – 5000 m	1	1	2	1	1	2	1	1	2
	5. > 5000m	1	1	1	1	1	1	1	1	1
11. proximity to residential areas	1. 0 – 500 m	1	5	5	1	5	5	1	5	5
	2. 500 – 1500 m	2	4	4	2	4	4	2	4	4
	3. 1500 – 3000 m	5	3	3	5	3	3	5	3	3
	4. 3000 – 5000 m	4	2	2	4	2	2	4	2	2
	5. > 5000m	3	1	1	3	1	1	3	1	1
12. proximity to industrial areas	1. 0 – 500 m	5	1	1	5	1	1	5	1	1
	2. 500 – 1500 m	5	3	2	5	3	2	5	3	2
	3. 1500 – 3000 m	3	5	5	3	5	5	3	5	5
	4. 3000 – 5000 m	2	4	4	2	4	4	2	4	4
	5. > 5000m	1	2	3	1	2	3	1	2	3
13. proximity to urban centres	1. 0 – 500 m	1	5	5	1	5	5	1	5	5
	2. 500 – 1500 m	3	4	4	3	4	4	3	4	4
	3. 1500 – 3000 m	4	3	3	4	3	3	4	3	3
	4. 3000 – 5000 m	5	2	2	5	2	2	5	2	2
	5. > 5000m	3	1	1	3	1	1	3	1	1
14. proximity to commercial areas	1. 0 – 500 m	1	5	5	1	5	5	1	5	5
	2. 500 – 1500 m	3	4	4	3	4	4	3	4	4
	3. 1500 – 3000 m	5	3	3	5	3	3	5	3	3
	4. 3000 – 5000 m	4	2	2	4	2	2	4	2	2
	5. > 5000m	2	1	1	2	1	1	2	1	1

I = industry, C = commercial, R = residential

In all cases, the same constraints were used i.e. the following areas are not available for development:

- built-up areas;
- environmentally sensitive areas;
- areas where the slope is greater than 35 degrees; and
- land above 150 metres.

However, the actual land available for development for each land use will also depend on the choice of land use types that can be converted for such development. For example, not all agricultural land use types are available for certain types of development.

Standardising the Factors: Usually factors are measured on a continuous scale, but in different units. For instance, slope is measured in degrees while all the proximity measures are in distance units, i.e. metres. In order to combine these factors into a single composite index, they need to be standardised and given certain ratings or scores. There are a number of ways of doing this. The simplest is a linear scaling approach using the minimum and maximum values as the scaling points (Eastman 1999). The other option is to categorise the factor into different types and each type is given a certain rating or score. For example, slope could be categorised into five classes: 0-2, 2-6, 6-10, 10-15, and above 15 per cent. Each of these classes is given a rating based on its suitability for certain purposes, lets say 5, 4, 3, 2, and 1 respectively.

This is the most common approach to standardising factors in the suitability analysis literature. This latter approach is the approach adopted in *KV-RPSS*. The ratings range from 0 to 5, where a 5 indicates highly desirable while 1 means least desirable. A rating of 0 is used to exclude the area from development, regardless of its rating on other factors. For example, the user can specify that development is to be excluded from areas with slope greater than 10°, regardless of those areas' suitability with respect to other suitability factors. Factor ratings must be specified for all of the land uses considered in the analysis; otherwise the system will use the default (or previously defined) values.

Establishing the Factor Weights: Each factor is given a weight to reflect its relative overall importance or contribution in determining the outcome of an evaluation. In the simplest case, we can assume equal weight among the factors (this is default option). However, in practice the importance of factors varies. For example, under the transit-oriented development scenario, a high weighting is given to factors related to the transit station; a high score rate may be assigned to give preferences for residential location to areas in close proximity to transit station. A wide variety of techniques exist for the development of weights. In the present version of the *KV-RPSS*, the weight for each factor is given based on an arbitrary judgement from discussions with planning officers from the Klang Valley Planning Division, with preliminary weights been revised. The total weighting for each set of factors is 100. Since the maximum rating for each factor is 5, the overall maximum suitability score will be 1,000, i.e.500 points for each set of factors. Factor weights must be assigned for all land uses considered in the analysis, otherwise the system will use the default (or previously defined) values.

Identify Land Use Type Considered for Conversion: Not all land use types are available for development. For example, only Class 1 and 4 agricultural lands may be used for industrial land uses. For each land use to be allocated, users may choose different land use types for conversion. The choices of land use types for conversion determine the amount of land available for development. A user must identify what land types are available for conversion for all land uses considered in the analysis, otherwise the system will use the default (or previously defined) values.

Undertaking the Multi-Criteria Evaluation: Once the criteria maps (factors and constraints) have been developed and the weights were established, the next step is to undertake the multi-criteria evaluation to combine the information from the various factors and constraints. This process uses the weighted linear combination method to produce the overall composite *suitability scores* that indicate the relative suitability of different locations for each land use. The location with the highest suitability score is the most suitable for that land use. The resulting set of suitability scores are then used to allocate projected demand to alternative sites. The suitability analysis results can be displayed in map or report form.

Land Use Allocation Sub-module

KV-RPSS provides two options to carry out the land use allocation: unconstrained or constrained approaches. In the unconstrained allocation process, land demand can be allocated freely within the study area. However, as stated earlier, the study area is administratively divided into five different districts and the land requirement was also projected accordingly. Thus, the user can choose to constrain the allocation process by district. However, this option will increase the complexity as well as slow down the allocation process, especially if certain districts do not have enough land available to satisfy the projected demand.

No matter which allocation procedure is selected, *KV-RPSS* allocates the projected demands for each land use in each projection year to different locations on the basis of their relative suitability for that land use. Presently, land use is allocated sequentially based on user-specified allocation priority. For example, if industrial land use is chosen to be the first priority; then lands are allocated to that land use first until all of the projected demand for a projection year has been satisfied. The land use is assigned first to the most suitable sites, then to the second most suitable sites, and so on. The user is notified if not enough land is available to satisfy the projected demand. If this occurs, the user must modify the growth assumptions (e.g. increase density) or suitability assumptions (e.g. release more land types).

In the case of constrained allocation, the user is also allowed to move the unmet projected demand to neighbouring districts or other districts (if any) that have extra land available for development. A land accounting model incorporated into the system provides the checking mechanism. In addition, the user can also control the growth allocation by opting to respect the existing structure plan(s). Most local authorities in the study area already have their own structure plans. In this case, the projected land use demand will first be allocated to areas in which the land use is planned under the structure plan. Projected demand not met by the structure plan will be allocated according to the procedure described above. The allocation analysis results can be displayed in map, tabular and graph form.

KV-RPSS interface

Figures 22.3 to 22.6 show examples of the interface for the scenario generation module. Some of the interfaces have been translated to English for the purpose of this chapter; the original interfaces are in Malay language. Figure 22.3 shows the main menu of the *KV-RPSS*, which provides user with four main options – Scenario Generation, Impact Analysis, Monitoring and Data Input.

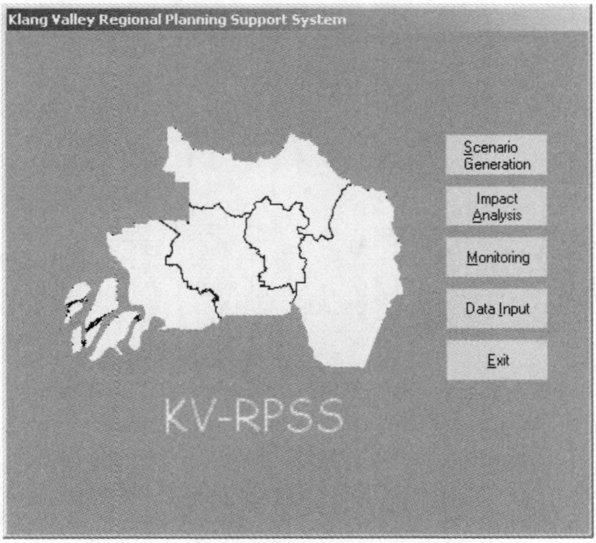

Fig 22.3: The main interface of the *KV-RPSS*

Figure 22.4 shows the initial interface for the land use projection sub-module. In general, this interface consists of two rows of tabs and a spreadsheet to display or input the relevant data. The first row of three tabs corresponds to the three major land use categories: residential, commercial and industrial. The second row of tabs corresponds to the major variable types that the user specifies for a particular land use, thus might change with land use.

The initial interface for the residential land use projection is shown in Figure 22.4. This screen consists of four tabs: Population, Housing factors, Percent of other development and Percent of redevelopment. The first tab is used to enter the population projection data. The user has the option to use population data calculated exogenous of the system (the default option) or input his/her own data. The second tab allows the user to specify the density and the five housing parameters i.e. backlog, immediate replacement, household increase, normal replacement and vacancy requirement. The Percent of other development tab allows the user to specify the percentage of other development to be added to the residential land use projection. These include road and utility reserve, social facilities and open space. The Percent of Redevelopment screen allows the user to input the percent of redevelopment for each of the five districts in the study area. The system assumes that redevelopment will fulfil a certain percentage of the new land use demand.

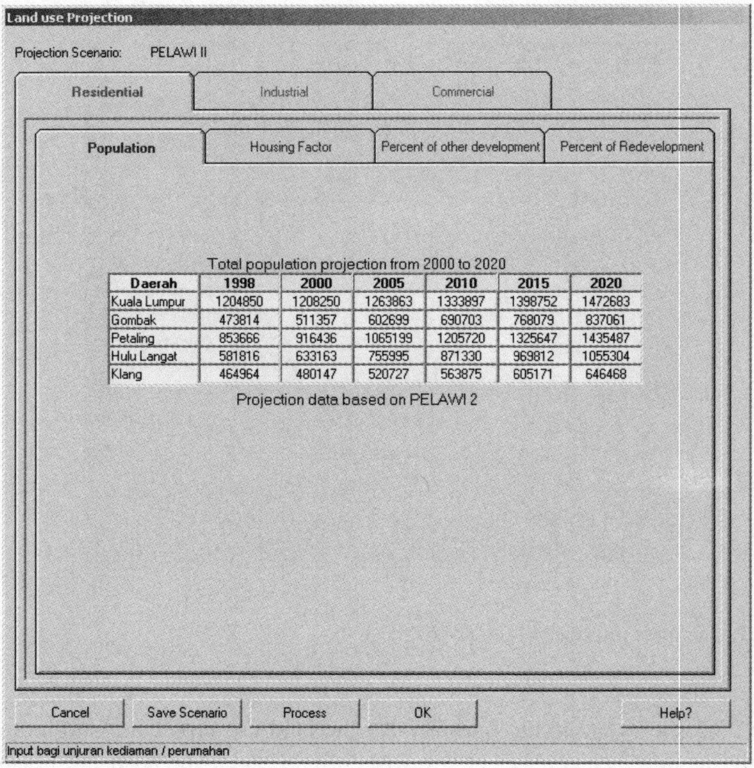

Fig 22.4: The initial interface of the land use projection sub-module

Figure 22.5 shows the basic interface of the suitability analysis sub-module. This interface allows user to specify the parameters for the Site Suitability, Locational

Suitability, and Land use Conversion. If the user selects the Site Suitability or Locational Suitability tab, the second rows of tabs will become active and consists of three tabs: Factor Considered, Factor Weight and Factor Rating. For each land use, these options allow the user to select the factors to be considered, their weights and ratings. There are seven possible factors for each set of suitability (as shown in Table 22.1).

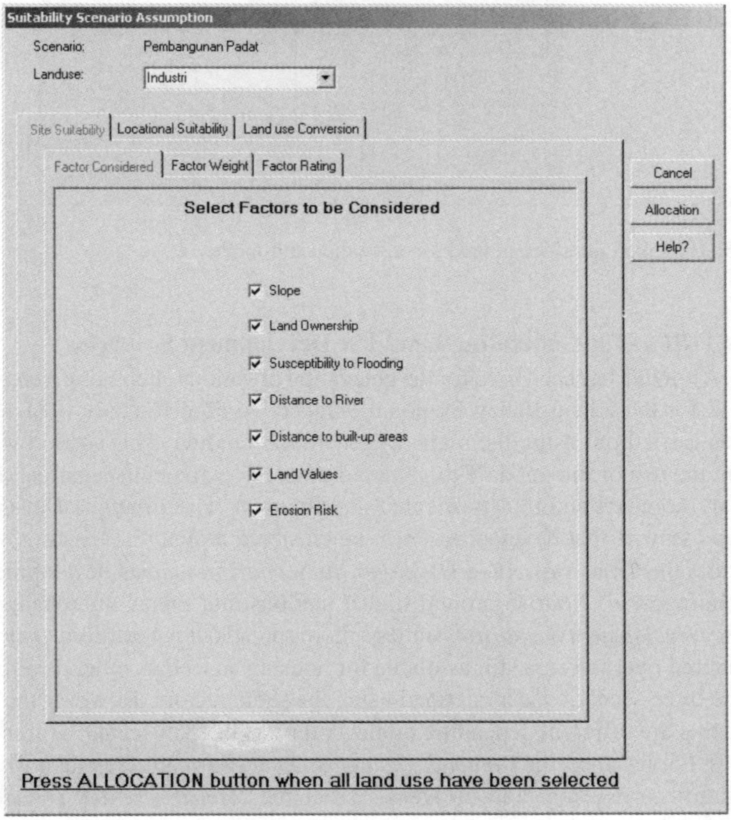

Fig 22.5: The initial interface of the land suitability analysis sub-module

The user can specify the land use allocation procedure using the interface shown in Figure 22.6. The interface comprises two major options (tabs): Allocation Priority, and Development Control. The Allocation Priority screen allows the user to determine the order in which the land use allocation process is carried out. The default is industrial land use as the first priority, followed by commercial and residential land uses. The user can also specify additional constraint(s) to the allocation process through the Development Control screen. In the present version, the user has two options – consider existing structure plan, or/and to allocate by district.

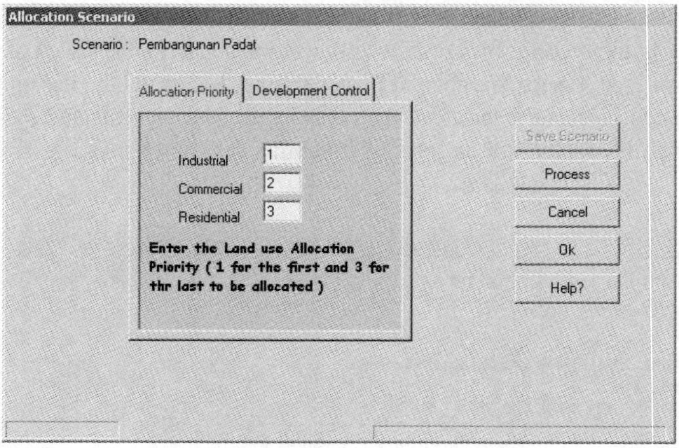

Fig 22.6: The initial interface of land use allocation sub-module

Using *KV-RPSS* for Generating Land Use Development Scenarios

To date, *KV-RPSS* has been used for the generation of various alternative development scenarios for the Klang Valley Regional Perspective Plan II Study (PELAWI II). This was carried out using the scenario generation module. This section will only focus on the use of this module to generate three development scenarios: disperse (as usual), compact and transit-oriented development. The *Compact Development Scenario* assumes that development will be clustered around the existing build up area, while the *Transit-oriented Development Scenario* assumes development will (be encouraged to) flourish around transit stations and major interchanges. The *Disperse Development Scenario* assumes a more spread out nature along major roads and scattered over the area still available for housing as well as other development.

All the three scenarios were defined using the same factors. However the ratings and weights are different depending on the emphasis of each scenario (Tables 22.1 and 22.2). For instance, the *Compact Development Scenario* gives greater weighting to proximity to existing built-up areas. Under the *Transit-oriented Development Scenario*, a high weighting is given to factors related to the transit station.

Results from the analysis indicate that a few of the districts in the region did not have enough land to accommodate the projected demand until 2020. These unmet demands were allocated to neighbouring districts that have a surplus of land suitable for development. Figure 22.7a to c show the distribution of future land uses under the three scenarios for the Klang district. The Klang district is the only district that has more land supply than required. For this reason most of the unmet demands of the neighbouring districts were allocated to this district. As expected, new development will be more dispersed under the *Disperse Development Scenario* (Figure 22.7a). The *Compact Development Scenario*, on the other hand, produces development that will be more concentrated around existing built-up areas (Figure

22.7b). Meanwhile, new development is expected to be concentrated around transit nodes under the *Transit-oriented Development Scenario* (Figure 22.7c).

Table 22.2: Weighting for each factor, land use and development scenario

	scenario								
	disperse			compact			transit		
site factor	I	C	R	I	C	R	I	C	R
slope	20	20	20	10	10	10	10	10	10
susceptibility to flooding	10	10	10	10	10	10	10	10	10
land ownership	20	20	20	10	10	10	10	10	10
erosion risks	10	10	10	10	10	10	10	10	10
distance to river	10	10	10	10	10	10	10	10	10
distance to existing build areas	10	10	10	40	40	40	40	40	40
land values	20	20	20	10	10	10	10	10	10
total	100	100	100	100	100	100	100	100	100
locational factor									
proximity to major roads	30	30	30	15	15	15	20	20	20
proximity to major road intersections	20	20	20	5	5	5	0	0	0
proximity to transit/LRT station	5	5	5	10	10	10	30	30	30
proximity to residential areas	10	20	20	15	15	25	10	10	15
proximity to industrial areas	20	10	10	25	10	10	20	5	5
proximity to urban centres	15	15	15	20	20	20	10	20	20
proximity to commercial areas	10	10	10	10	25	15	10	15	10
total	100	100	100	100	100	100	100	100	100

I = industry, C = commercial, R = residential

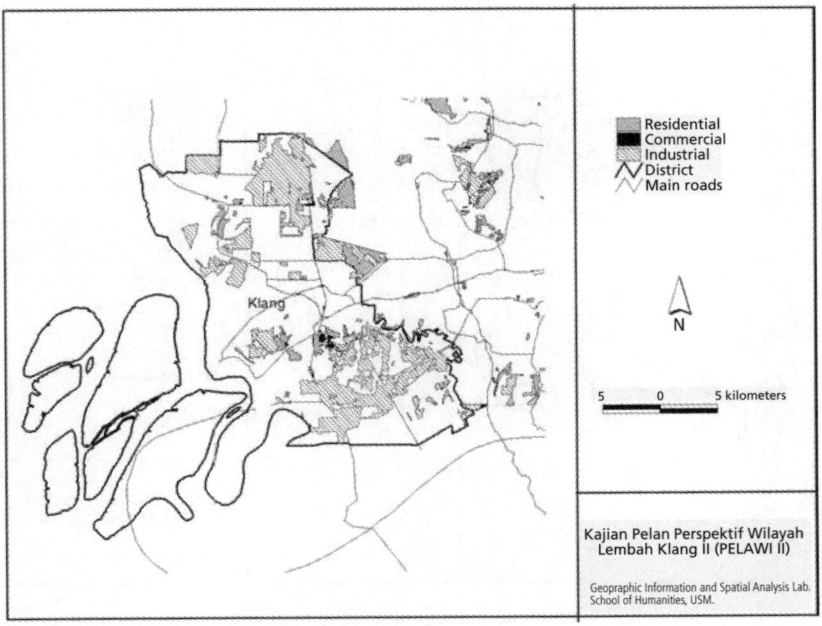

a)

Fig 22.7. Future land use pattern under (a) Disperse Development Scenarios for year 2020

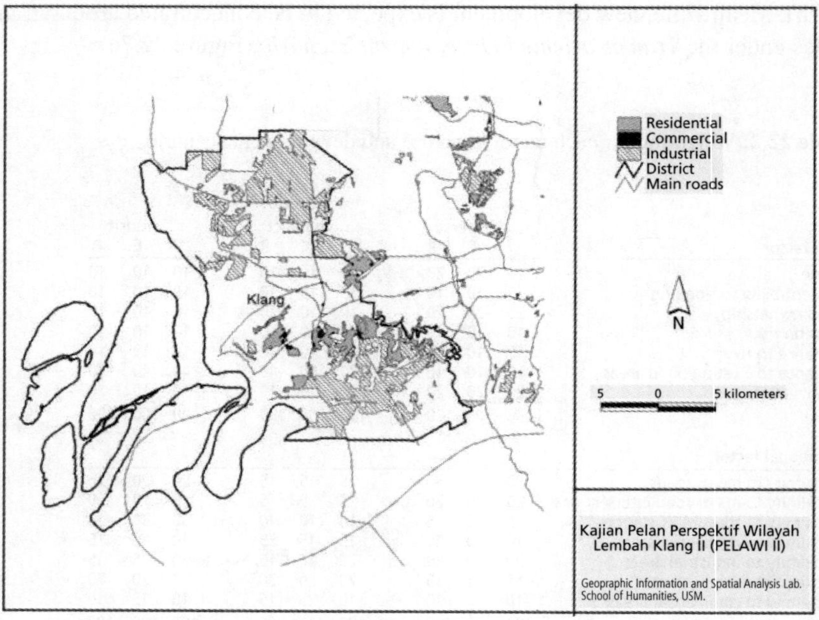

b)

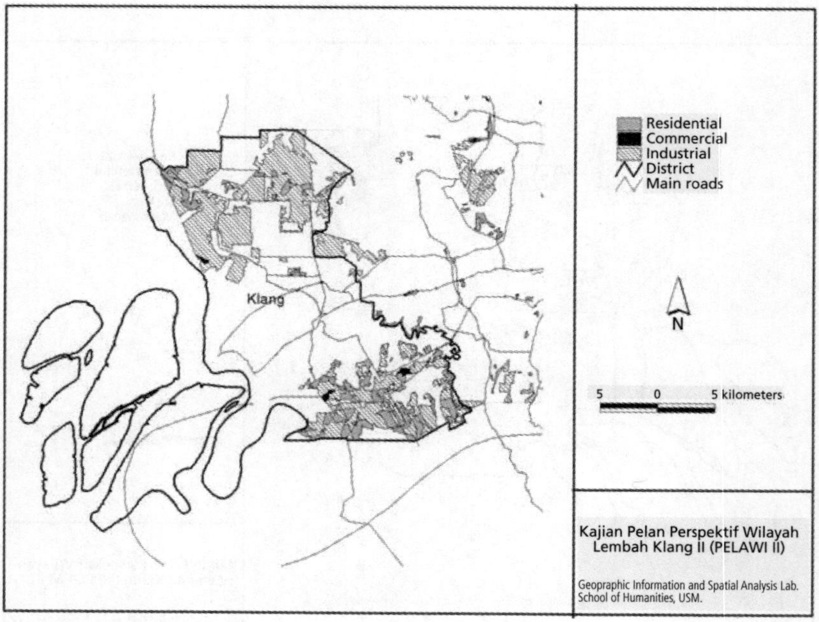

c)

Fig 22.7. Future land use pattern under (b) Compact and (c) Transit-oriented Development Scenarios for year 2020

Preliminary evaluation of the three scenarios indicates that none of them is suitable for all of the five districts. This lead to the creation of a hybrid scenario whereby each district adopts the development scenario that suits it owns characteristics and growth target. For instance, Kuala Lumpur and Petaling are highly developed districts, thus the *Transit-oriented Development Scenario* is more appropriate for these two districts. Meanwhile, the *Disperse Development Scenario* is deemed appropriate for Klang and Hulu Langat districts because they are less urbanised and have more land available for development.

22.4.2 The Impact Analysis Module

The impact analysis module has dual purposes. First it could be used for impact analysis of the alternative development scenarios generated by the scenario generation module. Secondly, it could also be used for monitoring purposes. At present, the module is limited to analyses related to three major sectors as follows:

Social and Recreational Facilities Including Police Station, Fire Station, Primary and Secondary Schools: The module analyses the availability and impact of existing and new development on the requirement for the various facilities.

Environment Including Water Quality and Environmental Sensitive Areas: Water quality is analysed using the water quality index (WQI) as developed by the Department of Environment (DoE). This DoE-WQI is calculated based on six chemical parameters: dissolved oxygen, biological oxygen demand, chemical oxygen demand, suspended solid, nitrogen and pH. Using the index (which ranges from 0 to 100), rivers in the study area will be classified into three categories: clean, dirty and very dirty. In terms of environmental sensitive areas, the module calculates the amount of sensitive areas directly affected and within certain distance from the proposed development.

Transportation: The general transportation analysis is based on logit modelling to examine the impact of land use changes on the aggregated need for movement for each of the four transportation modes – car, motorcycle, bus and rail. The Klang Valley region and its surrounding areas were divided into nine transportation zones of which five are internal zones (the five districts within the region) and four external zones (that surround the region). The analysis consists of two major parts: traffic generation and trip distribution, and modal split modelling. The origin-destination matrix for the four transportation modes is generated based on numerous data including housing density, average household size, commercial plot ratio, industrial plot ratio, travel time, frequency of bus services, average parking rate for car, average parking rate for motorcycle, average car ownership, average motorcycle ownership, average distant to rail/transit station, car operating cost, motorcycle operating cost, average travelling cost for bus, average travelling cost for rail, average travelling distance for car, average travelling distance for motorcycle and toll rate.

Depending on the analysis sector, the output of the module could be in the form of a map (thematic or travel desire lines), a cross-tabulation, a matrix table or a chart. Figure 22.8 shows a map of travel desire lines for car transport between the eight traffic zones in the study area.

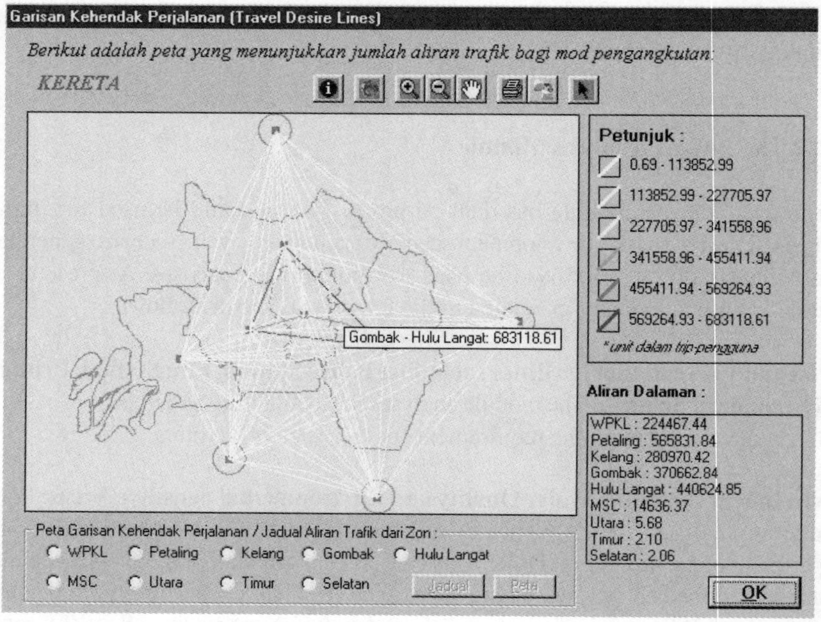

Fig 22.8: Example output from the impact analysis module – trip generation between zones

22.4.3 The Planning Monitoring Module

The monitoring module is developed to assist the coordinating agency to regularly monitor the implementation and performance of the perspective regional plan as well as other development (structure and local) plans in the region. Relevant monitoring or performance indicators were developed for each of the ten important sectors in the study: regional framework, population, economic, urban development, land use, housing, social and recreational facilities, environment, transportation and utility. These indicators were used in order to:

- measure the performance and consistency of the various programs and policies;
- examine development trends;
- monitor the urban or regional condition;
- (re)define the development objectives;

- (re)assess the development targets;
- compare localities in space and time;
- guide the selection of a strategy; and
- act as an early warning system.

The output of the module is either in the form of table, chart or map. Figure 22.9 shows the distribution of population growth between 1980 and 1991 for the five districts in the study area.

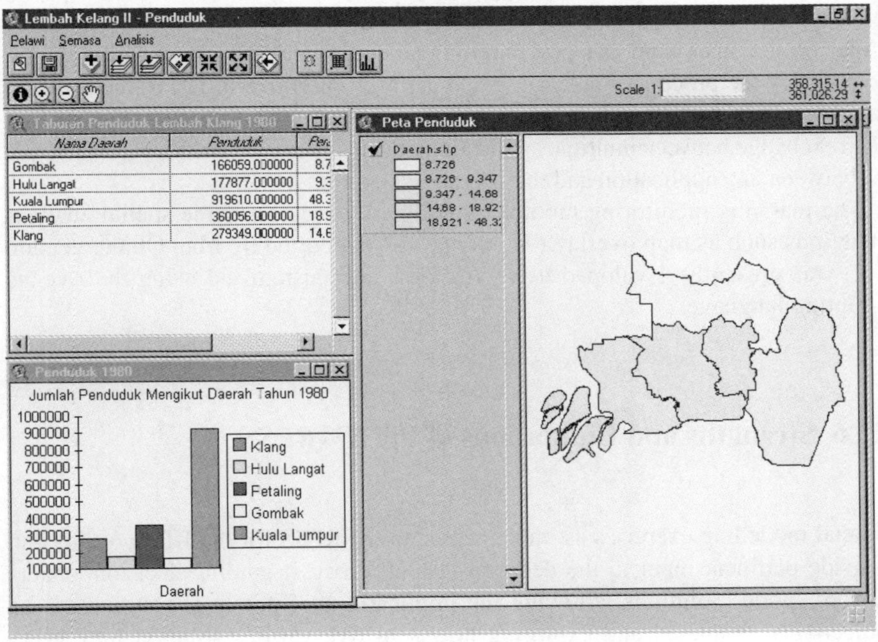

Fig 22.9: The general output of the monitoring module

22.5 *KV-RPSS* Application Development

Developing a stand-alone customised application such as *KV-RPSS* is necessary especially when 'out-of-the-box' software application packages such as ArcView or Arc/Info are either overly complex or too generic to support users' needs efficiently. The existing software must either be altered or completely rewritten with a programming language like Visual Basic or C++ to automate certain repetitive procedures as those analyses found in the *KV-RPSS*, as discussed above.

KV-RPSS may probably be the first attempt to design and develop an in-house planning analysis machine specifically tailored for local (Malaysian) applications. The scenario generation and impact analysis modules were developed using Map Object and Visual Basic. Using Map Objects and Visual Basic, *KV-RPSS* was programmed to read, analyse, and display spatial data stored in ArcView shapefile format and attribute data table stored in Microsoft Access.

Map Object, by Environmental Systems Research Institute (ESRI), is a collection of programming tools that extends the Microsoft Visual Basic language to include dynamic GIS maps. It is an ActiveX control (OCX) with more than 45 programmable automation objects that can be plugged in to many standard window development environments such as Visual Basic and C++ (ESRI 1998). Microsoft's Visual Basic (VB), on the other hand, is a programming language intended, among other things, to simplify the process of creating graphical user interfaces (GUI) in the standard windows style. In GIS environments, VB provides an effective means of managing data exchange between multiple applications, such as between ArcView and Access, or between any application and the operating systems.

The planning monitoring module, which partly requires some spatial analysis functions such as map overlay (which was not supported by Map Object version 1.2) was presently developed using ArcView and customised using the Avenue scripting language.

22.6 Strengths and Limitations of the *KV-RPSS*

Spatial modelling exercises as implemented in the RPSS for the Klang Valley can provide pertinent input to the decision-making process: guiding decision-makers toward 'good' solutions, allowing the implications of potential solutions to be objectively evaluated and exploring how sensitive various solution may be to assumptions or estimates incorporated into the analysis.

Although site suitability analysis and activity/land allocation implemented in *KV-RPSS* by its nature focuses on optimisation of pre-defined criteria, *"the real value of this modelling exercise is not entirely in its solution space, but rather their ability to explore and compare solutions"* (Berry 1993). The system allows the end user to test 'what-if' situations – e.g. if we allow agricultural land to be converted to other development types, how will this change the total land stock and the resulting development plan? The end user may also change the relative importance of the 14 factors incorporated within the system in terms of their weighting (or degree of importance). This in effect is equivalent to changing parameters of site suitability and thus will generate a different land use pattern.

The analysis can be replicated, repeated and re-tested. With the step by step graphic user interface (GUI), the system guides the end user to a more structured analysis of existing land use, defining land requirements, automating site suitability analysis and subsequently allocating the various land uses based on site suitability. Hence *KV-RPSS* integrates a number of analytical models into a workable framework. Similar analytical structures have been discussed by Kaiser *et al* (1999) and by Stillwell and Scholten (2002).

The current *KV-RPSS* needs to overcome a number of methodological and technical limitations, which hopefully could be overcome with continuous research and technological advancement in the GIS industry. One rather serious limitation of the system is its lack of a data input facility. Presently spatial data are pre-processed in GIS software – in this case an ArcView shapefile and Arc/Info for Windows NT.

KV-RPSS adopts a deterministic modelling approach. Variable inter-relationships are either known or taken as given. As an example, the model assumes land requirement for public facilities as 10 per cent of the total housing area. Instead of allocating to suitable sites, land requirement for public facilities is incorporated into the housing requirement. Land use allocation is solely based on predefined suitability criteria. No profitability, development cost or price information was incorporated into the model. Specific use of a land parcel is driven by the profitability of alternative development outcomes on the land, which in turn derives from the trade off between development costs and the consumer's willingness to pay for the development.

The vector data structure is rather rigid for the purpose of simulation and modelling. Accessibility to transit nodes, for example, has been defined as a series of buffer zones at 0.5 km intervals. As these buffer zones have been predefined and the data has been pre-processed, there is little possibility to automate variation in the interval, into say, 0.6 kms.

Computer processing time ironically appears 'slow'. The regional dataset contains about 30,000 DLUs polygons. Due to massive number of calculation iterations, reclassifications of spatial data, object and database searching, and linking of relational database, the *KV-RPSS* scenario generation module takes about 45 minutes to process, making it less suitable for impromptu analysis. This may be improved if a raster data model is used. However, Map Object version 1.0, which was used as the backbone of the system, does not support raster data processing (although Grid module of ArcInfo does support it).

The present model is also spatially static. The new land use configuration from each iteration (time interval) was not used as input to the subsequent iteration as originally conceptualised. This was mainly due to the limitations of the software (Map Object) and data model (vector) being used. Map Object version 1.0 does not have advanced GIS analysis functions (a newer version 2.0 does have both functions). Such vector data format requires longer processing time and is not well suited for overlay and neighbourhood analysis of large data set.

As described earlier, the study area is administratively divided into five different districts. Each district has its own characteristics and set of growth targets. Potentially, each district could have its own development scenario. However, in the present system, a development scenario can only be defined for the entire study area. It is not possible to define a separate scenario for each district. Experiences of using the system indicate that there is a need to have a capability to generate a hybrid development scenario.

Another limitation of *KV-RPSS* is concerned with model calibration. The models incorporated into the system are readily available and thus taken as given. At the moment, these models have not been properly tested nor verified. It thus requires calibration which is a matter for future research.

22.7 Implications for Planning Practice and Future Development

Discussion of plan generation inevitably opens up a Pandora's box of past and existing practices in Malaysia. The norm was to take short cuts thereby avoiding undertaking a multitude of factors and rules of land use allocation processes mentioned in this chapter. Manual sieve mapping, as carried out by private consultants for various planning studies, will have to move beyond colouring contour maps for producing suitability maps; it has to migrate to more relevant techniques when generating future or alternative options. Drawing circles and giving urban functions to future centres needs to be supplemented with relevant spatial tools.

Structure and local planning studies from the year 2002 onwards are expected to go fully digital with regard to data usage and it is expected that spatial analysis tools for plan generation will be used by planning consultants. The Federal Town and Country Planning Department has produced a manual on the application of GIS tools for development planning studies and this would hasten application of GIS technology for planning studies in the country.

How might the *KV-RPSS* modelling approach be of value to Malaysian planning activities? The Klang Valley RPSS is a beginning and is probably the first attempt to use such a modelling approach within the plan generation process. One possible scenario would be to improve and adapt the model as a general-purpose planning tool that blends the planning functions of developing, monitoring and revising regional, structure and local plans. However, this will entail a synthesis of the functions of land use modelling and monitoring in a way never previously attempted.

It would also require a more active multi-participant collaboration that involves local, municipal and regional as well as state agencies as partners. The collaborative aspect in plan generation is crucial for successful considerations of all relevant factors. The potential technical components of an integrated system to support such a collaborative planning system exist, but the political and organizational obstacles would have to be overcome first. AGISwlk, the GIS system at the Prime Minister's Department has taken this lead, and progress is good considering prior to this they had to break down barriers to opening up individual departmental interest; but the journey is long and full of pitfalls. This aspect of collaboration (within critical theory) has been widely discussed elsewhere and is becoming a potent communicative tool for successful GIS implementation, not only for plan generation but for other aspects of development planning, evaluation and tracking purposes (see Kaiser *et al.* 1999, Innes 1998, Klosterman 1999, and Nedovic-Budic 2000).

Within the Malaysian context, the development and usage of GIS for plan generation, impact analysis and monitoring has still a long way to go, as almost all planning consultants are only beginning to use GIS as a mapping tool. Perhaps five to ten years from now will we see that applied GIS is implemented in plan generation and other planning processes. The availability of packages tailored for planning application (for example Klosterman's 'What-If') and GIS software providing decision-making functions (such as Clarke University's IDRISI for Windows) are already able to undertake these tasks nowadays and it is expected that their usage by Malaysian planners will only be a matter of time.

Initially we intended to develop a spatially dynamic model based on cellular automata modelling approach. Cellular automata is becoming a popular modelling approach in urban modelling as well as various other fields (Couclelis 1987, Itami 1988). Actually Chapin *et al.* (1962) used a similar approach in the 1960s (but the term cellular automata was probably not yet well-known at that time). In land use modelling, the model basically attempts to simulate the dynamic of land use development. The land use evolves from a few cells to a complex urban development. The potential of an area for new land use development is influenced or controlled by a set of transition rules. These rules comprise of factors that hinder or encourage new development to occur from time t to t+m. Therefore, the major component of the model is the formulation of a function for the transition rules. The function can be generated by various methods such as heuristic (rule of thumb) or statistical approaches such as discriminant analysis (Rainis and Narimah 1996) or multi-logit modelling (Landis and Zhang 1997, Rainis *et al.* 2001). This modelling approach is easier to implement using raster data. Raster-based modelling is good in terms of speed and dynamics, but we would need to resolve the size of resolution and how that would affect information. It certainly would be suitable at regional level. The next version (generation) of *KV-RPSS* might follow a similar path. Funding to improve and enhance the present system for the year 2001/2002 is presently being requested from the Treasury Department, Ministry of Finance.

The Analytical Hierarchy Process (AHP) was earlier given some consideration for the study in order to come up with weighting/priority consistently (i.e. for measuring the consistency of weights of all factors). But as the study uses 14 factors, it was deemed too time consuming to conduct pairwise comparison. AHP would ideally be managed with around seven factors. With the decision to sub-divide the factors into two sets, each set comprising of seven factors, such an approach is becoming feasible.

The impact analysis module is presently limited to three sectors: social and recreational facilities, transportation and environment. It should also include other pertinent issues in the region such as flooding. Existing models in *KV-RPSS* such as transportation could be improved further from the current logit modelling capability. In addition, physical site factors could be further improved by incorporating an environmental model (including flooding and other geohazard factors in particular since AGISwlk has reasonably good data on these from its latest development).

Last but not least, suitability could be calculated based on profitability/economic return. It may also be based on a consumer preferences model (bid function) – similar to John Landis' CUF model I or a multi-logit (statistical) model – and also based on John Landis' latest version of California Urban Futures Model II.

22.8 Conclusions

This chapter has highlighted a number of areas showing the relevance of a GIS-based regional PSS capability and its limitations (especially the plan generation module) in the planning process. We believe that much more effort in this direction should be encouraged, thereby improving choices in alternative plan selection by planning periods. These further 'test' the viability of growth strategies that require re-evaluation over time. Together with impact models and also other supporting tools, development tracking and timely adjustment to development policies and strategies could be effectively undertaken. Perhaps the concomitant advent of GIS technology in plan generation is its encouragement towards more fruitful collaborative input from other stakeholders in the whole development planning scene – a phenomena yet to be fully harnessed in Malaysia although our hopes are high.

Acknowledgements
The development of *KV-RPSS* is funded under the Klang Valley Regional Planning Study (PELAWI II). The authors would like to thank the staff of the Federal Territory Development and Klang Valley Planning Division, Prime Minister Department and the various consultants for their input, assistance and involvement in the project.

References

Berry, J.K. (1993) Beyond mapping: concepts, algorithms, and issues in GIS, *GIS World*, Fort Collins, Colorado.

Chapin, F.S., Weiss, S.F. and Donnelly, T.G. (1962) *Factors Influencing Land Development: Evaluation of Inputs for a Forecast Model*, An Urban Studies Research Monograph, University of North Carolina at Chapel Hill and U.S Department of Commerce Washington D.C.

Couclelis, H. (1987) Cellular dynamics: how individual decisions lead to global urban change, *European Journal of Operational Research*, 30: 344-346.

Eastman, J.R. (1999) *Idrisi32: Guide to GIS and Image Processing,* Clark Labs, Clark University, MA, USA.

ESRI (1998) *Building Applications with Map Object,* Redlands, California.

Forbes, J. (1969) A map analysis of potentially developable land, *Regional Studies*, 3.

Innes, J.E. (1998) Information in communicative planning, *Journal of the American Planning Association*, 64(1): 52-63.

Itami, R.M. (1988) Cellular automatons: as a framework for dynamic simulations in Geographic Information Systems, in *Proceedings GIS/LIS*, San Antonio, 2: 590-597.

Kaiser, E., Godschalk, D.R. and Chapin, F.S. (1999) *Urban Land Use Planning*, University of Illinois Press, Urbana Champaign, Illinois.

Kamalruddin, S. and Nuha, M. (1999) Usage of GIS Technology for Regional Planning and Monitoring: The Klang Valley Experience, Paper presented at the XIIIth Malaysian Senior Town Planning Officers Conference, Perlis, Malaysia.

Klosterman, R.E. (1997) The What if? Planning Support System, *ESRI Annual User Conference*.

Klosterman, R.E. (1999) The What If? collaborative planning support system, *Environment and Planning B*, 26(3): 393-408.

Landis, J. (1994) The California Urban Futures model: a new generation of metropolitan simulation models, *Environment and Planning B*, 21:399-420

Landis, J (1995) Imagining land use futures: applying the California Urban Futures model, *Journal of the American Planning Association*, 61(1). 438-457.

Landis, J. and Zhang, M. (1997). Modeling urban land use change: the next generation of the California Urban Futures model, Paper presented at the *Land use Modeling Workshop*, Sioux Falls, South Dakota, USA.

Nedovic-Budic, Z. (2000) Geographic information science implications for urban and regional planning, *URISA Journal*, 12(2): 81-93.

Rainis, R. (2000). Spatial and Temporal Analysis of Land use Changes due to Urbanisation and Infrastructure Development, On-going research program, Urban Dynamics Research Group, Geoinformatic Unit, Universiti Sains Malaysia.

Rainis, R., Mohd Nazari, J. and Hadi Herman, S. (2001) *KV-RPSS Technical Documentation,* A consultancy report submitted to the Federal Territory Development and Klang Valley Planning Division, Prime Minister Department.

Rainis, R. and Narimah, S. (1996) Modelling urban residential land use development using GIS, *Journal HBP,* USM, 3:10-18.

Stillwell, J.C.H. and Scholten. H.J. (2002) *Land Use Simulation for Europe*, Kluwer, Dordrecht.

23 The LiNC Viewer: An Information System Designed to Assist with Land Reform

Marilene Heunis[1], Tania Smith[1] and Elizabeth Gavin[2]

[1] Department of Land Affairs, Private Bag X954, Pretoria 0001, South Africa
[2] ISLinkup, PO Box 13371, Hatfield, Pretoria 0028, South Africa

Abstract: The National Department of Land Affairs received a mandate in 1994 to develop a land reform programme aimed at redressing the legacy of unequal land distribution in South Africa while contributing to national reconciliation, growth and development. The LiNC Viewer is a web-based, spatial information system developed for use by the Department of Land Affairs who plan and manage land reform. The major advantage of the LiNC Viewer to the land reform officials is immediate access to spatial information concerning the location of the property or properties in relation to infrastructure and services, relating to the feasibility of the project. Furthermore, costs and time spent travelling to the areas can be saved. This chapter explains the steps followed to establish the LiNC Viewer, the data that are available in the system, the system's functionality and the future development plans. In general, the response of users to the system has been positive and this enthusiasm must be exploited in order to make the use of spatial information an integral part of the business of land reform. Difficulties were encountered in developing the LiNC Viewer, as well as in integrating its use into the land reform process. The difficulties are discussed and the steps taken to address these are outlined.

23.1 Introduction

The *LiNC Viewer* is a web-based, spatial information system developed for use by the National Department of Land Affairs (DLA) in South Africa in planning and managing land reform. Around 800 people are employed within the Department who are involved in land reform projects, together with the Land Claims Commission. Roughly US$63 million (R694 million at an exchange rate of R11.50 to US$1) are presently (2001) budgeted annually for advancing land reform. The development and future plans for the *LiNC Viewer* are presented in this chapter. Prior to describing the *LiNC Viewer* and its history, the context within which it was developed and purpose underlying its development, are sketched out.

23.2 Contextualising the *LiNC Viewer*

23.2.1 Background to Land Reform in South Africa

Land ownership is a sensitive issue, especially in countries of Southern Africa due to the history of colonisation. This is particularly true in South Africa, where policies of 'apartheid' and agriculture created a skewed distribution of land ownership, with a minority owning more than 80 per cent of the agricultural land (Kirsten and Van Zyl 1999). The aims of apartheid land laws were, *inter alia*, to define and physically separate various race groups, as well as to provide a legal framework for administrative and political control over black population movements and concomitant land rights (Van der Walt 1995).

As stated during the opening address to the Conference on Land Redistribution Options held in October 1993, this unequal distribution of land is not just the unfortunate legacy of apartheid; it also represents the totally unacceptable continuation of apartheid: *"Whoever owns the land is in effect the master or mistress of the people on the land. Unless we solve the land question, we cannot solve the human question, we cannot de-racialise the economy and we cannot achieve a secure legal regime for property rights respected by all"* (Ramaphosa 1993).

The close connection between land law and apartheid policies made land reform one of the major issues in the reform negotiations since 1990 (Kirsten and Van Zyl 1999). Furthermore, it is argued that the above reality provides sufficient motivation for land reform in South Africa, and that land reform, so long as it is implemented correctly, can contribute to increased efficiency and equity, increased growth and poverty reduction. The DLA received a mandate in 1994 to contribute to the Reconstruction and Development Programme by developing a comprehensive and far-reaching land reform programme (DLA 1997). The Land Reform Programme is aimed at redressing the legacy caused by apartheid, while contributing to national reconciliation, growth and development (DLA 2001c).

23.2.2 Elements of the South African Land Reform Programme

The Land Reform Programme comprises three main pillars, namely restitution, redistribution and tenure reform. One key objective of the Land Reform Programme is the restitution of land rights to people prejudiced by racially discriminatory legislation passed since 19 June, 1913. The process of restitution is guided by the Restitution of Land Rights Act of 1994, as amended. Restitution comprises the restoration of the original land, the provision of alternative land or financial compensation or other relief as negotiated (DLA 1997, DLA 1999). Facing the challenges to contribute to the Presidential Rural Development Strategy, the

Commission on Restitution of Land Rights re-prioritised its work (DLA 1999). The aim is to implement restitution in such a way that reconciliation, reconstruction and development is supported (DLA 2001b).

For land reform to succeed in South Africa, it needs to be implemented in an integrated manner with other relevant departments such as the Departments of Agriculture and Housing (DLA 2000). As a result, the Land Redistribution Programme was redesigned during the year 2000 under a ministerial review process, with a more pro-active approach in supporting the Integrated Sustainable Rural Development Strategy of Government (DLA 1999, DLA 2000). The new Integrated Programme of Land Redistribution and Agricultural Development in South Africa has the following objectives:

- to redistribute approximately 30 per cent of agricultural land over the next 15 years;
- to improve the nutrition and income of the rural poor;
- to overcome the legacy of past racial discrimination in ownership of land;
- to assist previously-disadvantaged who want to establish small and medium-size farms;
- to stimulate growth from agriculture;
- to create stronger linkages between farm and off-farm income-generating activities; and
- to empower beneficiaries to improve their economic and social well-being.

Tenure reform is about the improvement of terms and conditions through which people hold, use, occupy and access land. The legacy of apartheid for many South Africans is that most have only informal land rights – they have been living on land, in many cases for decades, that they do not own. In most cases, they are on the land with the permission of the landowner, but as there are only informal occupation agreements, most of these people are vulnerable to eviction. Furthermore, not only do they live in a constant state of insecurity, but they are also unable to develop and improve their homesteads without clear land rights. The wide range of informal tenure rights that have developed over time has complicated the question of securing formal land rights for such people. To deal with this the DLA has developed a differential approach to the variety of situations that exist (DLA 2001c).

23.2.3 Information Requirements for Effective and Efficient Land Reform

Land related information is vital for the execution of the Land Reform Programme. For restitution, information about past and present ownership of land is needed, as well as the value of the land and its current land use. The sustainability of potential redistribution projects needs to be addressed in planning such projects, i.e. information concerning the agricultural potential of the land and transport infrastructure is needed. Further, for management purposes, there is a need to know the location and details of existing and planned redistribution projects, as well as where land claims have been made.

Ideally access to the above information should be immediate in the interests of efficiency. It should not require the intervention of a 'geographical information specialist' to source and package the information required. For land reform practitioners operating from many district and provincial offices distributed across the country, such expertise may, in all likelihood, reside in a different location.

23.3 Reasons for the Development of the *LiNC Viewer*

For a number of years the DLA used so-called 'traditional' geographical information systems (GIS) software that could only be operated by a few GIS specialists at the national office of the DLA, the main output focus being map production. After the DLA had decentralized, the implementation of land reform via the establishment of various provincial and district offices, the number of map requests exceeded the capacity of the staff available to produce them. Many map requests were for location and orientation purposes only. Further, the national office is a remote location to most land reform officials, so hard copy maps have to be couriered to the offices at which the land reform officials are based.

Fortunately, technology allowing web-enabled GIS systems became available. The major backlog in map production, backed by a simultaneous concern about the lack of access to relevant spatial information by land reform officials, led to a decision to investigate alternatives such as the implementation of a web-based GIS and/or GIS viewer. It was recognised that managers and users in the land reform environment could benefit from a user-friendly system that would provide an integrated information, planning and decision-making tool. Subsequently, the DLA embarked on a joint project, under the co-ordination of the Directorate: the National Spatial Information Framework (NSIF), then situated in the office of the Chief Surveyor General, to address the development of a web-enabled land reform viewer, appropriately named the *LiNC Viewer*. The aim of the project is to link disparate data sets within the DLA, as well as external data sources, to provide management and operational level information to all officials within the DLA to assist in effective and efficient land reform in South Africa.

Great emphasis is placed on the new Integrated Programme of Land Redistribution and Agricultural Development in South Africa, requiring that an integrated approach be followed to ensure sustainable economic development, especially in rural areas. A web-enabled GIS such as the *LiNC Viewer* can serve as an integrated information and planning tool to empower land reform officials to pursue the above goal of the new programme.

23.4 Development of the *LiNC Viewer*

23.4.1 Needs Assessment

The most critical part of the project was to determine the needs of the users. The success of the implementation of the GIS system and/or GIS viewer depends on the correct identification of what the users need in order to do their work. The emphasis was on the users needs for information and not on the data content of the system, although the data content followed as a direct result of the information required by the users.

A pilot GIS viewer interface was developed in liaison with a supplier of web-based GIS software and the relevant role players at the DLA. The pilot interface was based on the State Land Information System (SLIS), maintained by the Chief Surveyor-General) and the Critical Project Data (CPD) system. The SLIS contained cadastral and some topographical data, whereas the CPD system was the operational land reform database used at that point in time in the DLA to capture and maintain attribute land reform redistribution data. This pilot GIS-viewer interface was utilised only as a demonstration of what could be provided in such a viewer and was not intended to satisfy or pre-empt any existing needs. This enabled the team undertaking the needs assessment to present each line functionary with a tangible and understandable presentation of the capabilities of such a viewer. Many functionaries would not have ever encountered a GIS before.

During the visits to national and the provincial offices, a short demonstration of the pilot GIS viewer interface was held, followed by a small workshop with the personnel. One main question was asked during these workshops, i.e. 'What spatially related (geographical referenced) information would you urgently/really need in such a viewer?', keeping in mind that the proposed information linked to such a viewer had to correlate with the core business of each line functionary (office). The use of an actual demonstration to show how such an information system could possibly be constructed proved effective in triggering responses.

As might have been expected, the information list was extensive and, in many cases, no national data set existed to address these needs. In view of the complexity of the information needs presented by the land reform officials, it was decided to initiate a basic system that could be expanded incrementally.

23.4.2 Data Included in the *LiNC Viewer*

The data currently available on the system are summarised in Table 23.1. Due to lack of standardisation of a unique identifier for land parcels, it was not possible to address the biggest need of the land reform officials, namely to link the land parcels from the cadastre to title deed information (maintained by the Registrar of Deeds in the DLA). This lack of standards to support integration of data in South Africa is currently being addressed by the National Spatial Information Framework (NSIF) created in 1997. The NSIF aims to facilitate geo-information driven decision-making, planning and service delivery, for an improved quality of life for all in South Africa. The NSIF builds on a National Spatial Data Infrastructure (SDI) or Framework that feeds into the Africa SDI and the Global Spatial Data Infrastructure (GSDI) (DLA 2001d for more information on the NSIF).

Table 23.1: Data currently available in the *LiNC Viewer*

data set	data source
cadastral data	captured and maintained by the offices of the Surveyors-General, Department of Land Affairs
1:50 000 and 1:500 000 topographical data, e.g. roads, rivers, railway, dams, pans, built-up (urban) areas	captured and maintained by the Chief Directorate: Surveys and Mapping, Department of Land Affairs (Mowbray, Cape Town)
magisterial district boundaries	captured and maintained by the offices of the Surveyors-General, Department of Land Affairs
district councils and local council boundaries	demarcated by the Municipal Demarcation Board. (The Board is an independent authority established in terms of the Municipal Demarcation Act (Act 27 of 1998) responsible for the demarcation of municipal boundaries since February 1999. For more information, visit their web-site at http://www.demarcation.org.za).
a live link to the operational land reform database, Landbase	national, provincial and district offices, Department of Land Affairs
state land information	maintained by the Directorate : Public Land Support Services. Department of Land Affairs. Public Land includes: • state land registered in the name of the Minister of Land Affairs and land held in trust by the Minister; • land registered in the name of the Republic of South Africa (RSA) and controlled by the Minister of Public Works and in the names of the nine provinces; and • land used for public purposes, comprised of land used by parastatals and local government authorities, e.g. commonage land (South Africa 1999).

23.4.3 Technical and Organisational Infrastructure

The DLA consists of one national office with nine provincial offices and various smaller district offices. In each provincial office there is a local server utilised to update the land reform project data in the Landbase database. This information is replicated to a central server at national office. The *LiNC Viewer* reads the land reform project data directly from this central server. The land reform project data is stored and managed in an Oracle database. Currently, the spatial data is converted to Autodesk MapGuide format, but it is envisaged that the spatial data will be stored and read directly from Oracle 9i in the near future. The users of the *LiNC Viewer* need to download a 'thin client' to enable them to view the spatial data via their web browser. Due to the limited bandwidth in South Africa, various complaints are received concerning the speed of the system. The distribution of spatial information via the Internet is a great technological achievement, but the communication networks in developing countries are often not adequate enough to support use of these systems to their full potential.

23.5 Functionality available in the *LiNC Viewer*

Officials are now able to plan the implementation of land reform projects with more information at their disposal than they had previously. The *LiNC Viewer* provides officials within the DLA the ability to overlay data according to their needs and preferences, by simply checking boxes to indicate which information layers they require. The deeper a user zooms into the data, the more detail is displayed on the screen and the more detail is reflected in the legend. In Figure 23.1, surveyed land parcels overlaid with topographical data (vital for orientation purposes in land reform) are shown, for example.

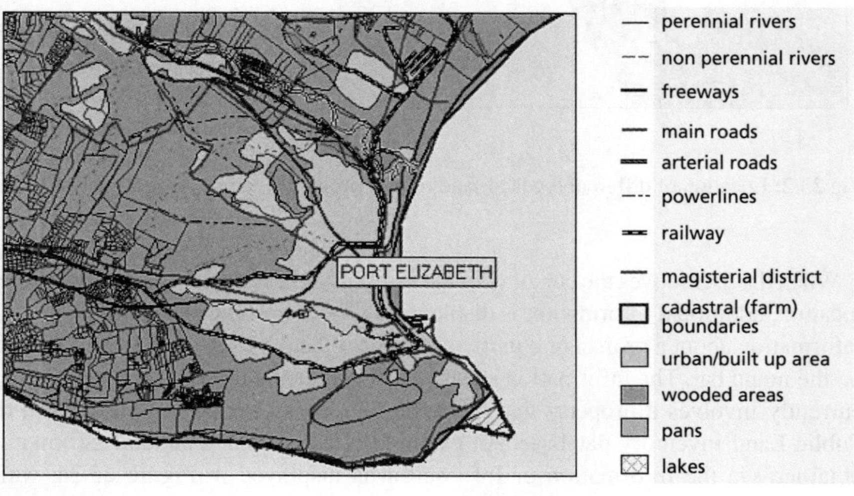

Fig 23.1: Cadastral and topographical data

The system has been developed in such a way that there is a linkage between the spatial data, particularly cadastral boundaries, and the operational land reform data in the Landbase database, providing data on properties involved in land reform projects and/or restitution claims. These are indicated with symbols (diamonds) of different colours reflecting the different project types, as illustrated in Figure 23.2.

With the assistance of such a map as illustrated in Figure 23.2, it is possible for managers to see in which areas land reform projects are focused. In the future, socio-economic information (e.g. data deriving from the population census conducted nationally every five years) will also be included on the system, thereby allowing managers to view which areas need assistance and could benefit from land reform projects most. The *LiNC Viewer* will thus serve as a decision-making tool to plan and re-direct future land reform projects in accordance with the aims of the new Integrated Programme of Land Redistribution and Agricultural Development.

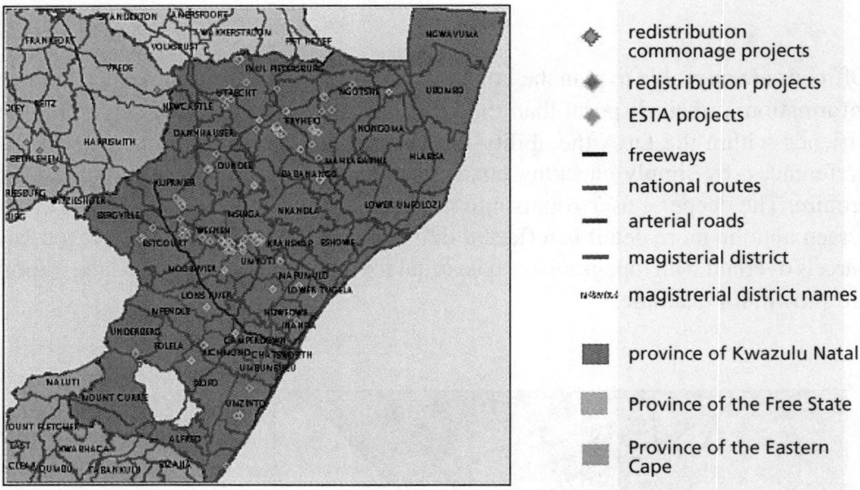

Fig 23.2: Location of different types of land reform projects

When the user moves the cursor over the symbols indicating the land reform project location, additional information is displayed as a tool tip. The user can obtain more information about a project or a particular piece of land by selecting the Info-button on the menu bar. The information sheet will also indicate if the land reform project currently involves a property that has been entered as public or state land in the Public Land Inventory database. An example of additional state land information obtained via the information or Info-button is displayed in Figure 23.3a, while additional project information is displayed in Figure 23.3b.

a)

Province name	GAUTENG
Magisterial district name	PRETORIA
Registration division	JR
Town name	JR
Registered owner	STADSRAAD VAN CENTURION
Area (Ha)	6.9539
Deeds number	T53696/1989
SG diagram	T53696/989
Property name	DOORNKLOOF
Parcel	00000391
Portion	00163
Vested	
SG key	T0JR00000000039100163

b)

Project Name	VUKILE/IMPALA
Project Status	MTEF Project
Property Name	Hope
Property SG key	N0GT00000000568400000
Planner	Peter Sapsford
Land Size	706.47000000000003
Group Name	Vukike / Impala Community Land Trust
Number of beneficiaries	110
Number of female beneficiaries	0

Fig 23.3: Additional information provided with the Info-button

A few basic reports on the operational land reform data, Landbase, have been developed. Fifteen standard reports are available which provide summarized, statistical as well as detailed project information. The facility to draw one generic report on state land has also been included. Examples of some of the reports available can be seen in Figure 23.4. Figure 23.4a displays management information retrieved from the operational and reform system, Landbase. Cumulative statistics are calculated for a specific land reform programme, in this case redistribution – commonage. The statistics are calculated and displayed for one province, namely the Free State as selected by the user.

Figure 23.4b displays project level information from Landbase. In this example the properties involved in the Alicedale (CPD 179) project are displayed. The farm names as well as their unique property identifiers are shown. Had the number of hectares been entered for each province, the total number of hectares would have been calculated.

Figure 23.4c is an example of a report indicating specific properties that are owned and / or administered by the state. It provides the unique identifier of the property as well as the farm name, area in hectares, the title deed number, the Surveyor-General diagram number as well as the name of the registered owner.

cumulative statistics

report is for redistribution - commonage and contact with DLA

page number : 1
print date : 18/09/2001

by cumulative the following is understood:
all the projects to have been completed for a specified milestone

province	No of hectares	No of projects	No of beneficiary households	No of female headed households
Free State	35675,2891	36	647	0

a)

project name
Alicedale CPD 179

no of beneficiary households

magisterial district : Albany

farm name	old prov. code	major code	minor code	parcel	portion	no of hectares
Domkerhoek	C	002	0000	00000276	00000	0.00000
New Years Drift	C	002	0000	00000274	00000	0.00000
					total hectares for project	**0.0000**

b)

public land national / provintial

public land catagory	property description	farm name	area (ha)	deed number	SG diagram number	registered owner
Bloemfontein						
Bethlehem RD						
rural						
stateland (national/provintial)	F0010000000000004400000	Kleinbegin	34.28	T4166/1987	T29067/1910	Republiek van Suid-Afrika
stateland (national/provintial)	F0010000000000005200005	Bramleyshoek	337.20	T6595/1990	T8300/1977	Republiek van Suid-Afrika
stateland (national/provintial)	F0010000000000005200006	Bramleyshoek	337.20	T6595/1990	T8302/1977	Republiek van Suid-Afrika

c)

Fig 23.4: Examples of reports available

Various search facilities have been developed to enable land reform officials to locate the land parcel or land parcels that the users need to focus on. A location map with orientation information of their choice can then be printed. They can search for urban or rural land parcels by narrowing down the search criteria from a national to a more local level, or they can search directly for the farm number if they already have that information. The result of a property search is illustrated in Figure 23.5. In the cadastral data, various codes are utilised, so searches on these code tables have also been included. All effort has been made to make it as easy as possible for officials not familiar with spatial information, and specifically the cadastre, to locate the property or properties they require. During training, this proved to be one of the biggest challenges, namely educating land reform officials regarding spatial data and the key that is utilised to identify every property uniquely in South Africa.

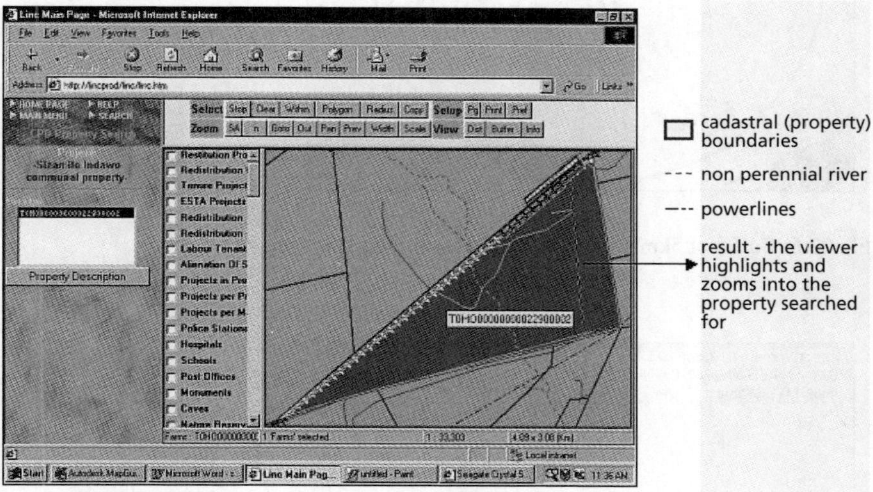

Fig 23.5: Result of property search

The *LiNC Viewer* provides the land reform officials with a buffer function. The user is able to select an object, such as a particular piece of land identified for land reform and then specify a distance to be buffered around that selected feature or object as illustrated in Figure 23.6. This functionality further enables a land reform official to locate other features within the buffer in the vicinity of the selected feature, for example state land, water or services such as schools and towns.

The land reform official can also measure distances with the *LiNC Viewer*, by indicating and drawing lines on screen with a mouse. This functionality assists in determining the feasibility of the location of a proposed land reform project in relation to the availability of other services and infrastructures in the surrounding area. In Figure 23.7, distances are measured between a land reform project and a school, police station and clinic.

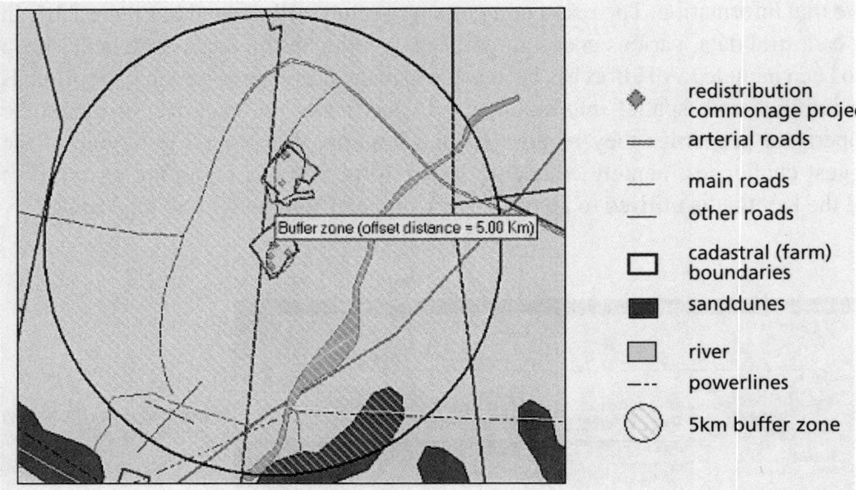

Fig 23.6: Buffer of 5km around a selected land reform project

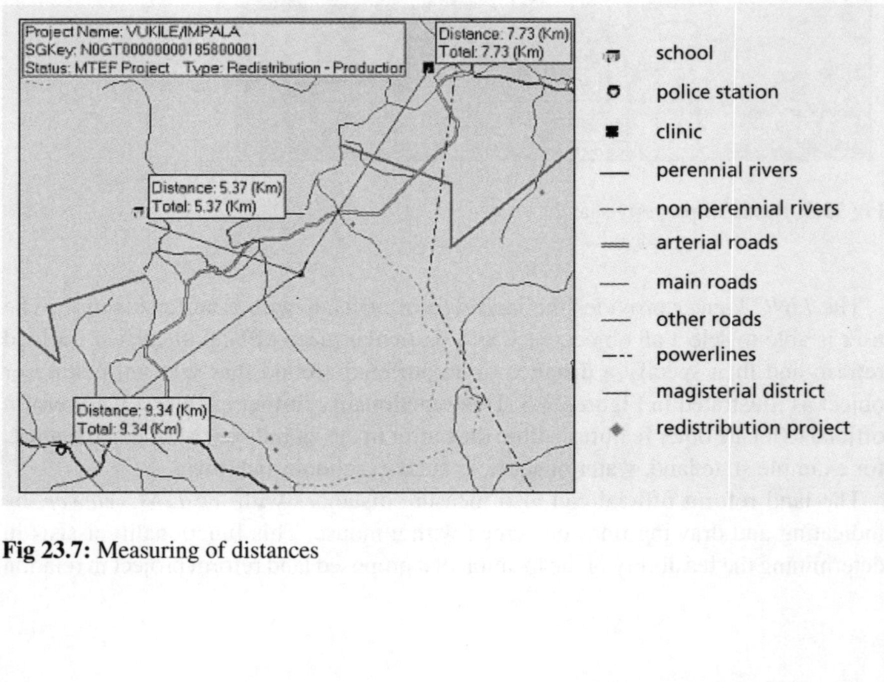

Fig 23.7: Measuring of distances

The functionality is available to copy an image of the map area displayed on the screen and then paste it into a word processor to be used as part of a report. Any report or map displayed on the screen can be printed to any local printer available to the land reform officials. A land reform official has the choice of specifying a heading for the map, before the map is automatically printed with a legend, scale bar, north arrow and frame.

23.6 Implementation of the *LiNC Viewer*

The DLA consists of offices across all nine provinces of South Africa. This makes the successful implementation of a system rather difficult. It was decided to identify 'champions' in every provincial office to assist with the implementation of the system. The criteria for choosing champions or super-users were that they should be experienced in the land reform process and acquainted with new (computer) technologies. These officials were nominated by their provincial directors and were the first to give input to the development of the system at a workshop. The champions were trained on the system as well as the data included on the system. They assisted officials from national office with the organisation of the Provincial *LiNC Viewer* roadshow by organising the date, venue and attendance. The utilisation of champions proved to be of real value as they would report any problems on the system. The personnel in the provincial offices know whom to contact in case of problems and they need not speak to an unknown official at a remote location at National Office for assistance. When new updates are made to the system or training is organised, this information is sent to the LiNC champions who subsequently inform their provincial and district offices.

Training is provided at a special IT venue at national office where departmental officials (champions and users) are given the opportunity to get hands-on experience on the live system being hosted on the departmental intranet. Courses are scheduled continuously and LiNC champions notified accordingly. Training is also available on site on request. The *LiNC Viewer* is also marketed to managers at management meetings in the DLA.

23.7 Practical Applications and Advantages for Users of the *LiNC Viewer*

The biggest advantage to the land reform officials is that they have immediate access to spatial information with which they can view the location of the property or properties in relation to infrastructure and services. Users can measure the distance to the nearest schools, clinics and other services and infrastructures in the vicinity. Access routes to the land in question can also be determined. This all relates to the feasibility of the project. It is also possible to get a quick overview of the area and determine if there is state land in the vicinity, which could be utilised for land reform. When land is identified for redistribution, officials can immediately see if there is already a restitution claim on the same property and therefore not available for the redistribution programme.

Due to the immediate access to valuable information, land reform offices can save resources in terms of travelling budget and man hours spent on the road to visit the areas (in many cases, fairly remote from their base) identified for land reform. They now have the ability to do research in-house and then plan trips with the aim of obtaining specific, detailed information.

The Quality of Life Project is a national study undertaken by the Monitoring and Evaluation Directorate within the DLA, where a service provider is appointed to administer a questionnaire (largely quantitative) amongst land reform beneficiaries. The *LiNC Viewer* is used to search and identify the magisterial districts in which projects lie for the Quality of Life Project. Furthermore the location of the specific project that is being monitored or evaluated by this Directorate can be viewed on-screen before the land or project is visited by fieldwork teams.

The Directorate: Tenure Reform is reviewing communal property associations (CPAs) that will be registered nationally. The *LiNC Viewer* is once again used to locate these sampled projects. The *LiNC Viewer* is also used to answer requests from various officials or persons from external organisations. Non-governmental organisations (NGOs) that assist disadvantaged communities had been assisted in the past by showing them the location of certain land reform projects in a specific province in relation to surrounding socio-economical services, such as roads, schools and clinics. They were even provided with simple maps produced with this viewer by land reform officials themselves. Previously such orientation maps would have had to be compiled by a GIS expert at the national office.

The *LiNC Viewer* is particularly useful with respect to ESTA. ESTA is an acronym for the Extension of Security of Tenure Act, 1997 (Act 62 of 1997). This Act protects more than six million people who live on someone else's land, with the permission of the owner providing a secure legal right to live on and use that land. It protects occupiers against unfair or arbitrary evictions but provides for legal evictions under certain circumstances. It sets out the right and duties of owners and occupiers, as well as allowing for the government to release funds to create more permanent solutions, i.e. on-site or off-site housing developments (DLA 2001c).

The *LiNC Viewer* is used to observe the land under question in order to determine whether the Act is applicable or not, before an extensive and time-consuming Deeds Office search is performed with regard to that particular piece of land. This implies that the *LiNC Viewer* is used to determine the landowner status, as ESTA is not applicable to land that is classified as township land. Given the urgency of responding to ESTA cases, the viewer is a very useful facility, as it saves the DLA financial and human resources that would be wasted if ESTA was not applicable to a particular piece of land.

23.8 Problems of Integrating the *LiNC Viewer* into the Land Reform Process

Although the system provides officials with access to information that could assist them to a great extent, there are problems of getting officials to integrate the use of the system into their business practices. There are three major reasons why officials are reluctant to utilise the system to its full extent. Firstly, fear of, as well as lack of exposure to, the technological world causes reluctance to take ownership of information systems in general. This is even more severe with the *LiNC Viewer*, which is a web-enabled GIS. Most users have had no or little exposure to computers, not to mention GIS. Secondly, officials and managers do not seem to understand that if they do not enter their information correctly on their land reform system, then it will not display correctly on the *LiNC Viewer*. As always, the system is blamed and not the incorrect data entered into the system. The lack of a culture toward information-driven decision-making as well as a culture that does not take the capturing or maintenance of information seriously has an extremely negative effect on any information system in the DLA. The information entered into the system is not quality checked and thus many errors occur on the system, which may make it impossible to display the land parcels involved in land reform on the *LiNC Viewer*. The negative cycle created with regard to information capture is illustrated in Figure 23.8.

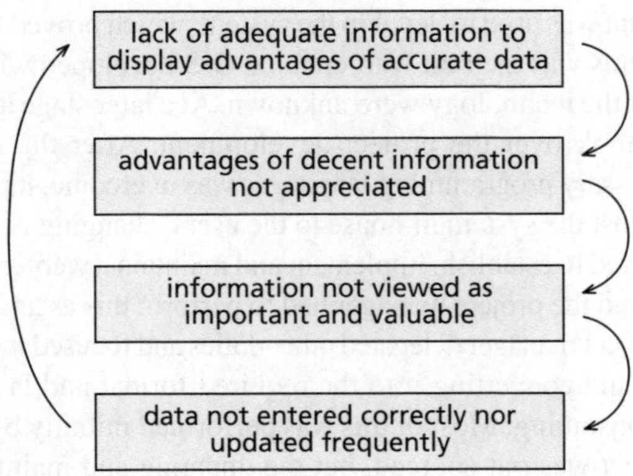

Fig 23.8: Present situation regarding information capture in the DLA

Thirdly, the recent restructuring of the DLA has also had an impact on the integration of the *LiNC Viewer* into the land reform process. Personnel who had become convinced of the importance of information management and hence were advocating better information management and the importance of the *LiNC Viewer*, have left the DLA and it is now necessary to convince new managers and staff of the importance of information systems.

Various steps have been undertaken to address these problems. First of all, a workshop with managers of the land reform sections of the DLA is to be held. During this workshop the advantages of spatial information integrated with accurate land reform project information as well as other relevant information will be demonstrated. Attendees will be given the opportunity to specify their frustrations as well as additional and specific information needs. The main aim is to determine the gap between user needs and what is available on the system, as well as any other factors that contribute to the users not utilising the system to its fullest extent. Secondly, it is also necessary to investigate the dynamics of 'computer human interaction'. Staff members will attend a conference dealing extensively with this subject and thereafter make suggestions as to how to address the problem of users not having had exposure to the technological world or GIS. Other measures include a map reading/awareness course which is to be integrated with the training provided on the *LiNC Viewer*; follow-up training sessions; and the promotion of awareness of the system on an ongoing basis in order to keep track of the changing needs of the users.

23.9 Problems Experienced with the Development of *LiNC Viewer*

Initially a consultant was used to develop the system, which proved to be very difficult to manage, since this was the first web-enabled GIS developed within the DLA and the possibilities of the technology were unknown. At a later stage a data technologist within the DLA took over the project development. After the initial obstacle of mastering the necessary programming languages was overcome, it proved to be easier to develop and adjust the system in-house to the users' changing needs. The extent of commitment required to establish, implement and maintain a web-enabled GIS system was not foreseen and the project manager had to perform this as an additional task. At a later stage the project manager delegated other duties and focused more on this system.

Obtaining data and converting it to the required format and layout proved to be extremely time-consuming. Most of this was performed initially by the project team to which the project was out-sourced, but the updating and maintenance of data is done internally within DLA. Conversion of spatial data especially is an extremely slow process. It is hoped that most of the data will be in Oracle Spatial in the near future and that the web-enabling GIS software will read the data directly from the Oracle Spatial database, making any data conversion unnecessary.

The speed of internet/intranet connections in South Africa is impacting extremely negatively on the performance of the system. Users are dissatisfied with the response time, even if it is much faster than waiting for a hard copy map. Many people who experienced having to wait for up to six weeks in the past for hard copy maps no longer work in the DLA and the new staff do not fully appreciate how privileged they are in being able to access the spatial data directly themselves.

23.10 Future Plans

The *LiNC Viewer* is being developed and implemented incrementally. Further modules are researched and included as directed by the users' needs. For example, at this stage it is not possible for users to add text onto the map area, but this is under investigation as there have been various requests for this facility. Most officials want to add the farm name or other text to specify the location of the property involved.

The updating of data sets is a continuous process, as updates become available, although it is recognised that the process needs to be streamlined. Educational aspects need to be integrated, namely the *LiNC Viewer* training, education on the land reform system as well as map reading workshops, that are currently provided by different components in the DLA.

In order to market the *LiNC Viewer* to more users in the land reform environment, functionality is to be added to simplify the software and to enhance access to the disparate data sets. A pilot project is to be planned to show the value that the system can have to the land reform process if data are entered accurately and timely. Additional data relevant to the land reform process and as requested by users is to be included in further modules, such as rainfall, drainage, soil types, demographics and general socio-economic statistical data. Users are very impressed with viewing a map on their own computers and feel that they learn something as individuals if they are able to utilise the system. This enthusiasm must be capitalised upon in order to make the use of spatial information an integral part of the business of land reform.

References

Department of Land Affairs (1997) *A Guide to the Department of Land Affairs' Land Reform Programme*. Communication Services, Department of Land Affairs, Pretoria.

Department of Land Affairs (1999) *Annual Report 1999*, Communication Services, Department of Land Affairs, Pretoria.

Department of Land Affairs (2000) *Discussion Document on Strategic Planning by the Minister of Agriculture and Land Affairs*, Unpublished document, Pretoria, February.

Department of Land Affairs (2001a) *Integrated Program of land Reform and Agricultural Development in South Africa*, Unpublished document, Pretoria.

Department of Land Affairs (2001b) *Introduction to Restitution*, *http://land.pwv.gov.za/BACKGROU.RES.htm* (2 May).

Department of Land Affairs (2001c) *Land Reform in South Africa*, *http://land.pwv.gov.za/briefin2.htm* (2 May).

Department of Land Affairs (2001d) National Spatial Information Framework (NSIF), *http://www.nsif.org.za*.

Kirsten, J.F. and Van Zyl, J. (1999) Approaches and progress with land reform in South Africa, *Agrekon*, 38, Special Issue, May. *http://www.iaae-agecon.org/saconf/sa23.htm* (26 April).

Ramaphosa, C. (1993) *Land Reform in South Africa,* Opening address to the Conference on Land Redistribution Options held in Johannesburg, 12-15 October. *http://jinx.sistm.unsw.edu.au/~greenlft/1993/124/124p21.htm*

Van der Walt, A. J. (1995) Land reform in SA since 1990 – an overview, *South African Journal of Public Law*, 10: 1-31.

24 The SPARTACUS System for Defining and Analysing Sustainable Urban Land Use and Transport Policies

Kari Lautso

LT Consultants Ltd., Melkonkatu 9, 00210 Helsinki, Finland

Abstract: SPARTACUS, the System for Planning and Research in Towns and Cities for Urban Sustainability, is a strategic urban planning support system based on a state-of-the-art land use and transport model called MEPLAN. It is combined with a set of urban sustainability indicators, a GIS-based Raster method to calculate values for some of the indicators, a database and presentation module as well as a decision support tool called USE-IT. The latter module is used for evaluating the results of policy options based on weighted indicator values representing social, environmental and economic effects of urban policies.

The SPARTACUS system has been used for assessing options for urban sustainability policies in a systematic way in three cities, Helsinki, Bilbao and Naples. The test results show that the level of sustainability can be increased in the test cities in a cost-effective way by using a variety of policies, especially pricing and regulatory policies and their combinations. The negative side effects of individual policies can be mitigated through combining policies in an innovative way. Recommendations based on the above studies have been given for sustainable urban policies and future research.

24.1 Introduction: The *SPARTACUS* Project

The *SPARTACUS* project has developed a methodology for assessing the sustainability implications of urban land use and transport policies. The original main objectives of the project were:

- to design and specify a system for analysing and providing forecasts of the interactions between land use, transport, the economy, the environment and social factors;
- to identify, on the basis of European initiatives and guidelines, a set of key policy actions which can be translated into development strategies and tested in the pilot city applications;
- to develop strategies for urban sustainability by using combinations of land use, transport and environmental policy instruments (regulation, pricing and investment);

- to simulate and assess the long-term effects of introducing policy actions in each of the selected pilot cities, and to compare and explain any differences in the results attained;
- to specify common sets of policy actions that are economically, environmentally and socially sound and viable, and that could successfully be used in other cities within the European Union having similar sustainability problems; and
- to give general recommendations and to disseminate the results in an effective way.

SPARTACUS is a research and development project within the Environment and Climate Programme of the Fourth Framework Programme of the European Community. The project has been funded by DG XII of the European Commission and national authorities in Finland, Spain and Italy. The research has been carried out by a consortium comprising the following: LT Consultants Ltd, Finland (Project Co-ordinator); Marcial Echenique & Partners Ltd, UK; Marcial Echenique y Compania SA, Spain; TRT Trasporti e Territorio srl, Italy; Institut für Raumplanung, Universität Dortmund, Germany (for further information see *www.ltcon.fi/spartacus*).

24.2 The *SPARTACUS* Approach to Urban Sustainability

The three components of sustainability (Figure 24.1) generally referred to in the literature have been taken as the points of departure for the *SPARTACUS* system. They are interpreted as follows: *environmental* limits are posed by the exhaustion of natural resources and exceeding natural carrying capacities; *social* sustainability includes the analysis of the distribution of policy impacts and access to resources and services, the justice of those distributions, and the state of the community with regard to health determinants and personal opportunities; and *economic* efficiency of the use of resources examines whether most benefits are achieved with the least use of resources.

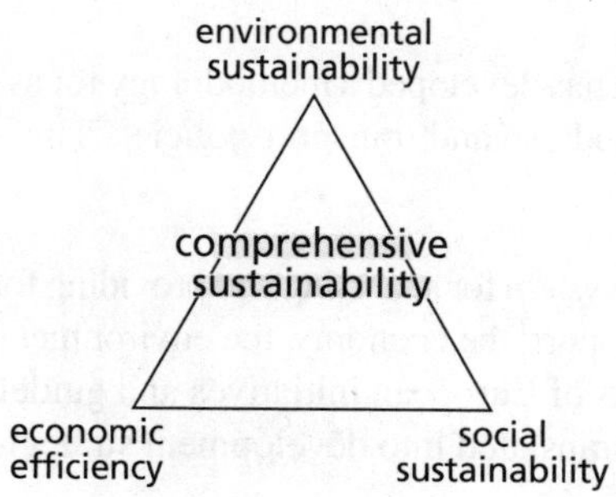

Fig 24.1: The approach to urban sustainability

For a city to be sustainable it needs to use resources in an efficient and equitable way within the carrying capacities of the environmental and social systems it is dependent on.

24.3 Indicators

The *SPARTACUS* indicator system (Table 24.1) consists of three components in accordance with the three components of sustainability: environmental, social and economic. The components are further divided into themes. The indicator values can be aggregated to the theme-level using indicator-specific weights, and further to component-level indices. The component-level indices are not combined, because no substitution is assumed possible between them. Indices are used to compare policy alternatives in one city or in studying effects of a specific policy in different cities.

Table 24.1: The components, themes and indicators of urban sustainability in the *SPARTACUS* system

environmental indicators			
air pollution	emissions of greenhouse gases from transport	EAGT	t/1000 inh./a
	emissions of acidifying cases from transport	EAAT	Meq/1000 inh./a
	emissions of organic compounds from transport	EAOC	t/1000 inh./a
consumption of natural resources	consumption of mineral oil products	ERMO	1000 t/1000 inh./a
	land coverage	ERLC	% points
	consumption of construction materials	ERCM	1000 m2/1000 inh./a
social indicators			
health	exposure to particulate matter in the living environment	SHEP	% points
	exposure to nitrogen dioxide in the living environment	SHED	% points
	exposure to carbon monoxide in the living environment	SHEC	% points
	exposure to noise in the living environment	SHEN	% points
	traffic deaths	SHTD	#/1000 inh./a
	traffic injuries	SHTI	#/1000 inh./a
equity	justice of exposure to particulates	SEJP	-
	justice of exposure to nitrogen dioxide	SEJD	-
	justice of exposure to CO	SEJC	-
	justice of exposure to noise	SEJN	-
	segregation	SES	% points
opportunities	total time spent in traffic	SOTT	hour/1000 inh./a
	level of service of public transport and slow modes	SOPT	min/trip
	vitality of city centre	SOVC	
	accessibility to the centre	SOAC	min
	accessibility to services	SOAS	min/trip
economic indicator (index)			
	total net benefits		(Euro/capita/a)

The social theme equity deals, among others, with the issue of justice. The distributions between socio-economic groups of various impacts, e.g. of noise exposure are analysed using optional theories of justice, and the evaluated indicator values are calculated on the basis of how well a given distribution matches the chosen theory of justice.

The economic index is expressed in one single value (EURO/capita/annum) describing the net benefits of any policy. It is composed of transport user and transport operator benefits as well as resource and external cost savings.

One of the central principles in the selection of the indicators has been to follow the 'causal chains of impacts' as far as possible. The system also has to be able to forecast the indicator values into the future.

In addition to indicators, a set of background variables is also used when analysing policy options. Examples of background variables are public transport share, car mileage and number of inhabitants or employees in different parts of the urban area. Some indicators can be studied in spatially disaggregated form at zone level or even in 100metre x 100metre raster cells.

24.4 Derivation of Indicator Values

The analytical framework of the *SPARTACUS* system is shown in the Figure 24.2. Policy packages are user-defined policy options or their combinations. They may present new transport links, new pricing schemes or changes in land use or regulatory actions, such as speed limits in the transport network. Some of the indicator values are calculated using the MEPLAN land use transport model. The transport and land use models estimate the changes in transport conditions, which may affect the land use patterns. A new land use pattern changes the transport demand and conditions on the network, which again may change the land use, *et cetera*. This interactive process is followed until equilibrium is reached on both the land use and the transport sides.

MEPLAN model output data is further processed in the MEPLUS GIS, Raster and USE-IT modules to derive values for the indicators. The MEPLUS module calculates values for certain indicators. These indicators include those under the themes of air pollution, consumption of natural resources and opportunities. MEPLUS also extracts data from the MEPLAN model output for calculation of the economic characteristics of the policy options. It also produces all the background variables of a given policy option.

The Raster module is based on the output of an aggregate urban model. Figure 24.3 shows the basic structure of the Raster module and its main sub-models. There are two main sources of input information for the Raster module. Firstly, there is a spatial database containing zone boundaries and main land use categories recorded as polygons and the network coded as vectors. Secondly, there are the results from the MEPLAN model, of which households by socio-economic group, employment and floorspace, all by zone, and link load by network mode, are used. This information is then converted to a raster data structure, i.e. to raster cells. Subsequently, environmental sub-models (emission and dispersion models) are employed to assess land coverage, air quality for several pollutants from road transport and noise levels for the raster cells.

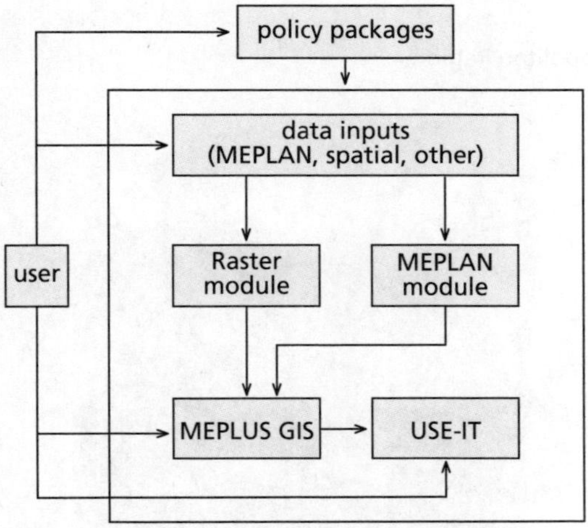

Fig 24.2: The *SPARTACUS* analytical framework showing the main modules of the system and their interaction

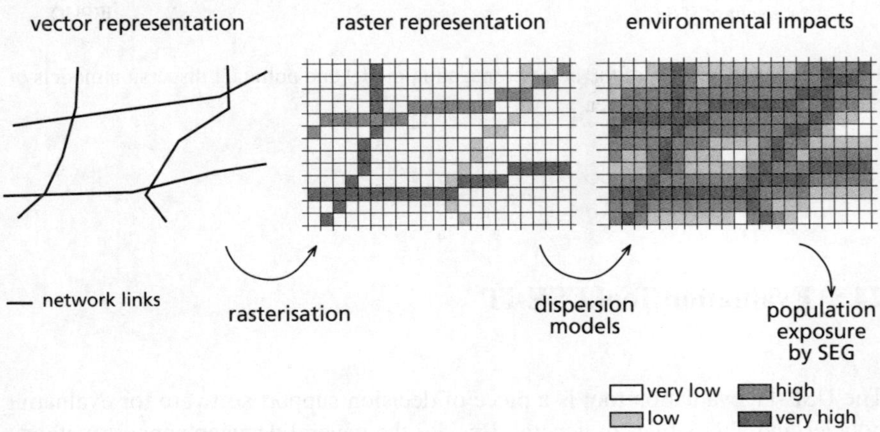

Fig 24.3: The structure of the Raster module

The estimation of the social impacts of these environmental impacts is the task of the next level of sub-models in which the exposure of population to air pollutants and to traffic noise is estimated. Figure 24.4 shows some sample output from the noise propagation model and pollutant dispersion models of the Helsinki Metropolitan Area. The output shows the noise map and the population by socio-economic groups disturbed by noise.

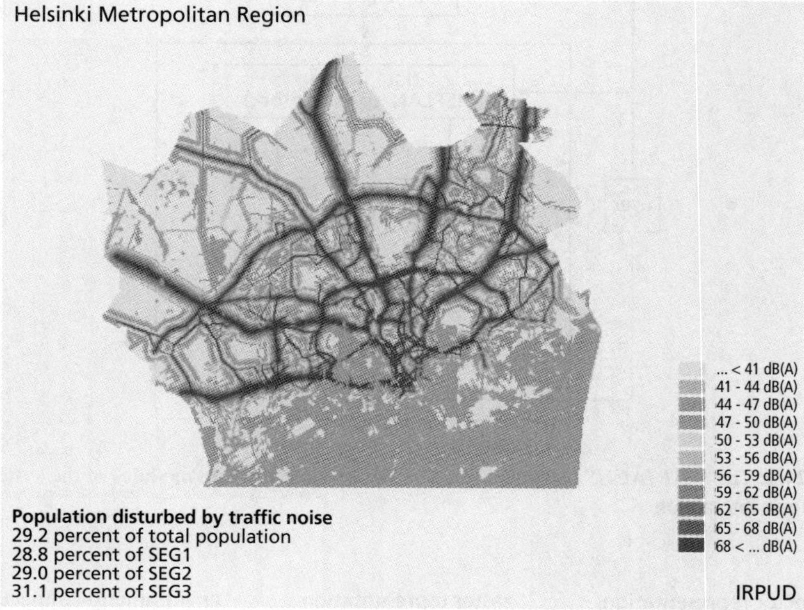

Fig 24.4: Sample output of the noise propagation model and pollutant dispersion models of the Helsinki Metropolitan Area

24.5 Evaluation Tool USE-IT

The USE-IT evaluation tool is a piece of decision support software for evaluating policies and alternatives in general. Besides the general decision support methods, direct weighting and the analytical hierarchy process (AHP) pair-wise comparing system, it has several features that help standardised testing of a wide variety of alternatives. In USE-IT, two methods can be used for giving weights. In the *direct weighting method,* proportional weights are given directly for each sub-goal. In the *pair-wise comparison method,* weights are calculated with an eigenmatrix method from the matrix created in the AHP process, where the indicators are compared pair-wise with each other.

In connecting the alternative policies to the indicators, a value function is given to calculate the degree of desirability of the indicator values (Figure 24.5). The input of the function contains a measurable indicator, such as the annual emissions per inhabitant, for a certain policy. The value of the function is a measure of the relative 'goodness' of this indicator value.

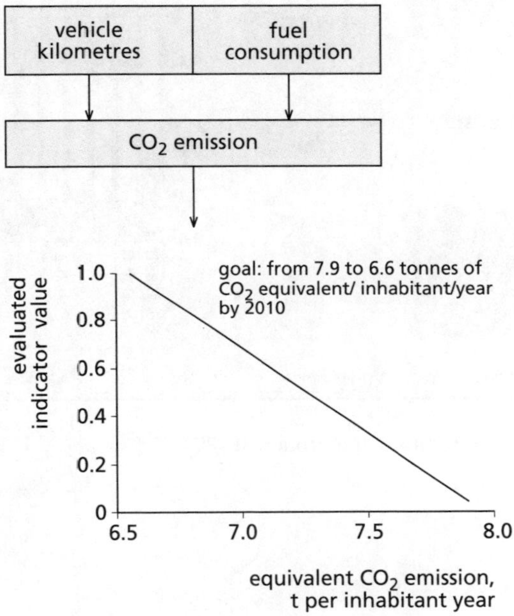

Fig 24.5: From indicator values to evaluated indicator values by using value functions

Four theories of justice are operationalised and are available for the user as options. According to the *equal shares principle,* benefits should be distributed as evenly as possible. According to the *utilitarian approach,* only absolute net improvement is important and no weight is put on the distribution of changes. According to the *egalitarian principle,* an equal distribution of welfare between different groups should be aimed at. Finally, according to the *Rawlsian difference principle,* the condition of the worst-off group should be improved most, in absolute terms.

The use of the tool is illustrated in Figure 24.6. On the left of Figure 24.6 are the weights for social themes (health, equity and opportunities) and social indicators (see Table 24.1 for explanations). On the right hand side are the social index values for the existing situation (1995B), base forecast situation (2010B) and a set of different policy options tested.

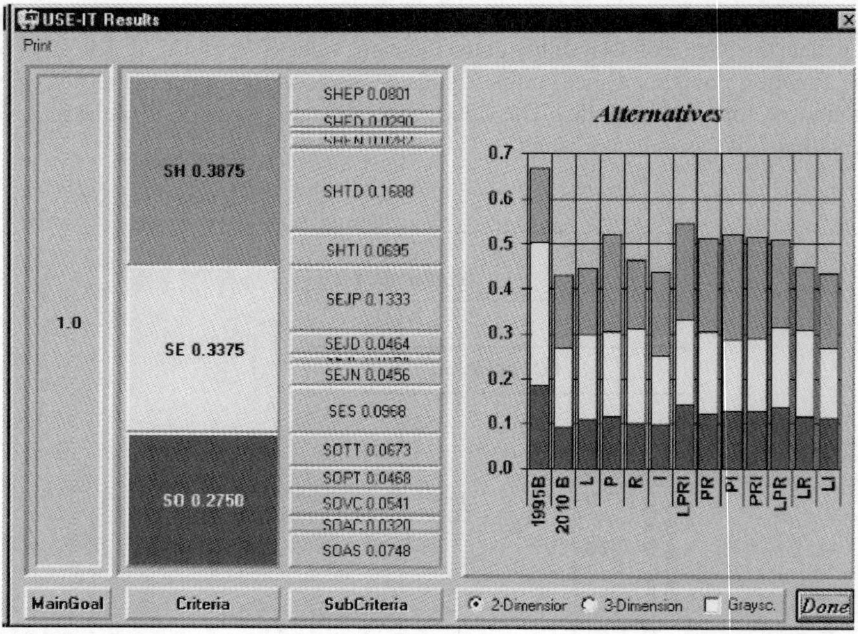

Fig 24.6: An example of results output from USE-IT

24.6 Use of the *SPARTACUS* System

The *SPARTACUS* system has been used to identify sustainable urban policies. The best policies could simultaneously improve each of the three components of sustainability. The identification process of sustainable urban policies started with the definition of the *policy element* tests. The main emphasis was on strategic measures. This is why many of the tests are related to region or city-wide measures instead of individual local measures (e.g. investment programmes instead of individual investment measures).

The policy element tests consisted of different subtypes. *Scenario tests* change some of the exogenous factors of the model (e.g. population). *Land use tests* change some of the land use related factors in the model (e.g. location of inhabitants or employment). *Transport tests* change some of the transport related factors or parameters in the model. Transport tests were further divided into several subgroups. Most attention was paid to the different ways of *pricing (car operating costs, parking costs, public transport fares)*, some of the elements are of *regulative* type (speed limits) and some of the policy elements test the effects of investments (new or improved road and rail links).

The combination of two or more policy elements is called a *policy* and the combination of two or more policies is called a *policy combination.* After analysing the policy element tests, a set of policies and policy combinations was defined and tested. Based on the results of the above tests, new policy elements were defined and tested before another set of policies and policy combinations was formed. To form a firm basis for comparisons, the *present situation* (1995 or 1996 depending on the city) and the base forecast which is called the *reference scenario,* (2010/11) are presented.

Many of the policy elements tested had positive effects from an urban sustainability point of view but, at the same time, also some negative side effects. By combining some of the policy elements, it was possible to mitigate the negative side effects. An example of the effects of a pricing policy on the three components of sustainability is presented on the Figure 24.7. A clear optimum level of pricing can be identified from the economic point of view. From the social point of view, the optimum level is less clear. From the environmental point of view, the index value keeps rising with the increase of car operating costs.

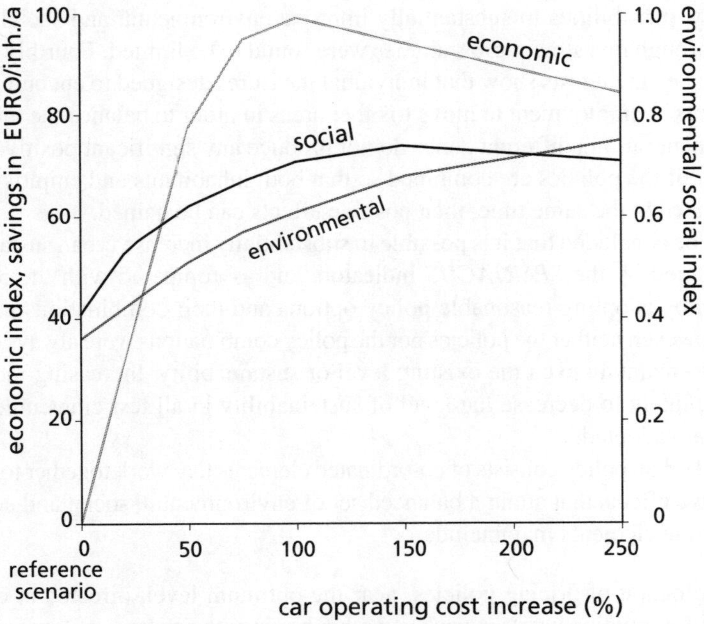

Fig 24.7: Illustration of the effect of the different car pricing levels on the environmental, social and economic index values of sustainability compared with the reference scenario

24.7 Conclusions

The *SPARTACUS* system produces large amounts of information, but it also makes possible a drastic stepwise and transparent aggregation of the data – down to three sustainability index values per policy based on the preferences of the user or client of the system.

Based on the *SPARTACUS* approach, methodology and system, the following types of conclusions have been achieved when using the system in practice. Firstly, pressure for urban sprawl is very strong. One per cent growth in urban population leads to more than one per cent growth in car mileage (in Helsinki 2 per cent). Many measures intended to decrease travel demand may in the long-term lead to increases in private car mileage. Secondly, the results clearly show that the most effective policies from an urban sustainability point of view are the pricing policies. Also, regulation policies, especially lowering car speeds, have significant effects. The pricing policy of increasing car operating costs by 50 per cent reduced the total private car mileage by 16 per cent in Helsinki and Naples, the regulation policy (car speeds down on main roads) by 4 per cent and the combined policies by 21 per cent. Simultaneously, positive effects on emissions, noise, accidents, total travel times and modal split could be achieved.

Thirdly, possibilities to substantially improve environmental and social sustainability through investment programmes were found to be limited. Fourthly, the land use policy element tests show that individual measures designed to encourage either inhabitants or employment to move to other areas in order to balance the inhabitant/employment ratio in different zones do not produce any significant positive effects. However, if the policies are combined so that both inhabitants and employment are redistributed at the same time, then positive effects can be gained.

It may be concluded that it is possible to substantially increase urban sustainability (as measured by the *SPARTACUS* indicators and as compared with the reference scenario) by adopting reasonable policy options and their combinations in the test cities. However, neither the policies nor the policy combinations were always effective enough to maintain even the existing level of sustainability. Increasing mobility is therefore likely to decrease the level of sustainability in all test cities unless strict policies are adopted.

A good urban policy consists of co-ordinated elements that work together to produce cumulative effects that attain a balanced set of environmental, social and economic goals. These elements may include:

- a combination of pricing policies, near the optimum level, directed at car users with differentiation between peak and other hours with appropriate level of pricing of public transport fares and parking;
- speed reductions taking into account the type, role and location of the road/street;
- investment programmes supporting the changes in demand caused by the above policies;

- land use plans that support the new need for people to live near central areas or in satellite cities and their increased need and opportunity to use public transport; and
- measures necessary to mitigate the negative side effects of the above policies.

The *SPARTACUS* approach, methodology and tools have proven to be operational and are, therefore, used by local authorities in the test cities as well as by the EU in order to search for new sustainable and acceptable urban policies. The general approach is well accepted and will be maintained but the models and prototype tools need further development, especially to improve their user-friendliness and capabilities to be interactively available for the general audience through the Internet. The general *SPARTACUS* approach can be applied in most cities and transport models normally available in larger cities can calculate many of the indicators. Some indicators can only be calculated by a land use transport model and some of the *SPARTACUS* indicators need special methodologies and tools, such as Raster and USE-IT, which are available on normal licensing basis.

24.8 Further Research

The PROPOLIS project will continue the research on sustainable urban policies using a variety of different land use and transport models (Tranus, Dortmund) in a number of new case cities (Dortmund, Vicenza, Brussels, Inverness). This will show the degree to which the results may be model or city-specific and the degree to which they can be generalised for European cities. It also develops some new indicators addressing e.g. the economic competitiveness of the city and the quality of open space, as well as changes in energy consumption and emissions from land use. New tools are being developed for more illustrative and understandable presentation of results as well as for communicating with the public through the Internet environment. Also environment feedback is aimed at when developing the existing land use and transport models.

The PROPOLIS project is part of the Key Action *City of Tomorrow and Cultural Heritage* of the Energy, Environment and Sustainable Development Thematic Programme of the Fifth Framework Programme funded by the EC. PROPOLIS is carried out by the same partners as *SPARTACUS*, together with Stratec, Belgium and UCL, University College London, United Kingdom. It is expected that PROPOLIS will be completed by the end of 2002 (for further information, see *www.ltcon.fi/propolis*).

Web Sites

SPARTACUS: *http://www.ltcon.fi/spartacus*

PROPOLIS: *http://www.ltcon.fi/propolis*

25 An Infrastructure Potential Cost Model for Integrated Land Use and Infrastructure Planning

Sharon Biermann

CSIR Building and Construction Technology, P O Box 395, Pretoria 0001, South Africa

Abstract: A strategic bulk infrastructure potential cost model has been developed as a decision support tool for planners and development decision-makers to ensure the proactive incorporation of bulk infrastructure cost considerations in the early land suitability assessment phase of the integrated planning process, to enhance the integration of land use and infrastructure planning. Not only does the model inform the early stages of the planning process prior to any proposals being made, but can also be adapted and applied in the plan evaluation stage, at a strategic level in situations where time, technical capacity, optimisation models and funds are limited or even lacking. Bulk engineering services infrastructure relating to water, sanitation and electricity have been included in the model. Potential costs are calculated on the basis of capacity in the existing system and include additional cost factors such as geotechnical, land use and environmental conditions to further enhance cost accuracy. The output of the model is in the form of potential cost contours which facilitates the relative comparison of infrastructure costs for different density scenarios.

Within a rapidly changing planning environment, it was found that although the model output was accepted at the time, before the output could be incorporated into the larger planning process, planning officials, decision-makers and the process itself had changed. This reinforces the importance of planning support systems (PSS) being integrally related to the planning process they are supporting if the outcomes are to have more than just an academic result. If the implementation structures are not in place and functioning, the PSS in isolation is not useful in influencing decision-making.

25.1 Introduction

25.1.1 Lack of Integrated Land Use and Infrastructure Planning

Infrastructure planning has generally tended to follow land use planning and infrastructure costs play no role whatsoever in the generation of land use strategies.

At most, the literature indicates the use of infrastructure cost assessment, particularly transport costs, at the evaluation stage of the planning process. In practice in South Africa, however, no evidence has been found for the strategic assessment of alternative strategies from the perspective of bulk engineering services costs (water, waste water and electricity). Historically, the preferred plan is selected and only then are the engineers tasked with optimising engineering services infrastructure for that particular plan.

In South Africa, infrastructure investment has historically been allocated in an imbalanced manner in accordance with racial policy, with some areas receiving massive investment and others, very little, resulting in areas with spare capacity that could be utilised at minimal additional cost. It is, however, not only political factors that make the incorporation of infrastructural aspects important but also the nature of infrastructure provision. From an economies of scale perspective, it is seen not to be cost effective in the longer term to expand infrastructure facilities in a continuously incremental manner but rather to construct facilities on an almost one-off basis, at considerable initial cost, but with more capacity than is currently required to ensure adequate capacity for future development. If the anticipated rate and extent of development does not occur, these areas are left with spare capacity in the system, theoretically already paid for. Knowledge of the location and amount of existing capacity is an essential input into the planning process. Not only is the additional capacity important but also an indication of the relative cost of extending and expanding the existing infrastructure.

25.1.2 Bulk Infrastructure Potential Cost Model

Purpose

A strategic bulk infrastructure potential cost model has been developed as a decision support tool for local government planners and development decision-makers to ensure the proactive incorporation of bulk infrastructure cost considerations, in the form of spatially comparative potential costs, in the early land suitability assessment phase of the integrated planning process, to enhance the integration of land use and infrastructure planning. Bulk engineering services infrastructure relating to water, sanitation and electricity have been included in the model. The output of the model is in the form of potential cost contours that facilitate the relative comparison of infrastructure costs for different density scenarios. Not only does the model inform the early, or plan generation stages of the planning process, prior to any proposals being made, but can also be adapted and applied in the plan evaluation stage, at a strategic level in situations where time, technical capacity, optimisation models and funds are limited or even lacking.

Relevance

Decision-makers need to be able to make informed decisions and be aware of the implications of decisions before implementation. Development planners and engineers need to understand the critical interrelationships between their disciplines. Stakeholders need to understand how technical factors impact on their choices and how their choices impact on issues such as cost and how their own situation relates to the wider area.

There is thus a need to transform data into decision-making material with 'what-if' testing capabilities and graphic and spatial presentation facilities. With an integrated planning approach, complexities in the urban process relating to the lack of familiarity between disciplines need to be simplified and the data inputs and outputs between various disciplines need clarification and integration. The bulk infrastructure potential cost model is a means to achieve this desired mutual understanding and facilitates the determination of impact before actually embarking on a specific plan of action.

The purpose of this chapter is to describe the bulk infrastructure potential cost model and how it has been applied in planning support for plan generation and evaluation. After referring to the theoretical basis of the model, which is threshold analysis, the three underpinning elements of the model, threshold, density and cost, are presented. The methodology inherent in the model is subsequently summarised and the application of the model in two instances of planning support, described. In both cases, the model was applied in the Greater Pretoria Metropolitan Council area. In the first case, the application of the model in determining bulk infrastructure costs as input to land suitability assessment, as part of plan generation, is explained. In the second case, the use of the model in the evaluation of alternative development scenarios from a bulk infrastructure cost perspective is presented. The chapter concludes by contextualising the model firmly as a Planning Support System (PSS) and refers to proposed future extensions of the model in relation to bulk services contributions.

25.2 Theoretical Basis

25.2.1 Threshold Analysis

The theory and technique of *threshold analysis* originated in Poland from the work of Malisz (1969) in the 1960s and was elaborated upon and refined by Kozlowski (1971) and the United Nations (1977) and applied in Scotland (Kowslowski and Hughes 1967, Kowslowski and Hughes 1972, Scottish Development Department 1973). According to Malisz (1969), threshold theory originated in response to three physical planning problems of the day. The first problem related to the lack of inter-disciplinary understanding and co-operation between physical planners and economic planners. Threshold theory attempts to translate qualitative urban design into quantitative factors. The second problem concerns feedback between the different levels of planning. Threshold analysis at the more detailed local level serves as a feedback into regional plans as to the feasibility of regional proposals. Thirdly, threshold theory goes some way to solving the problem of time. Instead of setting artificial time horizons for plans, threshold times, or times at which various thresholds are reached, can be used to plan for (Malisz 1969).

The theory of threshold analysis was based on the observation that towns encounter some physical limitation to their spatial growth in the form of natural limitations such as topography or in the form of man-made facility limitations such as public utility network systems (United Nations 1977). These limitations have been called the thresholds of urban development. They are not irremediable but can be overcome at a high capital investment cost or threshold cost (Kozlowski and Hughes 1967). Threshold analysis concerns those physical characteristics of an area that would cause significant differentiation in the unit cost of future urban development, dealing only with costs that vary with location (Lichfield *et al.* 1975). A threshold occurs when new units of development cannot be constructed and serviced at their previous unit cost levels and significant additional outlays are required. The presence of a threshold is indicated either by a steep rise in the gradient or a discontinuity in the marginal cost curve of urban development. Changes in the unit cost of development could be caused by a variety of physical factors ranging from the topography or the physical capacity of the public facility (Lichfield *et al.* 1975).

The methodology of threshold analysis concentrated initially on three groups of capital costs. Firstly, those costs due to physical constraints imposed by factors such as slopes, liability to flooding and subsistence were taken into account. Secondly, capacity constraints in engineering infrastructure services were considered. Thirdly, structural constraints were noted when the existing urban structure limited further development such as when increased population made the construction of a new public transport system necessary (Simpson 1977). In further applications of the technique, an attempt was made to include a range of social facilities (Intermet 1973). The United Nations (1977) includes environmental and social considerations.

25.2.2 Threshold Levels, Density and Cost

Three levels of threshold are identified in the literature: first, boundary and intermediate thresholds. Each level can be indicated spatially as a line on a map but would also have some quantity attached to it, e.g. population or capacity. The first threshold is the limit to which the town can expand without incurring any threshold cost. According to the literature, the developed area would be included in this first threshold. Boundary threshold lines are lines on a map that bound areas that are to be excluded from urban development due to natural or man-made environmental conditions that need protection. Intermediate thresholds are those lying between the boundary and first thresholds that require some cost expenditure in order to overcome the threshold (United Nations 1977:14, Scottish Development Department 1973:21).

There are three essential elements of the bulk infrastructure cost model: threshold, density and cost (Figure 25.1). It is the manner in which these three pillars are incorporated into the model and relate to each other within the model, which forms the basis for the model. Engineering services infrastructure is, by its very nature, a utility to people, particularly to urban people, necessary for health, access and industry. As the number of people increases within an area, so the infrastructure needs to expand to accommodate additional requirements. Due to the fact that infrastructure has physical dimensions (i.e. pipe diameter or reservoir capacity), limits or thresholds, directly proportional to size of infrastructure, exist. The relationship between density and threshold dictates whether or not additional infrastructure expenditure is necessary to accommodate additional people.

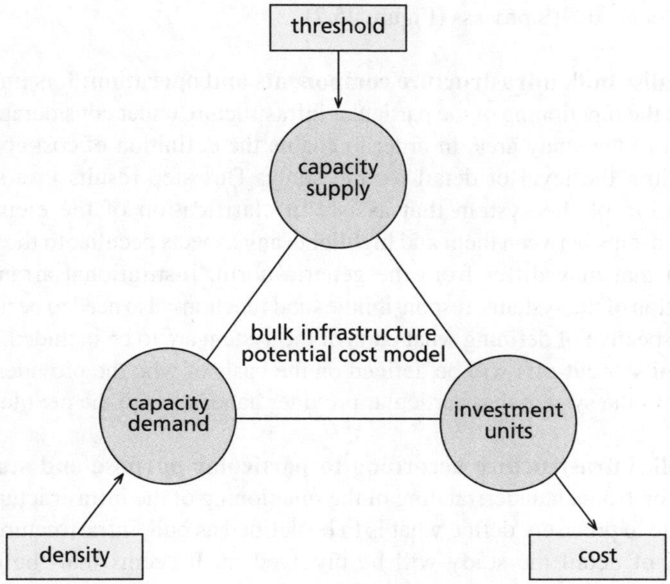

Fig 25.1: Mechanisms of interaction between the three pillars of the bulk infrastructure potential cost model

The manner in which the three pillars are incorporated into the model is largely through capacity analysis. The density levels set convert into the number of additional person units required per sub area, which in turn can be translated into infrastructure capacity demand figures. The resulting capacity demand figures are influenced by the desired level of services. The lower the level of service, the less the demand for infrastructure. In the present study, levels of service are kept constant in order to calculate relative costs. Infrastructure thresholds are equivalent to the design capacities of the various components of the infrastructure system. The existing infrastructure network and facility design capacities are compared with the current utilisation of infrastructure in order to quantify the capacity supply situation. The quantitative comparison of the capacity demand situation with the capacity supply situation determines firstly, whether or not additional infrastructure is required and secondly, if infrastructure is required, how much, or how many infrastructure investment units are required. Investment units can be in the form of metres of conduit or number of facility modules. Base cost per investment unit is influenced by additional cost factors relating to local conditions. The resulting relative costs are mapped and incorporated into a wider land suitability assessment model.

25.2.3 Generic Bulk Infrastructure Potential Costing Methodology

The various methodological steps involved in the bulk infrastructure potential cost model are explained with reference to a schematic representation of the costing process and the associated GIS process (Figure 25.2).

Conceptualise bulk infrastructure components and operation: It is important to understand the functioning of the particular infrastructure under consideration within the context of the study area, in order to enable the definition of cost components and to delimit the level of detail for the study. This step results in a schematic representation of the system that assists in clarification of the elements and interrelationships between them and highlights any aspects peculiar to the particular study area that may differ from the generic norm. Institutional arrangements, administration of the systems, responsibilities and functions also need to be understood from a perspective of defining what parts of the system are to be included. It may be that the limits or cut-offs will be defined on the basis of who the provider is and to what point in the system that particular provider hands over to the next level.

Define bulk infrastructure according to particular purpose and study area: Following on from an understanding of the functioning of the infrastructure system, it is possible to precisely define what is to be defined as bulk infrastructure and thus what level of detail the study will be involved in. It seems that 'bulk' cannot categorically be defined, say in terms of size of pipe, but rather needs to be defined for a particular study area, depending on the requirements of the particular study and on the basis of physical and institutional factors.

Subdivide study area according to intermediate threshold areas: The entire study area needs to be delimited in terms of supply or catchment areas. Each section of network and each facility has some area which it serves in terms of either supplying water or electricity or taking away waste. A major distinction needs to be made between those areas with existing networks and facilities and those areas with no infrastructure. For those areas with no existing infrastructure, two situations need to be identified. Firstly, for areas where there is planned infrastructure, the supply and catchment areas and networks need to be included. Secondly, for those areas where no planning has yet been undertaken, coarse level assumptions regarding possible future infrastructure provision in these areas need to be made in consultation with engineers familiar with the area and from an assessment of natural drainage in the area. Assumptions must be made as to whether or not the area will be served by existing infrastructure, i.e. if it falls within the same drainage basin as an existing facility or if the possibility exists for pumping over a watershed, or if an entirely new system is more likely. If absolutely no development is envisaged in those areas, they can be excluded from the cost model.

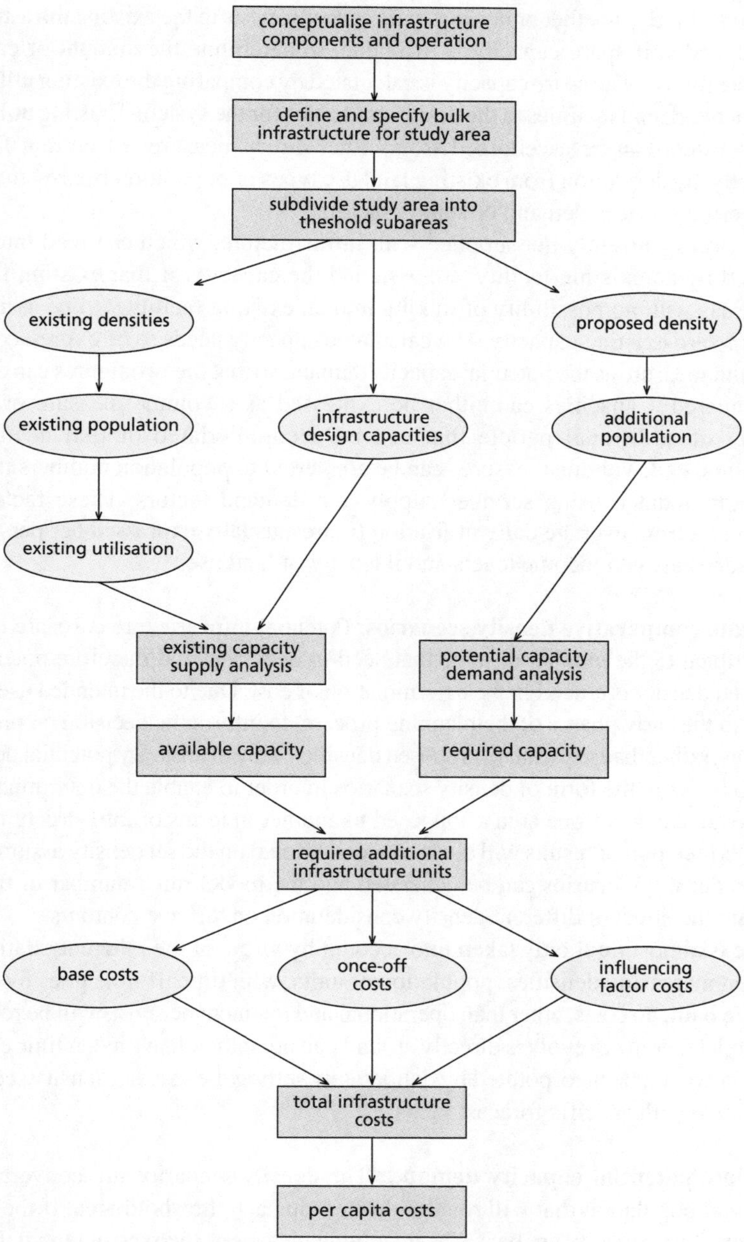

Fig 25.2: Overall bulk infrastructure potential cost model methodology

Undertake existing capacity supply analysis: For each threshold area defined in the step above, the current capacity situation needs to be quantified in order to determine, firstly, whether or not any spare capacity exists in the existing infrastructure and, secondly, if spare capacity is available, to determine the amount of capacity available for use. The spare capacity is calculated by comparing the existing utilisation of the network and facilities to the design capacities for the system. Existing utilisation of the system can be ascertained from either direct measures of current flow or indirectly, by derivation from existing land use types or population figures and using engineering services demand or supply factors.

New areas, currently not supplied with infrastructure, which can feed into or be supplied by an existing facility, are assigned the capacity of that existing facility. New areas with no possibility of linking into an existing facility will be assigned a value of zero existing capacity. The calculated capacity needs to be expressed in the same unit measure as the potential capacity demand so that the two figures can directly be compared. Capacities can either be expressed as a volume measure or as the number of additional people that can be accommodated or that need to be accommodated. Volume measures can be converted to population numbers that can be accommodated using services supply and demand factors. These factors are determined from average daily utilisation figures, usually expressed per person and often vary between income levels and intensity of land use.

Generate comparative density scenarios: Potential infrastructure costs are directly proportional to the number of units that need to be served and therefore potential or proposed density is crucial in the determination of cost. Due to the intended use of this model in the early phases of the planning process, to inform the decision on proposed densities, rather than to evaluate proposed densities, almost arbitrary potential densities need to be set in the form of density scenarios in order to enable the determination of relative advantage of one area as opposed to another in terms of infrastructure costs. The interpretation of results will always be conditional on the set density assumptions. Various density scenarios can be generated and the model run a number of times to ascertain the effect of different density consideration on the cost contours.

Time is almost implicitly taken into account by virtue of the inherent relationship between increasing densities, population or units, with time. If time goes by and no units are built, no costs, other than operational and maintenance costs will be required. The model therefore revolves directly around density rather than time. A time element could, however, be incorporated by deliberately setting the various density scenarios to coincide with specific forecast years.

Calculate potential capacity demand: The density scenarios are converted into additional population that will require services in each threshold area. If the output of the existing capacity analysis was in volume measures, then the potential demand output also needs to be converted from population figures to volume figures using different per capita services demand and supply factors for low, medium and high income groups. Residential populations are not the only users of bulk infrastructure – industry, agriculture, social and commercial uses also have infrastructure requirements. In the urban context, industrial and commercial activities are most pertinent. For the purposes of this model, usage of infrastructure by social facilities, such as schools and libraries, and by business areas, is not taken into account since the usage is relatively minor and does not peak at the same time as the main residential peak (National Housing Board 1994). It is assumed that for large water and electricity dependant, high effluent producing industrial uses, separate infrastructure provision arrangements are made and are thus not included in the model.

Determine required additional infrastructure: A comparison of the output of the existing supply analysis with that of the potential capacity demand analysis, allows the determination of what additional infrastructure is required to meet the potential demand. The additional requirements specified would relate to both networks and facilities. Required network infrastructure specification would involve the stipulation of location, length and diameter or related characteristic of additional pipes or cables needed. For the purposes of this model, where additional capacity is required in sections of the existing network, the assumption is that a new section of conduit will be laid alongside the existing section so that the locational position and length of the new conduit is the same as the old. For totally new conduits in the form of links to the current network or new networks, the route is assumed after a consideration of topography, current development and existing infrastructure, i.e. a route of least cost as calculated by the GIS. Size of the conduit, as measured by diameter for water and sewer pipes or resistance for electricity cables, is determined according to tables prepared by engineers for the specific purposes of this model, which relate density, slope (where applicable) and diameter in accordance with required flow, pressure or electricity output requirements.

Required facility specification involves the determination of the location and number of new or additional facility modules that need to be constructed to serve the additional potential population. Tables and graphs, relating numbers of persons to amount of storage required and size of facilities, have been prepared for this purpose using engineering design principles.

Calculate threshold costs: For each additional section of network or facility required, as specified in the step above, the cost of overcoming those thresholds of limits in the physical infrastructure needs to be calculated. There are two broad cost components to be considered: base costs and additional costs. Base cost for the network refer to the per metre cost of conduit under ideal conditions. Infrastructure facility base costs are defined as the average cost per facility module related to size of module. Additional costs take account of costs involved due to non-ideal conditions and include what have been termed, influencing factor costs and once-off costs. Influencing factor costs take into account the impact of locality on the per metre network base cost, in terms of natural and built environmental conditions. One-off costs refer to the additional network cost of traversing a particular physical, man-made obstacle such as a railway line or road.

Determine base costs: For network costs, all cost components relating to laying an infrastructure conduit need to be identified. The components can be broadly categorised as preliminary and general costs (insurance, tools site office, *et cetera*), labour costs (project management, earthworks, bedding) and materials costs (conduit, valves, meters, manholes). The contribution of each component to the per metre cost needs to be ascertained and a total base cost calculated. The reason for breaking the base cost down into component parts is that different influencing factors influence different components of the cost and only that specific portion of the cost is factored up to take account of influencing factors.

The facility module average costs need to be determined and can be ascertained from an assessment of contract documentation for the construction of such facilities. For water infrastructure, facilities include reservoirs and water towers, for sanitation, waste water treatment works and pump stations and for electricity, electrical substations.

Determine influencing factor costs: Those factors which cause an increase in the per metre base cost of conduit need to be determined and the cost implications calculated. The manner in which it has been conceptualised that the influencing factors be included for the purposes of this study, is that each influencing factor is categorised into one of three categories. Category 1 conditions are ideal conditions and only the per metre base cost applies. Category 2 conditions are such that some additional cost is necessary. Category 3 conditions are least favourable and significant additional cost needs to be added to the per metre base cost. Additional costs are included by means of percentage increases which need to be applied to specific cost components of the base cost, e.g. to the excavation or materials components (Table 25.1).

Influencing factors include aspects of land use, geotechnical conditions, natural environmental factors and topography. The influence of land use is in terms of intensity of development. Ideal conditions refer to undeveloped land where there is no impediment to installing infrastructure. Generally, costs increase as intensity of development increases with the increasing possibility of obstacles and digging up existing construction. The highest cost is associated with central business district type development. Geotechnical conditions influence the per metre base cost largely according to hardness of rock and amount of clay in the soil, requiring blasting or special protection of pipes (Table 25.1). In the GIS, all influencing factors are overlaid and the cost influence of each, in terms of percentage increase on the base cost, added, to create an influencing factor cost surface (Figure 25.3).

Determine one-off costs: One-off costs account for additional expenditure as a result of traversing an obstacle that can be man-made such as a road or railway line or natural, such as a river. The per metre cost is not affected; rather a once-off cost is involved. As in the case of influencing factors, one-off costs are categorised in terms of size of obstacle. Single carriageways and railway lines would cost less to traverse than highways, multiple railway lines and large rivers (Table 25.1).

In the GIS, all one-off costs are overlaid and a one-off cost layer created indicating the total cost effect of overcoming man-made and natural obstacles.

Calculate total relative potential cost: For each particular density scenario, and for each defined threshold area, the network and facilities cost need to be calculated. Network costs are calculated by multiplying the distance to be traversed with the appropriate per metre cost (base costs and influencing factor costs included), adding any additional once-off costs required. Facility costs are simply the number of modules necessary to service the additional population multiplied by the module cost. The location of facilities is assumed to be the same as that of existing facilities in the case of upgrading and the location of new facilities is assumed from an assessment of the current network and topography.

Table 25.1: Costs of influencing and one-off factors

Sewer base cost (1995 R/m): 206.76
Water base cost (1995 R/m): 92.83

factor	category	classification	sewer % cost inc	sewer R/m	water % cost inc	water R/m
influencing factors						
land use	undeveloped	agricultural	0	206.76	0	92.83
		farmland	0	206.76	0	92.83
		government	0	206.76	0	92.83
	developed	commercial	8	223.30	1.5	94.22
		industrial	8	223.30	1.5	94.22
		informal	8	223.30	1.5	94.22
		residential	8	223.30	1.5	94.22
		education	8	223.30	1.5	94.22
	CBD/high density	activity area	40	289.46	3	95.61
		high rise res	40	289.46	3	95.61
	environmental	nature reserves	10	227.44	12.5	104.43
		open spaces	5	217.10	6.5	98.86
geotechnical	geology*	1 AC	0	206.76	0	92.83
		1 BEG	0	206.76	0	92.83
		2 ADE	12.5	232.61	12.5	104.43
		2 ADE	12.5	232.61	12.5	104.43
		2 B	10	227.44	10	102.11
		2 E	12.5	232.61	12.5	104.43
		2 EG	12.5	232.61	12.5	104.43
		3 B	20	248.11	20	111.40
		3 BCI	20	248.11	20	111.40
		3 EGH	12.5	232.61	12.5	104.43
		3 F	40	289.46	40	129.96
	slope	2 - 6 degrees	0	206.76	n/a	n/a
		6 - 12 degrees	25	258.45	n/a	n/a
		< 2 degrees	25	258.45	n/a	n/a
		> 12 degrees	50	310.14	n/a	n/a
one-off factors				R		R
man-made obstacles	roads	national		20000.00		20000.00
		major		20000.00		20000.00
	railway lines	single track		10000.00		10000.00
		multiple track		30000.00		30000.00
natural obstacles	rivers	perennial		15000.00		15000.00
		non-perennial		5000.00		5000.00

*** code description**

A collapsible soil
B active soil
C highly compressible soil
D erodability of soil
E difficulty of excavation
F instability in areas of soluble rock
G steep slopes
H areas of unstable natural slopes
I areas subject to flooding
1 most favourable for urban development
2 intermediate favourability
3 least favourable for urban development

In the GIS, required water and sewer pipes are overlaid with the influencing factor cost surface and the once-off cost surface and the total cost for each pipe determined according to the distance covered by each pipe under differing environmental conditions.

Calculate per capita cost: In order to relatively compare costs between locations, the infrastructure costs need to be distributed amongst the people who use that infrastructure. Some costs, i.e. the link costs would only be distributed amongst the people in the particular threshold area whereas the costs of say, a major outfall sewer, would need to be distributed amongst the potential population of all the threshold areas that utilise that particular section of pipe. Thus, there needs to be a per capita assignment of costs across the entire study area to facilitate the relative comparison of costs between locations. As is currently the case in most local authorities, the local authority carries the cost of the infrastructure until it can be redeemed, as and when development occurs, in the form of bulk services contribution levies.

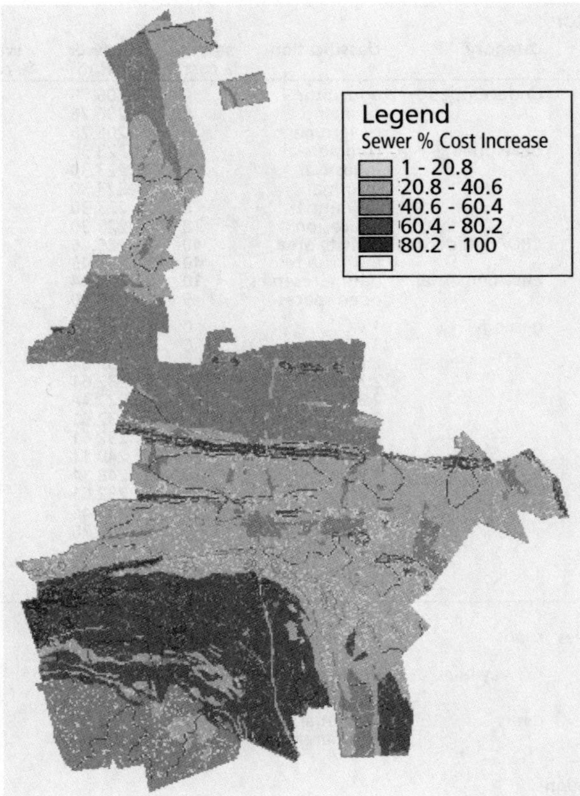

Fig 25.3: Locality map of the Greater Pretoria Metropolitan Council area

Prepare thematic maps for incorporation into multicriteria evaluation model: For each density scenario, thematic maps, displaying cost contours or relative cost categories, need to be prepared in order to make the results explicit and visual to decision makers and planners. All the various engineering services costs can be overlaid in the GIS and total cost layers derived. In addition, the mapping of the bulk infrastructure cost results can be combined with other multicriteria layers, into an overall land suitability assessment model.

25.3 Application of Model

In this chapter, the application of the model in the plan generation phase and also the plan evaluation phase of the integrated strategic planning process is illustrated. In both cases, the model was applied in the Greater Pretoria Metropolitan Council (GPMC) area. The study area is about 130,000 hectares in extent, stretching over a distance of 65 kilometres from north to south and over 50 kilometres from east to west (GPMC 1997:28). The study area comprises three metropolitan substructure areas of the Town Council of Centurion in the south, the City Council of Pretoria in the centre and the Northern Pretoria substructure in the north (Figure 25.4).

25.3.1 Application in Plan Generation: Determining Bulk Infrastructure Costs as Input to Land Suitability Assessment

The generic bulk infrastructure potential cost methodology described earlier was applied in the GPMC area. Water, sanitation and electricity capital costs, at 1995 prices, were determined and represented spatially, for various population density scenarios. Bulk water costs comprise link costs (cost of linking the development with the distribution reservoir), distribution reservoir costs (including pressure towers), feeder main costs (pipes linking distribution reservoirs with system mains or Rand Water mains), system main costs (pipes linking distribution reservoir feeder mains to receiving reservoirs) and receiving reservoir costs. Waste water treatment works (WWTW) costs, bulk outfall sewer network costs and sewer link costs (linking the development with the outfall sewer network) are included in sanitation cost. Electricity costs incorporate 132/11kV substation, cabling and transformer costs. Infrastructure costs vary with location according to local density, infrastructure capacity, land use, geotechnical, environmental and built conditions, making the role of GIS in the model appropriate and important.

All individual costs are ultimately integrated into a single cost for each of water, sanitation and electricity which are overlaid to achieve a total infrastructure cost overlay (Figure 25.5). Costs are calculated for three different density patterns: high, medium and low. For illustration purposes, only the medium density results are discussed. The spatial pattern of cost for the medium density scenario indicates highest costs in the vicinity of Klipkruisfontein, the central northern areas, the far east, the southern parts of the central area and in the south western area (Figures 25.4 and

25.5). The pattern is dominated by electricity costs but WWTW thresholds in the central areas have been exceeded, contributing to increased costs in the central regions.

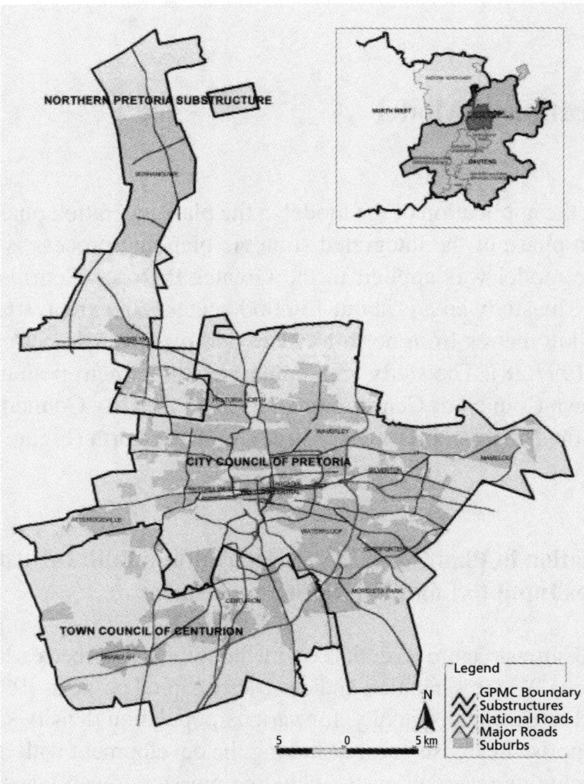

Fig 25.4: Combined water, sanitation and electricity per capita costs for the medium density scenarios

Higher costs in new peripheral areas are usually due to a combination of all services, whereas in the developed area, there is usually only one or two cost factors, predominantly water factors, responsible for the higher cost (Figure 25.5). This can be attributed to the fact that in developed areas not all engineering service thresholds are reached at once, whereas in newly developing areas, all services need to be provided from scratch at the start of development – there is no condition of existing spare capacity.

Relatively low combined costs in the far northern Soshanguve area and far eastern Mamelodi area can be attributed to lower usage factors and the way in which the model deals with situations where existing density exceeds scenario density (Figures 25.4 and 25.5). Available spare capacity can therefore go much further in meeting the demand than the same available capacity in areas of higher usage. In situations where existing population exceeds scenario population, the model takes the existing population as the scenario population and as long as existing population equals scenario population, no additional costs are incurred. Due to the fact that these are areas of existing high density, the model generally allocates no additional people to these areas and thus costs are zero.

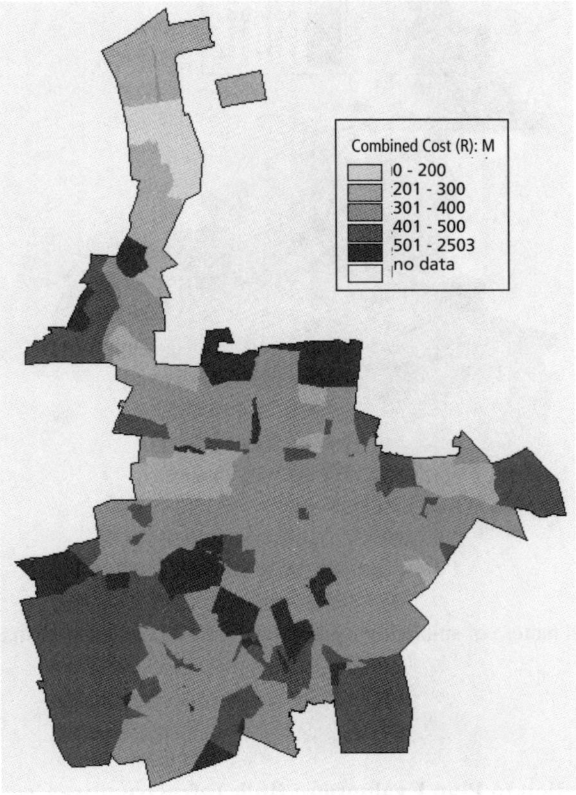

Fig 25.5: Multicriteria suitability assessment including bulk infrastructure potential costs

Using GIS-based multicriteria evaluation, the bulk infrastructure cost layer is combined with other suitability layers: access to employment; access to health, education and recreation facilities; agricultural potential; geotechnical suitability and land values. The overall spatial pattern of suitability obtained by including bulk infrastructure potential costs indicates higher suitability dominating the more central, developed areas with the highest suitabilities occurring in the western central areas and lower suitability in the peripheral areas (Figure 25.6).

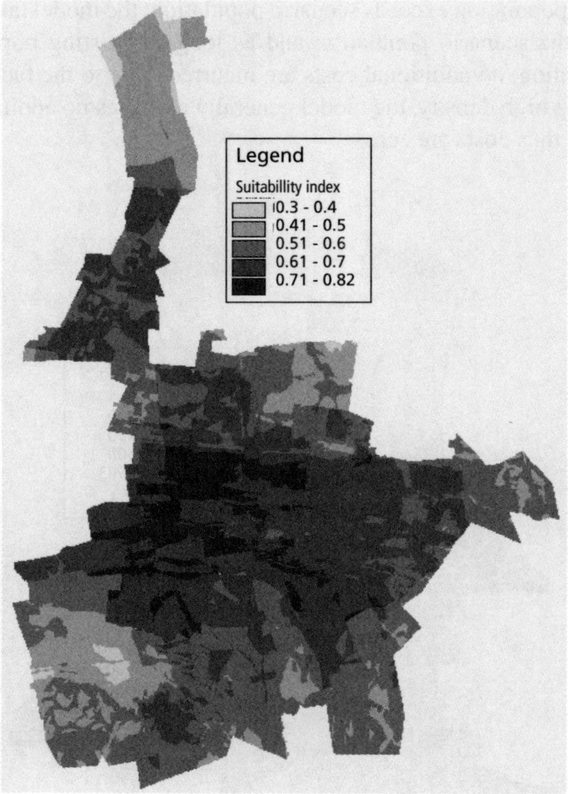

Fig 25.6: Overall pattern of suitability including bulk infrastructure potential costs

25.3.2 Application to Plan Evaluation: Bulk Infrastructure Cost Implications of Two Alternative Urban Density Scenarios

The aim of this application was to develop a general costing and multi-criteria evaluation model to be used in evaluating two residential development scenarios, qualitatively and quantitatively. The 'Trend' scenario was prepared based on current development trends and approaches. The 'Densification' scenario, or 'Integrated Urban Densification Strategy', was prepared on the basis of sustainability precepts.

The emphasis was on comparative costs not necessarily actual costs and every effort was made to ensure that precisely the same factors, methodologies and calculations were used in calculating the costs associated with the Trend and Densification scenarios. Bulk engineering services costs, including waste water treatment works, outfall sewers, distribution reservoir, water feeder main and electricity costs, were calculated. Local reticulation costs, including sewer, water, electricity and roads and stormwater costs were also included in the cost calculation. Land costs, building costs, social facilities costs and transportation costs were further included. For the purpose of this chapter, the following discussion revolves around the bulk infrastructure cost component as the calculation of bulk infrastructure costs relied heavily on the methodology developed in the bulk infrastructure potential cost model.

The generic methodology and cost tables developed for the bulk infrastructure cost model described earlier were adapted in order to evaluate the two scenarios. Required capacity was calculated by converting proposed scenario population numbers per planning area into volume capacity requirements per catchment or supply area. GIS technology was used extensively in the process. Available capacity was determined from a comparison between existing utilisation and design capacity. The difference between available capacity and required capacity yielded an estimation of required additional infrastructure units, either as number and size of infrastructure facility modules or as length of conduit. Costs were allocated to each unit of infrastructure required on the basis of average cost data determined from records and engineering experience. Total costs for each service were then expressed per additional person added per scenario so that relative comparison between scenarios was facilitated.

In interpreting the results it is essential to realise that the results are unique to a particular place at a particular time under a particular scenario. The costs that have been calculated arise from a unique combination of factors relating to where and how many additional people have been allocated in relation to the unique situation of existing capacity conditions in the engineering services infrastructure. The results would therefore be very different from a simple factorising exercise where the number of additional people is multiplied with a cost factor for each service.

The total bulk infrastructure cost required to support the Densification scenario is 63 per cent greater than for the Trend scenario (Table 25.2). This is largely as a result of almost 2 million more people being added in the Densification scenario than in the Trend scenario due to higher densities. The per capita costs reveal that the Densification scenario costs 17 per cent less per person, than the Trend scenario (Table 25.2). The most costly infrastructure components are reservoirs and electricity costs. The least costly items are water and sanitation conduit costs.

Table 25.2: Summary of bulk infrastructure costs, 1997

cost component	cost element	cost (per person) TREND Rand	TREND %	DENS Rand	DENS %	total cost TREND Rmill	TREND %	DENS Rmill	DENS %
bulk water	reservoirs	340	45	252	40	840	49	1276	46
	feeder mains	17	2	15	3	43	2	75	3
bulk sanitation	WWTW	29	4	27	4	71	4	139	5
	outfall sewers	11	2	8	1	27	2	39	1
bulk electricity	substation, line, transformers	357	47	323	52	728	43	1265	45
total cost		754	100	625	100	1709	100	2794	100

25.4 Practical Experience in Applying the Model

The application of the model in the Greater Pretoria metropolitan area demonstrates the benefits of the model as a PSS in incorporating bulk infrastructure cost aspects during both the plan generation and plan evaluation phases of the planning process. Although the model was successful, technically, in terms of achieving the objectives, the institutional context within which the model was applied had significant implications on the impact of the model outcomes on the planning process.

Political changes in South Africa over the past few years have resulted in massive changes in local government institutional structures and processes which have been implemented through a series of transitional phases involving local government elections, organizational restructuring and transformation and new legislation. The planning process itself has undergone major changes. Within this rapidly changing planning environment, it was found that although the model output was accepted at the time, before the output could be incorporated into the larger planning process, planning officials, decision-makers and the process itself had changed. This absolutely reinforces the importance of planning decision support systems being integrally related to the planning process they are supporting if the outcomes are to have more than just an academic result. If the implementation structures in the planning process are not in place and functioning, the PSS in isolation is not useful in influencing decision-making.

25.5 Bulk Infrastructure Potential Cost Model as a PSS

GIS technology is relevant to any spatial system and is thus largely independent of the system of interest. Spatial Decision Support Systems (SDSS), although incorporating a focus on a particular decision-making problem, have arisen out of DSS from management science or operation research modelling rather than being focussed on large-scale planning and urban systems theory (Batty 1994:13). It has been argued that the gap between SDSS technology and its use in strategic planning and design is so wide that planning support systems (PSS) are required to achieve an effective incorporation of planning decisions into SDSS (Harris 1989:85).

The spatial planning arena is very different to the management arena in that the concerns of planning are far longer term than those of management. An important concern of planning is the avoidance of unintended consequences while pursuing intended goals. In order to assess these consequences, planning needs techniques for making conditional predictions based on alternative hypothetical 'what-if' decisions (Harris and Batty 1993: 184). These techniques require extensive computation resources and simulation modelling and although GIS plays an important role, in itself it is not sufficient to be used as the main tool of analysis in planning (Harris and Batty 1993:185). PSS require analytical and design functions that incorporate goals, objectives, costs and benefits. For GIS to be truly beneficial in the spatial planning domain, it needs to be constantly and strongly linked to problem definition and the systems models used to inform the planning process. The framework within which this linkage between computation, social and functional theories of the system being planned, the theory of planning and spatial representation takes place, is termed PSS (Harris and Batty 1993:187). It has been concluded from an assessment of the problems and opportunities of linking GIS and PSS, that systems and models are not easily embedded within one another and that more rapid progress can be made if PSS were to be developed independent of but in parallel with and loosely coupled to, GIS (Harris and Batty 1993:197).

SDSS emerged out of the combined traditions of DSS and GIS, applied in the context of the spatial planning tradition, using urban modelling and spatial analytical techniques. The methodology described for the bulk infrastructure potential cost model, complies with the requirements of a model-oriented DSS, and the strong spatial component, with multicriteria evaluation and the cost model, being loosely-coupled with GIS, means it could be referred to as a model-oriented SDSS.

In addition, with the focus of the models on a specific planning problem, the bulk infrastructure potential cost model complies with the requirements of a PSS. There are two essential requirements for planning and thus, in turn, for PSS. The first requirement is that, since urban system optimisation is impossible (this would equate to the automatic generation of plans), the search for good plans must be by means of an informed process of trial and error that generates plans and prepares them for testing (often referred to as sketch planning). Secondly, planning and policy making need tools for determining the consequences of alternative courses of action or decisions so that choices can be made in terms of costs and benefits or adjustments and improvements to alternatives can be made (Harris and Batty 1993: 194). The cost model generates potential bulk infrastructure costs under different density conditions that informs the plan generation process but can also be used to determine the engineering services cost consequences of certain development decisions.

25.6 Conclusion

The bulk infrastructure potential cost model provides a means for incorporating bulk engineering services cost consideration into the strategic planning process at an early stage during land suitability assessment. Potential costs are calculated on the basis of capacity in the existing system and include additional cost factors such as geotechnical, land use and environmental conditions to further enhance cost accuracy. The model can also be used in evaluating alternative spatial development proposals in terms of engineering services cost implications. Rather than a particular density scenario being set artificially, as is the case in applying the model in the early, generation phase of the planning process, actual proposed densities, which may vary with location, are used as model input to determine demand.

It has been shown that the model can be considered as a PSS. The problem with the term 'planning' in PSS is that, particularly in South Africa, with recent moves to enforce an 'integrated planning' approach, strategic planning is seen to be far broader than spatial and physical planning, incorporating the entire spectrum of development, including the economic, social, institutional, financial, environmental and spatial realms. PSS should accordingly, theoretically support much more than only spatial decisions. It is proposed that the term spatial planning support system (SPSS) would better describe SDSS applied to the spatial component of strategic integrated planning.

A possible future application of the model is in the determination of bulk service contributions that local authorities levy on developers. Infrastructure cost considerations, as with all the other suitability criteria considered, only serve to better prioritise land for development. No land is excluded from development due to infrastructure cost factors, but rather land, which is cheaper to service, is promoted for development first. This is of significance in both public and private funding of housing. In the case of public provision, the lower the development cost, the more housing can be provided for the same amount of money. For private development, local authorities can recover the real bulk infrastructure costs by charging developers real costs, which developers would need to build into the price of the land. If private developers wish to develop in areas which are not cost effective from a bulk infrastructure cost point of view, they can proceed with development as long as they pay the local authority the real cost of provision. The manner in which service contributions are levied at present in many local authorities, is often on an averaged basis. It is argued that this promotes sprawl because on the periphery, where infrastructure costs are usually the highest, development costs to the developer, and eventually to the consumer, are the same as infill development. If real costs were levied, this would contribute to development occurring where it is economically sensible to do so, thus reducing the negative impacts of sprawl. The model can assist in determining a differential levy for various locations, more reflective of real costs for that particular location.

Ongoing research is investigating linking the bulk infrastructure cost model with other existing models to incorporate transportation costs, local reticulation cost models and financial cost models so that affordability aspects of local authorities and end users can be brought into the equation. The model only considers capital costs and the possibility of incorporating running and maintenance costs is also being investigated.

References

Batty, M. (1994) A chronicle of scientific planning: the Anglo-American modelling experience, *Journal of the American Planning Association*, 60(1): 7-16.

Greater Pretoria Metropolitan Council (1997) *Strategic Metropolitan Development Framework: Status Quo Report*, The Council, Pretoria.

Harris, B. (1989) Beyond geographic information systems: computers and the planning professional, *Journal of the American Planning Association*, 55(1): 85-90.

Harris, B. and Batty, M. (1993) Locational models, geographic information and planning support systems, *Journal of Planning Education and Research*, 12: 184-198.

Intermet (1973) Threshold Analysis as a Management Aid, International Association for Metropolitan Research and Development, Glasgow.

Kozlowski, J. (1971) The place and role of threshold analysis in the model planning process, *Ekistics,* 192(32): 348-353.

Kozlowski, J. and Hughes, J.T. (1967) Urban threshold theory and analysis, *Journal of the Town Planning Institute,* February: 55-60.

Kozlowski, J. and Hughes, J.T. (1972) *Threshold Analysis*, The Architectural Press, London.

Lichfield, N., Kettle, P. and Whitbread, M. (1975) *Evaluation in the Planning Process*, Pergamon, Oxford.

Malisz, B. (1969) Implications of threshold theory for urban and regional planning, *Journal of the Town Planning Institute,* March.

Scottish Development Department (1973) *Threshold Analysis Manual*, HMSO, Edinburgh.

Simpson, B. (1977). Some theoretical developments on threshold analysis and testing, *Urban Studies*, 14: 79-88.

United Nations Department of Economic and Social Affairs (1977) *Threshold Analysis Handbook*, United Nations Publications, New York.

PART 5

SUPPORT SYSTEMS FOR ENVIRONMENTAL PLANNING

26 The Environmental Information System: A Data Analysis and Presentation Tool

Flavio Bonfatti[1], Paola Monari[1], Andrea Aime[2] and Silvia Ascari[2]

[1] Dip. Scienze dell'Ingegneria, Universitá di Modena e Reggio Emilia, Via Vignolese 905, 41100 Modena, Italy
[2] Municipality of Modena, Via Santi 40, 41100 Modena, Italy

Abstract: The ISOLA project aims to provide methods and tools to support control and planning in urban environments with a strong focus on exportability and adaptability. This chapter describes the Environmental Information System, a set of software tools that enables end-users to create, maintain and analyse geographical and tabular data, create reports and maps and share data and processing power among several users.

26.1 Introduction: The ISOLA Project

The ISOLA project, funded through the framework of the European LIFE/ Environment programme and co-ordinated by the municipality of Modena, is aimed at providing methods and tools to support environmental control and planning for a medium-sized town. Modena, with a population of 180,000 inhabitants, is located in the centre-north part of Italy and is characterised by a strong presence of agricultural, industrial and service activities. Other partners in the project are: ARPA, the regional agency for environmental protection; ERVET, the regional agency for economic development; and Autodesk, a well-known company in the field of computer-aided design and GIS.

The project is based on three fundamental activities referred to as eco-balance, eco-plan and eco-management, and a software package, the *Environmental Information System* (*EIS*), that is designed to support them. Here follows a brief description of each activity and how it is supported by the *EIS*.

- *Eco-balance*. This deals with assessment of risks in the components of the urban environment and measurement of the pressure exerted by human activities and settlements. Eco-balance evaluations are based on a proper set of indicators, in particular: *status indicators*, showing the condition of a phenomenon at a particular point in time; *trend indicators*, providing a dynamic picture of a phenomenon as it changes; and *sustainability indicators*, synthesising the concurrent impact of different phenomena. Each indicator refers to a specific phenomenon, such as pollution, energy or transportation. It takes the form of a time series of one or more non geo-referenced values, or that of a fully fledged geographical data set, such as interpolated rainfall precipitation. Eco-balance also requires the calculation of indices for comparison. The *EIS* is required to store and give access to both raw and derived data with associated meta-information, provide a means for data representation (tables, diagrams, thematic maps), provide data analysis functions, and support the automatic production of Eco-balance reports, i.e. integrated documents containing data, tables, diagrams and maps to represent indicators and indexes ready for the decision-maker to consult.

- *Eco-plan.* This deals with integration of environmental factors in urban planning according to proper methods, criteria and procedures. The main objectives are to: ascertain the degree of suitability of each area for specific purposes or uses; assess the exposure levels to potential or actual sources of environmental, natural and human degradation; and estimate the response levels of each area and its environment when subject to a specific use. This process requires field data acquisition, elaboration, comparison with the current area use and the presentation of results in the form of thematic maps, diagrams and data tables. The *EIS* is required to store raw and derived data, perform repeatedly pre-defined analyses by minimising user effort, and create thematic maps and reports for every product concerned with the Eco-plan.
- *Eco-management.* This deals with support for decision-making by means of models and simulation tools that allow evaluation of the impact of alternative intervention scenarios on the urban environment. Considering that simulation tools are specialized applications that are outside the scope of a standard GIS tool, the *EIS* is required only to exchange (import and export) data with them.

The project extends and generalises the experience acquired by the Modena municipality from several years of work in this field, with the objective to set-up an integrated approach that can be replicated in urban contexts differing in geographical, cultural and economic conditions. The expected project results are:

- a sound methodology, expressed in the form of guidelines addressed to those municipalities that plan to adopt a disciplined way of facing the environmental problem;
- a software package (*EIS,* the *Environmental Information System*), to support the methodology and make its application directly feasible and controlled by end users not expert in GIS technology; and
- an experiment, obtained by applying the methodology to the real case of the Modena municipality. The main experiment goals are validating the methodology, testing user friendliness and performance of the *EIS*, and providing material and application examples for dissemination actions.

26.2 *EIS* Requirements

The ISOLA software package is required to be exportable to realities other than the Modena municipality and to improve GIS tool useability, while supporting the ISOLA project's three main activities. The *EIS* is based on the following requirements:

- Spatial analysis must be developed directly by the domain expert, and every technological aspect must be hidden under a user-friendly interface based on a graphical approach, called the 'process model'. In current practice, spatial analysis development requires at least two experts: a *domain expert*, who understands the phenomenon being studied and knows how to process raw data to obtain new synthetic indicators that can support decision-making; and a *GIS expert*, who knows the GIS technology and is able to translate domain expert indicators into a sequence of commands to the adopted GIS tool. The presence of a GIS expert is needed because the domain expert is commonly unable to use the GIS engine, mainly because of the detailed nature of the GIS technology. In fact, coding a complex spatial analysis procedure requires knowledge of data formats, database operations, scripting languages, *et cetera.* Unfortunately, the different cultures of domain and GIS technology experts make information sharing difficult and result in time and money consuming collaborations. *EIS* is required to eliminate the problem by hiding GIS tecnology details let the user focus on what he wants, and be concerned about 'the how' only from a conceptual point of view (as a sequence of steps).

- Spatial analysis processes cannot be a set of hard-coded routines; they must be easy to create, modify and to store for later execution. This is a direct consequence of the exportability requirement, since planning practices can vary from one municipality to another. Such variations are related to several factors:
  - *input data availability:* field data acquisition is usually expensive; some organisations simply cannot afford to get all the data needed for a sophisticated analysis;
  - *environmental issues:* related to the peculiar features of the local environment, even if there is an obvious set of issues related to human activities that does not change from place to place;
  - *legal constraints:* legal pollution limits can vary from country to country, as well as the list of chemicals whose level control is enforced;
  - *domain knowledge:* the knowledge depth about the studied phenomena may vary from organisation to organisation; and
  - *GIS tools available:* last but not least, the position of the single municipality is affected by the GIS tools, if any, it already uses.

  With respect to the single organisation the above-mentioned factors can also change in time, leading to possible modifications in the spatial analysis process.

- Tools must support effectively an iterative style for spatial analysis development. This is a key factor for actual tool utilization; a shorter development cycle enhances user experience and lays down the basis for further and more sophisticated investigation about the phenomenon under study. The tools must effectively support the standard development cycle depicted in Figure 26.1, in which we see the fundamental steps and the unavoidable feedback flows that one gets when evaluating spatial analysis results.

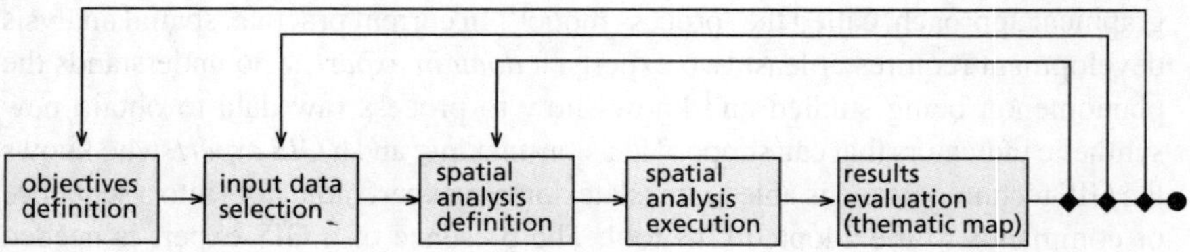

Fig 26.1: Spatial analysis development cycle

- Thematic maps must be easy to create and the map structure must be stored to support comparison between the results of periodic analyses. The same applies for tabular data investigation, diagram creation and report-building.
- The system must be independent of the GIS engine and a default configuration must be based on cheap (possibly free) products. This stems again from the portability requirement, in that some organizations may already have settled on a GIS package, while others need a full installation but cannot afford to buy a full blown commercial product.
- The processing power must be shared among users, and data, processes and thematic map definitions must be held in a common repository to foster knowledge sharing. The ISOLA project activities area carried out by several different people in a single organisation, these people need to share data and analysis processes in order to carry out their tasks efficiently. A central processing server is considered fundamental: when different users perform analyses on it, they use its full processing power for the needed time while splitting the investment cost.
- The system must support different kinds of users. Not every single user of the information system can create spatial analysis procedures. In fact, users can be subdivided into three categories: *domain experts with knowledge of geographic tools* who know both the phenomenon and GIS operators well enough to manage spatial analysis problems; *simple domain experts,* who are not able to handle the GIS technology but are interested to use already defined processes; and *viewers*, usually decision-makers, who only need to visualise results in proper forms.
  EIS supports this classification by providing different access level to system functionalities.

26.3 System Architecture

EIS is based on the traditional client-server architecture, where the client is in charge of the interaction with the user and the server manages data and executes computations (Figure 26.2). This separation ensures the independence of the client functions from the underlying GIS technology, and allows implementation of the ISOLA environment on different GIS platforms.

The client side of the architecture includes the process editor, the thematic map builder, the explorer, the tabular data viewer and the report generator. The server side includes the DMBS, one or more GIS engines in charge of spatial analysis, a virtual machine that allows for GIS engine independence and a lock manager. Adopting a client/server architecture means that several users can view, analyse and modify the same data at the same time. This is a well-known problem in computer science, and it is usually solved by putting data into a database management system (DBMS), under transaction and lock management. Unfortunately, most GIS engines still use files and file systems as the preferred way to store information; moreover they usually adopt a workstation approach, without considering multiple users of the same data. To enable user concurrency, we use a lightweight lock manager, adopting a simple reader/writer policy on shared files, essentially preventing any other access while a file is being modified.

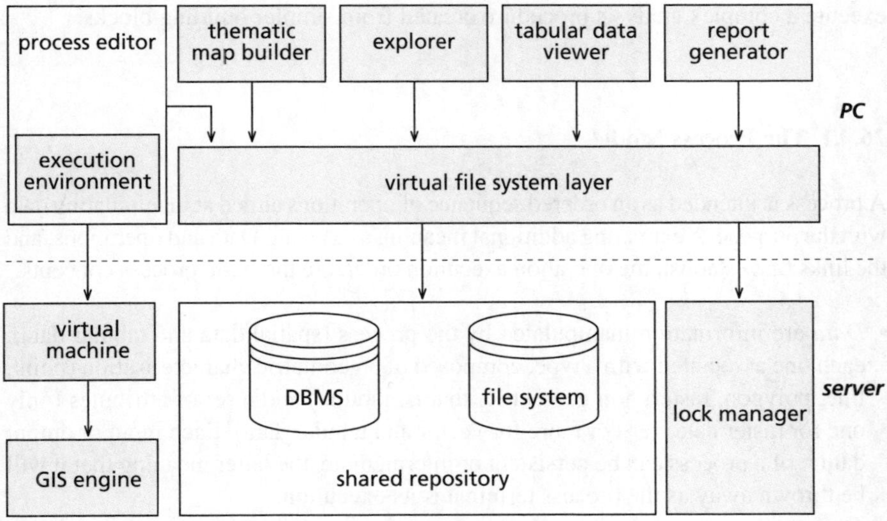

Fig 26.2: System architecture

GIS engine independence is obtained by layering the system architecture and letting only the lower level cope directly with the GIS engine (API) and data formats. In particular, process execution is made independent of the GIS engine by using a virtual machine, that is, a software component that exposes a standard interface to the process execution environment and translates requests in direct API calls to the GIS engine. If the GIS engine has to be changed, only the virtual machine needs to be rewritten, preserving all the code that manages process construction and user interaction. DBMS independence from the vendor is obtained by using only SQL standard features and by using access libraries that are built to assure client independence from the particular DBMS (ODBC).

Fig 26.2 depicts the client/server architecture and provides some details about client/server interaction: every access to the shared repository is made through the virtual file system (VFS) layer, the only component dependent on the real data structure; every call to the GIS engine is made through the virtual machine (VM), the only component that is aware of the GIS engine used by the implementation.

26.4 The Process: Modelling and Execution

The process model is at the base of the exportability feature found in the *EIS* environment. It is based on a graphical tool that enables the user to specify and execute a complex analysis procedure created from simpler building blocks.

26.4.1 The Process Model

A process is intended as an ordered sequence of operations aimed at manipulating data with the purpose of extracting additional meaning as a result. Data and operations, and the links that establish the operation execution order, are the basic process concepts.

- *Data* are information manipulated by the process (spatial data and tabular data), each one associated with a type, composed of a geometric characterisation (point, line, polygon, raster, non geometric, that is, tabular) and a set of attributes (only one for raster data, zero or more for vector and tabular data). Each input or output datum of a process can be persistent or intermediate, the latter meaning that it will be thrown away as the process terminates its execution.
- *Operators* represent the operation performed on available data in order to derive new information; they may be spatial operations (overlay, buffer, interpolation), synthesis operations (that create a tabular report starting from spatial data) and tabular operations (such as selections or joins). Each operator has a set of input roles (each one with a specific meaning) that will be linked to data, a set of

parameters to decide how the operator computes the output data, and, of course, a set of output data.
- *Links* represent input/output relations between data and operators; in other words, they connect operators to their input and output data.

The process can be thought of as an oriented, acyclic digraph whose nodes are, alternatively, spatial data and operations. The graph extreme nodes are spatial data, in that every process starts from its input data and reaches the resulting output data. The process model is a set of primitives that allow the user to express his/her intentions at the conceptual level, that is, without being distracted by implementation issues.

The advantages of such an approach are several:

- the user can specify a sequence of operations without using a programming language, but still giving a formal and executable specification;
- the user can work in an interactive fashion, modify the process again and again in order to refine the analysis procedure and obtain the desired results; and
- the process enables the reuse of the same spatial analysis on different input data sets, with the same meaning but coming from different location (spatial reuse) or from the same location in different times (temporal reuse).

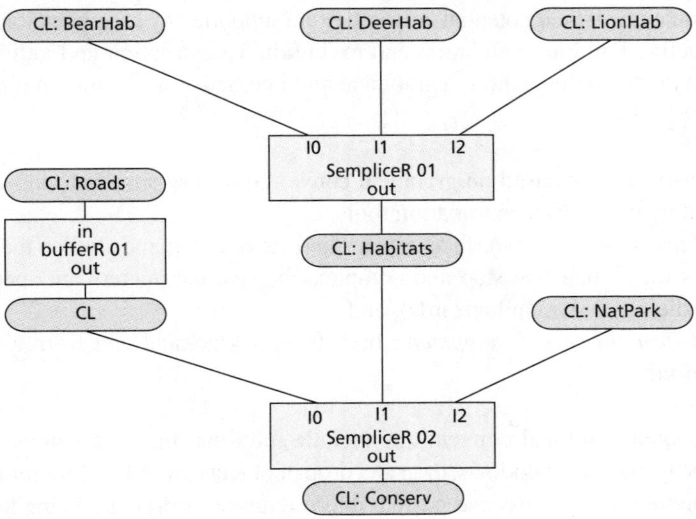

Fig 26.3: Selection of candidates for conservation areas

An example of process model is depicted in Figure 26.3. Here we want to select some candidates for a new conservation area that will protect wild animals, particularly bears, mountain lions and deer. We are looking for areas that fall into wild animal habitats that are not too close to roads and that are not already included in national parks. Supposing to work with raster data, cell layers (CL), the desired information can be obtained by performing the following steps:

- calculate a buffer around roads (bufferR 01), possibly by using a different buffering distance according to road class;
- merge habitats by means of a multiple data reclassification or by a Boolean expression, in this case performed with 'SempliceR 01', that is, simple calculation over raster data (mathematical formula, also known as raster algebra); and
- delete from the merged habitats both road buffers and national parks, again with a multiple reclassification or by using a Boolean expression (SempliceR 02).

A process represents a compound operation that can be thought of as a building block for more complex analysis procedures (macro-operators). The use of macro-operators is very useful to express complex procedures by refinement steps and whenever a certain analysis is planned to be executed over and over with different data.

26.4.2 The Process Editor

Graphical processes are useful as a conceptual tool to depict process structures, but they can unleash all their potential only if they are supported by an application allowing for interactive building, validation and execution. Construction and validation are delegated to the process editor, a graphical and interactive tool whose main features are:

- *familiarity*, as it is based on graphical conventions very similar to those already adopted by most office automation tools;
- *ease of use*, as the user interface uses a small set of commands (when the operator requires more than one step and complex choices, parameters are specified by simple dialogs boxes and wizards); and
- *control over process development*, both from a graphical and from a semantic point of view.

The adopted graphical conventions include multiple undo/redo steps, multiple selections by means of windows, drag and drop of objects, multilevel zooming, ability to get a picture of a process and paste it in presentations or reports being developed with an office automation tool. Obviously, moving items around does not change the process structure, as every link between data and operator is moved, thus keeping data hooked to operators that use/generate them. Considering the interactive use of the process editor, there are some features that improve its ease of use and help to obtain a straightforward representation. First, the editor enforces the placement of items onto a predefined grid, helping the user to develop a clear and well-organised layout. Second, if a link between data and operators intersects other items, it is possible to use a 'place card', called a reference, which allows the user to get a better representation.

The process editor is not a simple flowcharting tool since it assists the user in the creation of correct and executable processes by means of semantic control features. Considering the nature of a data flow structure and the broad editing possibilities granted to the user, some errors are underlined but allowed, while some others are forbidden. In particular, linking two data values to the same role (which is nonsense) and creating a loop in the graph (which would lead to an infinite loop during execution) are both forbidden.

Figure 26.4 shows an example of semantic validation. Correct operators are depicted with a white background, whilst the others are not correct or complete and thus not executable (operators that lack inputs are depicted in cyan, 'SempliceR 01'; operators that lack parameters in yellow, 'BufferR 02'; operators that are connected to unsuitable input in red, 'SempliceR 02'). Controls are executed in real time, as the user creates and modifies the process, allowing for an immediate evaluation of the effect of a change on the whole process.

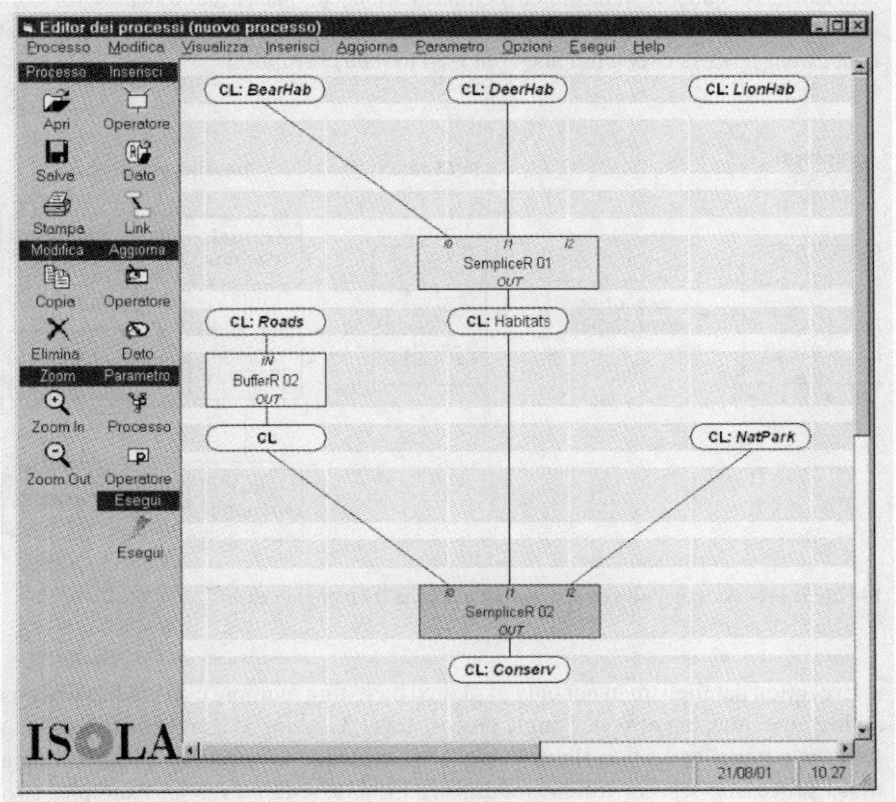

Fig 26.4: Process editor, semantic validation

26.4.3 Process Execution

As already stated, the user interacts directly with the process editor in order to build a spatial analysis process in the form of a graph. To allow execution, this graphical representation needs to be translated into a sequence of commands to the GIS engine. Process translation is performed in two steps: first, the execution environment built into the process editor translates (compiles) high level operators into calls to a virtual machine whose interface is still independent of GIS engine. Second, the virtual machine translates this intermediate language into specific GIS engine calls.

This architecture aims to minimise the quantity of code dependent on the GIS engine, by making the porting activity simple and not expensive. Moreover, the execution environment and the virtual machine communicate through sockets, a quite standard communication interface that enables the software distribution to be hidden (software does not need to be changed in order to use a one-machine approach) and lets the VM programmer choose the implementation language that best suits the GIS environment, server operating system and his particular programming skills. Figure 26.5 shows how the graphical process is executed by translating graphical objects into low level calls. If multiple GIS engines are used, one virtual machine for each will be used. Each virtual machine works along with the lock manager to ensure that concurrent process execution does not lead to data corruption.

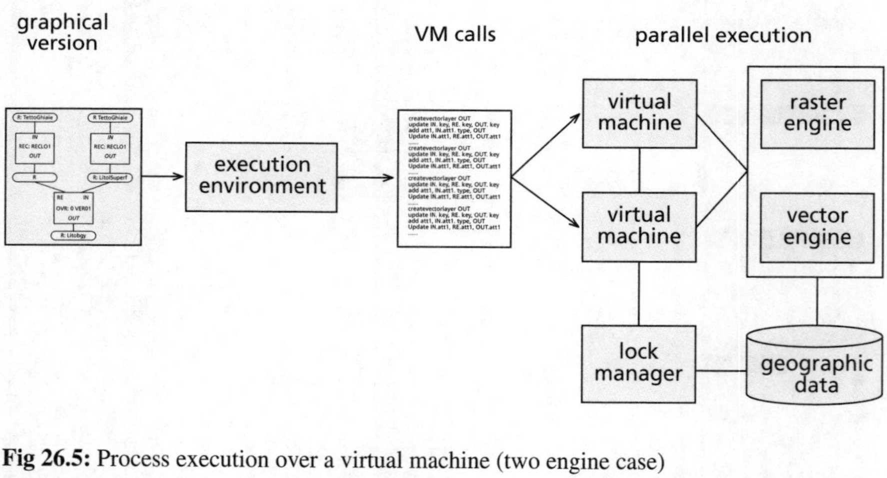

Fig 26.5: Process execution over a virtual machine (two engine case)

Execution parallelism in not only exploited by letting multiple users make analysis at the same time, but also at a single process level. Looking at a process, it is easy to find out parallel execution flows that can be exploited to speed-up execution on a multi-processor server without modifying the GIS engine (as an example, two concurrent execution flows can be identified in Figure 26.4). A great advantage would be obtained if every single command could be implemented in a data parallel fashion, but this involves radical changes in the GIS engine, which are beyond our control. By now, parallel execution is made possible by creating more than one virtual machine per single process, one for each independent execution flow. As a rule of thumb, the server side creates one virtual machine for each client, allowing for further virtual machines depending on current load and hardware characteristics, in order to optimise the overall execution time. Since process execution can even take long times, depending on the number and nature of operators used, the execution environment provides a visual feedback on the execution progress, highlighting operations that have already been done; as a side effect, the user will be able to notice parallel execution too.

26.4.4 Viewer and Report Generator

EIS includes some modules dedicated to data viewing, querying and report generation, that enable the user to evaluate the products generated by the process execution and to produce printed output. The thematic map builder has a quite standard implementation, inspired to the most common commercial viewers. It is aimed to make thematic map building as simple as possible by using a legend metaphor to organise the building and visualisation process. The main point of interest about this module is its capability to store thematic map structure rather than building a project comprising data and rendering information. In fact, only the latter is stored, thus enabling the user to build several thematic maps with little storage space consumption by sharing data and to build maps with the same structure, but related to different data, leading to easy comparison between different locations or at different times.

Fig 26.6 shows the *EIS* thematic map builder in our current implementation, based on Autocad Map. The main points of interest are the user interface simplification (if compared to the standard map user interface), the presence of a legend that acts both as a description of the thematic map and as a way to manage it, and a number of specialized dialogs that enable the user to associate data with its thematic representation (Fig 26.6 shows a dialog that enables the user to specify continuous colour shades to represent raster data).

The tabular data viewer is a quite standard grid viewer that enables the user to perform filtering and ordering, and to get some statistical information about data (such as maximum, minimum, average, variance). This is coupled with the ability to generate simple diagrams (histograms, line charts, pie charts) to better evaluate a data distribution or evolution (in the case of time series). Tables, charts and metadata can be put togheter to form a printed report that is generated in Word and HTML formats; that enables the user to create a structure for an eco-balance, and to obtain the current eco-balance based on the latest indicator and computed indexes. Both the tabular data viewer and the report generator share the thematic map approach; they only store rendering information and re-generate the output each time input data changes.

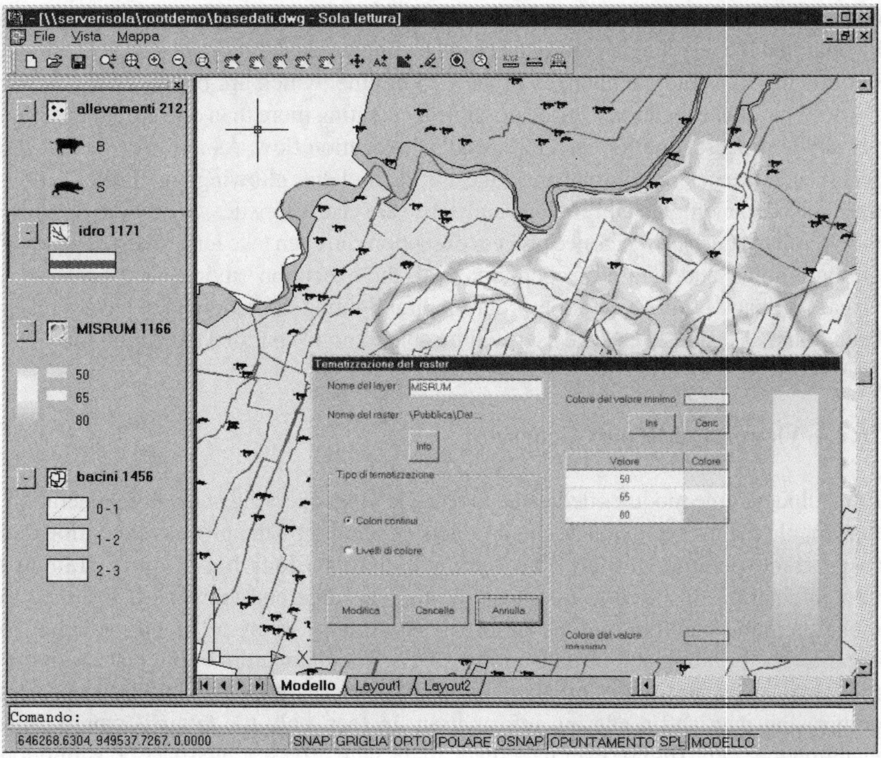

Fig 26.6: *EIS* thematic map builder

26.4.5 Shared Repository

The information system is based on a client/server approach, thus enabling information sharing among different users by means of a shared repository. Since we are striving for a GIS engine-independent solution, we also want to avoid retaining a single GIS data organisation. Thus we have created an abstract layer to handle this issues that we call the virtual file system (VFS). Users perceive the VFS as a standard file system in which every user gets a home folder, plus a common folder that every user can access to store shared information. Depending on the access level, a user may have write grants limited to their own folder, or may be enabled to write to the common folder, thus avoiding untrained people damaging important analysis processes while letting them try process development on their own home folder. Having to deal with user capabilities and access rights requires the user to authenticate himself/herself at start-up by providing a username and a password.

Every entity managed by the system, be it a process, a datum or a thematic map definition, is treated as a single file, independent of its real nature. In the ISOLA prototype, most information is stored in the file system, but with a different structure than the one perceived by the user, while other information is stored in a centralised database (mainly metadata and vector data attribute tables). Figure 26.7 shows *EIS* Explorer and the way shared the repository is proposed to the user, by means of a standard directory tree and a file view. In particular, Figure 26.7 shows the common folder ('Pubblica') and a user folder ('Silvia'); the file view comprises all kinds of data (point, line, polygon and raster) and some processes as well.

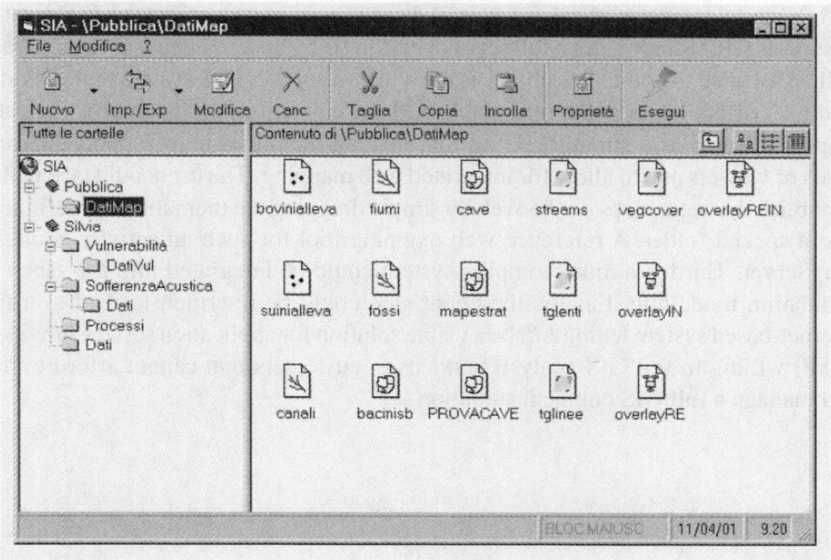

Fig 26.7: Shared repository managed through *EIS* Explorer

26.5 Project Status and Foreseen Improvements

As already stated, *EIS* aims to be GIS engine-independent. Anyway, within the ISOLA project we had to provide a reference implementation that is as cheap as possible, using at the same time software that is likely to be found in other local administrations. The reference implementation is based on:

- Linux as the server operating system;
- Windows as the client operating system;
- PostreSQL 7.0 as the relational database management system;
- GRASS 5.0 as the raster GIS engine; and
- Autocad Map 2000 as the viewer and the vector GIS engine.

Autocad Map 2000 works on the client side, but at the same time does vector computation, leading to an imperfect distribution of tasks and to a non-full client/server implementation. That was necessary since GRASS 5, at the actual stage of evolution, is not capable of handling vector analysis operations. On the other side, we have to notice that Autocad Map 2000 does not support at all raster data, so we had to add this support ourselves with some additional software modules. Also, as we need to work with an incompatible GIS system, we had to develop data format routines that made it possible to exchange raster and vector data between Autocad Map and GRASS.

As a modular system, there are several directions of improvement for *EIS*. Firstly, as soon as GRASS provides a full set of operators, the standard client/server approach will be restored. Another possibility is to switch to another GIS system that provides a full set of operators and a reasonable API for customisation. Secondly, thematic maps are stored as a structure definition, making them usable to support different kinds of viewers and to allow for integrated web mapping. The user could be enabled to publish thematic maps on the Web by simple dragging the thematic map definition into a special folder. A reference web mapping tool for such an activity could be MapServer. Thirdly, a more complex system could be integrated into *EIS*, such as simulation modelling. Finally, the client side could be rewritten to lead to a full internet-based system letting *EIS* be a viable solution for application service providers (ASP) willing to sell GIS analysis services to customers that cannot afford to buy and manage a full GIS engine installation.

27 DESIMA: A Decision Support Tool for Integrated Coastal Zone Management

Marieke Eleveld[1], Philippe Carreau[2], Wolfram Schrimpf[2] and Ardy Siegert[3]

[1] Institute for Environmental Studies (IVM) – FALW, Vrije Universiteit (VU), De Boelelaan 1087, 1081HV Amsterdam, The Netherlands
[2] Joint Research Centre of the EC, Institute for Environment and Sustainability – Inland and Marine Waters Unit, T.P. 272, I-21020 Ispra (VA), Italy
[3] Covalent Technologies, Inc., 706 Mission Street (2nd Floor), San Francisco, CA 94103, USA

Abstract: Evaluating engineering solutions to coastal erosion and assessing the potential impacts of oil spills are some of the tasks that management authorities in Europe's coastal regions are concerned with. The Inland and Marine Waters Unit of the European Commission's Joint Research Centre aims to support these authorities through the development of methods, tools and systems that integrate distributed coastal data and information. This chapter introduces a distributed information system, called DESIMA, that has been specifically created to support (European) integrated coastal zone management. The DESIMA-Demonstrator was designed to show the possibilities of such a distributed system. A key requirement for its design and implementation was that large data sets remain with data providers and that their data formats did not have to be changed. Furthermore, users had to have real-time access to data and models through a system architecture and user interface that allow interpretation of various formats on a single PC or workstation using a regular browser. To meet these requirements, the Common Object Request Broker Architecture (CORBA) was selected as a basis for the design of the DESIMA-Demonstrator, whereas Java was used for the user interface. Two applications, a coastal protection scenario and an oil spill scenario, demonstrate how a user can benefit from the DESIMA concept by using data and software that are implemented and maintained elsewhere. The coastal protection scenario enables the appraisal of three coastal management options in relation to flood risk for the town of West Bay (UK). The oil spill scenario simulates the evolution of an oil slick in the Gulf of Lion (France) and provides land cover data that allows the user to assess possible onshore impacts.

27.1 Introduction

Effective coastal management requires a combination of data from different sources. It is, however, often difficult to access the necessary data and tools. A distributed system, called *DESIMA* was developed to support coastal management by allowing users to access, process and integrate various data and information sources.

DESIMA stands for DEcision Support for Integrated coastal zone MAnagement, which is one of the activities of the Inland and Marine Waters Unit of the Joint Research Centre of the European Commission. This chapter introduces the on-line *DESIMA*-Demonstrator (*http://desima.jrc.it*), which aims to show the possibilities of such a distributed system. It was developed by a consortium consisting of organisations whose names and responsibilities are listed in Table 27.1.

Table 27.1: Full names and functions of the participating organisations

abbreviation	full name	function
JRC	Joint Research Centre of the European Commission, EU	project co-ordinator; project definition and funding; host of the DESIMA server
Matra	Matra Systèmes & Information (currently a subsidiary of the EADS Systèmes & Information Group), France	technical development of the system
HR Wallingford	HR Wallingford, UK	developed the coastal protection scenario
ACRI	ACRI mechanique des fluides, observation de la terre, sciences de l'environnement, France	developed the oil spill scenario
SOS	Satellite Observing Systems, UK	data source

The function of the demonstrator is to show how new technology can help in solving real problems of coastal management. From a technical point of view, its main components are CORBA and web technologies (Java, HTML), which are discussed in the next section of the chapter. From a thematic point of view, the demonstrator comprises a fully implemented coastal protection application and an oil spill application which are elaborated in the section on applications. *DESIMA*-Demonstrator users are involved in coastal management (mainly in Europe). Their experiences with the *DESIMA*-Demonstrator are described in a section on feedback from users. This chapter concludes with an 'Outlook' section, which outlines our future plans for the *DESIMA*-Demonstrator.

27.2 System Components

The general system architecture (Siegert and Schrimpf 2000) is illustrated by Figure 27.1. The *DESIMA*-Demonstrator system consists of three components: it has a client side, a server side and a data source side. The system's client side comprises a regular browser such as MS Explorer or Netscape, and uses Java applets for the user interface.

The server side is located on a UNIX Solaris machine. Some data are local and can be accessed through an intranet. The other data sources are distributed over different remote servers that are working on different operating systems. In principle, all data sets can remain at their original place, and are transferred through the Internet if access is required. Furthermore, the *DESIMA*-Demonstrator system is designed to launch, monitor and stop remote applications on a source server such as an oil slick propagation simulation model (the so-called TIDAL model from ACRI).

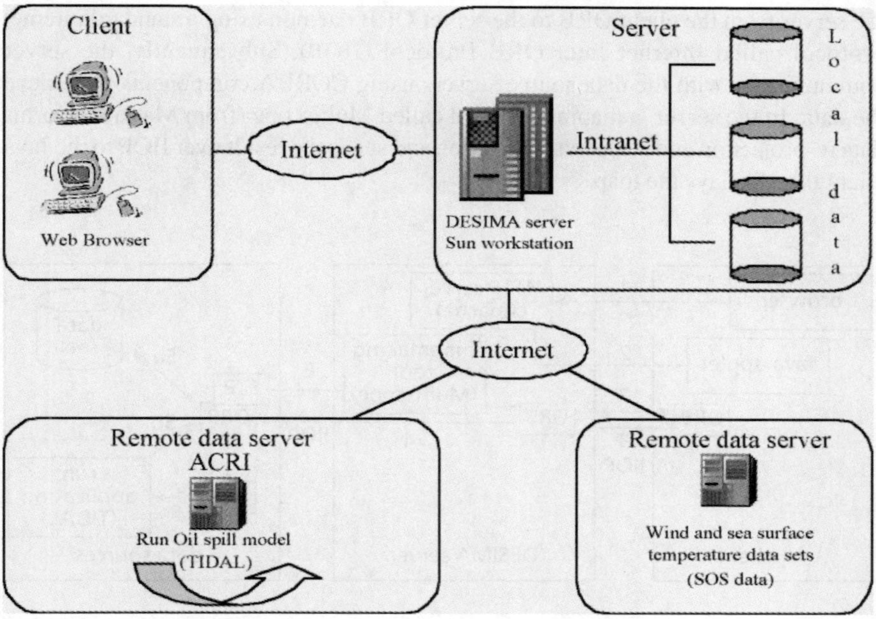

Fig 27.1: General *DESIMA* architecture

The three components (client, server and data sources) communicate through the use of Java and CORBA technologies (Figure 27.2). On the client side, the user interacts with a Java interface to build queries that are sent to the server. At the end of the process, this Java client receives the resulting map from the server and displays it. Java technology is used for the user interface because its cross-platform language (UNIX and Windows) and applets can be developed and integrated in web pages. The graphical user interface (GUI) for the applications consists of sheets with, for example, Display, and Legend tabs. Within each sheet, there are checkboxes, radio buttons and pull down menus that enable users to specify their choices.

On the *DESIMA* server side, the queries sent by the client are handled and communicated with the relevant data source. It uses an Apache HTTP server that transfers HTML pages and communicates with the Java client. This server was selected because it is robust, open source and widely used. The core of the *DESIMA*-Demonstrator architecture is the Common Object Request Broker Architecture (CORBA). CORBA is an open standards cross-platform and cross-language distributed framework that enables it to deliver applications through the Internet (Curtis 1997). Its central component is the Object Request Broker (ORB). It encompasses the entire communication infrastructure necessary to identify and locate objects, handle connection management and deliver data. As soon as the Java applet on the client side is started, the ORB handles all communications between the web browser, the server and the sources. The request made by the user is transferred to the server from the client ORB to the server ORB daemon using a standard internet protocol called Internet Inter ORB Protocol (IIOP). Subsequently, the server communicates with the data source servers using CORBA components to retrieve the data. In the server, a mapmaking tool called Multiscope (from Matra) performs data re-projection and scale transformation and sends the result over IIOP to the Java client that displays the map.

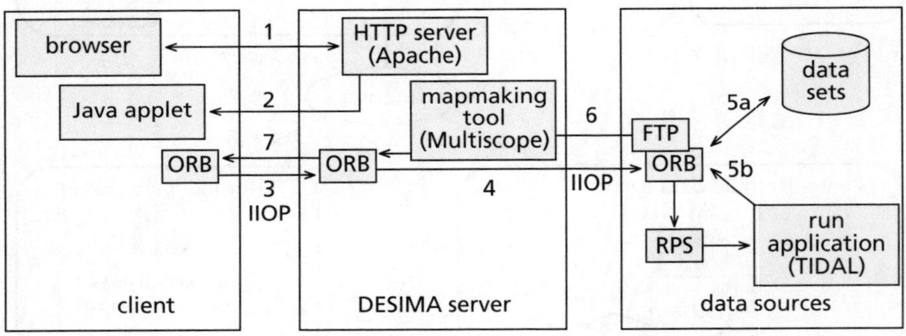

Fig 27.2: Interactions between the *DESIMA* components

Notes: (1) Get the HTML page; (2) Get and load the Java applet; (3) The Java applet sends a request to the *DESIMA* server; (4) The *DESIMA* server makes an ORB to ORB connection to the data sources server; (5) The data are retrieved from the data sets (5a) or from the TIDAL model (5b); (6) The data are retrieved from the data sources server and processed by a mapmaking tool on the *DESIMA* server; and (7) The Java applet gets and displays the resulting map.

Several CORBA components of the *DESIMA*-Demonstrator have been developed for remote access to the data sources. Parts of the communication modules use the file transfer protocol (FTP) to retrieve and manage external data. The TIDAL model is managed by a dedicated component called remote processing server (RPS). The RPS launches a process, monitors the model progress and returns the results. It can easily be applied to other models.

27.3 Applications

Two applications, a coastal protection application and an oil spill application, demonstrate the practical use of the techniques described in the previous section.

The coastal protection application is based on an appraisal of coastal management options for the town of West Bay in the UK. This is a result of a project carried out by HR Wallingford for the Environment Agency of England and Wales. This application aims to illustrate how coastal defence management can benefit from the use of a system such as *DESIMA*, because it facilitates the combination of data from different sources. Furthermore, the application allows the user to evaluate the effect of different scenarios (alternative engineering strategies) on flood risk in the region.

The user can display heterogeneous data sets such as maps, images (either georeferenced or not), vectors (points, lines and polylines) and graphs through interaction with the user interface by clicking on tabs, checkboxes, radio buttons and pull down menus. These data sets provide, amongst others, information on wind and wave characteristics, bathymetry and topography and on the coastal protection structures that are currently present in the area. This information also serves as input into a numerical modelling tool that comprises a comprehensive suite of wave, beach, and flood modelling software. The software by HR Wallingford is based on a range of widely accepted equations and methods. The modelling tool allows the user to predict the impact of several predefined engineering solutions on this region.

The modelling tool predicts the effects of three management options. These are (i) current management (the present day scenario), (ii) construction of a 250 berth marina, and (iii) an extended revetments scenario – on water motion, sediment transport and morphological change, and the subsequent effects on flooding risk and its financial consequences. The underlying modelling incorporates many interactions and feedback mechanisms, which are further elaborated in an intricate process diagram that can be displayed when running the application. Because of this complexity, the user of the–*DESIMA*-Demonstrator has no access to the underlying models, but is allowed to choose for one of the three predefined management options, and to subsequently run the modelling tool through the *DESIMA* interface (Figure 27.3).

After running the modelling tool the results can be obtained and visualised within the system. Running the modelling tool for several design options enables the user to compare the effects of alternative engineering strategies. Figure 27.4 is a screen dump that illustrates flood risk results and economic appraisal, for the present day scenario. In addition, a listing of output parameters from the modelling, such as offshore wave rose, beach width, storm response of beach, flood risk map and economic appraisal, is present on the right hand side of the interface.

The oil spill application provides tools to simulate the spreading of an oil slick in the Gulf of Lion (France) and data on land cover in order to evaluate its effects onshore. This application is based on SOS wind and wave data and an ACRI model, and was developed with support from CEDRE (CEntre de Documentation, de Recherche et d'Expérimentations sur les pollutions accidentelles des eaux). This application aims to demonstrate the remote access and remote processing possibilities of the *DESIMA* system.

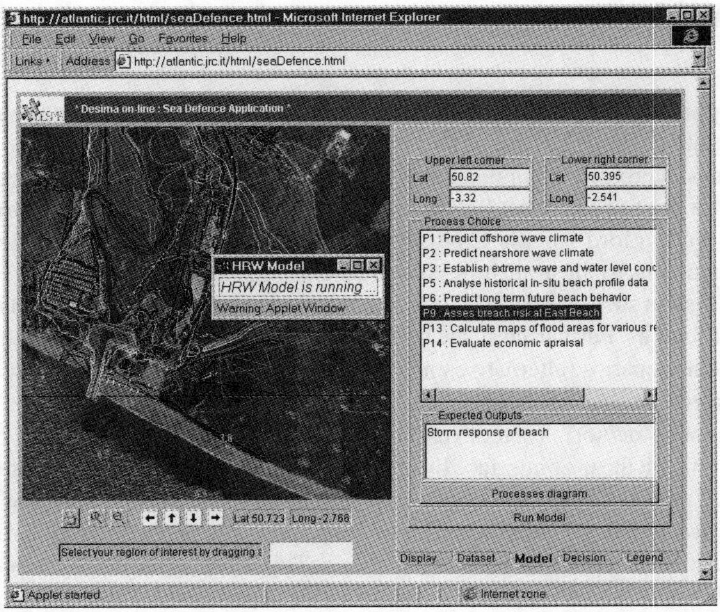

Fig 27.3: Running the coastal protection modelling tool, with an output parameter to be calculated (storm response of beach) selected on the right-hand side

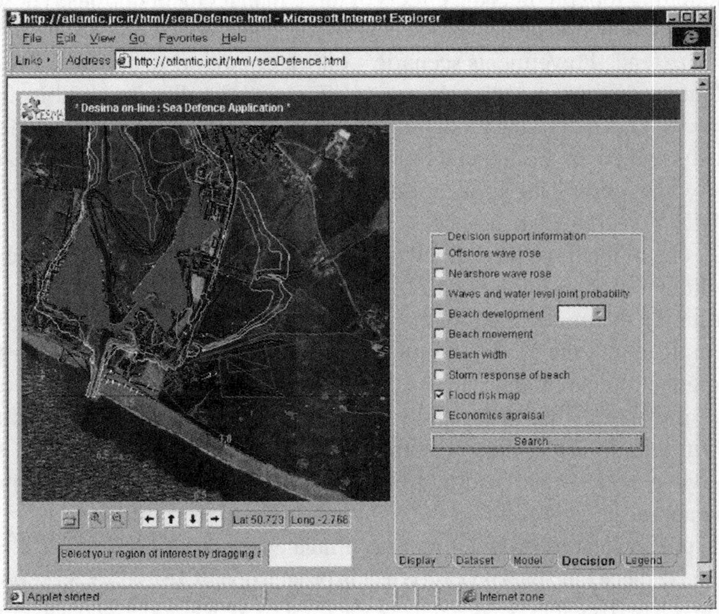

Fig 27.4: Predicted flood risk (uniformly dark grey areas indicate the flooded parts) for the present day scenario of the coastal protection application

Users can obtain and display information about monthly mean data sets including wind speed and sea surface temperature. These are actually some of the input parameters that need to be specified for the oil slick evolution simulation model. TIDAL, the model from ACRI, is a relatively simple flow model that can simulate the spreading of an oil slick under certain wind and water conditions. To run the model, the original slick position, date, time and duration of the evolution of the oil slick (simulation time) need to be given. A certain type of oil has to be selected, and either a punctual or a continuous leak can be chosen. Users can either specify a constant wind speed and direction, or invoke a file that contains time-series of wind speeds and directions (originally from ECMWF, the European Centre for Medium-range Weather Forecasts). Surface water temperature also has to be given. Figure 27.5 shows a screen dump of the interface made during the running of the model. An example of values for the previously mentioned input parameters is provided on the right-hand side of the interface.

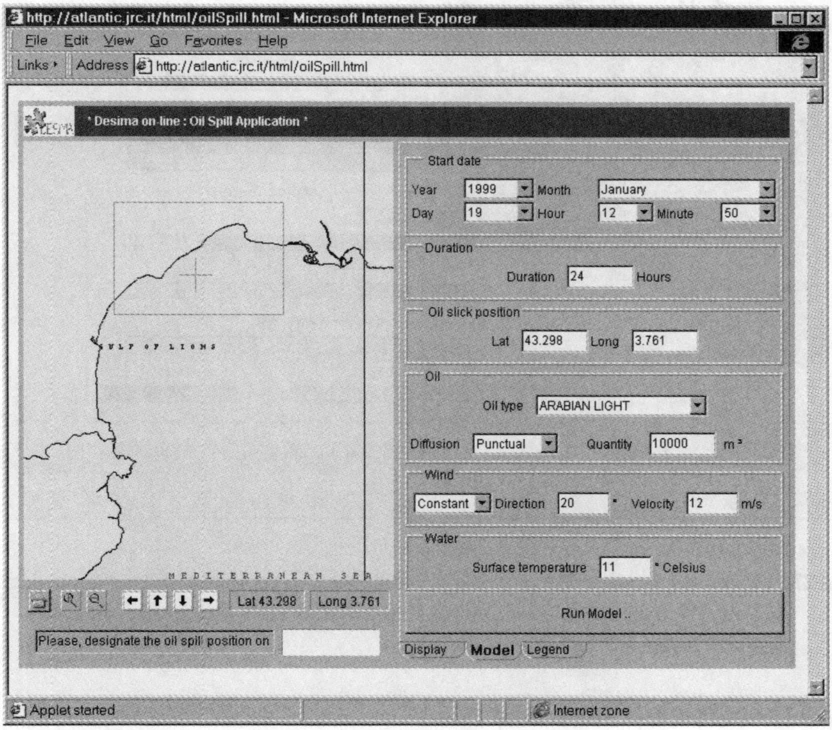

Fig 27.5: Running the oil spill model with the input data that the user provided on the right-hand side

As output, the model generates animated images that dynamically show the evolution of the oil slick superimposed on a background image that shows the coastline. Additionally, the composition and physical properties of the remaining oil (e.g. evaporation rate, density, emulsion rate, viscosity) at the end of the simulation are given. To assess the possible onshore impact of the spill, information on land cover (LaCoast data) of the threatened onshore area can be obtained (Loudjani *et al.* 1999). Figure 27.6 gives an example of the latter. It shows an oil slick approaching shore with a land cover consisting mainly of wetlands and semi-natural areas.

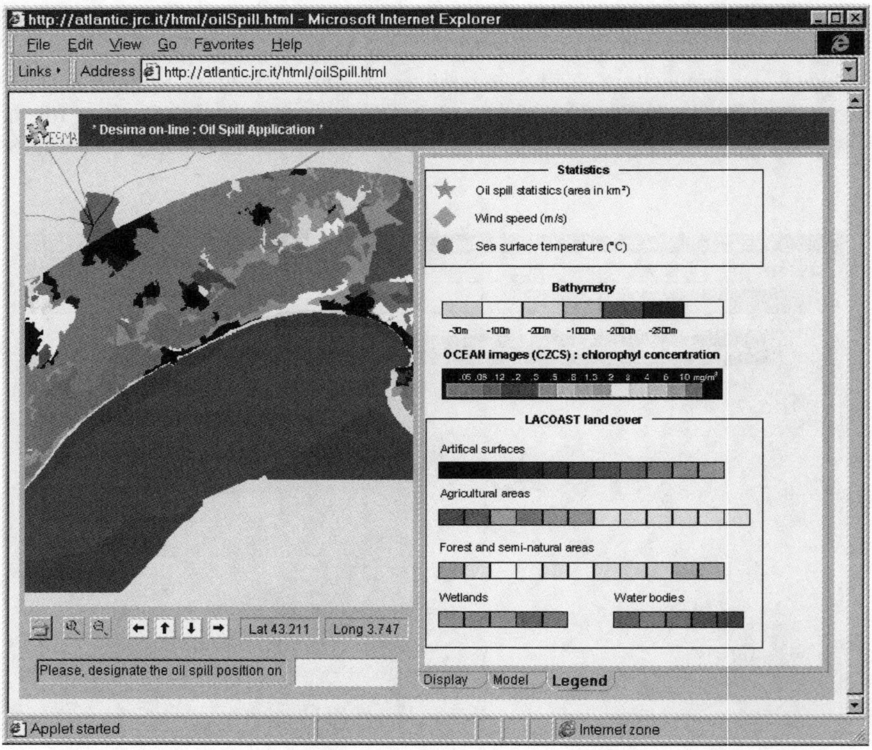

Fig 27.6: Predicted location of the oil slick (middle grey) combined with land cover data

27.4 Feedback from Users

The *DESIMA*-Demonstrator has about a hundred users within Europe and also a number of individual users from other parts of the world. A user is defined as an entity (for example, a person or an organisation) that requested a user name and password, which enable the two applications of the system to be run interactively.

DESIMA-Demonstrator users are involved in coastal management on all levels: European, regional, national and local. Regional, in this case, refers to a so-called 'Regional Sea', an entity larger than an individual 'European Sea', and managed by the joint Oslo and Paris Commission (OSPAR), the Helsinki Commission (HELCOM), or by the United Nations Environment Programme – Mediterranean Action Plan (UNEP-MAP). Many of the users function as policy-makers, or as policy advisors and project managers, but there are also researchers, university teachers and students registered.

Because users have to register before they can run the *DESIMA*-Demonstrator applications, some contact with them can be maintained in case of technical problems. Feedback from users has shown that interaction with the system proved to give some problems for first time users. Therefore, on-line tutorials consisting of commentated screenshots were added to the web site for each application (Siegert and Schrimpf 2000). The tutorials show the full functionality of *DESIMA* (and can be accessed without registering).

Both applications were originally developed in collaboration with coastal managers and scientists, and have proven their value by attracting a large and enthusiastic additional user group. The applications demonstrate how users can benefit from data and software developed and implemented elsewhere, and how they can limit their own investments in hardware and software without limiting their requirements. Users indicated that they are aware that the web site offers only a demonstrator of a distributed information system, and most users are positive that the technique enables them to actually access distributed data and to run models, but they would like to apply such a system to their own problems. Therefore, we are aiming at extending the demonstrator to a system that can cover a wider range of topics and locations.

27.5 Outlook

For the future, we intend to expand our *DESIMA* activities. From a technical point of view, the *DESIMA*-Demonstrator provides data management and GIS functionality (spatial analysis) through a GIS tool (Multiscope from Matra). In parallel with this, several of our current internet-related activities also consist of the components data management and/or GIS functionality. For example, CoastBase – a European virtual coastal and marine data warehouse (CoastBase 2001) like the *DESIMA*-Demonstrator, also involves a distributed system based on a CORBA core, and also incorporates data management and GIS functionality. In addition, we are setting up a (meta)data management system for our own Inland and Marine Waters Unit (Donlon 2001). It is intended to combine these activities into a new system called *DESIMA+* that will be based on XML technology. It will allow users to access metadata from these various projects, to display the actual data on a map, but also to run new models, such as a pollution spreading model for the Adriatic Sea.

From a thematic point of view, we are working on water quality and eutrophication issues, which would allow a future *DESIMA*-Water Quality (DESIMA-WQ) application, and coastal change, which would allow a future *DESIMA*-Coastal Change (DESIMA-CC) application. In addition, another coastal planning scenario will be implemented for the development of the CoastBase prototype (Eleveld *et al.* 2000). The latter comprises two examples: a pre-feasibility study for an international airport on an artificial island off the Dutch coast, and a project on sustainable management of a gold mining area in the coastal zone of Strymonikos Bay, Greece. It illustrates how to access the multitude of sources and retrieve information on the variety of topics that are needed for integrated coastal planning.

Acknowledgements

We would like to thank all those who have been involved in *DESIMA*. We thank Craig Donlon and the editors of the book for kindly checking the manuscript. CEO (Centre for Earth Observation) and SAI (Space Applications Institute) are acknowledged for funding the DESIMA activities.

References

CoastBase (2001) CoastBase: The European Virtual Coastal and Marine Data Warehouse. An open system architecture for integrated, distributed coastal and marine information search and access, CoastBase, (*http://www.coastbase.org*).

Curtis, D. (1997) Java, RMI and CORBA, Object Management Group (OMG), (http://cgi.omg.org/library/wpjava.html).

Donlon, C.J. (2001) A data management framework at the Marine Environment Unit, *EUR 19804 EN*, EC JRC, Ispra, (Available from JRC, IES-IMW).

Eleveld, M.A., Siegert, A.G. and Schrimpf, W.B.H. (2000) CoastBase. Information definition and user requirements, *EUR 19732 EN*, EC JRC, Ispra, (Available from JRC, IES-IMW).

Loudjani, P., Meyer-Roux, J., Schmuck, G., Annoni, A. and Perdigao, V. (1999) The LACOAST project: land cover changes survey of European coastal zones, in Nieuwenhuis, G.J.A., Vaughan. R.A. and Molenaar, M. (eds.) *Operational Remote Sensing for Sustainable Development*, Proceedings of the 18th EARSeL Symposium on Operational Remote Sensing for Sustainable Development, Enschede, Netherlands, 11-14 May 1998.) Balkema, Rotterdam, pp. 185-186.

Siegert, A. and Schrimpf, W. (2000) DESIMA Evaluation, verification and improvement, *EUR 19555 EN*, EC JRC, Ispra, (Available from JRC, IES-IMW).

Web Site

DESIMA: *http://desima.jrc.it*

28 WadBOS: Integrating Knowledge to Support Policy-making for the Wadden Sea

Guy Engelen[1], Inge Uljee[1] and Kees van de Ven[2]

[1] Research Institute for Knowledge Systems bv, P.O. Box 463, 6200 AL Maastricht, The Netherlands

[2] National Institute for Marine and Coastal Management, Ministry of Transport, Public Works and Water Management, Directorate-General of Public Works and Water Management, P.O. Box 207, 9750 EX Haren, The Netherlands

Abstract: WadBOS version 2 is a decision support system featuring an integrated model representing the ecological and the economic functions of the Wadden Sea in an explicitly dynamic and spatial manner. Sub-models represent processes operating at three time scales: the tidal cycle, monthly and yearly; and spatial processes operating at three spatial scales: the whole sea, 12 relatively homogeneous compartments within the sea, and small cellular units of 25 hectares each. The WadBOS system relies heavily on GIS information for its inputs, but its models need economic, demographic and ecological data from other sources. WadBOS has been developed for use by policy-makers working in isolation or in a group setting. A lot of effort has gone into its user-friendly character, its interactive capabilities, the (geo)graphical representation of its input and dynamic output, and its transparency. The system is most instrumental in analytical exercises, but its use for communication, learning, and storage of knowledge is nearly as important.

28.1 Introduction

The Wadden Sea is an important coastal system in the north of the Netherlands. It is part of a larger system extending into northern Germany and western Denmark. In the Netherlands, the sea has the status of a protected natural area because of the important ecological functions that it fulfils. At the same time, the sea has important economic functions. Fishing, recreation, transportation and mining are among the main economic activities. The management of the different activities and functions of the sea is distributed over a great number of institutions, ranging from the local to the European. When new decisions are required or policies need to be developed relative to the exploitation or protection of the area, incompatible views tend to slow down the decision-making process.

In order to streamline this process, CUBWAD, the organization of governmental institutions responsible for the management of the Dutch Wadden Sea, decided in 1996 to initiate the development of an information system, named *WadBOS* (*Wad*denzee *B*eleids *O*ndersteunend *S*ysteem, Dutch for Wadden Sea planning support system). Their general thinking was that such an information system representing the Wadden Sea in a holistic manner, integrating the ecological functions and human activities at the appropriate temporal and spatial scales, would enhance policy-making in the region.

The intended end-users of *WadBOS* are policy-makers actively involved in the Wadden Sea. Their profile is best described as high-level technicians actively involved in the design and evaluation of policies. They perform policy work of a formal/analytic nature in support of the administrator or politically appointed person responsible for taking the actual decision and starting the actual policy implementation process. This policy-maker is not a politician; rather, he is a technician usually stressed by time constraints and rapid changes in the emphasis and relevance of the policy work itself.

In this chapter, we will discuss the technical choices made during the development process. From this it will become clear that *WadBOS*-like systems are very demanding on modelling and information and computer technologies, but that their successful development and use in the decision-making organization is even more dependent on the intricate interplay of different actors: end-users, domain specialists and developers.

28.2 Functions of *WadBOS*

Based on the definition of the end-users and on the functional specifications of CUBWAD, *WadBOS* had to become a planning support system (PSS): an information system providing expertise and knowledge in support of the preparation of, and possibly the implementation of, integrated spatial policies for the Wadden Sea. The functions of *WadBOS* are as follows:

- *Analysis* is certainly the most important task of *WadBOS*. It is to provide the end-user with good insights into the autonomous dynamics of the Wadden system and should facilitate evaluation of the effects of policy intervention. It should assist in reaching decisions relative to precise policy measures to be implemented.
- *Learning* is an inherent result of analysis, since solutions are predicated on sound understanding. A *WadBOS* featuring a holistic representation of the Wadden system becomes a very valuable learning instrument in that it provides insights, not only into the individual domains but also into the linkages between the domains represented. Users that are experts in the dependencies in their own field of interest may want to use *WadBOS* to learn about linkages to the fields of others and the processes unknown to them.

- *Communication* or the lack thereof is what renders integrated policy-making as difficult as it often is. A facility that visualises the anticipated effects of policies will not necessarily lead to consensus, but at least it makes things more tangible and thus facilitates greatly discussions among policy-makers, stakeholders and the public at large.
- *Knowledge management.* A WadBOS that orders and links knowledge can serve as a knowledge management infrastructure and can fulfil the function of a dynamic library. It may reveal knowledge gaps, and thereby give impetus to further research and data collection. The more this knowledge is available in a formal, quantitative manner, the more it can be made operational and thus be immediately useful to a large group of users.

The above is an ambitious set of functions, not easily attained by traditional information systems such as databases or GIS. Rather, an information system is envisaged with a high level of intelligence (Catanese 1979): an instrument with the ability to manipulate and aggregate data as the result of statistical, mathematical, heuristic or algorithmic operations and to catch from this complex information and data the essential factors explaining the functioning and the dynamics of the Wadden system. To achieve this, this information system is to be equipped with formal models and methods able to address complicated, complex problems. This is where decision support systems (DSS) become useful instruments.

28.3 Definition of a DSS

DSS are computer-based information systems developed to assist decision-makers to address semi-structured (or ill-defined) tasks in a specific decision domain. They provide support of a formal nature by allowing decision makers to access and use data and appropriate analytic models (El-Najdawi and Stylianou 1993). The terms 'semi-structured' and 'appropriate' in this definition refer to the fact that DSS are typically applied to find answers for problems that, due to their specific nature and complexity, lack an unambiguous solution method. Rather, usage of the 'most appropriate' analytical solution methods available approximates the unique answer. Thus, the DSS provides the decision-maker with a suit of 'analytic models' considered appropriate for the decision domain and capable of representing the complexities of the system and the problems studied.

But a DSS is more than a *model base* alone. Typically, three more components can be distinguished (Engelen *et al.* 1993): a *user interface* enabling easy interaction between the user and the instrument; a *database* containing the raw and processed data of the domain and the area at study; and a *toolbase* with the methods, analytical techniques, decision models, statistical and operations research methods and software instruments required to work in an effective manner with the domain models and the data. Each of the four components fulfils specific tasks within the DSS and has a complex internal structure. In the following sections we will discuss briefly their role and nature in *WadBOS*. First our attention will go to the development, the contents and the implementation of the model base.

DSS can only be successful if they are developed as part of an effort involving both the intended end-users and the DSS developers. This is mostly so because they are still highly technical, novel products with a high degree of sophistication, typically built by those not using them and used by those not building them. The development of *WadBOS* involved a relatively small team of professionals (around 8), specialized in the development of DSS applications, and formally trained in computer science, knowledge engineering, systems analysis and modelling. This team worked in close collaboration with end-users, representing member organisations of CUBWAD. In particular a small group of representatives (around 6) from the Ministry of Public Works, Water Management and Transport participated in the development process. They were very much the initiators of the project; they clarified the functionality expected from the instrument, but took responsibility for a number of administrative, technical and networking tasks equally well. Finally, when specialised knowledge was needed, collaboration was sought with a group of domain experts and scientists (around 25), knowledgeable about the problems and the area studied.

The life cycle model known as *evolutionary delivery* (McConnell 1996) is used to develop and upgrade *WadBOS*. As a result, a series of prototypes is developed in an iterative manner. Each is developed according to a predefined set of specifications and with particular objectives in mind. Different from 'throwaway' prototypes, they evolve during the project towards the intended final product. The first prototype, *WadBOS-1*, available in October 1997, was mostly developed with a view to prove the feasibility of the approach and the usefulness of the instrument before engaging in its stepwise completion. To that effect, the Wadden system was represented in a general way, and a few human activities and ecological functions were represented in detail. In particular, recreational activities and shell mining as well as the nutrient, algae and filter feeder dynamics got a more detailed representation. The second prototype, *WadBOS-2*, available in November 1999, completed the systems representation. Fishing and recreational boating got a more encompassing representation and assessment indicators at the landscape level were introduced. At the same time the toolbase was extended with additional analytic instruments. *WadBOS-2* can be downloaded from *http://www.riks.nl/RiksGeo/proj_wad.htm*. Finally, the third prototype, *WadBOS-3*, available in December 2001, improves the instrument with a view to its application in an environmental impact assessment study and a practical policy-making context.

28.4 Developing the Model Base

To develop the model base of *WadBOS*, the approach taken is bottom-up: based on a reasonable understanding of the characteristic processes and problems typifying the Wadden Sea and based on a fair amount of complementary knowledge and modelling material available from a large number of organizations, an integrated model is designed and constructed. This integrated model, consisting of linked sub-models, represents the Wadden system as completely as possible with a view to facilitate the design and evaluation of integrated policies. Thus, the user of the DSS gets immediate access to very rich and operational knowledge of the decision domain.

28.4.1 Knowledge Acquisition and Modelling

In order to gather and structure the available data, the information and knowledge, techniques from both the fields of systems analysis and knowledge engineering were used (see, for example, Firley and Hellens 1991, Gonzalez and Dankel 1993, Kok *et al.* 1997). The developers of *WadBOS* organised three consecutive knowledge-acquisition sessions. Prior to these, a preliminary requirements analysis had been carried out focussing strongly on the use and user of the instrument.

In the first series of sessions, an extensive group of potential end-users, all actively involved in policy-making, were interviewed. Answers were obtained to the following questions:

- What and where are the boundaries of the Wadden system?
- What knowledge is available, from whom or where can it be obtained, and what knowledge should be included in *WadBOS*?
- What are desirable levels of detail and accuracy for *WadBOS*?

Clearly the expectations of the instrument were diverse, but consensus grew over a representation covering the entire Wadden Sea, not the land adjoining, and including all the major human activities that take place on the water. The appropriate representation of the processes in their complexly coupled multi-spatial and multi-temporal context was considered very fundamental.

With this information, complemented with material obtained from literature research and the analysis of policy documents, an initial systems analysis was conducted; the main processes were identified and their definition in terms of measurable characteristics and state variables was carried out. Next, an initial set of system diagrams and conceptual models were drafted, and, for each process considered, experts were identified.

In a second knowledge acquisition round, the developers set up structured interviews with a large group of selected domain experts. Most of the work with the experts was carried out in sessions involving 2 to 5 people only. Often a member of the end-user group would participate. During the interviews, 'visual modelling' was applied; a graphical modelling tool was used to draft, discuss, upgrade and complete the system diagrams and conceptual models according to the experts' knowledge. At the end of these sessions a rather complete description of the Wadden system was available in the form of graphical, qualitative models. There was a very reasonable consensus relative to this representation. Ambiguity and difference of opinion had been resolved to the degree possible in a number of consecutive visits and discussions in which conflicting views were clarified and resolved.

Parallel to the qualitative representation, a number of existing mathematical models had been gathered and evaluated for incorporation. With this material, the developers began the translation of the qualitative models into a mathematical representation. To the extent possible, existing models were incorporated or adapted for this purpose. However, a lot of knowledge was not available in the form of readily usable mathematical models and needed to be developed.

Partway through this modelling phase, a third knowledge acquisition round was organised, in which the developers confronted end-users, domain experts, scientists and model developers with the mathematical representation of the system. These were very intense exercises in which consensus was sought relative to the representations chosen, the level of integration accomplished and the simplifications introduced. With a view to meeting the general objectives of the *WadBOS* system, much effort went into separating the detail from the essence in the domain representations.

With the information thus obtained, the modelling phase was completed, the technical integration of models was started, and so was the technical implementation of the system. For the development of successive prototypes, *WadBOS-2* and *WadBOS-3*, similar knowledge acquisition sessions were organised with the appropriate specialists. With more elaborate versions of *WadBOS* in operation, these have become more focussed and effective to the extent that improved representations of processes can now be introduced in a time span of a few days.

28.4.2 Technical Integration

Technical integration of models is very much a software problem. Two aspects are decisive: the *architecture* chosen for representing the model base and the integrated model and secondly, the *software technology* used to implement the DSS and its model base.

As for the *architecture*, Figure 28.1 shows four possible solutions to choose from (Hahn and Engelen 2000). For each solution, the end-user is shown on the left. He interacts with the DSS via a user interface and has access to an integrated model. The integrated model consists of more or less strongly coupled component sub-models available from the model base of the DSS. On the right is shown where the sub-models reside and how they are accessed and integrated. In the solution named loose and distributed models, use is made of existing models residing on the machines of the developers or owners. They are accessed and run via a network in a synchronous or an asynchronous manner. A human intervention might be required to run particular sub-models and exchange input and output data. In the solution named existing models linked in a single system, existing models are equipped with interfaces that enable them to be accessed and run in a synchronous manner.

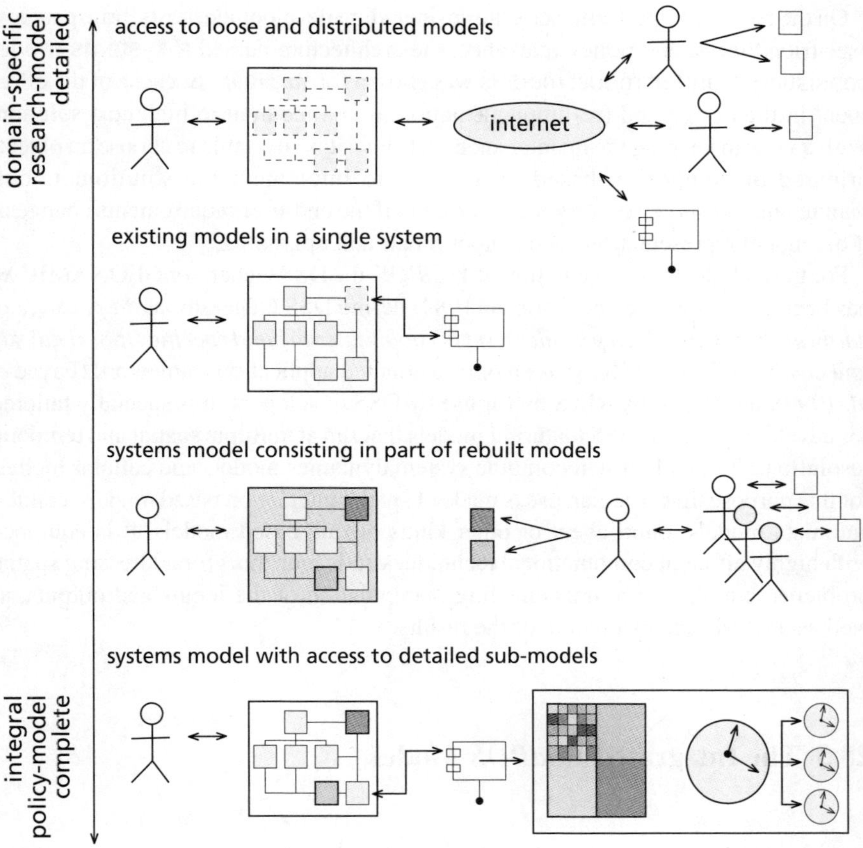

Fig 28.1: Approaches to model integration in four possible DSS architectures

Contrary to both previous cases, in the solution named a systems model consisting in part of rebuilt models, most sub-models are reformulated as parts of an integrated model, fully tailored to the precise role of the DSS and the needs of its end-users. Hence, a clear problem definition, an in-depth user requirements analysis, and a precise user profile are essential elements to decide on the exact depth and the extent of the integrated model and its constituent sub-models. All sub-models reside and run on the machine of the end-user. More than in the solutions linking existing models, this integrated model is very strongly coupled. It is by design a truly complex model representing all relevant processes at the same level of abstraction and detail. However, it will not necessarily produce the most detailed or most accurate output possible. The final solution, named systems model with access to detailed models, is in fact a hybrid form of the previous solution. It is different in that it allows for the additional incorporation of existing, more detailed models. In most cases, these are sophisticated algorithms that carry out particular calculations required in the integrated analysis.

On the basis of 15 performance criteria, including development costs, transparency, user-friendliness, and policy relevance, the architecture named a systems model consisting in part of rebuilt models was chosen for *WadBOS*. Because of the effort spent in the design and (re-)implementation, it is a medium to high cost solution. With a view to keeping the maintenance costs low, it is advisable to choose an object-oriented or component-based technology to implement the solution. If not, maintenance costs can be very high, certainly if the end-user requirements change or if the model representation needs major repair and upgrading.

For the technical implementation of *WadBOS,* the DSS Generator GEONAMICA® has been used. Sprague and Carlson (1982) define DSS Generator as *"a package of hardware/software which provides a set of capabilities to build specific DSS[s] quickly and easily"*. GEONAMICA® is an object-oriented application framework (Fayad *et al.* 1999), developed by RIKS bv for use by DSS developers. It is specially tailored for developing spatial DSS featuring models that run at multiple spatial and temporal resolutions. Typically it will combine system dynamics models and cellular models for this purpose. In particular, use is made of spatial interaction based models, cellular automata models, multi agent or other kinds of rule-based models. It is equipped with highly efficient computational techniques and algorithms for addressing spatial problems, but also with tools enabling manipulation of the inputs and outputs, as well as analysis and evaluation of the results.

28.5 The Integrated *WadBOS* Model

In its version 2, the Integrated *WadBOS* model has the following main characteristics:

- It represents strongly coupled social, economic, ecological, biological, physical and chemical processes (Figure 28.3). They are grouped into three main sub-models: economy, ecology and landscape.
- It integrates models that run at three spatial scales (Figure 28.2):
  - the Wadden sea as a whole;
  - 12 compartments of the sea; and
  - cellular representation of the sea: some 11,000 cells of 25 hectares each.
- It integrates models that run at one of three possible time steps:
  - one tidal cycle;
  - one month; and
  - one year.

It typically performs a run of 10 years and shows new outputs every month. The tidal cycle outputs are integrated over the month. This model is sufficient to generate the output required for most relevant policy questions. It meets the performance criteria specified and performs a 10-year simulation run in less than 10 minutes on a state of the art PC with Windows installed.

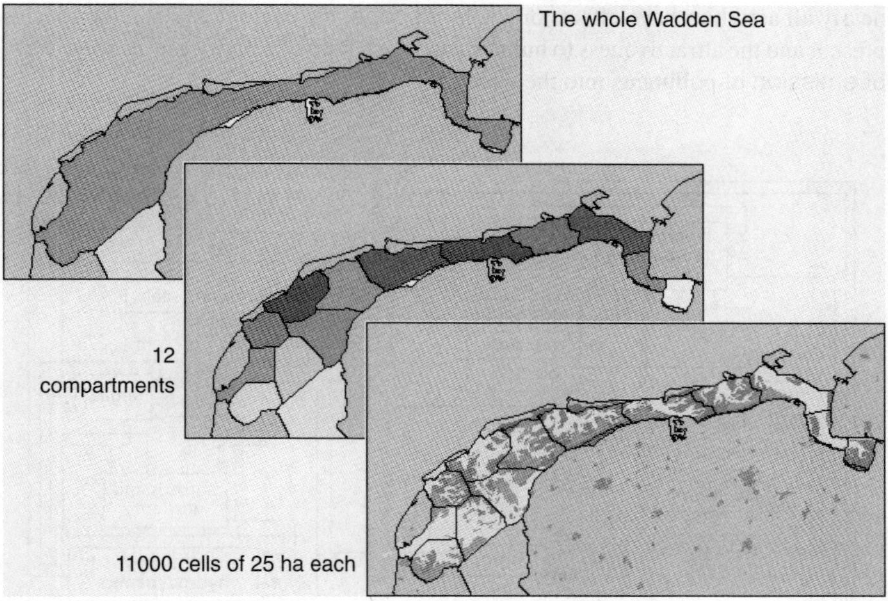

Fig 28.2: *WadBOS* spatial scales

Most economic processes run on a monthly or yearly time step, while most ecological processes are represented at the compartment level and run on a time step equal to a tidal cycle, but there are many exceptions to this general rule. Figure 28.3 shows the system diagram of the integrated model at the highest level of abstraction. This diagram is also the user interface of the model (in particular the *System Diagram View*, see Section 28.7). The boxes represent sub-models and the arrows between the boxes show the main data flows in the model. When a box is clicked, the details of the underlying model are shown, which is either a dialog window relative to the model, or a new system diagram of a sub-model at the next hierarchical level.

28.5.1 Economic Sub-model

In the economic sub-model (Figure 28.3), all the major economic activities present in the Wadden Sea are represented at some level of detail. shell mining, fishery and recreation have been worked out in more detail. Most activities carried out at sea are an input into the local industry and cause directly or indirectly the need for shipping of goods and people. shell mining and fishery directly extract biological resources from the Wadden Sea, while recreation, electricity generation by means of wind turbines, and defence activities use the open space as their prime resource. Finally, gas-mining taps underground resources and may thus have an impact on the morphology of the bottom of the sea. The presence and the noise generated by nearly all activities have effects on the landscape: the ecological state, the species present and the attractiveness to humans; and each kind of activity causes some form of emission of pollutants into the waters.

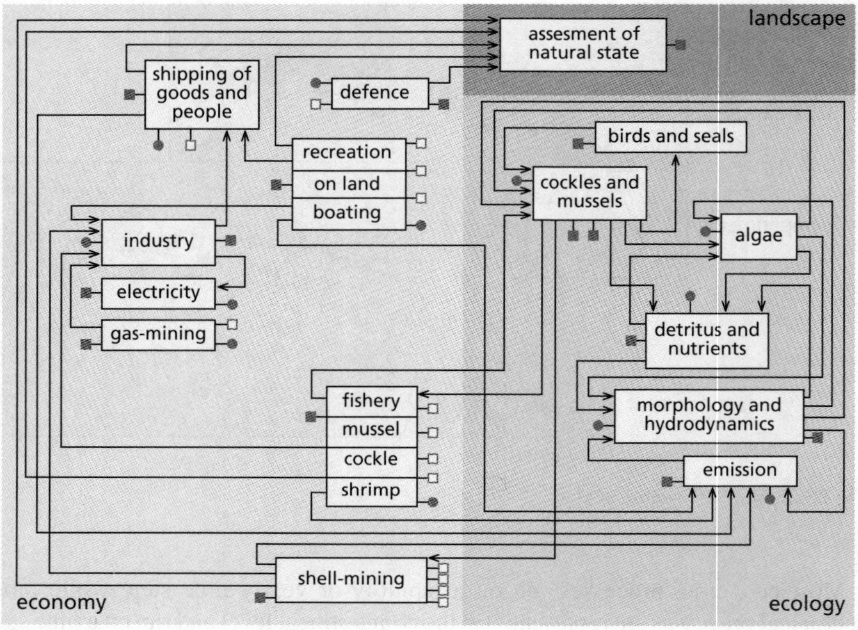

Fig 28.3: System diagram of the integrated *WadBOS* model represented at the highest hierarchical level of abstraction

Each of the economic activities is modelled according to the same general scheme, represented in Figure 28.4. This increases greatly the consistency, the learnability, and the transparency of *WadBOS*. The models calculate through a series of interlinked relations on a monthly basis. The mussel fishery is used as an example. The boxes in grey are part of sub-models different from the one represented. The effort of an economic activity is determined by the available infrastructure. Infrastructure for mussel fishery, for example, is the amount of fishing equipment available on board of vessels, while infrastructure for the recreational sector is the number of beds available. In order to operate the infrastructure, a workforce is required, which determines the employment in the sector. The deployment of effort and the availability of clients (in the recreation sector), or biomass in the fishery sector (as calculated in the Ecology sub-system), determine the extent of the extraction: catch for the fishery sector and number of recreants present in the area for the recreation sector. At a fixed price, the catch generates a particular turnover.

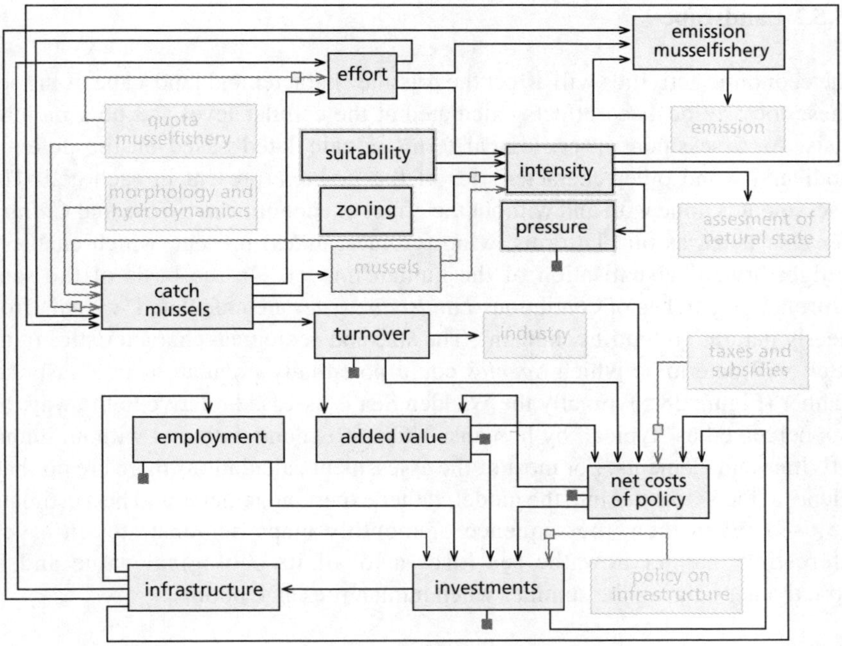

Fig 28.4: The model of an economic sector in *WadBOS*

In *WadBOS*, prices are externally determined constants. They are part of economic scenarios that can be tried out with the system. Debits, labour costs and technical costs are deducted from turnover to calculate value added. If the latter is positive, investments may be made to increase the infrastructure. Policies on infrastructure can intervene directly and set a maximum limit on the infrastructure allowed, or indirectly through taxes and subsidies. Another way of intervening into an economic sector is by setting a quota. These will possibly reduce the catch and will prevent the exhaustion of the species or other resources (i.e. shells).

The above relations are calculated for the entire Wadden Sea or per compartment (i.e. quota can be set per compartment). Further to these there are calculations relative to the precise location of the activity at the cellular level and 25 ha resolution. This is because effort is not equally distributed over the entire sea. It is the cell's suitability for an activity that determines where the intensity is high or low. So does the availability of the resource – mussels to be fished or mined in the example of Figure 28.4. However, the distribution of activities can be controlled through zoning policies by closing areas permanently or during particular months in the year. The latter is an important policy lever, because a high intensity may cause disturbance of particular species of bird or sea animals locally and add to specific forms of pressure.

28.5.2 Landscape

The economic activities will affect the pristine character and landscape of the sea. These too are localised effects calculated at the cellular level and on a monthly basis. An assessment in ecological terms is calculated based on the potential biodiversity and other characteristics of the *Ecotope* present in each cell. This assessment is done with and without the effect of economic *activities* and *Cultural elements* (such as oil platforms, wind turbines, dikes) present, which enables a straightforward visualisation of the human impact. On the basis of the same information, patches of contiguous *Landscape types* are calculated, varying from 'nearly natural' to 'entirely human'. The size and ecological characteristics of the patch will determine which *Species* could potentially populate it in a sustained manner (Figure 28.5). Finally the Wadden Sea is assessed relative to the way that it is perceived as 'typical' by humans. This too is done with and without human activities and elements. For most of the assessment calculations there are no 'hard scientific facts' to enter into the model, rather expert judgements and best estimates are used. Even then, the sequence of monthly maps demonstrates in a very interesting manner how the sea loses a lot of its ecological value and its attractiveness during the summer when human presence booms.

28.5.3 Ecological Sub-system

Where activities are carried out, some form of pollution will be generated and particular emissions (copper, TBT, PAK, and Oil) will end up in the water (Figure 28.3). River-borne pollutants will enter the sea mainly from Lake IJssel and the river Eems. A simple hydrodynamic model calculates how the water moves from one compartment to the next to eventually end up in the North Sea. Not only pollutants will move with the water, but so will other dissolved matter – detritus and nutrients – including: phosphates, silicon and nitrogen. In a food chain model derived from the EcoWasp model (Brinkman 1993), the nutrients serve as an input to the *Algae* dynamics. Simply stated, the growth of the latter is determined by the availability of nutrients, the right climatic conditions, light, or influx from the North Sea and sweet water systems. They are grazed by filter feeders, including the commercial species cockles and mussels. The latter are prey for birds and the fishery. More precisely, the policy-makers will decide every year the quantities that are needed as food for birds and for the survival of the species. The rest can be fished commercially. Cockles are fished and cooked at sea and their shells are returned into the sea. Thus, they become a resource for the shell mining industry. The food chain dynamics are calculated on a tidal cycle time step and mostly on the spatial level of the compartments. Some information, such as the mussel and cockle biomass and the consumption rates of birds is calculated on the cellular level. The latter serves as a cellular input into the fishing activity.

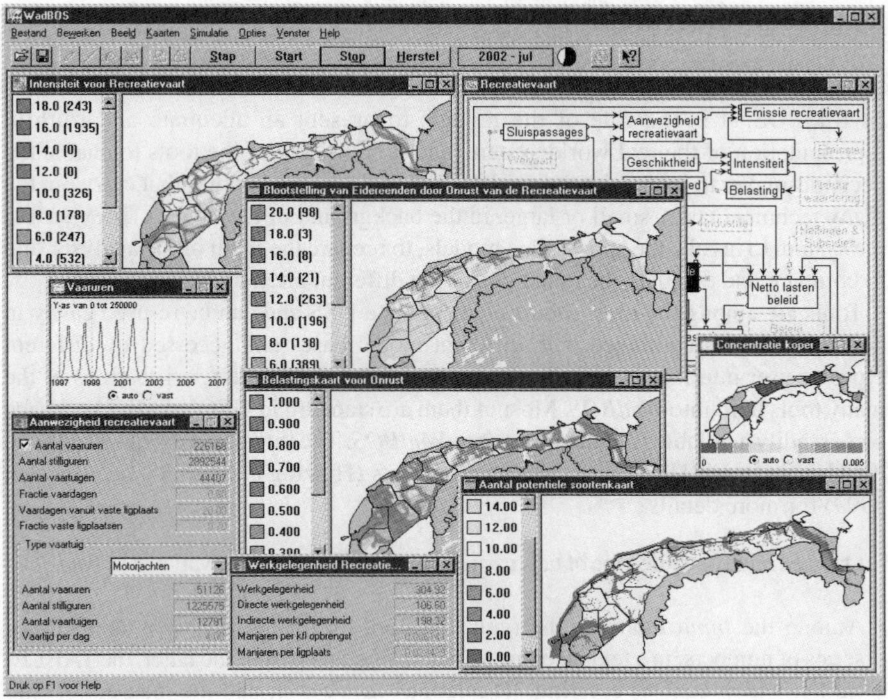

Fig 28.5: Screen dump of *WadBOS*

Figure 28.5 shows a typical output relative to recreational boating in the month of July, including dialogs with textual information, a graph of the boating activity, and maps with: intensity, stress caused by boaters on eider ducks, disturbance (partly due to boating), number of red list species potentially present, and the concentration of copper (partly due to boating).

The above description is kept very superficial. More details can be obtained from the technical reports (Huizing *et al.* 1998, Engelen 1999, Engelen 2000 and Uljee *et al.* 2000) or from the on-line documentation of *WadBOS* itself. However, it should be clear how the different sub-models are linked to each other in a network of mutual, reciprocal influence. Outputs are presented in the form of text, time graphs, and linked MS Excel tables. All the variables calculated at the compartment and the cellular levels are available as dynamic maps, updated on a monthly basis (Figure 28.5). Dynamic maps generated during a simulation can be stored on file for interactive comparison and further analysis by means of the tools available from the toolbase of *WadBOS*.

28.6 Tool Integration

In *WadBOS*, it is the role of the models to present an adequate and truthful representation of the real world system, and it is the role of the tools to enable the decision-maker to work with the models. The tools are the gnomes that carry out the many technical tasks, small or large, in the background of the system. They enable the system to invoke the appropriate models, to prepare the input of the analysis, and to compare and evaluate the outcomes of the different alternatives generated.

Tools are among the most robust elements in a DSS and can be re-used easily in new applications combined with different model bases and accessed by different types of user interfaces. We will not dwell extensively on the specifications of the many tools built into *WadBOS*. Most of them are standard in GEONAMICA®, hence were readily available for integration into *WadBOS*. The interested user is referred to the User Manual (Uljee *et al.* 2000) and reports (Huizing *et al.* 1998 and Engelen 1999) for more details.

Ordered relative to the type of task carried out, the following are available in *WadBOS*:

- Among the *Input tools* are the typical editors for changing single numbers, or series of numbers, in a textual or graphical manner. As part of the latter, the TABLE-Editor, which enables entering 2D relations as a curve, and the map editors are very essential and powerful instruments. But so are the tools to open, close, import and export files, scenarios and policy exercises.
- The *Output tools* fulfil the difficult task of presenting the massive amounts of data generated interactively in as concise and precise a manner as possible. For presenting dynamic spatial data, fast dynamic mapping facilities are available. For storing simulation results and enabling in depth analysis at a later stage, *WadBOS* features map recorders and players of (map)animations, as well as tools for storing dynamic map output in so-called .LOG files, and tools for writing results in a linked MS Excel spreadsheet.
- *Exploration tools* enable the user to interactively search the solution space. The capacity to generate scenarios, produce map overlays interactively (OVERLAY-Tool) (Figure 28.6), compare maps interactively (ANALYSE-Tool), and carry out sensitivity analysis puts great analytical power in the hands of the end-user.
- *Evaluation tools* help the user to decide on a 'best' solution. To that effect score tables and interactive multicriteria analysis are available. A built-in goal-seeking feature, fed with the results of sensitivity analysis, presents the dependency of state variables on policy relevant parameters. Thus, the risk can be avoided that the 'best' solution is never attained when series of 'what-if' analyses are carried out.

Not all the tools in *WadBOS* have the same importance for all its users. According to the function the system is to carry out (Section 28.2) they play a more or less pronounced role. The analyst uses the near complete set extensively. He needs both the very down to earth instruments for entering data and reading input files, as well as the sophisticated instruments for analysing output. When the main purpose of the DSS is communication, output instruments are very essential. The learner will be most interested in transparent input and output tools and the documentation systems. Finally, for the knowledge management function, a good documentation system is paramount, as are facilities to quickly enter, update and retrieve the information stored.

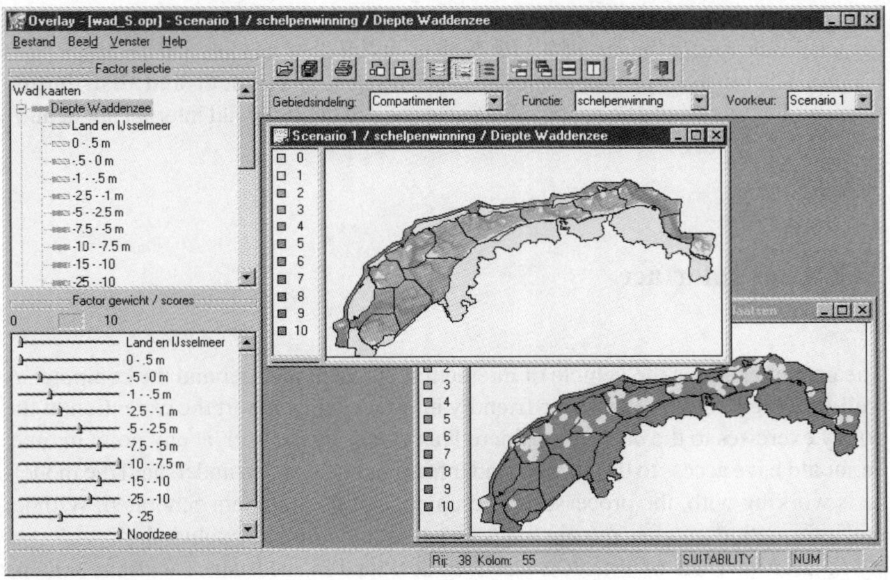

Fig 28.6: Screen dump of the OVERLAY-Tool of *WadBOS* used for the interactive generation of suitability and zoning maps e.g. for shell mining

28.7 Data Integration

The integration of the data required to run the *WadBOS* models is implemented by means of an offline connection to an external database for the GIS data and by means of a stand-alone database for the other data. This solution is in line with the general usage of the instrument and is possible because none of the models momentarily incorporated in *WadBOS* requires data from an online connection to an external database. The original geographical information is generated, maintained and stored in the GIS of the National Institute for Coastal and Marine Management. It is retrieved when better quality data become available to be distributed with new versions of *WadBOS*. The economic and ecological data are obtained from a wide variety of sources, some of which are not in digital format. When in use, *WadBOS* takes care of the retrieval and storage of its data: new or additional information is entered via the dialogs of the instrument and is stored in a hierarchically organised directory structure. In addition, all the files are in a readable format, hence can be viewed and edited directly by the end-user if he wishes to do so.

In the future, when *WadBOS* is installed on the machines of the many end-users of the member organisations of CUBWAD, there will be an absolute need for a dedicated central site from where the management and distribution of the databases is organised. In fact, this is true for the rest of *WadBOS* also – its models, its tools and the application proper. Currently, options are evaluated for a management and distribution organisation that would be accessible from a page in the Interwad internet application (*www.Interwad.nl*).

28.8 User Interface

The user interface is the vehicle of interaction between the user and the computer. A well-designed, intuitive and user-friendly interface will support the execution of the policy exercises to the degree considered necessary by the user, at any point in time, he should have access to the background information needed to understand the models he is working with, the processes represented, and the numbers generated. Without this information, models remain black boxes and learning is excluded.

The user interface of *WadBOS* has been designed in accordance with the scheme shown in Figure 28.7. The Wadden system (System box) is represented by means of the integrated dynamic model discussed. Because of its autonomous dynamics and the policies in force, the Wadden system evolves into new states as time goes by (Resulting state box). The policy relevant aspects of new states are captured and synthesised in a number of indicators. The policy-maker should begin any policy exercise, aimed at selecting ways of altering the autonomous dynamics or future state of the Wadden system, with a clearly defined set of criteria defining the desired state at a particular point in time (Preferred state box). In order to bring the actual state of the Wadden system closer to the preferred state, the policy-maker can intervene in the instrument and impose policy measures (Policy measures box). In an iterative process, he can tune the latter in an attempt to approach the preferred state with a minimum amount of effort and costs. A chosen policy measure can be tested for robustness by trying out scenarios (Scenarios box), which are effects on the system beyond the control of the policy-maker.

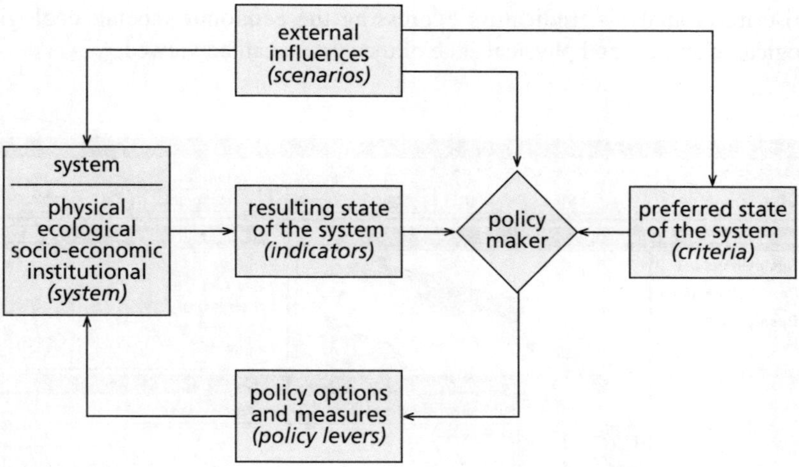

Fig 28.7: *WadBOS* system that enables the policy-maker to work with the models in a structured manner

The user-interface of *WadBOS* features for each of the boxes in the above scheme a specific window and view on the underlying models. Each view shows in a graphical manner how policy relevant features relate to the processes modelled.

The *System Diagram View* contains an overview of the structure of the integrated model at the most synthetic level (Figure 28.3). It represents the linkages between the ecological, landscape and economic processes typifying the dynamics of the Wadden Sea. The *Scenarios View* shows the parts of the model that are most subject to external influences. These are elements not under the control of the policy-maker, including: atmospheric and climatologic variability; economic growth trends; development of prosperity levels in the Netherlands; and exchanges of water and dissolved matter between the Wadden Sea, the North Sea and the IJsselmeer.

The *Policy Options View* shows the parts of the model that are subject to policy interventions (Figure 28.8). These are elements that are under the control of the policy-maker. The user can experiment with different combinations of policy measures and set values for: zoning policies, including closure of parts of the sea (*Sluiten gebieden*); policies on infrastructure (*Infrastructuur*); taxes & subsidies (*Heffingen & Subsidies*); and quotas set on fishing and mining. The costs and benefits of these Government (*Overheid*) policies can be assessed.

The *Impacts View* shows the parts of the model containing the summarised information and policy indicators required to evaluate the success of scenario and policy options tried out on the system. In general, the indicators are taken from the management plan of the Wadden Sea. For most indicators, norms and criteria are available which can be entered as reference values in the built-in score tables or multi-criteria analysis. Indicators expressing the economic, social, ecological, biological, chemical and physical state of the system can be viewed.

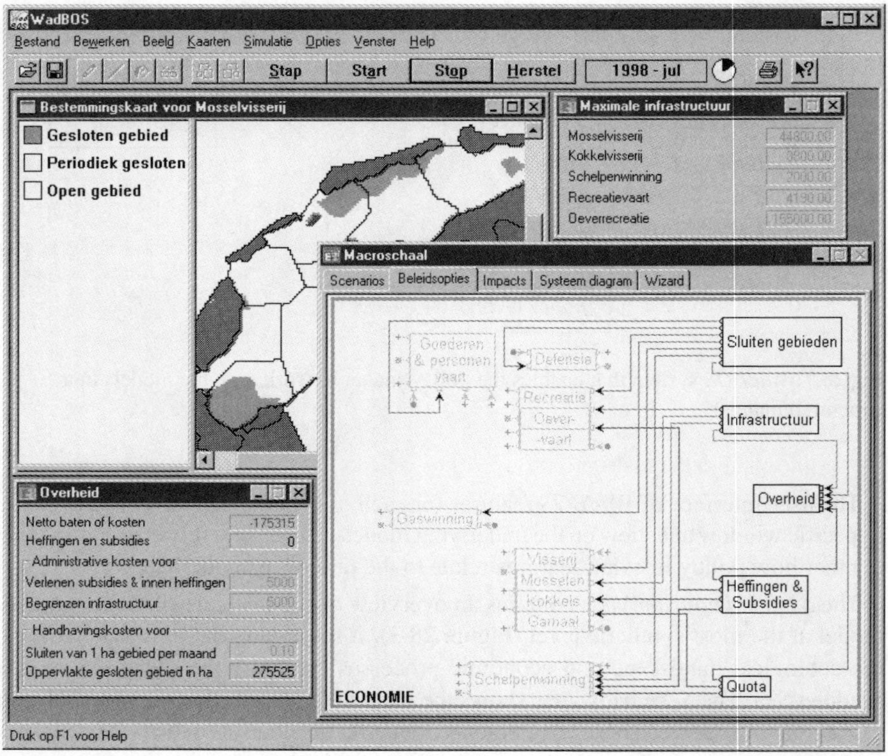

Fig 28.8: Screen dump of the Policy Options View, which is one of the four main views from the user interface of *WadBOS*

28.9 Using *WadBOS*

The development phase described in the previous sections is only the first one in the lifetime of a DSS. It needs to be followed by an evaluation of the technical contents of the instrument: its constituent models, its coverage of the decision domain(s), its flexibility and adaptability relative to changing issues and policies. Even more important is an evaluation on behalf of the end-users as the success of any DSS is ultimately decided upon by its end-users. How do they believe their decision domain has been represented? Does the instrument speak their language? Does it work in a manner that they find useful and pleasant? Will they want to use the instrument and make it part of their professional toolkit? With the answers to these questions a more or less major redesign of the DSS might be possible or necessary.

Following the implementation phase of both *WadBOS* versions 1 and 2, workshops were organised to test the appropriateness of the instrument and to get user feedback relative to the contents of the instrument, its role relative to policy-making in the Wadden Sea and its usability and user-friendliness. The results served as inputs for go no-go decisions relative to the further development of (parts of) the DSS. In particular, *WadBOS* version 2 was used and tested extensively (Reijngoud and van de Ven 2000). Two 2-day workshops were organised. Each of which gathered 12 to 16 participants to work on two cases, shell fishing and recreational boating, which were in need of new policy decisions at the time. The workshops were not intended to produce any particular usable policy, rather were meant for learning, dissemination and evaluation purposes only. The participants consisted of intended end-users actively involved with recreational or fishing policies, but also four developers of *WadBOS*. The role of the latter was to provide technical information when asked for it and otherwise to assess and evaluate the role of *WadBOS* throughout the exercise. The end-users usually discovered the instrument for the first time. Some had been involved in the knowledge acquisition phases, but most had never seen or worked with the actual instrument before. After a short guided tour through the instrument, the general objectives of the exercise were explained and the actual policy work started.

The objective of the recreational boating exercise consisted in defining a new policy allowing for a maximum amount of recreational activity and associated economic output but with a minimum impact on the bird colonies and on the pristine natural character of the Wadden landscape. Bird colonies can be protected best by zoning policies that displace the boating activity from bird-inhabited areas to parts of the sea with fewer birds. More generally, the ecological value of the sea and the assessment of the landscape are highest when large contiguous areas can be kept free of any human activity.

The objective of the shell fishery case study consisted in developing a new policy aimed at a sustainable exploitation of the resources (cockles and mussels). A fishery that leaves sufficient food for the birds and that enables a stable and high work force without exploiting the stocks beyond the regeneration capacity of the species is the ultimate goal. Further criteria for this exercise consisted in minimising the effects on the gear, the fleet, the economic output of the fishery, the employment in the fishery and linked sectors, and minimising the level of disturbance of other environmental or economic values of the Wadden Sea.

In fact, part of the exercises consisted of defining the set of policy criteria and entering norm values into the score tables of the impact screen of *WadBOS*. Next, finding a 'sufficient' or 'best' policy consists in carrying out 'what-if' simulations followed by an evaluation of the score tables. *WadBOS* has, in principle at least, all the processes and the actual data built-in to develop the policies described. All that the policy-maker needs to do is set and fine-tune the policy sensitive parameters relative to for example the employment, the exploitation of the shell fishery, and the expected '(re-)generation' capacity of the species. He can open and close areas for exploitation, and set quota on the totals to be fished. He can link the quota to the regeneration capacity of the species in the whole sea or parts thereof. The exact timing of a policy intervention is to be analysed and decided. Some policies will meet the criteria at an early stage while others will take more time to result in the desired effect. Some cause economic damage, while others damage the environment. Hence, compromises need to be sought. Policies need to be evaluated on the longer run anyway, because many processes have stochastic components. For instance, the regeneration of mussels and cockles is very dependent on the weather and changes in the climate as cold winters, storms and turbidity of the water will have an important effect on the species itself or its food (algae). A good and robust policy will allow for a number of consecutive bad years. Spatial policies can be tried out in *WadBOS* by changing zoning maps interactively. Closure of areas can be year-round or for particular periods only, for instance while birds are nesting.

As part of the exercises, the policy-makers were asked to compare policy-making with and without *WadBOS*. From their answers it could be concluded that in the absence of *WadBOS*, they generally spent less time defining the set of criteria for the evaluation of alternatives prior to their elaboration. They develop fewer alternatives and home in much more quickly on 'politically acceptable', 'likely', 'no-regret', or 'easy to implement' policies. Domain experts are hired to carry out the job. This way of working involves high costs, uncertainty relative to the outcome of the analysis, and problems relative to the timing and availability of the results. But, the policy-makers expressed a lot of faith in this process and its results.

According to the participants of the sessions, there is a distinctive role for *WadBOS* in future policy exercises. Its added value resides in:

- its integrated approach to policy-making and the capabilities of the instrument as a means for communication;
- the fact that *WadBOS* incorporates the information and knowledge required to carry out a lengthy, expensive, and cumbersome policy analysis in a manner of hours; and
- the role of the instrument in testing sensitivities and constraints as it explicitly reveals which aspects affect the dynamics and state of the Wadden system in a more or less major way. The fact also that it shows immediately the gaps in the (formalised) knowledge about the Wadden system in relation to the policy field at stake.

WadBOS is also criticised on a number of points. Firstly, the DSS is not complete. The policy-makers were missing particular information or policy criteria, which they would have liked to incorporate in their analysis. For example, effects of human activity (such as shell fishing) on the morphology of the bottom of the sea are not as yet available in the instrument. Some of the information is missing because the instrument is not finished yet, but other knowledge simply does not exist in a formal manner at this moment, hence cannot be integrated.

Secondly, some experts among the users had doubts about the validity and correctness of the scientific knowledge incorporated in the models of the DSS. These remarks require special attention and analysis because the DSS should indeed incorporate the best scientific knowledge available at any time. If not, the outcomes generated have little or no value in the policy-making process. However, 'best available' is not the same as 'indisputably correct' or free of 'scientific controversy' and 'value judgement'. The finger is put on a very hard scientific problem that cannot be easily solved.

Thirdly, while the geographical resolution (500 metre grid) at which policies can be formulated and analysed was considered sufficient for the shell fishing, this was not the case for the recreational boating. More detail was wanted. For example, some policy-makers wanted to test the effects of an increased docking capacity in particular harbours and the impacts caused by the additional number of boaters operating from these harbours. To that effect they wanted to create detailed designs of docking areas for yachts. *WadBOS* was never intended for this particular kind of detailed analysis, but the remark shows clearly that each policy-problem has a spatial and temporal dimension of its own. It is technically not possible to build an instrument that serves all purposes.

Fourthly, the policy-makers were not entirely satisfied with the user-friendliness and transparency of the instrument. They liked the fact that results are presented in the form of graphs and dynamic maps and liked the way data are entered and changed interactively. They are also impressed with the on-line documentation of all the models, variables, and parameters present in *WadBOS*. But, they indicated that concepts, definitions, units, and assumptions needed to be stated in a clearer manner. Some policy-makers considered the instrument to be too complex and too demanding because of the level of knowledge needed to use it. The latter did not think of themselves as being the end-users, rather suggested that the instrument should be used by technical people facilitating sessions in which policy-makers would take part.

These four points contain sufficient food for a long process of further improvement. Some remarks are very fundamental, they have been mentioned in similar exercises and show that the developers of *WadBOS*-like instruments have a long way to go still. For example, the first three points are a plea for more detail and complexity in *WadBOS*, while the fourth states that the instrument as it is, is already too complex and too cumbersome to be used by exactly those users formulating points one to three. How can this contradiction be resolved? Further to this, some end-users have expectations relative to DSS systems that are beyond the capabilities of the information technology at this moment and others seem to forget that a DSS is only a technical tool enabling users to address complex issues more easily and systematically, but, that in return some investment in time and effort is required to learn how to work with it. Driving a car from A to B is faster than walking, but requires learning how to drive the car.

28.10 Discussion

The development cost for *WadBOS* versions 1 and 2 amounts to some 400,000 . It took some 20 calendar months to build, and involved a team of eight part-time developers. Other people including scientists, domain experts, and end-users invested time in the development at no costs. From this, it can be concluded that an elaborate policy support tool can be developed within very reasonable constraints relative to budget, human resources and development time. But, this is only true when good base material and expertise is available and when a stimulating collaboration between visionary end-users and competent DSS developers is propelling the development. From the workshops reported on in Section 28.8, we must conclude that we have only gone part of the way: systems supporting public policy-making and planning are very complex instruments that will need a lot more technical development, experimentation and experience before they will meet the promises that are made so easily and so frequently.

We have not dealt with the quality of integrated models in terms of their calibration and predictive capabilities. The technology presented in this chapter is not an alternative for calibration and good modelling practices. As long as a model is not calibrated, it should stay in the hands of the model developer and should not be used for actual policy-making purposes. Calibration remains a very difficult and time-consuming task. This is more so for a complex integrated model than for a simple and small model. However, linking individual sub-models into an integrated model does not necessarily increase the level of uncertainty and error. Rather, the strong linkages and loops between the individual models and the fact that variables are passed from sub-model to sub-model at every time step of a simulation, brings to the surface mistakes in the model formulation much more easily than when the same models are applied in isolation. Nevertheless, it should be emphasized that even an extensively calibrated *WadBOS* model will only generate potential developments rather than predictions of future states or changes. This is not only due to the problem of calibration, but because of the inherent uncertainty in the processes represented too. Consequently, the model and the encompassing DSS should be used for exploratory rather than predictive purposes.

Introducing and institutionalising the DSS in the end-user organisation is the ultimate factor determining its success (see, for example, Alter 1980, Klein and Methie 1995). Even the best DSS will go unnoticed or will fail if the wrong implementation and introduction strategy is chosen. Generally speaking, the development and acceptance of the DSS will be much easier if the development is initiated by its end-users right from the beginning. The more they feel the need for a change and the more they are involved in defining the precise role of the instrument in the organisation, the more likely it is that the DSS will be accepted and used. The more the product itself fulfils tasks that are perceived as real and in ways that solve problems in an obviously better, more transparent manner than before its introduction, the more the product will have a chance to survive. It is therefore very crucial to determine the right moment during the development of the DSS to have it change state from a prototype into an operational version. This is when a gradual introduction, a good technical documentation, hands-on training, and prolonged technical assistance during usage, become of paramount importance. A hasty, unprepared introduction should be avoided under all circumstances.

The methodology developed in this chapter and the technical components that constitute *WadBOS* are transferable to other coastal zones in the Netherlands, Europe and the world beyond. There are examples to prove this point:

- RamCO is a coastal zone management application in Sulawesi, Indonesia developed for the National Institute for Marine and Coastal Management, Ministry of Transport, Public Works and Water Management (*http://www.riks.nl/RiksGeo/proj_ram.htm*);
- SimLucia is a coastal zone management and land use planning application in St. Lucia, West Indies, developed for the United Nations Environment Programme (*http://www.riks.nl/RiksGeo/proj_sim.htm*);
- Murbandy is an urban land use planning application generically applicable for European urbanised regions developed for the Joint Research Centre of the European Commission (*http://www.riks.nl/RiksGeo/proj_mur.htm*);
- MODULUS is an integrated watershed management and land use planning application applied in Mediterranean coastal watersheds subject to land degradation and desertification developed for DG12 of the European Commission (*http://www.riks.nl/RiksGeo/proj_mod.htm*); and
- Environment Explorer is an integrated land use planning application in the Netherlands, developed for Ministry of the Environment and the Ministry of Transport, Public Works and Water Management (*http://www.riks.nl/RiksGeo/proj_lov.htm*).

Such transfer could take place on a site-by-site and application-by-application basis. However, a more systematic approach in which databases and model bases are developed and maintained containing generic building blocks applicable in more than a single region or for more than one type of application, holds much more potential. In the Netherlands, such initiative has started as part of the GEOMOD project of the Ministry of Transport, Public Works and Water Management. One of the aims of this initiative is to find out precisely what the contents would be of the national database, model base and toolbase enabling the implementation of a *WadBOS* like instrument in other coastal zones of the Netherlands, and possibly of Europe.

Acknowledgements

The *WadBOS* system presented in this chapter is developed by a consortium consisting of Research Institute for Knowledge Systems bv (Maastricht), Infram bv (Zeewolde), Delft Hydraulics (Delft), Institute for Environmental Studies (Amsterdam), Resource Analysis (Delft), and *DHV* (Amersfoort). The work was supported by the Land Water Information Technology Programme of the Dutch Government as well as the Ministry of Transport, Public Works and Water Management. A more elaborate and earlier version of this text was produced for the National Institute for Coastal and Marine Management/RIKZ, The Hague, The Netherlands as part of Contract 42002555 (Engelen 2000).

References

Alter, S.L. (1980) *Decision Support Systems: Current Practices and Continuing Challenges*, Addison-Wesley, Reading, MA.

Brinkman, A. (1993) Biological processes in the EcoWasp ecosystem model, *IBN-DLO report 93/6*, Texel.

Catanese, A.J. (1979) *Information for Planning. The Practice of Local Government Planning*, International City Management Association, Washington D.C.

El-Najdawi, M.K. and Stylianou, A.C. (1993) Expert support systems: integrating AI technologies, *Communications of the ACM*, 36 (2): 55-65.

Engelen, G., White, R. and Uljee, I. (1993) Exploratory modelling of socio-economic impacts of climatic change, in Maul, G. (ed.) *Climate Change in the Intra-America's Sea,* Edward Arnold, London, pp.306-324.

Engelen, G. (ed.) (1999) *BOS Integraal beheer van Estuariene en Waddensystemen. Deelproject: Case Waddenzee*, Stichting LWI, Gouda.

Engelen, G. (2000) *The Development of the WadBOS Decision Support System. A Bridge Between Knowledge and Policy in the Wadden Sea*, National Institute for Coastal and Marine Management, The Hague, pp. 38.

Engelen, G. (ed.) (2000) *MODULUS: A Spatial Modelling Tool for Integrated Environmental Decision-making, Final Report*, EU-DG12 (Contract ENV4-CT97-0685), Brussels.

Fayad, M.E., Schmidt, D.C. and Johnson, R.E. (1999) *Building Application Frameworks: Object-Oriented Foundations of Framework Design*, Wiley Computer Publishing.

Firley, M. and Hellens, D. (1991) *Knowledge Elicitation. A Practical Handbook,* Prentice Hall, New York.

Gonzalez, A.J. and Dankel, D.D. (1993) *The Engineering of Knowledge-based Systems. Theory and Practice,* Prentice Hall, Englewood Cliffs.

Hahn, B. and Engelen, G. (2000) Concepts of DSS Systems, Decision Support Systems for river basin management, Proceedings of an Internationl Workshop in Koblenz, 6 April, Bundesonstalt für Gewässerkunde, Koblenz-Berlin, Germany, pp. 9-44.

Huizing, J.J., Engelen G., van de Ven, K., Pothof, I. and Uljee, I. (1998) *WadBOS. Een Prototype van een Kennissysteem voor Beleidsanalyse van de Waddenzee*, Eindrapport, Directie Noord Nederland, Rijkswaterstaat, Leeuwarden.

Klein, M.R. and Methie, L.B. (1995) *Knowledge-Based Decision Support Systems,* Wiley, New York.

Kok, J.L. de, Arifin, T., Noor, A., Wind, H.G. and Augustinus, P.G.E.F. (1997) Systems analysis as a methodology for sustainable coastal-zone management in tropical countries, *Torani, Marine Science and Technology Bulletin*, 8: 31-41.

McConnell, S. (1996) *Rapid Development. Taming Wild Software Schedules,* Microsoft Press, Redmond, Washington.

Mulligan, M. (1999) Adapting research models for policy application: revise vs. rebuild, *Journal of Modelling and Assessment,* Submitted.

Reijngoud, T. and van de Ven, K. (2000) Beleidsoefeningen schelpdiervisserij en recreatievaart met WadBOS, National Institute for Coastal and Marine Management/RIKZ, Haren.

Uljee, I., Hahn, B., van der Meulen, M. and Engelen, G. (2000) *WadBOS Gebruikershandleiding*, National Institute for Coastal and Marine Management/RIKZ, Haren.

Web Sites

WadBOS Version 2: *http://www.riks.nl/RiksGeo/proj_wad.htm*

Interwad: *http://www.Interwad.nl*

RamCO: *http://www.riks.nl/RiksGeo/proj_ram.htm*

SimLucia: *http://www.riks.nl/RiksGeo/proj_sim.htm*

Murbandy: *http://www.riks.nl/RiksGeo/proj_mur.htm*

MODULUS: *http://www.riks.nl/RiksGeo/proj_mod.htm*

Environment Explorer: h*ttp://www.riks.nl/RiksGeo/proj_lov.htm*

29 A Planning Support System for Policy Formulation in Water Resources Rehabilitation

Mohammad Ali Sharifi[1] and Erasmo Rodriguez[2]

[1] International Institute for Geo-Information Science and Earth Observation, ITC, Enschede, The Netherlands
[2] Department of Civil Engineering, Universidad Nacional de Colombia, Bogota, Colombia

Abstract: In order to support policy formulation for rehabilitation of the natural environment in the Western Mancha region of Spain, a planning support system was developed and applied. The system is based on a framework developed for planning and decision-making and includes three main components, namely, a water balance model of the groundwater basin, a planning model and an evaluation model. The water balance model, which makes use of GIS and remote sensing, simulates the average yearly recharge of the aquifer system in relation to the land use changes for average meteorological conditions, to help understand the current situation. The planning model which makes use of mixed integer programming, simulates the reaction of farmers towards the changes in the present subsidy schemes and helps formulate proper policy instruments. Finally, the evaluation model makes use of multicriteria decision analysis to support the evaluation of developed policies and the selection of attractive scenarios based on the identified criteria and the preferences/opinions of various decision-makers.

29.1 Introduction

The Western Mancha is a low populated region characterized by high degree of aridity. The economy is mainly agrarian. Water resources are limited to the groundwater in Aquifer 23. The aquifer has a surface outlet in a wetland area called Las Tablas de Daimiel Natural Park. Many years ago, the low return from rain-fed crops, like cereals and vineyards, encouraged farmers to switch to irrigated crops and to construct wells and *'norias'* to extract groundwater from the aquifer. Pumping has dramatically increased since 1970, and in some years, the volume of extractions (discharge) has been greater than the recharge of the aquifer system. The progressive lowering of the water table in the aquifer has reduced the wetland by more than 60 per cent of its original area and has jeopardized the supply of drinking and irrigation water.

To improve the situation, a plan to reduce the rate of extraction was carried out, and limits on the use of groundwater for irrigation were imposed on the farmers. The European Community and the Spanish Government have initiated a programme for the rehabilitation of the natural environment in the Western Mancha region. To support this effort, a planning support system (PSS) was developed to help in formulating and evaluating the impacts of different policy instruments.

This was a typical policy decision problem, which involved choices at two levels. At one level, the policy-makers were trying to decide on the policy that would have most effect on rehabilitation of the region. This was undertaken in the face of uncertainty about the impact of each policy as far as the farmers' responses were concerned. At another level, farmers had their own decision problem: how best to respond to the new policy environment, given their own objectives and constraints. In order to solve such a problem, the uncertainty about the farmers' reactions needed to be reduced. This was achieved by modelling farmers' reactions towards various policy decisions and their impacts on the environment. Using this system, the existing problem was substantiated; then attractive policies were generated, appraised and evaluated for their performance.

The system was presented to the *Fundación Municipal para el Desarrollo Económico y el Empleo*, in *Alcázar de San Juan*. At local level, it was well received and used as a promising tool for formulation and evaluation of the policies that intended to make sustainable use of the groundwater resources in Aquifer 23. At the governmental level, however, it was not well adopted as an additional tool for policy formulation and evaluation. This was due to the large number of different authorities with local, regional and basin competence on the management of the system. As an example, at basin level, the inter-basin transfers from Tajo basin (hydraulic and hydrological solution) have been used as a solution to the environmental problem at Las Tablas de Daimiel Natural Park, rather than through a more sustainable scheme like the one formulated through the PSS.

As reported by Beaufoy (2000), the scheme of subsidies in the area has had some success: many farmers have signed up and there has been a notable shift in the cropping pattern in the area towards those crops that use less water and a corresponding reduction in the extraction from the aquifer. However, the scheme has also been criticised for several reasons: firstly, as a result of repeated droughts during the last years, and in spite of the reduced extractions, the system Aquifer 23 is far from being restored to a sustainable state; secondly, the widespread problems of farms with illegal bore-holes outside the scheme have not been addressed; and thirdly, the cultivation of high-water demanding crops like maize, beet and melon, are heavily subsidized by the Common Agricultural Policy of the EU, which is in contradiction to the spirit of conservation of the groundwater system.

Nowadays, a Committee of independent experts, established by regional and state governments, is proposing new studies of the system. The objective of this Committee is to develop a Plan for the Management of Water Resources and for Sustainable Development of the Upper Guadiana River. At this stage of analysis, the PSS may be again used as a tool for the development of new policy instruments not only for managing water consumption, but also for the identification of a sustainable strategy for water resources management. In this chapter the basic principles, components and functions of the system are described.

29.2 Basic Principles of Planning and Planning Support Systems

29.2.1 Planning Support Systems Definition and Components

There are innumerable definitions of planning, of which we prefer the definition given by Conyers and Hill (1989) that suggests that planning is *"a continuous process, which involves decisions, or choices, about alternative ways of using available resources, with the aim of achieving particular goals at some time in future"*. This definition attempts to incorporate the main functions included in most other definitions, e.g. as a means to choose, to allocate resources, to achieve goals and to plan for the future. On the other hand, planning is a specific type of decision-making. Therefore, it should comply with the definition and phases of decision-making and problem-solving processes that include the following main phases (Simon 1960, Sharifi 1999):

- intelligence: examining the environment to identify problem or opportunity situations;
- design: initiating, developing, and analyzing the possible courses of action. This involves application of decision models that, generate solutions, test their feasibility and analyze different alternatives;
- choice: evaluating alternative options and selection of specific course of action;
- implementation; and
- monitoring and evaluation.

In this context we define PSS as a class of geoinformation systems composed of data/information, models and visualization tools that are primary developed to support different phases of the planning process and its functions. PSS contributes to rationalizing planning and related decision-making processes by providing necessary support to systematically structure and formulate problems, develop alternative plans or policy scenarios, assess and evaluate their impacts (considering objectives of the relevant stakeholders), and to select of a proper policy or plan. Underlying the development of a PSS is the assumption that planning is a dynamic process, and therefore requires the relevant support for continuous updating of data, generating and evaluating plans and policies based on the updated data and assumptions. Naturally, a greater degree of access to relevant knowledge and information will lead to the consideration of more effective number of alternative scenarios, which will result in a better informed planning and public debate.

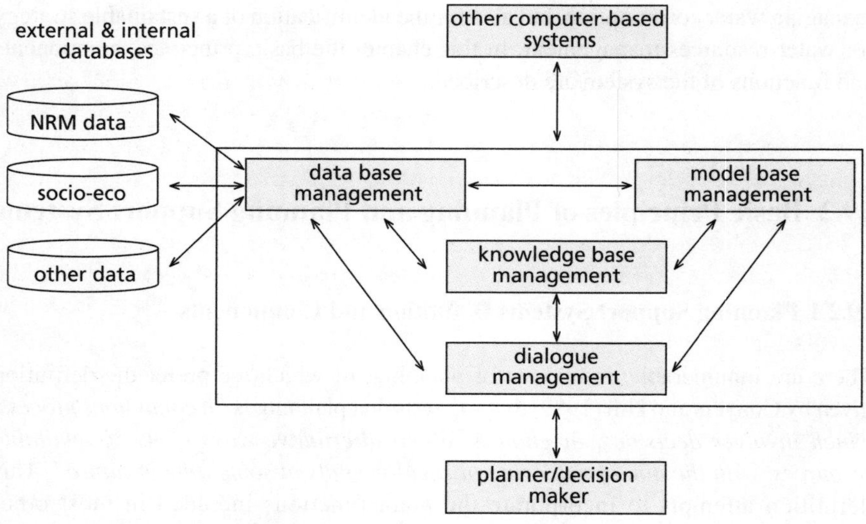

Fig 29.1: Overall architecture of a planning support system (adapted from Turban 1995)

The PSS, as a specific type of decision support system (Figure 29.1), includes the following main components:

- A *database management system* that includes databases designed to accommodate and organize the basic spatial and thematic data, provide facilities for selection and manipulation of data as well as interrelating data from various sources.
- A *model base management system* that includes quantitative and qualitative models that support resource analysis, assessment of potential and capacities of resources at different levels of management. This is the most important component of the system, which forms the foundation of the model-based planning support (Sharifi 2000). It includes three classes of models (Figure 29.2), which make use of the existing data, information and knowledge for identification of the problem, and formulation, evaluation and selection of the proper solution. These models are as follows:
  - A *process/behavioural model* describing the existing functional and structural relationships among elements of the planning environment to help analyze the actual state of the system and identify the existing problems or opportunities. This also supports resource analysis, which clarifies the fundamental characteristics of land/resources and helps understanding of the process through which they are allocated and utilized (Sharifi and van Keulen 1994).
  - A *planning model*, which integrates the potential and capacity of the resources (biophysical), socio-economic information, goals, objectives and concerns of the different stakeholders to simulate behaviour of the system. Conducting experimentation with such a model helps to understand the behaviour of the system and allows generation of alternative feasible scenarios to address the existing problems.

- An *evaluation model*, which allows evaluation of impacts of various strategies/scenarios and supports selection of the most acceptable solution, which is acceptable to all stakeholders and improves the management and operation of the system.

• A *knowledge base that* provides information on data and existing processing capacity and models which can be used to identify problem, to generate solutions, to evaluate and appraise their impacts, and finally to communicate the results to the decision-makers.

• A *user-friendly interface*, that allows smooth and easy communication with the system, visualization and communication of the results of analysis to the decision-makers in a manageable and understandable form.

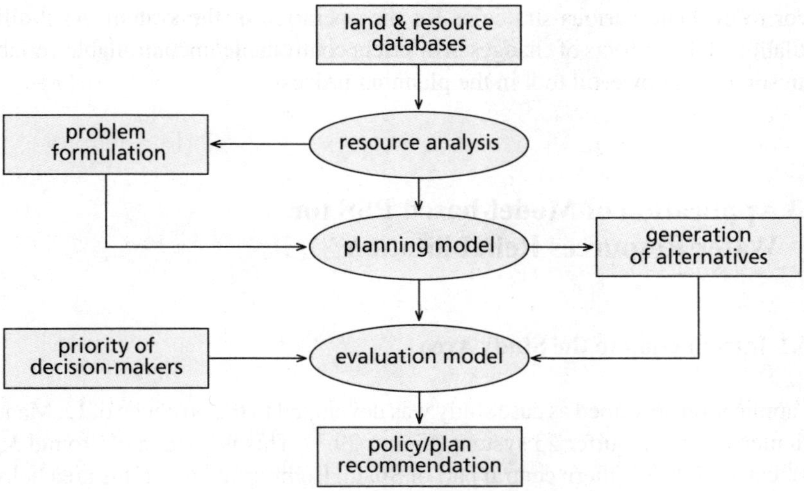

Fig 29.2: Basic principles of model-based PSS

29.2.2 Land Use Planning Process

Tinbergen (1956) and Thorbecke and Hall (1982) considered land use planning as part of agricultural sector and/or regional planning, where the effects of economic policies on patterns of and changes in land use are studied. In this approach, changes in land use are considered as the result of the interaction between policy variables (like infrastructure, investments, prices, credit facilities) and exogenous parameters (resource endowments) that lead to realization of a number of defined goals (welfare, equity, rehabilitation of environment) and possible (undesired) side effects (environmental pollution).

In this context, land use decisions involve choices of at least two levels, e.g. regional and farm levels. At the regional level, a policy-maker is trying to decide how best to allocate resources or lead the agricultural development process in the desired direction, in the face of uncertainty about the impact of the allocation process on the other systems (economic, cultural and ecological). This uncertainty is related to the way that farmers in the economic system will respond to the new policy. At the farm level, farmers have their own decision problem: how best to respond to the new policy, given their own resources and objectives that are influenced by socio-cultural values and impacts of the other systems. In order to reduce the uncertainty about farmers' reactions to and support for an effective decision on a proper policy measure at the macro-level, their impact at farm level has to be evaluated.

There are several types of empirical research which can support this type of analysis. Simulation is used most frequently, especially when 'satisficing' is selected as a principle for choice. Following Shannon (1975), we define and use a simulation model for conducting experiments in order to understand the behaviour of the system and/or to evaluate various strategies for the operation of the system. As it allows simulation of the impacts of changes in different controllable/uncontrollable variables, it can serve as a powerful tool in the planning process.

29.3 Application of Model-based PSS for Water Resources Rehabilitation

29.3.1 Introduction to the Study Area

The application described as case study was developed in the province of La Mancha, particularly in the Aquifer 23 system (Figure 29.3). This is an area of around 5,500 km^2 located in the southern central part of Spain. Eighty per cent of this area belongs to the province of Ciudad Real and 20 per cent to the provinces of Albacete and Cuenca. The climate is semi-arid. According to Koppen's classification, it should be catalogued as continental Mediterranean, with warm summers and cold winters. Average annual precipitation is around 450mm; average annual temperature is between 11.5 and 14.5^{o}C; annual sunshine is around 2,800 hours, and annual evapotranspiration is around 950 mm.

Two main aquifers can be distinguished in the Aquifer 23 system: the upper one, composed of limestone of Miocenic origin with an average and maximum thickness of 35 and 200m respectively (Niñerola 1976); and the lower one comprising Jurassic and Cretaceous limestones and dolomites. The upper unit extends all over the plain. It is a free aquifer, very heterogeneous due to sedimentation factors. The lower unit connected to lateral aquifers (Aquifer 19 in the north and Aquifer 24 in the south) extends in the eastern half of the plain (around 3,500 km^2). It is a series of confined aquifers that locally could be considered as free or semi-confined.

Due to the total lack of surface water, groundwater from the aquifer is used for drinking and mainly for the irrigation water supply. Farmers have made huge investments in pumping and irrigation systems. The major problem in the area is caused by the overexploitation of the groundwater reserves, which has lead not only to jeopardising the aquifer, but also to big environmental problems, resulting in the continuous decline of wetland in the aquifer's discharge zone at Las Tablas de Daimiel Natural Park (Rodriguez 1994).

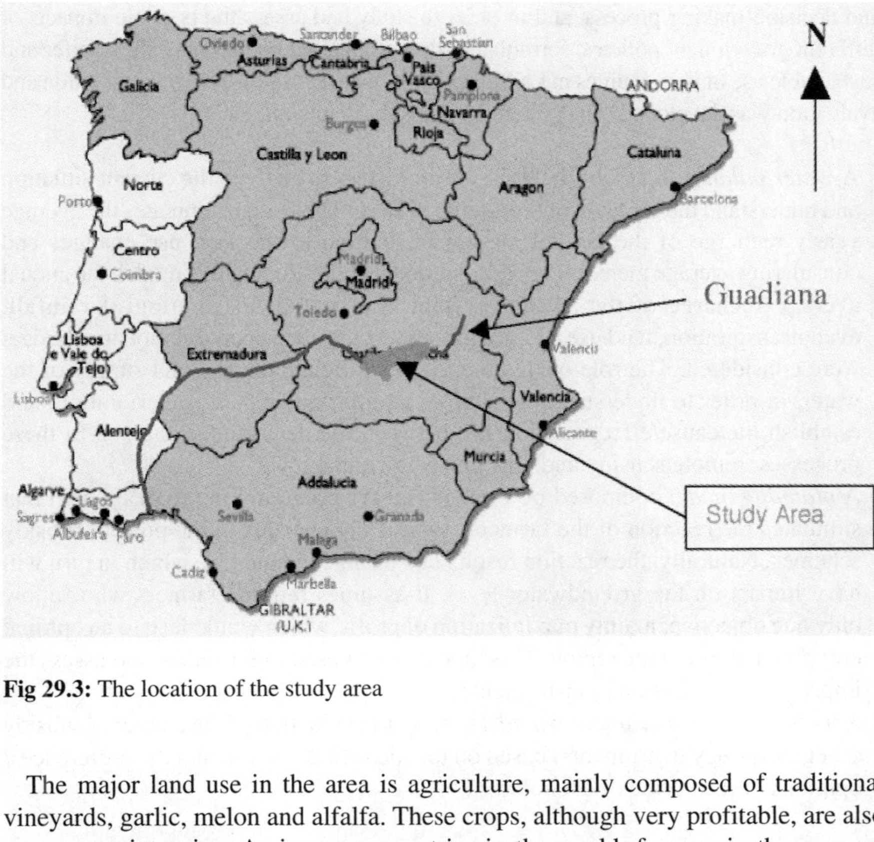

Fig 29.3: The location of the study area

The major land use in the area is agriculture, mainly composed of traditional vineyards, garlic, melon and alfalfa. These crops, although very profitable, are also very water intensive. As in many countries in the world, farmers in the area are receiving subsidies from the Government. But because of the lowering of the water table in the aquifer, the Ministry of Agriculture offered the farmers additional subsidies for reduction in the consumption of irrigation water. Besides, the *Confederacion del Guadiana* (CHG) established maximum pumping water volumes from the aquifer depending on the size of the farm and the crops planted.

Due to the big extension of the aquifer system, and based mainly on data availability and counterpart support, the *Alcázar de San Juan* district was selected as the study area. This district is located in an area where the groundwater level was relatively high (Getachew 1994). *Alcázar de San Juan* is located in the central northern part of the aquifer. It has an extension of around 68,000 hectares, from which approximately 52,000 hectares are within the aquifer's limits.

29.3.2 PSS Concept

Based on the framework presented in Section 29.2 (Figures 29.1 and 29.2) for the planning and decision-making process, and in order to study and assess the possible impacts of different government policies, formulated for a gradual recuperation of the aquifer and sustainable use of its resources in La Mancha province, a PSS for policy formulation and evaluation was developed. The system includes three main models.

- A *water balance model* of the groundwater basin to analyze the current situation and understand the problem of the aquifer system. This model simulates the average yearly recharge of the aquifer system in relation to the land use changes and considering average meteorological conditions. In developing this model, the annual average recharge of the aquifer system, the spatial distribution of rainfall, evapotranspiration, land use and soil properties, and the water-soil moisture budget were considered. The role of this model was to relate the use and supply of the water, in order to understand the state of aquifer, quantify its deterioration, and establish the cause/effect relationship between the determining factors. In these processes, remote sensing and GIS played a crucial role.
- A *planning model* composed of a mixed integer-programming (MIP) model that simulates the reaction of the farmers towards the changes in the present subsidy schemes. Naturally, the reaction results in a change in land use, which in turn will have impact on the groundwater level. It assumes rational farmers who follow only one objective, mainly maximization of profit, which would lead to an optimal cropping pattern in the region. This model is also used to formulate and assess the impacts of various policy instruments.
- A *multiple criteria evaluation model* to support evaluation of the attractive subsidy schemes (policy instruments) based on the identified criteria, and the preferences/ opinions of the various stakeholders.

In the following sections, each of the models is briefly explained and their application results are reported.

29.3.3 Water Balance Model

The purpose of this model was to understand the state of the aquifer and to study different components of the water balance at a local (cell) level, under different conditions of climate, soil and vegetation, in order to estimate the aquifer's average direct recharge. Therefore, it was proposed to estimate the annual average recharge of the aquifer system, coupling the spatial distribution of rainfall, evapotranspiration, land use and soil properties with a water-soil moisture budgeting method. In this process, a geographic information system (GIS) was used, because GIS facilitate the use of existing hydrological models. Besides, they provide a convenient medium for coupling hydrological models with spatial entities, permitting the use of lumped conceptual hydrological models at grid scale.

A typical computational element for the hydrological modelling at grid scale is shown in Figure 29.4. It illustrates the approach to quantify the spatial variability of hydrological parameters. The processes of rainfall, infiltration, inflow and runoff are computed for each element with each time step. A depth h_1 exists at the beginning of each time step and a depth h_2 at the end of each time step. A detention storage depth d is assigned depending on the land cover types, the soil characteristics and the present soil moisture. The partitioning of the watershed into computational elements, as shown in Figure 29.4, was made for the entire groundwater basin.

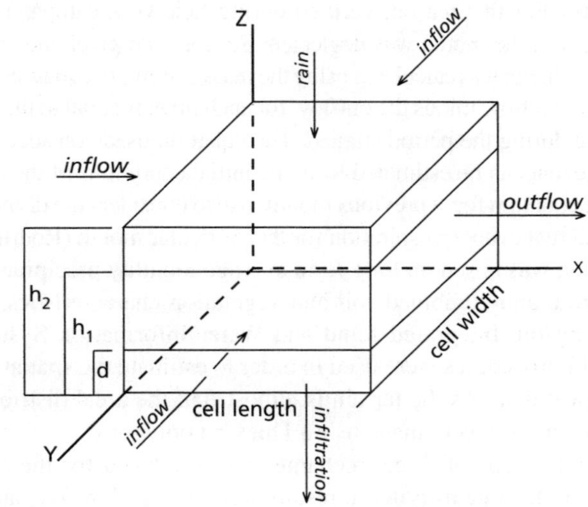

Fig 29.4: Elements in hydrological modelling at grid scale

Landsat TM data of 1993 was utilized to derive land use classes that were used to estimate the water use and a GIS was used to carry out the water balance for every grid (pixel) in the entire basin, and estimate the impact of land use changes. A soil map compiled by Carlevaris (1992), covering about 80 per cent of the aquifer, was digitized and used in order to establish the soils parameters considered in the water balance equation. For the remaining 20 per cent of the area (northeast), average soil conditions were established. The boundary of the groundwater basin was digitized from an analysis made by Lopez Camacho in 1989. Meteorological measurements, consisting of monthly data on rainfall and temperature for the period 1950-1991, were obtained for 10 stations evenly distributed within the aquifer's area.

The core algorithm for the calculation of the water balance was based on the Thornthwaite and Mather method (1957). It was applied on a cell basis and a monthly time step, making the model a mesoscale fully distributed hydrological model. The model required input series for rainfall, temperature, and soil data in order to estimate the water holding capacity. Potential evapotranspiration was calculated using the Thornthwaite method. Real evapotranspiration was estimated as the potential evapotranspiration times the consumptive coefficient of the different crops planted.

The water balance model takes into account the initial soil moisture conditions for every pixel in the grid (30km x 30km). Calculations start at the end of the rainy season that corresponds to March. It aims to have soils at water holding capacity (WHC) as initial conditions. However, due to the dryness of the area, some parts of the basin could not be saturated. For dealing with this, three different sets of WHC were analysed (100, 80 and 65per cent) in these areas.

The drainage density in the basin is very low. Besides, during the whole year, all the rivers remain dry. For this reason, verified during fieldwork campaigns, the surface runoff component in the model was neglected. So, for each pixel and each month the aquifer's direct recharge was calculated using the mass continuity equation. This equation establishes that the inflow minus the outflow for each pixel is equal to the change in soil moisture storage during the period studied. The equation used considers previous soil moisture storage that can be estimated from the initial conditions of the model or from a solution of the equation for a previous month. It also considers the dryness of the soil, precipitation and real evapotranspiration for the particular month (Rodriguez 1994).

The application was based on long-term average monthly precipitation, potential evapotranspiration and combined soil and vegetation characteristics. Calculations were made using the Integrated Land and Water Information System (ILWIS). Available ILWIS procedures were used in order to estimate the spatial variability of the meteorological data. As the terrain is almost flat, the areal distribution of both rainfall and temperature were made using Thiessen polygons.

For each month, maps of direct recharge were produced for the entire aquifer. Using the annual recharge as criteria, results were grouped in five categories from low to high. GIS visualization functions were used to identify the location of the main direct recharge spots. Results of the model showed an average annual direct recharge of around $43*10^6$m3/year or 8mm/year. Getachew (1994) estimated the indirect recharge of the aquifer at $40*10^6m^3$. According to this, total annual aquifer recharge should be around $83*10^6m^3$. This quantifies the trend of decreasing ground water level and triggers a decision.

29.3.4 Mixed Integer Programming (MIP) Model

One approach to manage the declining groundwater level in the aquifer system was to assign a fixed amount of water to the farmers. The *Confederación del Guadiana* has limited farmers to extract an average of 225mm of water per year (well above the average recharge), depending on their farm size. In fact, with efficient irrigation systems, efficiency and appropriate management water was not a constraining factor for some crops. However, for some other crops, water was a limiting factor.

In spite of the Government policies, the aquifer continued its decline. This means that less water had to be allocated in future years to sustain the aquifer. Crop selection should be based on water requirements and water availability. The selection must maintain economic income while less water is used. In order to make the best use of rainfall and limited supply of irrigation water, a more efficient water management is required.

From 1993, the European Economic Community (EEC) and the Spanish Government were implementing four kinds of subsidies in the study area: subsidies for type of crop; subsidies for reduction in water use for irrigation; set-aside subsidies; and subsidies for reforestation.

Subsidies for type of crops. One annual subsidy was given to stimulate the cultivation of specific crops either in irrigated or rain-fed conditions. These subsidies were established yearly and were based on the EEC Common Agricultural Policy. They were given on area basis (per ha). Farmers receiving these subsidies had to maintain fallow at least 15 per cent of the total area of the farm.

Subsidies for reduction in the use of water for irrigation. Both, the *Confederación del Guadiana (CHG)* and the Irrigators Association established the maximum yearly amount of groundwater available for irrigation (in m^3/ha). This endowment was calculated based on crop water requirements and farm size. The total water use for each farm was derived through multiplication of the area by its corresponding water requirement, as established by the CHG. An average value of consumption of water per hectare was obtained by dividing the total water used over the total number of hectares. The result was compared with pre-established figures for use of water. Subsidies were given when they were applicable. Depending on the amount of water used, three types of subsidies were granted. Subsidies were given for programs of reduction in water use of 50, 70 and 100 per cent.

Set-aside subsidy. Farmers in irrigated lands opting for set-aside subsidies received compensation. These subsidies are aimed at stimulating farmers to leave part of their land fallow. In rain-fed lands, this subsidy was lower. Farmers could receive simultaneously payments for water reduction and the set-aside subsidy.

Subsidy for reforestation. The subsidy for reforestation included the costs of reforestation, plus a premium of maintenance. For five years, farmers received an average annual payment and a compensatory premium for 20 years. Farmers could take simultaneously the subsidy for reforestation and the set-aside subsidy. It was not possible for them to take the subsidies for reforestation and reduction in water at the same time. For non-irrigated lands, subsidies for reduction in water use were not applicable. A schematic flow chart of the possible farmers' decisions with respect to the different subsidies is presented in Figure 29.5.

MIP model formulation. To simulate the behaviour of each class of farmer (big, medium and small) holding different types of land (irrigated and non-irrigated), and seek the cropping pattern that maximizes their profits considering various subsidies schemes, an optimization model was developed. The optimization was based on irrigation water availability, land availability, crop rotation, production policies and existing subsidy schemes. Subsidy are offered for crop type, reduction in water use (50, 70 or 100 per cent), set-aside and reforestation. The optimum cropping pattern for each type of farm and land was the basis for the calculation of the total cropping pattern in the *Alcázar de San Juan* district. The planning period has been assumed to be one year. However, the results can be considered as an average over a span of time of five years (program of subsidies for reduction in water consumption duration).

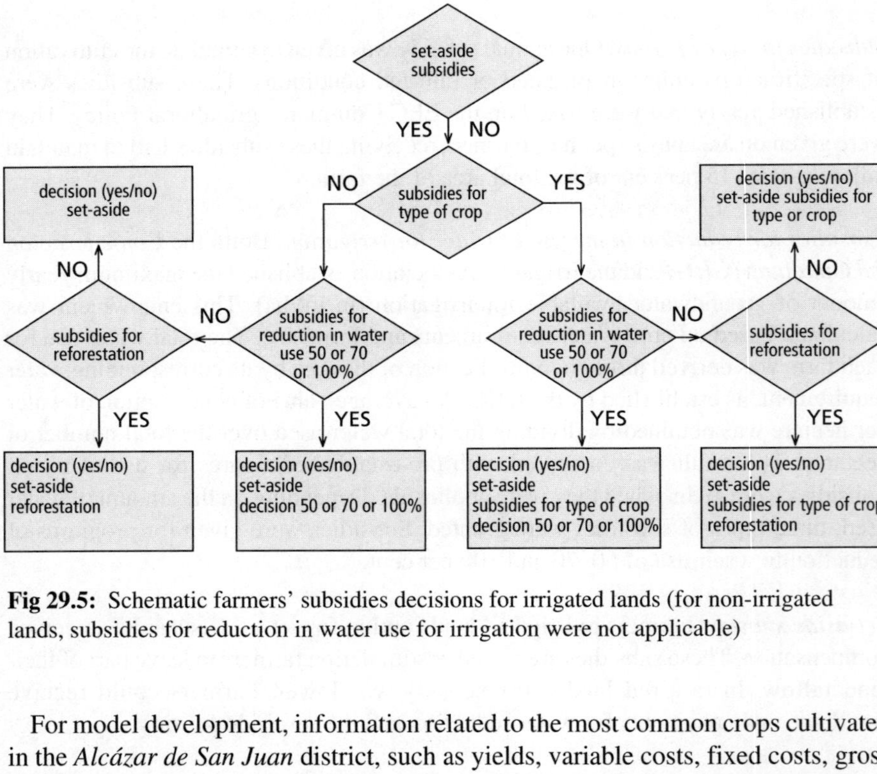

Fig 29.5: Schematic farmers' subsidies decisions for irrigated lands (for non-irrigated lands, subsidies for reduction in water use for irrigation were not applicable)

For model development, information related to the most common crops cultivated in the *Alcázar de San Juan* district, such as yields, variable costs, fixed costs, gross margins, sale prices and net incomes were estimated. Crop water requirements were obtained from estimations made by the CHG. Crop calendars were identified during fieldwork. For irrigated lands, 17 crop rotations were selected. In non-irrigated lands, six crop rotations were considered. Selection was made according to the more common rotations in the area listed by the Ministry of Agriculture.

The annual income, annual water requirements and monthly labour requirements for each rotation were calculated based on the assumption that where the rotation enters the 'optimum' plan, the crops considered in that rotation would be cultivated each year in equal proportions. For example, a farm following a rotation of four years of fallow and six years of crop would have 4/10 of its area in fallow and 6/10 in crops, rather than having the whole area under fallow for four years followed by six years under crops.

In order to study the farmers' decisions with respect to the different combination of subsidies, a MIP model was developed. It simulates the behaviour of farmers with respect to different subsidy policies and estimates their economic and environmental impacts. It was mixed because the farmers' decision outcome with respect to a particular subsidy was either positive 1 or negative 0. In the problem formulation, the objective function for optimization was profit maximization for each of the three farm sizes in both, rain-fed and irrigated conditions. Decision variables were either the rotation type or the crop type and the decisions for each subsidy scheme. Constraints included type of farm (irrigated/rain-fed), water and land availability, and policy and marketing constraints with regard to the traditional vineyard cultivation, garlic, cereals and tubers. Restrictions expressing the relationships between the different subsidy schemes were also considered.

With respect to rotations, two approaches were considered. The first one (Approach 1) took rotations as the decision variables. The second approach (Approach 2) considered each crop as an activity and included market and rotation constraints obtained from the analysis of the cropping pattern evolution in the case study area. The cost of the land was not included in the economic analysis. Due to a strong tradition for vineyard cultivation, it was assumed that the area under vineyard would remain the same. Studies made with respect to the evolution of this crop between 1984-1995 strongly supported this assumption.

The matrix structure of the model is shown in Table 29.1. When crop rotation is considered as activity (Approach 1), the model results in 246 variables and 133 constraints. The number of non-zero elements in the matrix equals 1,182. The matrix density is 2.97 per cent. When crops are considered as separate activities (Approach 2), the number of variables becomes 198 and the number of constraints 136. The number of non-zero elements in the matrix equals 1,390; the matrix density is 4.21 per cent. The optimization package for linear and mixed integer programming, OMP (Beyers and Partners 1993), was used for the problem solution. For solving the optimization problem, it uses the branch and bound method (Williams 1985).

Before experimentation, the model has to be calibrated and validated internally and externally. Internal validity refers to the existence of the casual relationship between variables, or absence of relationship, which implies the absence of cause. If internal validity is lacking, no statement about the cause and effect can be made. The external validity refers to the possibility of applying/extrapolating the result of the experiment to the real world cases. Validation of the simulation model is the investigation of the correspondence of the model with the real life situation; in other words, if model presents a realistic representation of the reality. The correspondence test was carried out by studying of the trends in various runs of the model, as well as comparison of the model results with an estimate of the actual land use map. The land use map, which was obtained from a supervised classification of the 1993 Landsat image (corresponding to the first year of the subsidies scheme), was used for the calibration and validation of the MIP model (Table 29.2). Compared to previous years, cropping patterns in 1993 showed a reduction in the irrigated area. Results of the model predicted a further reduction of around 40 per cent of the irrigated land after 1993.

Table 29.1: Matrix structure of the MIP model

	decision variables										
	crops in irrigated lands						crops in rain-fed lands				
	farm size big - medium - small						farm size big -medium - small				
	subsidized crops			no subsidized crops			subsidized crops		no subsidized crops		resource constraints
constraints	W	R	A	W	R	A	R	A	R	A	
water availability											
land availability											
policy constraints											
water subsidies											
crop subsidies											
subsidies for abandon											
subsidies for reforestation											
objective function	net income										max

W = water subsidies; A = set-aside subsidies;
R = subsidies for reforestation; MAX = maximize

Table 29.2: Comparison between results of the land use classification (1993) and the results of the two approaches

land cover	land use class. (ha)	approach 1 (ha)	approach 2 (ha)
rain-fed vineyard	9,729	9,729	9,729
irrigated vineyard	2,037	2,037	2,037
fallow	5,667	3,354	4,012
rain-fed (no vineyard)	14,098	-	15,764
irrigated (no vineyard)	20,674	11,671	10,635
others (reforestation and abandon of lands)	390	19,506	9,437
total	51,595	51,595	51,595

According to Government estimates, the enforcement of the program of subsidies for water reduction, big farmers opted for the 70 per cent reduction in water consumption, medium farmers for the alternative of 50 per cent reduction. Small farmers were expected to be not interested on this type of subsidy. The MIP model predicted the same results.

In rain-fed lands, results of the model indicated that big and medium farmers would opt for cultivation of sunflower. Small farmers would opt either for reforestation or for the traditional vineyard cultivation. In irrigated lands, medium farmers would opt for the cropping of sunflower while big and small farmers would decide to crop melon and irrigated cereals. It is important to note that the model results were very sensitive to the crops that received subsidies. In 1993, low subsidies were given to crops with high water requirements. With exception of melon, the model predicted a decline in the area of these crops. In 1993, high subsidies were given to the cultivation of sunflower and vecht; it explains the big trend of the farmers to the sunflower cultivation. However, almost 70 per cent of the net income for sunflower was due to subsidies, making the cultivation of this crop attractive.

With respect to vineyard, the model results indicated that this crop might not appear in the optimum cropping pattern strategy. However, because of the cropping tradition and its high social value, farmers were expected to continue with the vineyard fields. A policy stimulating the cultivation of vineyard under many varieties could be helpful to the economic improvement of vineyards. However, a policy like this would be against the EEC plans, which are in favour of the abandonment of vineyard fields.

Comparison of the land use classification from Landsat data of 1993 and the results of the model, as shown in Table 29.2, shows reduction in the irrigated area, combined with an increase in the reforested and abandoned lands. For each type of farm, average results of farmers' total net income were calculated (Table 29.3). In general, from the total average income per hectare, 30 per cent corresponded to subsidies given by the Common Agricultural Policy (subsidies for crops, set-aside and reforestation), 25 per cent to subsidies for reduction in water use and 45 per cent to crop income.

Table 29.3: Average income results (in pesetas/ha) for each farm size, two approaches analyzed

approach	farm type	big farm	medium farm	small farm	average all farm sizes
1	irrigated	146,320	152,924	140,564	146,602
1	rain-fed	48,792	48,795	31,083	42,890
2	irrigated	159,835	147,233	108,511	138,527
2	rain-fed	54,281	54,305	31,083	46,556

Small farms in both irrigated and rain-fed lands had lower income than the other farms, because of the assumption that they would continue cultivating vineyards. This explains the reason why small farmers in irrigated lands would only pay for water 7 pesetas/m^3, while medium and big farmers would offer around 30 pesetas/m^3.

The MIP model predicted the decline in cultivation of crops with high water needs (alfalfa, sugar beet and maize) and their replacement by crops with lower water demand. It also showed that some existing land use patterns (vineyards) continued to exist because they represent a long tradition. The model predicted a reduction in the area of irrigated land, because the farmers would set aside a part, and also showed that the applied subsidy policies would stimulate the reduction in use of groundwater for irrigation. However, if the average annual recharge of the aquifer ($83*10^6m^3$) is compared to the annual discharge ($135*10^6m^3$), it can be clearly seen that, although subsidies are effective, the water table in the aquifer would continue decreasing. Logically, the water table depletion would be reduced, but sooner or later, the groundwater reserves would finish if the mining of groundwater continued. However, it is still possible, for a number of more dry years, to allow the same level of agricultural production as in wetter years, through further mining of groundwater.

29.3.5 Multicriteria Evaluation Model

Results of the previous section showed that the MIP model is rather good at simulating the behaviour of the farmers with respect to different subsidy schemes. The next step in simulation is experimentation with the model in order to formulate different policies and study of their impacts from different perspectives. Using the MIP model 10 aspatial scenarios that were deemed to be attractive and include extreme variations of the present subsidy scheme were generated as follows:

- currently subsidies scheme implemented in the area (Scenario 1);
- policy without subsidy for crop type (Scenario 2);
- policy without subsidy for reduction in water use for irrigation (Scenario 3);
- policy without subsidy for abandon of lands (Scenario 4);
- policy without subsidy for reforestation (Scenario 5);
- policy without any kind of subsidies (Scenario 6);
- 25 per cent reduction in subsidy for sunflower and peas (Scenario 7);
- 50 per cent increase in subsidy for cereals (Scenario 8);
- Scenario 7 + Scenario 8 (Scenario 9); and
- water pricing policy (groundwater cost = 10 pesetas/m^3), without considering subsidy for reduction in water use for irrigation and without constraint about maximum water use (Scenario 10);

The next step in the planning process is the assessment and evaluation of different scenarios from different perspectives, considering different criteria. This is carried out through application of the multicriteria evaluation model. In this process, the overall utility of alternatives were evaluated on the basis of their impacts in terms of the decision criteria and the associated values that decision-makers assign to them. The latter are generally referred as weights. In this experiment, the following four criteria were considered for multicriteria evaluation:

- income generation (estimated in pesetas);
- total employment (estimated in average man/month);
- irrigation water use (estimated in m^3/year); and
- Government expenditures (estimated in Pesetas).

Within these criteria, income and employment are considered as a benefit criteria (the higher the better) and the irrigation water use and Government expenditure as cost criteria (the lower the better).

The evaluation model allows the planner to consider the priorities of different actors, e.g. farmers, environmentalists and Government. In this case, the preferences or weights related to each criterion were estimated through inquiries carried-out with the main interest groups concerned with the problem (politicians, farmers and environmentalists). Three sets of weights were derived, as presented in Table 29.4.

Table 29.4 Sets of weights considered in the analysis

criteria	W1=equal weights for all the criteria	W2=weights estimated using direct assessment	W3=weights estimated using AHP[1] method
profit	0.25	0.35	0.292
employment	0.25	0.15	0.110
water use	0.25	0.40	0.542
Gov. expenses	0.25	0.10	0.056

1 AHP refers to analytical hierarchy process which converts qualitative factors to quantitative scale (Satty 1980)

A project impact matrix represents performances of each alternative scenario on each criterion. Using MIP, impacts were estimated and presented in Table 29.5. As it can be seen in table, values of total income were very low in case of scenarios 6 and 7. For the remaining scenarios, income values were very similar at around 4,300 millions pesetas. With respect to employment, all the alternatives, with exception of Scenario 10, gave similar results. These values compared well to the figures obtained from the Agrarian Census 1991, 1,450 people were working in the agricultural sector. Scenario 10 gave a very high level of employment because of the vast area cultivated under melon. In relation to consumption of water for irrigation, almost all the alternatives showed figures around 43 million m^3/year (direct recharge of the aquifer). However Scenario 10 presented a very high water consumption, which is again explained by the large area cultivated under melon. Government expenditures were quite different in different scenarios. For multicriteria evaluation, scores that were measured in different scales (m^3/year, pesetas, men/month) had to be standardized to a common dimensionless unit (value judgment). In this case, an interval method was used to scales each score between the lowest value 0 and highest value 1. The weighted summation and an out-ranking method, ELECTRE II (Roy 1973), were chosen for the multicriteria analysis. They were selected based on the type of information available and because they are particularly useful when distinct alternatives are available. The software used for this application was the DEFINITE program (Janssen and van Herwijnen 1992, Janssen *et al.* 2000).

Results of the multicriteria analysis for the two methods are shown in Figures 29.6 and 29.7. Results were quite sensitive to the set of weights. When set W1 was selected, the weighted summation results indicated the scenario 10 to be the best alternative. However, using the ELECTRE II method, the same conclusions were not derived. This is because scenario 10 could not be ranked with the imposed thresholds. When other sets of weights were used, scenario 10 became poorly ranked. Results of the weighted summation method using sets of weights W2 and W3, indicated scenarios 8, 1 and 5 as the best. Scenarios 6, 7 and 10 could be discarded for further analysis as they rank as the worse. For the ELECTRE II method it was not clear which scenario could be classified as the best one (considering the selected thresholds). However it was evident that scenarios 4 and 7 always were outranked by other alternatives. Although scenario 10 could not be ranked it was also dominated by almost all the alternatives.

Table 29.5 Impact matrix for all the scenarios analysed for *Alcázar de San Juan* district IV

	scenario									
criteria	**1**	**2**	**3**	**4**	**5**	**6**	**7**	**8**	**9**	**10**
total income Alcázar (millions pesetas)	4,658	4,394	4,135	4,607	4,642	2,900	2,576	4,748	4,696	4,807
employment (avg. men/month)	1,318	1,271	1,465	1,063	1,320	1,463	1,348	1,201	1,493	3,029
total water for irrigation (m^3*1000/year)	43,282	40,218	42,279	43,282	43,282	42,279	43,282	38,505	49,061	105,580
Government expenditures (millions pesetas)	2,444	2,521	1,739	2,731	2,403	0	569	2,737	2,436	700

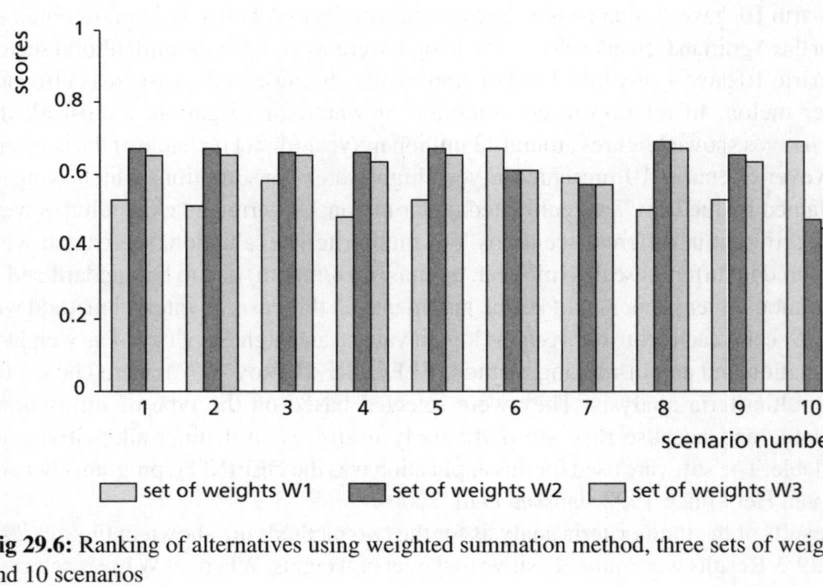

Fig 29.6: Ranking of alternatives using weighted summation method, three sets of weights and 10 scenarios

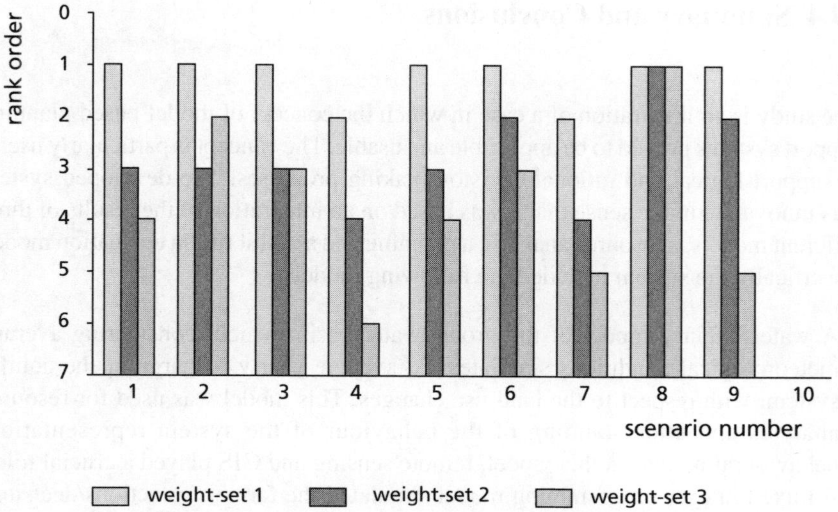

Fig 29.7: Ranking of alternatives using ELECTRE II method, three sets of weights and 10 scenarios

According to the criteria considered in the analysis, a policy of increasing cereals subsidies would be the best (scenario 8). The 1993 scheme of subsidies in the area ranked at an intermediate level. The scheme without subsidies for reduction in water use ranked as the second best alternative. Although in 1993 all the farmers were accepting this type of subsidy, the reduction in water use seems to be motivated by the maximum water endowment established by the Government and not by the given amount of subsidy. In this sense, it would be reasonable to maintain these subsidies only for a period of five years, as the Government planned. Alternatives involving a reduction in the subsidies for sunflower or pea cultivation or set-aside, were ranked as the worst alternatives. A policy of water market pricing with the cost of groundwater equal to 10 pesetas/m^3 would not be effective.

The multicriteria analysis indicated that a policy of increases of 50 per cent in subsidies for cereals would be better than the subsidy scheme implemented in the area in 1993. However, this strategy would not lead to a recovery in the aquifer's water table. Reduction in water use for irrigation is most effective through a policy of maximum groundwater endowment established by the Government, and not through subsidies set to discourage the use of groundwater. In order to re-establish the natural equilibrium discharge-recharge in the aquifer system, strong restrictions on the use of groundwater for irrigation have to be implemented. The cropping pattern in the area would change dramatically as a consequence of policies diminishing subsidies for sunflower, peas or set-aside. An adequate and planned policy for subsidies for crops has to be implemented and continued.

29.4 Summary and Conclusions

The study is an illustration of a case in which the concept of model-based planning support systems proved to be applicable and usable. The concept is particularly useful to support logical and rational decision-making processes. The developed system was innovative in the sense that it was based on an integration of the results of three different models: a resource analysis, a planning and a multicriteria evaluation model. Specifically the system included the following models:

- A water balance model of the groundwater basin, which, considering average meteorological conditions simulates the average yearly recharge of the aquifer system, with respect to the land use changes. This model was used for resource analysis and understanding of the behaviour of the system representation/ behavioural model. In this model, remote sensing and GIS played a crucial role.
- A mixed integer-programming model simulated the farmers' reactions/decisions to the different subsidy schemes. This model, which was based on rational farmers who wants to maximize his profit, was used as a 'planning model' to formulate and assess the impacts of various policy instruments.
- A multicriteria evaluation model, which allowed planners to consider the impacts of different policy instruments with priorities of different actors (farmers, environmentalists and Government) and select the most attractive options.

The results obtained with the system are specific for water resources rehabilitation in the study area of La Mancha, Spain. However, the concept of PSS as applied here, is general and can be applied in and extended to other disciplines and areas. The concept might turn out to be particularly useful in cases where decisions have to be made on the basis of an evaluation of various scenarios that are developed on the basis of a detailed resource analysis and an analysis of the resource allocation at a disaggregated level. Thematic areas where the concept can be further developed and applied include agriculture, land rehabilitation and forest resources.

Acknowledgements
The authors wish to thank, Drs E. Dopheide and Dr. M. McCall for their critical review and editing of the text. We are also thankful to *La Fundación Municipal para el Desarrollo Económico y el Empleo*, especially Julio Plaza Tabasco and his team and *La Comunidad de Regantes*, principally Pablo Valero, both at *Alcázar de San Juan,* for their collaborations in data collection and fieldwork support. The research was carried out with the support of the International Institute for Aerospace Survey and Earth Sciences (ITC) and the Netherlands Organization for International Cooperation in Higher Education (NUFFIC).

References

Beaufoy, G. (2000) Integrating water and agriculture through river basin management: a practical exercise. Las Tablas de Daimiel. The case of a protected area in the upper Guadiana Basin, Castilla-La Mancha, Spain. Introductory Note. European Freshwater Programme WWF.

Beyers and Partners NV (1993) *OMP Manual.*

Camacho, L., Camacho, B. *et al* (1989) Sobreexplotación del Acuífero de la Mancha Occidental. (Unidad Hidrogeológica 04-04), Almería (España). In Spanish.

Carlevaris, J. (1992) La fertilidad de los principales suelos agrícolas de la zona oriental de la provincia de Ciudad Real. La Mancha y Campo de Montiel. Consejo Superior de Investigaciones Científicas. Consejería de Agricultura. Junta de Comunidades de Castilla-La Mancha. Centro de Ciencias Medioambientales. In Spanish.

Conyers, D. and Hill, P. (1989) *An Introduction to Development Planning in the Third World*, John Wiley, New York.

Getachew, H, (1994) Hydrogeological study of an over exploited aquifer system using mathematical ground water flow simulation model. A case study in the La Mancha Occidental ground water basin. La Mancha, Spain. ITC.

Janssen, R. and Rietveld, P. (1990) Multicriteria analysis and GIS: an application to agricultural land use in the Netherlands, in Scholten H. and Stillwell, J. (eds) *Geographical Information Systems and Urban and Regional Planning*, Kluwer Academic Publishers. Dordrecht, pp 129-139.

Janssen, R. and van Herwijnen, M. (1992) *DEFINITE A System to Support Decisions on a Finite Set of Alternatives*, Kluwer Academic Publishers, Dordrecht.

Janssen, R., van Herwijnen, M. and Binat, E. (2000) Decision support system for a finite set of alternatives (DEFINITE), *Manual Institute for Environmental Studies*, Free University of Amsterdam.

Niñerola, S. (1976) El Embalse Subterráneo de la Llanura Manchega. I Simposio Nacional de Hidrogeología, Valencia (España). In Spanish.

Rodriguez, E. (1994) Decision support analysis for water resources management. A case study of Alcazar de San Juan district in the Aquifer 23, La Mancha, Spain. ITC.

Roy, B. (1973) How outranking relation helps multiple criteria decision-making, in Cochrane, Zeleny, J. and Zeleny, M. (eds) *Multiple Criteria Decision-making,* University South Carolina Press, Columbia.

Satty, T. (1980) *The Analytical Hierarchy Process*, McGraw Hill, New York.

Shannon, R.E. (1975) *System Simulation: The Art and Science*, Prentice-Hall, Englewood Cliffs, N.J.

Sharifi, M.A. and van Keulen, H. (1994) A decision support system for land use planning at farm level, *Agricultural Systems*, 45: 239-257.

Sharifi, M.A. (1999) Remote sensing and decision support systems, Chapter 14 in Stien, A. and van der Meer, F. (eds) *Spatial Statistics for Remote Sensing*, Kluwer Academic Publisher, Dordrecht.

Sharifi, M.A. (2000) Planning Support Systems to Enhance Sustainable Land Utilization. Resource Paper presented at the Seminar on Impact of Land Utilization Systems on Agricultural Productivity, Asian Productivity Organization & The Iranian Ministry of Agriculture, 4-10 November 2000, Tehran, Islamic Republic of Iran.

Simon, H.A. (1960) *The New Science of Management Decision*, Harper and Brothers, New York.

Steuer, R.E. and Schuler, A.T. (1981) An iterative multiple objective linear programming approach to a problem in forest management, *Operational Research Journal*, 26: 193-208

Thorbecke, E. and Hall, L.(1982) Nature and scope of agricultural sector analysis: An overview. In: Ed. E. Thorbecke, Agricultural sector analysis and models in developing countries, *FAO Economic and Social Development Paper No. 5*, FAO, Rome.

Thornthwaite, C.and Mather, J. (1957) Instructions and tables for computing potential evaporation and the water balance, Publ. Climatology Lab., Centeron, New Jersey, v. 10 No. 3. pp. 244.

Tinbergen, J. (1956) *Economic Policy: Principles and Design,* North Holland Publishing Company, Amsterdam.

Turban, E. (1995) *Decision Support and Expert Systems: Management Support Systems*, Macmillan Publishing Company, New York.

Williams, H. (1985) *Model Building in Mathematical Programming*, John Wiley & Sons, Chichester.

Subject Index

Figures

Tables

List of Contributors

Mohammed Ahmadul Hoque
Department of Geography and Planning, University of Akron, Akron, Ohio 44325-5005, USA

Andrea Aime
Municipality of Modena, Via Santi 40, 41100 Modena, Italy

Kheir Al-Kodmany
University of Illinois at Chicago, CUPPA Hall, 412 South Peoria Street, Chicago, IL 60607, USA

Silvia Ascari
Municipality of Modena, Via Santi 40, 41100 Modena, Italy

Vladimir Badenko
St. Petersburg State Technical University, 195251 Polytechnical 29, St Petersburg, Russia

Glen Barry
Land Information and Computer Graphics Facility, University of Wisconsin-Madison, B102 Steenbock, 550 N. Babcock Drive, Madison WI 53706, USA

Michael Batty
Department of Geography and Centre for Advanced Spatial Analysis, University College London, 1-19 Torrington Place, London WCIE 6BT, United Kingdom

Sharon Biermann
CSIR Building and Construction Technology, PO Box 395, Pretoria 0001, South Africa

Mark Birkin
School of Geography, University of Leeds, Leeds, LS2 9JT, United Kingdom

Peter Boden
GMAP Consulting, 15 Blenheim Terrace, Leeds LS2 9HN, United Kingdom

Flavio Bonfatti
Dipartimento Scienze dell'Ingegneria, Universitá di Modena e Reggio Emilia, Via Vignolese 905, 41100 Modena, Italy

Glen Bramley
School of Planning and Housing, Edinburgh College of Art/Heriot-Watt University, 79 Grassmarket, Edinburgh EH1 2HJ, Scotland, United Kingdom

Philippe Carreau
Joint Research Centre of the EC, Institute for Environment and Sustainability – Inland and Marine Waters Unit, TP 272, I-21020 Ispra (VA) Italy

Gerrit Jan Carsjens
Land Use Planning Group, Wageningen University and Research Centre, PO Box 9101, 6700 HB Wageningen, The Netherlands

Steve Carver
Centre for Computational Geography, School of Geography, University of Leeds, Leeds, LS2 9JT, United Kingdom

Tony Champion
Department of Geography, The University of Newcastle-upon-Tyne, Newcastle-upon-Tyne NE1 7RU, United Kingdom

Marieke Eleveld
Institute for Environmental Studies (IVM) – FALW, Vrije Universiteit (VU), De Boelelaan 1087, 1081 HV Amsterdam, The Netherlands

Guy Engelen
Research Institute for Knowledge Systems bv, PO Box 463, 6200 AL Maastricht, The Netherlands

Andrew Evans
Centre for Computational Geography, School of Geography, University of Leeds, Leeds, LS2 9JT, United Kingdom

Stewart Fotheringham
Department of Geography, The University of Newcastle-upon-Tyne, Newcastle-upon-Tyne NE1 7RU, United Kingdom

Cary Gaunt
Science Applications International Corporation (SAIC), 11521 Roger Bacon Drive, Reston, VA 20190, USA

Elizabeth Gavin
ISLinkup, PO Box 13371, Hatfield, Pretoria 0028, South Africa

Stan Geertman
URU/Nexpri, Faculty of Geographical Sciences, Utrecht University, Heidelberglaan 2, 3508 TC Utrecht, The Netherlands

Hadi Harman Shah
Faculty of Development Science, Universiti Kebangsaan Malaysia, Bangi, Selangor

Jan-Jaap Harts
URU Urban Research Centre Utrecht, Utrecht University, PO Box 80115, 3508 TC Utrecht, The Netherlands

Bo Huang
Department of Computer Science, Keele University, Staffordshire, ST5 5BG, United Kingdom

Marilene Huenis
Department of Land Affairs, Private Bag X954, Pretoria 0028, South Africa

Laura Jackson
U.S. Environmental Protection Agency, Office of Research and Development, Research Triangle Park, NC 27711, USA

Bin Jiang
Division of Geomatics, Institutionen för Teknik, University of Gävle, SE-801 76 Gävle, Sweden

Tom de Jong
URU, Faculty of Geographical Sciences, Utrecht University, PO Box 80.115, 3508 TC Utrecht, The Netherlands

Jung-Wook Kim
Department of Geography and Planning, University of Akron, Akron, Ohio 44325-5005, USA

Richard Kingston
Centre for Computational Geography, School of Geography, University of Leeds, Leeds, LS2 9JT, United Kingdom

Richard E. Klosterman
Department of Geography and Planning, University of Akron, Akron, Ohio 44325-5005, USA

Dmitry Kurtener
Agrophysical Research Institute, St Petersburg, 195220 Grazhdansky pr.14, St Petersburg, Russia

Kari Lautso
LT Consultants Ltd, Melknokatu 9, 00210 Helsinki, Finland

Arend Ligtenberg
Centre for Geo-information, Wageningen University and Research Centre, PO Box 9101, 6700 HB Wageningen, The Netherlands

Kees Maat
OTB Research Institute for Housing, Urban and Mobility Studies, Delft University of Technology, PO Box 5030, 2600 GA Delft, The Netherlands

James Macgill
Centre for Computational Geography, School of Geography, University of Leeds, Leeds LS2 9JT, United Kingdom

Tracy Miller
Land Information and Computer Graphics Facility, University of Wisconsin-Madison, B102 Steenbock, 550 N. Babcock Drive, Madison WI 53706, USA

Jukka Mikkonen
Regional Council of Päijat-Häme, Hämeenkatu 9A, FIN-15110 Lahti, Finland

Paola Monari
Dipartimento Scienze dell'Ingegneria, Universitá di Modena e Reggio Emilia, Via Vignolese 905, 41100 Modena, Italy

Mohd Nazari Jaafar
Department of Geography, Universiti Malaya, Kuala Lumpur, Malaysia

Kari Oinonen
Finnish Environment Institute, PO Box 140, FIN-00251 Helsinki, Finland

Henk Ottens
Urban Research Centre Utrecht, Utrecht University, PO Box 80.115, 3508 TC Utrecht, The Netherlands

Aziza Parveen
Department of Geography and Planning, University of Akron, Akron, Ohio 44325-5005, USA

Christopher Pettit
School of Geography, Planning and Architecture, University of Queensland, St Lucia 4072, Queensland, Australia

Ruslan Rainis
Geoinformatic Unit, School of Humanities, Universiti Sains Malaysia, 11800 Penang, Malaysia

Philip Rees
School of Geography, University of Leeds, Leeds LS2 9JT, United Kingdom

James Reilly
New Jersey Office of State Planning, 33 West State Street, PO 204, Trenton, NJ 08625, USA

Mika Ristimäki
Finnish Environment Institute, PO Box 140, FIN-00251 Helsinki, Finland

Erasmo Rodriguez
Department of Civil Engineering, Universidad Nacional de Colombia, Bogota, Colombia

Wolfram Schrimpf
Joint Research Centre of the EC, Institute for Environment and Sustainability – Inland and Marine Waters Unit, TP 272, I-21020 Ispra (VA) Italy

Kamalruddin Shamsudin
Development Program Division, Federal Department of Town and Country Planning, 20th Floor, Wisma Sambanthan, Jalan Sulaiman 500000, Kuala Lumpur, Malaysia

Mohammad Ali Sharifi
International Institute for Geo-Information Science and Earth Observation, ITC, Enschede, The Netherlands

Xun Shi
Department of Geography, University of Wisconsin, Madison, WI 53706, USA

Tung-Kai Shyy
School of Geography, Planning and Architecture, University of Queensland, St Lucia 4072, Queensland, Australia

Loren Siebert
Department of Geography and Planning, University of Akron, Akron, Ohio 44325-5005, USA

Ardy Siegert
Covalent Technologies Inc., 706 Mission Street (2nd Floor), San Francisco, CA 94103, USA

Tania Smith
Department of Land Affairs, Private Bag X954, Pretoria 0001, South Africa

Ken Snyder
PlaceMatters.com, 3469 W Hayward Place, Denver CO 80211, USA

Henning Sten Hansen
Helsinki, Finland National Environmental Research Institute, Department of Policy Analysis, Frederiksborgvej 399, DK-4000 Roskilde, Denmark

John Stillwell
School of Geography, University of Leeds, Leeds LS2 9JT, United Kingdom

Robert Stimson
School of Geography, Planning and Architecture, University of Queensland, St Lucia 4072, Queensland, Australia

Paul M. Torrens
Department of Geography and Centre for Advanced Spatial Analysis, University College London, 1-19 Torrington Place, London WC1E 6BT, United Kingdom

Inge Uljee
Research Institute for Knowledge Systems bv, PO Box 463, 6200 AL Maastricht, The Netherlands

Kees van de Ven
National Institute for Marine and Coastal Management, Ministry of Transport, Public Works and Water Management, Directorate-General of Public Works and Water Management, PO Box 207, 9750 EX Haren, The Netherlands

Ron van Lammeren
Landscape Architecture Group and Centre of Geo-Information, Wageningen University and Research Centre, PO Box 9101, 6700 HB Wageningen, The Netherlands

Gert van Maren
Asset Information Systems, PO Box 102 032 NSMC, Auckland, New Zealand, and Faculty of Civil Engineering and Geosciences, Department of Geodesy, Delft University of Technology, the Netherlands

Vit Vasek
Laboratory on Geomatics and Cartography, Masaryk University, Brno, Czech Republic

Steve Ventura
Land Information and Computer Graphics Facility, University of Wisconsin-Madison, B102 Steenbock, 550 N. Babcock Drive, Madison WI 53706, USA

Coen Wessels
Nexpri, Utrecht University, PO Box 80.115, 3508 TC Utrecht, The Netherlands

Julia Williams
GMAP Consulting, 15 Blenheim Terrace, Leeds LS2 9HN, United Kingdom

Anthony Yeh
Centre of Urban Planning and Environmental Management, The University of Hong Kong, Pokfulam Road, Hong Kong SAR, China

Titles in the Series

M. M. Fischer and *J. Fröhlich* (Eds.)
Knowledge, Complexity and Innovation Systems
XII, 477 pages. 2001. ISBN 3-540-41969-1

M. M. Fischer, J. Revilla Diez and *F. Snickars*
Metropolitan Innovation Systems
VIII, 270 pages. 2001. ISBN 3-540-41967-5

L. Lundqvist and *L.-G. Mattsson* (Eds.)
National Transport Models
VIII, 202 pages. 2002. ISBN 3-540-42426-1

J. R. Cuadrado-Roura and *M. Parellada* (Eds.)
Regional Convergence in the European Union
VIII, 368 pages. 2002. ISBN 3-540-43242-6

G. J. D. Hewings, M. Sonis and *D. Boyce* (Eds.)
Trade, Networks and Hierarchies
XI, 467 pages. 2002. ISBN 3-540-43087-3

G. Atalik and *M. M. Fischer* (Eds.)
Regional Development Reconsidered
X, 220 pages. 2002. ISBN 3-540-43610-3

Z. J. Acs, H. L. F. de Groot and *P. Nijkamp* (Eds.)
The Emergence of the Knowledge Economy
VII, 388 pages. 2002. ISBN 3-540-43722-3

R. J. Stimson, R. R. Stough and *B. H. Roberts*
Regional Economic Development
X, 397 pages. 2002. ISBN 3-540-43731-2